Springer Geography

The Springer Geography series seeks to publish a broad portfolio of scientific books, aiming at researchers, students, and everyone interested in geographical research.

The series includes peer-reviewed monographs, edited volumes, textbooks, and conference proceedings. It covers the major topics in geography and geographical sciences including, but not limited to; Economic Geography, Landscape and Urban Planning, Urban Geography, Physical Geography and Environmental Geography.

Springer Geography—now indexed in Scopus

More information about this series at http://www.springer.com/series/10180

Dahe Qin · Tandong Yao · Yongjian Ding ·
Jiawen Ren
Editors

Introduction to Cryospheric Science

Editors
Dahe Qin
China Meteorological Administration;
Northwest Institute of Eco-Environment and Resources
Chinese Academy of Sciences
Beijing, China

Yongjian Ding
Northwest Institute of Eco-Environment and Resources
Chinese Academy of Sciences
Lanzhou, Gansu, China

Tandong Yao
Institute of Tibetan Plateau Research,
Chinese Academy of Sciences
Beijing, China

Jiawen Ren
Northwest Institute of Eco-Environment and Resources
Chinese Academy of Sciences
Lanzhou, Gansu, China

ISSN 2194-315X ISSN 2194-3168 (electronic)
Springer Geography
ISBN 978-981-16-6427-4 ISBN 978-981-16-6425-0 (eBook)
https://doi.org/10.1007/978-981-16-6425-0

Jointly published with Science Press
The print edition is not for sale in China (Mainland). Customers from China (Mainland) please order the print book from: Science Press.
ISBN of the Co-Publisher's edition: 978-7-03-067738-9

This Springer imprint is published by the registered company Springer Nature Singapore Pte Ltd.
The registered company address is: 152 Beach Road, #21-01/04 Gateway East, Singapore 189721, Singapore

Foreword by Sun Honglie

The main research object of cryospheric science includes ice, snow and frozen ground. The cryosphere is an earth layer that is severely affected by climate warming. It is prominently manifested in the severe retreat of glaciers, the rapid reduction of Arctic sea ice extent and snow in the northern hemisphere, and the thickening of active layers of permafrost. At the same time, the cryosphere has a significant impact on the natural and the socio-economic systems through mass and energy exchanges with other earth layers. Chinese scientists have primarily clarified the status, evolution and change mechanisms of the Chinese cryosphere, and revealed the interaction between the cryosphere and other earth layers after last decades. Chinese scientists have also taken the lead in conducting exploratory research on the relationship between the cryosphere and sustainable development.

Considering its high surface reflectivity, huge cold storage and latent heat exchange in phase change, the cryosphere serves as a source and sink of greenhouse gases and a recorder of the climate environment, as well as huge freshwater reserves and other irreplaceable functions, and its change process, trend and the interaction with other earth layers, leading to the research on cryospheric science which has become one of the active fields in the current climate system and sustainable development research, and has received unprecedented attention. In 2007, China established the State Key Laboratory of Cryospheric Science for the first time. In the same year, the International Union of Geodesy and Geophysics (IUGG) upgraded its former International Committee on Snow and Ice (ICSI, the second-level society) to the International Association of Cryospheric Science (IACS, the first-level society), which become the only first-level Society added in 87 years since IUGG was founded. In 2016, the China Association for Science and Technology approved the establishment of the Chinese society for cryospheric science.

In recent years, the State Key Laboratory of Cryospheric Science, in conjunction with experts from other scientific research institutes and universities, has successively opened graduate courses on "Introduction to Cryospheric Science" at the University of Chinese Academy of Sciences, Beijing Normal University and Lanzhou University, and the book was compiled and published on the basis of these courses. This book is the third series on Cryospheric Science which was published by academician Qin Dahe and other scientists following on the "English-Chinese Dictionary of Cryospheric Science" in 2012 and "Dictionary of Cryospheric Science" in 2014. "Introduction to Cryospheric Science" is rich in content, covering the basic concepts and theories of cryospheric science, expounding the formation and evolution of various elements of the cryosphere, the interaction between the cryosphere and other layers of the climate system, and the impact of cryosphere changes on society and sustainable economic development. It can be used as a textbook for related majors in universities and colleges and a reference for scientific and technological researchers in the field of ecological environment. This book is the first monograph on cryospheric science; I believe the publication of this book will further promote the development of cryospheric science.

Beijing, China
March 2019

Sun Honglie
Academician of Chinese Academy of Sciences

Foreword by Fu Bojie

Since the older generation of Chinese geographers started scientific investigations on modern glaciers and permafrost in the 1950s, the glaciers, permafrost and snow were studied in parallel as sub-disciplines for more than half a century. In the past 10 years, the team led by academician of Chinese Academy of Sciences Qin Dahe has put forward and developed a new subject—Cryospheric Science—from a systematic and comprehensive perspective, established the State Key Laboratory of Cryospheric Science and successively published "English-Chinese Scientific Vocabulary of Cryospheric Science" and "Cryospheric Science Dictionary". Today, they launched "Introduction to Cryospheric Science", which systematically introduces the scientific content of cryospheric science. I was fortunate to witness and experience this process, and along the way, I was moved by their courage to pioneer and climb the peak. As an entourage, I first read "Introduction to Cryospheric Science".

Cryospheric science intergrated with geography, meteorology and climatology, hydrology, geomorphology, ecology, marine science, remote sensing science, environmental science and other disciplines, and it is closely related to the sustainable development of social and economic and even geopolitics. The depth and breadth of cryospheric science research has achieved considerable development with new technologies and new methods, and it has involved in global change, environmental change, sustainable development and other fields. "Climate warming ice prophet" cryospheric changes have an immeasurable impact on human society by affecting water resources, ecological environment, sea-level changes and extreme weather and climate events. Therefore, the development of cryospheric science is not only related to cryosphere research itself but is also related to the aspects of human environment, economic and social, and is closely related to our daily lives.

In this context, the launch of a book that comprehensively introduces the basic concepts, research overview and cutting-edge progress of the cryosphere has important practical significance for the establishment and development, education and popularization of the scientific system of cryospheric science. Cryospheric science includes multiple branches such as glaciers, permafrost and snow, and each branch is extremely rich in research content and they are varied. It is quite difficult to ensure the continuity of the content, which organically integrates the various elements of the cryosphere into the cryosphere system to address a certain aspect (such as physics, chemistry, observations and simulations) in one chapter or section. It can be seen that the book has put a lot of effort in this issue. It has made a good arrangement of the lecture content, the sequence of explanations and the theoretical depth, leading to the main elements of cryosphere well represented in the book. The chapters are arranged from the shallower to the deeper from step to step, which include both basic knowledge and mechanism exploration, and field work introduction and observational experiment analysis. It will not make one feel boring and is highly readable while acquiring theoretical knowledge.

Another feature of the book is its novel materials. Many of the materials in the book are the latest results and advanced technologies of the past 10 years of research. It not only illustrates the cryospheric observations in pictures and texts but also provides physical explanations, which greatly expands the audience of this book. Although the book still has the shadow of a monograph in a small part of content, it is suitable for undergraduates and postgraduates with a certain foundation in physical geography as a textbook for introductory study and later scientific research of cryospheric science. It is believed that it is of great benefit for broadening students' view and to continue scientific research. The book can also be used as a reference book by teachers, media and other public who are interested in cryospheric science.

Cryospheric science is a discipline that focuses on field observations. The progress of observation technology will greatly promote the development of the discipline. Therefore, cryospheric science will continue to develop rapidly with the advancement of future observation technology, and new research results will continue to emerge. I believe it will be more comprehensive when it is republished. For the development of disciplines, education must go first. The author team is far-sighted. On the basis of successively publishing a series of reference books on cryospheric science and personally teaching related courses in many universities, they formed "Introduction to Cryospheric Science" which systematically introduces the scientific content of cryospheric science, and this is the first time in the scientific field of cryospheric science. The publication of this book will definitely strengthen the backbone of scientific research, and the construction of subsequent talent teams of cryospheric science will definitely promote cryospheric science to make more progress and lead cryospheric science to a high level.

Beijing, China
February 2017

Fu Bojie
Academician of the Chinese Academy of Sciences, Chairman of the Chinese Geographical Society

Preface

In the current world, scientific and technological progress has prompted rapid economic and social development, which not only improved people's living standards but also brought about global warming and deterioration of the ecological environment. It has received widespread public concern, among which the cryospheric issues such as mountain glacier shrink, snowline rise, permafrost degradation, Antarctic ice sheet melting and the extent of Arctic sea ice reducing get more attention. Many researches from different fields have published many articles related to Cryospheric Science; the mass media published a large number of papers discussing the impact of climate change and cryosphere changes, and the audience was very wide; in various speeches and reports, the cryosphere was also used as an example. These phenomena indicate that Cryospheric Science is being integrated into different disciplines, the scientific knowledge of Cryospheric Science has been popularized, its influence has expanded and social benefits have improved. This has enabled the cryosphere to develop and spread.

It is our long-cherished wish to closely integrate the impact and adaptation of climate change and cryospheric changes with economic and social development, protect the earth's environment and achieve sustainable development. Realizing long-term aspirations and ambitions in a good environment is an exciting thing. But there are also worries. Due to the rapid expansion of the scientific research team and the slow development of professional training, some cryosphere scientific and technological workers have insufficient understanding of the updated development and trends of the discipline, and their structure of professional knowledge is flawed. This is not conducive to the in-depth development of Cryospheric Science, the popularization of related sciences, the protection of the earth and the environment, and the realization of sustainable development. With the development of human production abilities and the improvement of science and technology, especially the improvement of satellite and remote sensing monitoring technology, Cryospheric Science has been rapidly developed. "Introduction to Cryospheric Science" is a book which responds to the development demand for the science, and provides a more comprehensive introduction to the science.

The book includes 11 chapters, which comprehensively introduce relevant issues of Cryospheric Science. Chapter 1 is about cryosphere and Cryospheric Science, which give the definition and review the research history of Cryospheric Science, and introduce the role of cryosphere in global change and social development. Chapter 2 is about classification and geographical distribution of cryosphere, in which are introduced the global geographic distribution of various elements of the cryosphere and the classification system of each element. Chapter 3 is the formation and development of the cryosphere, in which are introduced the zonality of cryosphere development, and the formation mechanism and development conditions of each element of the cryosphere. Chapter 4 is the physical characteristics of the cryosphere, in which are introduced the physical characteristics of glaciers, frozen ground, snow, sea ice and other elements of cryosphere from the aspects of Mechanics, Thermodynamics, Electricity and Magnetism. Chapter 5 is the chemical characteristics of the cryosphere, mainly including the chemical characteristics of snow and ice in glaciers, the chemical characteristics of frozen ground and the chemical characteristics of sea ice. Chapter 6 is the climate and environmental records in the cryosphere, in which are systematically introduced the climate change records from ice cores, frozen ground, tree rings and other media records in cold regions. Chapter 7 is the evolution of the cryosphere at different temporal scales, in which are introduced the characteristics of changes in various elements of the cryosphere from the tectonic scale, orbital/suborbital scale, millennium scale, centennial-decadal scale to interannual-seasonal scale. Chapter 8 is about the interaction between the cryosphere and other earth layers, in which is discussed the interaction between the cryosphere and the other four major layers of the climate system, especially the close intersections. Chapter 9 is about cryosphere changes and sustainable development, in which are introduced the issues closely related to social and economic development, including the impact of cryospheric changes on society, cryospheric disasters and risk management, and major engineering construction in cryosphere areas. Chapter 10 is the techniques of observation and experimental technology of the cryosphere. The traditional in situ observation methods and laboratory analysis techniques of cryosphere are systematically introduced, and some latest technologies and methods which have accelerated the development of Cryospheric Science are discussed.

It can be expected that there will be more advanced progress in the research of Cryospheric Science in the next 10 years, and some cryospheric scientific issues mentioned in this book will also have more comprehensive conclusions, especially the current research hotspots such as interaction between the cryosphere and other earth layers, and cryospheric changes and sustainable development will have more reliable conclusions. From this view, this book can be served as an "introduction", which can be regarded as the "primary stage" and the basis for further research of Cryospheric Science. The readers of this book are teachers and students of related majors in universities and researchers in scientific research institutions. The readers of the book are stuffs of related majors in universities and the researchers in scientific research institutions. By reading this book, readers can understand the cryosphere and its elements from the perspective of the earth layers, understand the complexity and important impact of cryosphere changes, gain new knowledge and raise new research questions that need to be further studied. Cryospheric Science is very rich in research content, and it is impossible to elaborate on it with a single introduction. We have considered compiling a series of books on Cryospheric Science at the same time as publishing as supplementary teaching materials of "Introduction to Cryospheric Science". Since it is the first time for compiling "Introduction to Cryospheric Science", due to the wide range and rapid development of disciplines, there are unavoidable omissions in the book. Criticism and corrections are well welcomed. We will supplement or correct in the reprint.

The compilation and publication of this book are supported by the National Natural Science Foundation of China's Innovative Research Group Project (41421061), the National Major Scientific Research Project (2013CBA01800), the National Key Basic Research Development Program (2007CB411507) and the independent project of the State Key Laboratory of Cryospheric Science (SKLCS-ZZ-2016), and the Textbook Publishing Center of the University of Chinese Academy of Sciences, and the book is also supported by the subject of development strategy research by the Standing Committee of the Chinese Academy of Sciences. The authors express their sincere gratitude. At the same time, we would also like to thank all teachers, colleagues and friends who have shown concern, support and help for the publication of this book.

Beijing, China
December 2019

Dahe Qin

Contents

Chapter 1
Cryosphere and Cryospheric Science

Lead Authors: Dahe Qin, Tingjun Zhang

Contributing Authors: Ninglian Wang, Shichang Kang, Cunde Xiao, Yongjian Ding, Jiawen Ren

The Intergovernmental Panel on Climate Change (IPCC) Fifth Assessment Report (AR5) states that it is extremely likely that more than half of the observed increase in global average surface temperature from 1951 to 2010 was caused by the anthropogenic increase in greenhouse gas concentrations and other anthropogenic forcing together. The global warming is indicative in atmosphere, hydrosphere, cryosphere, biosphere and lithosphere. Among them, the cryosphere warming leads to melting and retreating of glaciers, degradation of permafrost, and reduction in extent of snow cover and sea ice.

The cryosphere change has a significant impact on global and regional climate, ecosystem and human well-being. On a global scale, the Antarctic and the Greenland ice sheets are very sensitive to climate, and their changes may affect ocean circulation and sea level rise. Changes in the extent of snow cover and sea ice play critical roles in energy and radiation balance on earth surface, in addition to key processes and feedback of atmospheric circulation.

On a regional scale, the future changes of mountain glaciers, river ice, and lake ice may have a dramatic impact on water resources and ecosystems along with the recurring and intensifying cryosphereic hazards and disasters. Meanwhile, the retreat of sea ice in the Arctic Ocean would create more opportunities for new shipping routes and exploration of subsea resources. However, it can also noticeably elicit more disputes for territory and resources among the circum-Arctic and pan-Arctic nations.

Locally, a freezing and thawing process of seasonally frozen grounds (including the active layer over permafrost) changes soil moisture, subsequently affecting vegetation and ecosystem as a whole. Additionally, the frost heaving and thaw settlement may cause serious damages to infrastructure in cold regions. Considering a global warming scenario, the frozen organic carbon in permafrost can be decomposed by microbial processes resulting from the increase of the active layer thickness and thermokarst activities. It increasingly releases methane (CH_4) and carbon dioxide (CO_2) into the atmosphere, adding to the atmospheric greenhouse gas concentration and further exacerbating global warming.

D. Qin et al. (eds.), *Introduction to Cryospheric Science*, Springer Geography,
https://doi.org/10.1007/978-981-16-6425-0_1

The above-mentioned various important roles that the cryosphere plays are amid the intensified inter-disciplinary and multi-disciplinary researches, aiming at more pragmatic applications in close relationship with human activities. Overall, the cryospheric processes are critical in the climate system, and complex when interacting with the other spheres. Improved knowledge about them will considerably advance the understanding of the cryosphere and the cryospheric science as a whole.

1.1 Cryosphere

1.1.1 Definition of the Cryosphere

The cryosphere is a sphere on the earth's surface with a certain thickness, where temperature is continuously at or below 0 °C. Therefore, water in the cryosphere generally exists in a frozen state. In interacting with the lithosphere, the key components of the cryosphere occur in ground from the earth surface down to a certain depth from tens of meters to thousands of meters. In relation to the hydrosphere, the cryosphere appears in the Southern Ocean and the Arctic Ocean surface at a depth from less than a meter to several meters. The subsea permafrost may even extend to several hundred meters deep into the continental shelf. The thickness of glaciers and ice sheets varies from a few meters to thousands of meters. In the atmosphere, the cryosphere exists above the layer of 0 °C isotherm in the troposphere and stratosphere.

In general, the cryosphere is composed of glacier (including ice sheet), frozen ground (including permafrost and seasonally frozen ground), snow cover, river ice and lake ice, sea ice, ice shelf, iceberg and subsea permafrost, and frozen water in the atmosphere. Spatially, the mid-latitude and high-latitude regions are the main areas for the cryosphere.

It should be noticed, on the surface of ice crystals and soil particles, there exists a quasi-molecular membrane of water with temperature below the freezing point. The unfrozen water also exists in frozen ground due to soil capillary and particle absorption. The unfrozen water is still considered as a part of the cryosphere. In contrast, although temperature of surface sea water or water underneath sea ice and ice shelves in the Southern Ocean or the Arctic Ocean may be at a few degrees below 0 °C, this part of sea water in a liquid state is not considered as part of the cryosphere.

1.1.2 Cryosphere Classification and Its Characteristics

Cryosphere can be divided into three major categories: continental cryosphere, marine cryosphere and aerial cryosphere, mainly depending on its geographical distribution, dynamics, and thermodynamic conditions.

Continental cryosphere consists of glaciers (including ice sheets), frozen ground (including permafrost and seasonally frozen ground, excluding subsea permafrost), snow, river ice and lake ice. Marine cryosphere mainly includes sea ice, ice shelf, iceberg, and subsea permafrost. Aerial cryosphere includes all frozen water in the atmosphere, such as snowflakes, ice crystals, and ice clouds. Both cryospheric science and atmospheric science cover aerial cryosphere with each having its own focus (Fig. 1.1).

Fig. 1.1 Distribution of the global cryosphere (after IPCC AR5 WGI 2013). *Note* 1. In the Northern Hemisphere, Arctic sea ice extent shows the minimum area of the Arctic summer sea ice (September 13, 2012). The south boundary of the average yearly minimum extent is based on the 30-year average sea ice extent during the period between 1979 and 2012 (in yellow lines), which considers ice concentration as less than 15%. In the Southern Hemisphere, the maximum sea ice extent and the north boundary of the average yearly maximum sea ice extent are described. The instereographic projection is applied to the Northern and Southern Hemispheres, and the information of low-latitude glaciers and snow extent cannot be displayed

The extent of continental cryosphere covers 52–55% of the land surface. Mountain glaciers and ice sheets cover approximately 10% (Antarctic ice sheet and Greenland ice sheet for 9.5%, mountain glaciers for 0.5%). Snow cover takes a range from 1.3 to 30.6%. An average maximum snow cover extent in the northern hemisphere holds about 49% of the northern hemisphere's total land surface. The global permafrost region (excluding permafrost underneath glaciers and ice sheets) accounts for 9–12% of the global land area. In the northern hemisphere, the maximum seasonally frozen ground takes 33% (excluding areas of the active layer over permafrost). As indicated, if the active layer over permafrost is included, areas of seasonally frozen ground occupies more than 56% of the northern hemisphere's land area, and as much as 80% in extreme cold years.

There is about 75% of the Earth's freshwater stored in the cryosphere, taking into account approximate 70% from modern glaciers, Greenland and Antarctic ice sheets. If all ice of the Greenland and Antarctic ice sheets would have melted to oceans, the global sea level would rise up to 7.36 m and 58.30 m (sea level rise equivalent), respectively. The sea level rise equivalent would be about 0.41 m as a result of the meltwater from mountain glaciers, and about 0.10 m from the excess ice in permafrost. As observed from 1993 to 2010, the ice melting in the continental cryosphere caused a global sea level rise at a rate of 1.36 mm per year.

On average a total area up to 5.3–7.3% of the Earth's ocean surface is covered by sea ice and ice shelves. The maximum extent of the Arctic sea ice in winter is about 15×10^6 km^2 and the minimum in summer is about 6×10^6 km^2. The maximum extent of the Southern Ocean sea ice in September is about 18×10^6 km^2, while the minimum is about 3×10^6 km^2 in February. According to the age of ice, sea ice is divided into the first year ice, second year ice and multi-year ice. The majority of sea ice drifts subject to the wind force and ocean surface currents. Distributions of the floating ice in terms of thickness and age as well as snow cover and open water areas are extremely non-uniform, with a spatial scale from several meters to hundreds of kilometers. The ice shelves around the edge of Antarctic ice sheet have a total area of approximate 1.617×10^6 km^2, or about 0.45% of the global ocean area, while the extent of global subsea permafrost accounts for about 0.8% of the ocean area (Table 1.1).

The water volume in the atmosphere is very low, which is about 114×10^3 m^3, the least among three types of cryosphere, The ice in the atmosphere also has the shortest life amid them.

1.1.3 Cryosphere Changes

The Soviet Union geographer, C. B. Kalesnick, claimed in 1939, "first of all, glacier is a product of certain climatic conditions." It was demonstrated after more research and better comprehension. Research indicated that each component of the cryosphere should be considered as a "natural climate indicator".

Table 1.1 Representative statistics of cryospheric components

Continental cryosphere	Percent of global land surface[a]	Sea level equivalent[b] (m)
Antarctic ice sheet[c]	8.3	58.3
Greenland ice sheet[d]	1.2	7.36
Glaciers[e]	0.5	0.41
Permafrost[f]	9–12	0.02–0.10 [g]
Seasonally frozen ground[h]	33	Not applicable
Seasonally snow cover (seasonally variable)[i]	1.3–30.6	0.001–0.01
Northern hemisphere freshwater (lake and river) ice[j]	1.1	Not applicable
Total[k]	52.0–55.0	~ 66.1
Marine cryosphere	Percent of global ocean area[a]	Volume[l] (10^3 km^3)
Antarctic ice shelves	0.45[m]	~380
Antarctic sea ice, austral summer (spring)[n]	0.8(5.2)	3.4(1.1)
Arctic sea ice, boreal autumn (winter/spring)[n]	1.7(3.9)	13.0(16.5)
Subsea permafrost[o]	~0.8	Not available
Total[p]	5.3–7.3	

Notes
[a]Assuming a global land area of 1.476×10^8 km^2 and ocean area of 3.625×10^8 km^2
[b]Assuming an ice density of 917 kg m^{-3}, a seawater density of 1028 kg m^{-3}, with seawater replacing ice currently below sea level
[c]Area of ice sheet not including ice shelves is 1.2295×10^7 km^2
[d]Area of ice sheet and peripheral glaciers is 1.801×10^6 km^2
[e]Includes glaciers around Greenland and Antarctica (see sources of SLE in AR5 WGI Table 4.2)
[f]Area of permafrost excluding that beneath the ice sheets is $1.32–1.8 \times 10^7$ km^2
[g]Value indicates the full range of estimated excess water content of Northern Hemisphere permafrost
[h]Long-term average maximum of seasonally frozen ground is 4.81×10^7 km^2; excluding Southern Hemisphere
[i]Northern Hemisphere only
[j]Areas and volume of freshwater (lake and river ice) were derived from modelled estimates of maximum seasonal extent
[k]To allow for areas of permafrost and seasonally frozen ground that are also covered by seasonal snow, with total area excluding seasonal snow cover
[l]Antarctic austral autumn (spring) and Arctic boreal autumn (winter). For the Arctic, volume includes only sea ice in the Arctic Basin
[m]Area is 1.617×10^6 km^2
[n]Maximum and minimum areas taken from IPCC AR5 WGI, Sects. 4.2.2 and 4.2.3
[o]Few estimates of the area of subsea permafrost exist in the literature. The estimate of 2.8×10^6 km^2 which is significantly uncertain, was assembled from other publications by Gruber (2012)
[p]Summer and winter totals assessed separately
Data Source IPCC AR5 (2013)

To investigate cryospheric changes, it is prerequisite to understand what the climate change is and its effects on cryosphere. Currently, the climate change is represented by the changes in five spheres in the climate system, and a change in any of the five spheres can be indicative of climate change. For example, the global warming is not only manifested as an increase in mean surface temperature, but also reflected in the increases in oceanic heat capacity, glacier retreat, thickening of the active layer, reduction in snow cover and sea ice extent, as well as biodiversity deterioration.

There are two kinds of definitions of the climate change. At first, defined by IPCC, "climate change refers to changes that can be identified (e.g., using statistical tests) for a longer period of time (typically for several decades or longer), including changes in average value and/or rate. The cause of climate change can be either internal processes of nature, or external forcing, such as solar cycle, volcanic eruptions, or artificially continuous changes in atmospheric composition, land use and land cover change." Secondly, defined by the United Nations Framework Convention on Climate Change (UNFCCC), climate change "observed besides natural climate variability in comparable period directly or indirectly attributable to human activities to change the global atmospheric composition." It is clear that in the UNFCCC's definition, the climate change caused by changes in atmospheric composition due to human activities is distinguished from the climate variability caused by natural factors.

Changes in the cryosphere are considered as an internal variability of climate system, instead of external forcing. Cryospheric change refers to the change in temporal and spatial distribution of heat energy and mass within the cryosphere. Specifically, it statistically refers to a long-term trend of changes in each component of the cryosphere, including changes of glaciers and ice sheet areas in thickness, volume, length and their terminus; frozen ground (including permafrost and seasonally frozen ground) area, extent, and thickness; snow cover extent and snow water equivalent; sea ice extent and thickness; river ice, lake ice freezing and melting dates, freezing days, and thickness. Changes in variables such as temperature, structure, geometry and volume are also indicative of cryospheric changes.

The cryosphere is one of the crucial research subjects in climate change science. The IPCC AR5 Working Group 1 (WGI) reported that the cryosphere is most sensitive to climate change in the climate system. Every component in the cryosphere system changes subject to global warming (Fig. 1.2). Situated in remote cold regions, the cryosphere has been observed warming and retreating, an undoubtful evidence of global warming.

As known, phenology pays a particular attention to the comparison of phenological observation, which has a dramatic impact on plant growth, such as, but not limited to, the first time of snowfall and frost in autumn and in spring, types of freezing injury of plant and frozen plants, the first date of river, lake and near surface soils frozen completely, the onset and ending dates of frozen soil thawing and ice melting, and the starting date of shaded areas frozen, as well as cross-observations and measurements of biological, agricultural, meteorological and cryospheric elements. The study of tree rings in high altitude and latitude regions has also advanced cryospheric science.

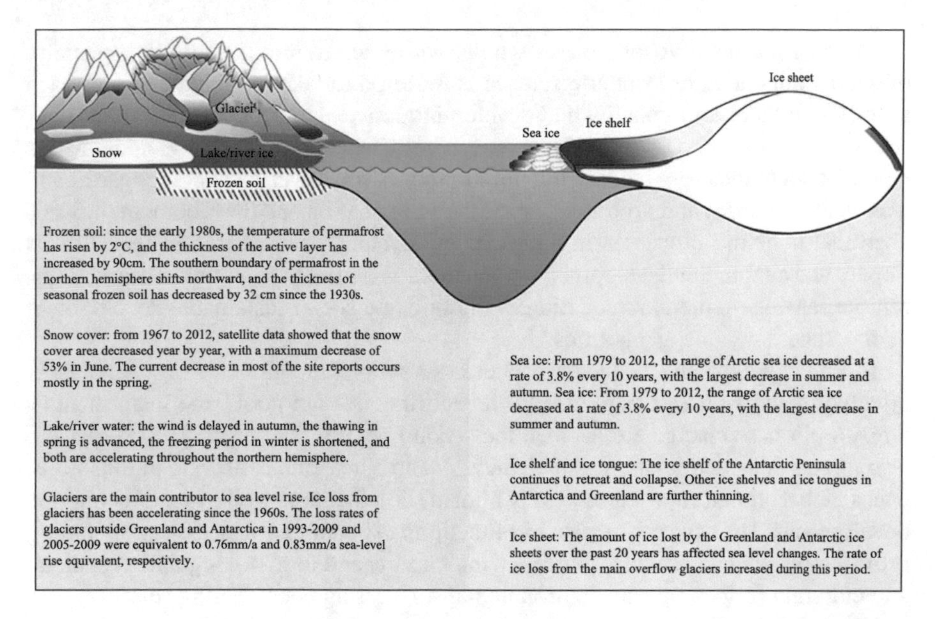

Fig. 1.2 Schematic summary of the observed major changes in the cryosphere (after IPCC AR5 WGI 2013)

1.2 Cryospheric Science

1.2.1 Definition and Scope of Cryospheric Science

Cryospheric science is a newly developed interdisciplinary science. It studies the formation of cryospheric components, evolution and internal dynamic processes, relationship between cryosphere and the other spheres in climate system, impacts and adaptation of cryospheric changes and climate change as a whole. The primary objectives of cryospheric science move forward to understand the physical dynamic processes in the cryosphere, and to develop the understanding in order to serve the society and contribute to the sustainable development. How the understanding contributes to national development goals is also part of research in cryospheric science.

The traditional cryospheric science is based on the study of each separated component in the cryosphere, considered as sub-disciplines, such as glaciology, geocryology, glaciers and periglacial geomorphology. The studies in those sub-disciplines have a relatively long history, which has laid a solid foundation and developed rich contents for the understanding of the cryospheric components. Nevertheless, their studies are relatively independent of each other and little interconnected. Facing the increasing challenges of global warming, the traditional approaches become difficult to keep pace with the needs for scientific development in adaptation to the increasing impact of global warming.

Considering the cryosphere as a whole, while taking into account the specialty of each component, cryospheric science is required to take an integrated approach to look into their commonality and develop inductive classification as well as analysis of the entire cryosphere. The cryosphere science should be considered as a whole and include those studies, but not limited to, such as cryospheric physical and chemical properties and processes, cryopsheric formation and development, biogeochemical dynamic processes in a cryosphere system, cryosphere change with its impact and adaptation development, cryosphere experiments and field observations, remote sensing, numerical modeling, economic and social sustainable development in cryosphere regions, geopolitics.

From the perspective of the cryosphere as a whole, the global energy balance is affected by cryosphere's high albedo. Its enormous "cold pool" and latent heat as a result of phase change are deemed the second next to the ocean. The cryosphere also affects ocean circulation as it changes ocean's heat and salinity, and thus influences global climate, and subsequently human settlements as well as socio-economic development. The cryosphere also has functions to benefit human society, including provisioning such as water resources, winter sports and tourism, regulation societal and cultural services, special environmental services and engineering services.

There is no inconsistency between the emphysis on the entity as a whole in cryospheric science and the individuality of each cryospheric components in subdisciplines. The cryospheric science is the outcome of the requirement to further advance the discipline, and the later is the foundation to continuously advance the cryopshere science.

From the aspect of one sphere, cryospheric science is mainly composed of aspects including cryospheric heat-mass transfer dynamic processes, changes in cryosphere, impacts of and adaptation to cryosphere change. Among them, the cryosphere formation processes, heat and mass transfer dynamic processes, cryosphere changes are the fundamental research; interactions and feedbacks between the cryosphere and the other spheres, adaptation and cryosphere services are the applied fundamental research; adaptation strategies and advance for sustainable economic and social development are the applied research (Fig. 1.3).

In general, the cryospheric science mainly includes the following:

(1) Dynamic processes of cryosphere formation and development

From both micro-scale and macro-scale points of view, it investigates the physical, chemical and bio-geochemical processes of the cryosphere with a particular attention to their thermodynamics. By using data and information from traditional and modern ground-based measurements, space-borne sensor monitoring, laboratory experiments, and numerical modeling, it analyses the changes of each cryospheric component at different temporal scales (i.e. daily, monthly, seasonal, yearly and interannual) and at different spatial scales (i.e. sites, localities, catchments, basins, regions, hemispheres and the globe). By these, it sheds light on the dynamic processes of the cryosphere, which lays a foundation for the prediction and projection of its future changes, and evaluation of subsequent effects or impacts.

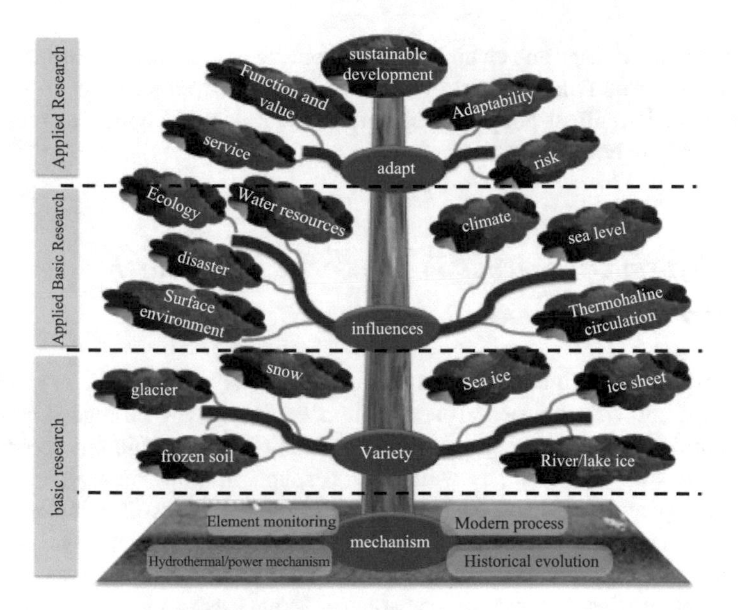

Fig. 1.3 Framework of cryospheric science

(2) Impact of cryospheric change

Cryospheric change impact refers to both positive and negative effects on nature and society imposed by changes in cryospheric components and cryospheric system as a whole. It can be considered as a result of interactions and feedbacks between the cryosphere and the other spheres in the climate system, i.e. effects on climate, ecology, water, environment and society, and even geopolitics.

(3) Adaptation to cryospheric change

Through the integration and crossing amid natural and social sciences, the adaptation to cryospheric change is based on the understanding of exposure, vulnerability and risks subject to the future change of cryosphere. It is established by integrating regional economic and social knowledge to reduce exposure, vulnerability and risks of cryospheric change, ultimately to identify strategies of adaptation and provide scientific and technological support for sustainable economic and social development at global, regional and local scales.

(4) Cryospheric evolution and geologic background

By looking into geologic deposits and geomorphological features, cryospheric evolution studies the formation, evolution processes, impacts of glacier, ice sheet and permafrost in various geological periods. Research is generally carried out by evaluating climatic and environmental records with different resolutions in cryosphere, for example, climate change and changes in atmospheric greenhouse gas concentrations recorded in the Antarctic ice sheet over the past 800,000 years and records of outer space events. Permafrost temperature gradient can also be applied to reconstruct climate change over the past millennium.

1.2.2 Cryospheric Science Framework and Research Approaches

Cryospheric science originates from numerous sub-disciplines, such as glaciology, geocryology, snow-ice physics, frozen ground engineering, cold-region hydrology and meteorology, periglacial geomorphology snow-ice microbiology, cold-region ecology and snow research, as well as cryosphere remote sensing, geographic information system (GIS) and other related high-technologies. It is initially driven by research needs in climate change and sustainable development. On thc aspect of the cryosphere in terms of momentum, energy, water as well as socioeconomic characteristics, the core of cryospheric science is to investigate the role of cryosphere and its interactions with the atmosphere, the hydrosphere, the biosphere, the lithosphere and human society. Therefore, cryospheric science has much broader contents, which interconnects with, but not limited to, geography, atmosphere, hydrology, oceanology, geology, ecology, environmental science, mathematics, physics, chemistry, biology, humanities and social sciences, economy, sustainable development, remote sensing, modeling, computation, big data analysis and other high-technological issues as well as tourism, culture and geopolitics. It emphasizes the interactions of the cryosphere with the other spheres and their changes and the influence in the climate system. The adaptation of cryosphere to the changes deeply involves the influences of all the other spheres including the anthroposphere. It is therefore also closely related to socio-economics, sustainable development, and societal needs. All of these would further improve scientific understanding and knowledge of the cryosphere.

Cryospheric science applies both natural and social science. On the aspect of its natural science, it is linked with approaches relevant to optics, thermology, mechanics, electricity, electromagnetism, chemistry, ecology, observational system and observational data of each cryospheric component, laboratory experiment, global and regional cryospheric modelling and simulation coupled with earth system models. Meanwhile, it also pays a great attention to the principles of economic and social science by investigating the relationship between the cryosphere and the society, for examples, studying the impacts of cryospheric change on societal vulnerability and adaptability.

Cryospheric research is generally fulfilled through field observations, in-situ measurements, aerial and satellite remote sensing, GIS and laboratory experiments, social investigations and so on, to obtain scientific data and information for each cryospheric component. By combining these data and information together with data analysis and modeling, we can further deepen the understanding of the cryosphere and its interactions with the other spheres in the climate system, explore the interaction of the cryosphere with the other spheres, and develop relevant adaptation measures, which are all part of cryospheric research.

1.2.3 Establishment of Cryospheric Science

In this book, we intends to establish the cryospheric science for the first time. Although the concept of cryosphere was proposed a long time ago, the path to shaping up cryospheric science shows very much difference at different stages of the establishment. Polish scholar A. B. Dobrowolski originally introduced the concept of cryosphere in 1923, Russian scientists P. A. Shumskii, O. Reinwarth and G. Stäblein further promoted the concept in the 1960s through 1970s. The World Meteorological Organization (WMO) confirmed the "cryosphere" as a separate sphere in parallel with atmosphere, hydrosphere, biosphere and lithosphere during the United Nations (UN) Conference on Human Environment in Stockholm in 1972. The UN conference also demonstrated the interactions and feedbacks among the five major spheres, established the concept of climate system, and further comprehended the importance of cryosphere. World Climate Research Program (WCRP) joint scientific committee established "Climate and Cryosphere" (CliC) program, aimed to quantitatively evaluate the impact of climate change on cryosphere and the role of cryosphere in climate system. Arctic Council and Japan also implemented cryospheric research program.

After enormous efforts of several generations, Chinese scientists eventually proposed and established the concept of cryospheric science with its theoretical framework shaped up by integrating climate change science and sustainable development theory.

In the early 1920s, Chu Coching introduced glaciers in his teaching lectures of "The General Theory of Geography". Later in 1943, he proposed to artificially accelerate the melting of mountain snow for increase of water supply in Hexi and southern part of Tianshan Mountains. In 1957, Shi Yafeng organized an investigation of glaciers in Qilian Mountain and Tianshan regions, then published "Investigation Report on Modern Glaciers in Qilian Mountain". He later established "Lanzhou Institute of Glaciology and Geocryology (LIGG)", Chinese Academy of Sciences (CAS), which has been a cryospheric research base in China. Under the LIGG, the long-term observational stations, including the Tianshan Station and the Cryospheric Research Station on Qinghai-Tibetan Plateau, were established. By the time, the former LIGG/CAS obtained the First Class Award of National Science and Technology Progress for its outstanding achievements in developing Qinghai-Tibetan Railroad project. In the 1980s, while the research of cryosphere on global change was playing more and more important roles, international research communities started to pay an increasing attention to cryospheric science as a whole. Chinese scientists took the opportunity and established "The State Key Laboratory of Cryospheric Science" in April 2007, the first international institute under the name of "cryospheric science".

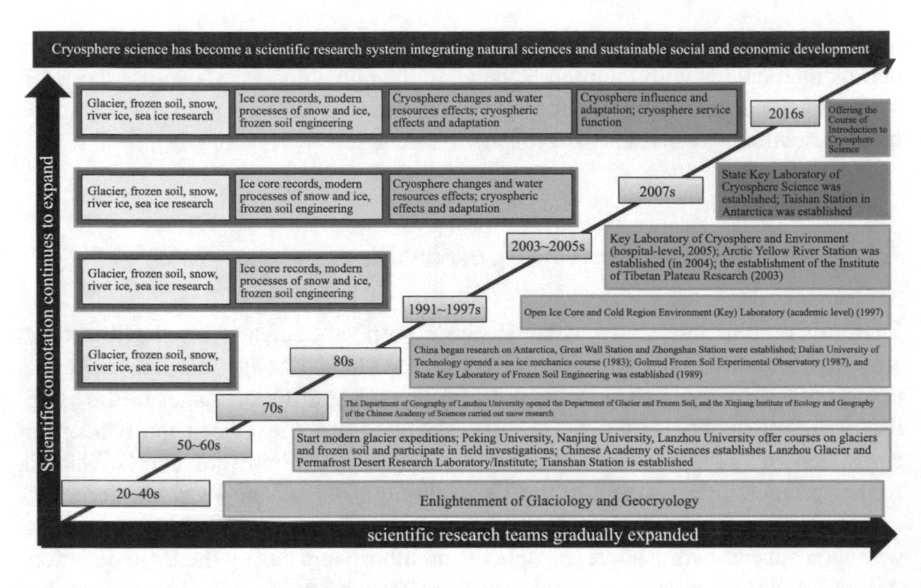

Fig. 1.4 Progress of cryospheric science in China

Meanwhile, the National Key Science and Technology program funded the newly established institute with several other key state projects, including "Cryosphere change, its impact and adaptation". It investigated cryospheric change and its impact on socio-economic and sustainable development, as well as adaptation at various scales. By taking inter-disciplinary and multi-disciplinary approaches to integrate cryospheric science with social sciences, economics, and sustainable socio-economic development, a system of cryospheric science as a whole was moved forward to a new era (Fig. 1.4).

International cryospheric research programs were developed along two main pathways. One is WCRP CliC program to advance the understanding of physical processes and feedback between cryosphere and climate system to improve the climate modelling and prediction for disaster prevention and mitigation. Another is the International Association of Cryosphere Society (IACS) under the International Union of Geodesy and Geophysics (IUGG), primarily to promote the establishment of cryosphere scientific system in order to better support socio-economic and sustainable development.

1.2.4 Cryospheric Science in International Science Programs

Over the past several decades, the International Council for Science (ICSU) has implemented four major global change science programs, including the International Geosphere-Biosphere Program (IGBP), the Global Change Human Factors Program (IHDP), the WCRP and the Biodiversity Program (DIVERSITAS). In 2002, the four programs merged into the "Earth System Science Partnership (ESSP)" program. ICSU and the International Social Science Council (ISSC) jointly launched a ten-year scientific plan of the Future Earth (FE) in 2014. At the same time, the four ICSU scientific projects became the core projects integrated into the FE program. In retrospect, cryospheric science was always a key component in those programs. WCRP/CliC has been the most representative international program for cryospheric science. Meanwhile, the Past Global Change Research (PAGES) in the IGBP program also has a major component of cryospheric science, including the integrated research program led by Chinese scientists—"Impact of cryosphere change on ecology and economic society in the arid regions of Asia". Additionally, there were seven associations such as the International Hydrological Association (IAHS) under IUGG. During the 24th General Meeting of IUGG in Perugia, Italy in July, 2007, IACS was established as its eighth association with a relevance to cryospheric science, the first added association over 87 years of the IUGG history. On 4 July 2018, ICSU and ISSC had merged to form the International Science Council (ISC).

During the fourth International Polar Year (IPY) in 2007/2008, 228 scientific programs have been implemented in the Arctic and Antarctic regions involving more than 50,000 scientists. After the IPY 2007/2008, WMO established a working group on Polar and High Mountain Observations, Research and Services (EC-PHORS), which promotes and coordinates programs in Antarctic, Arctic and high mountain regions. During the 17th Session of the WMO Congress at Geneva, Switzerland in 2015, it was decided that observation in polar and high mountain regions was one of the future seven key programs. These plans are directly linked to cryospheric research and services.

Recently, the "Third Pole Environment (TPE)" Program was launched and implemented by Chinese scientists in 2007, as an example for regional cryospheric and environmental change programs. The TPE program focused on cryosphere of the Qinghai-Tibet Plateau, carrying out observations and investigations with considerations of interactions among the five spheres in climate system over the Third Pole regions. The outcome of the TPE program contributes to the sustainable development of the Tibet Autonomous Region in China and its surrounding regions and countries.

1.2.5 IPCC's Focus in Cryospheric Science

The IPCC is the international body for assessing the science related to climate change. The IPCC was established in 1988 by the WMO and United Nations Environment Program (UNEP) to provide policymakers with regular assessments of the scientific basis of climate change, its impacts and future risks, and options for adaptation and mitigation. The IPCC embodies a unique opportunity to provide rigorous and balanced scientific information to decision-makers because of its scientific and intergovernmental nature. The key principles for the IPCC assessments are rigor, robustness, transparency, and comprehensiveness. The IPCC published five assessment reports and a series of special reports, technical papers, methodology reports, and supporting materials in 1990, 1995, 2001, 2007, and 2014, respectively. The IPCC assessment reports are written by hundreds of leading scientists who volunteer their time and expertise as Coordinating Lead Authors and Lead Authors of the reports. They enlist hundreds of other experts as Contributing Authors to provide complementary expertise in specific areas. The Sixth IPCC Assessment Reports (AR6) are under way and will be published by 2022.

The IPCC AR6 will continue to cover key issues related to cryospheric science. Of relevance are there three special reports to be published in 2018 and 2019, including Special Report on Global Warming of 1.5 °C (SR1.5), Special Report on the Ocean and Cryosphere in a Changing Climate (SROCC), Special Report on Climate Change and Land (SRCCL). Among those subjects, the impact and adaptation strategies for regional and global cryospheric change under different climate warming scenarios, especially related to sea level rise, water resources, ecosystems and socio-economy are important. Cryosphere scientists contributing to the effort should understand the causes, the impact and adaptation as well as mitigation considering climate changes, and are clear about the implications of outputs of earth system models, subject to different socio-economic pathways.

1.3 Cryosphere and Climate System

1.3.1 Dynamic Processes and Changes of the Cryosphere

The key research objectives of cryospheric science includes the formation, dynamic processes, changes, and monitoring of the cryosphere (Fig. 1.5 inner circle). They are also the main research for each sub-discipline in relevance to cryospheric science, provisioning the fundaments for cryospheric science.

The physics is a theoretical fundament to understand the formation and evolution of the cryosphere. Over different time scales, with appropriate temperature, precipitation and topographic conditions, glaciers, permafrost, snow and other components of the cryosphere are forming and then evolving with climate change. As a result of the involvement of a variety of medium, chemical elements and organics, it is inevitable to naturally involve bio-geochemical processes from the start to the end.

In relation to changes in environmental conditions, cryosphere may retreat under global warming and advance in a cooling climate. Technically speaking, with the physical properties of pure ice and ice with impurities into consideration, a macroscopic variation of cryosphere and cryospheric components can be described by mathematical and-physical models to project future cryospheric change given different global warming scenarios. However, in reality, the natural conditions of cryospheric processes are extremely complex. Adding with the consideration of biogeochemical processes, the physical models currently available are insufficient to describe the full processes within cryosphere and its interactions with the other spheres. Continuous

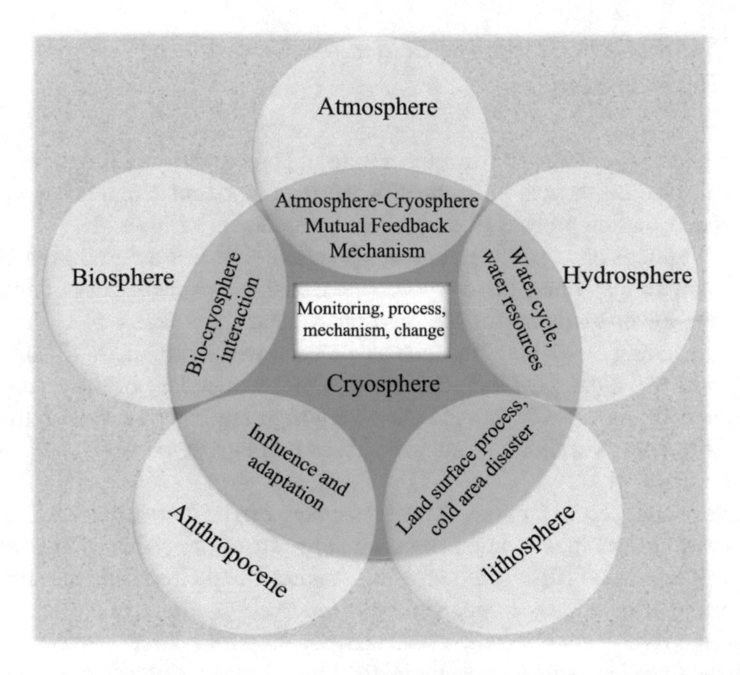

Fig. 1.5 Cryosphere and its interaction with hydrosphere, biosphere, lithosphere and anthroposphere

improvements of cryospheric models and their applications in earth system models are the urgent tasks for cryosphere scientists.

Observation and monitoring of cryospheric change are amid the major subjects of cryospheric process studies. They started for each component, such as glaciers (including ice sheets), frozen ground (including permafrost and seasonally frozen ground), and snow cover at specific locations. These observations include, but not limited to, physical parameters of each cryospheric component, hydrological and meteorological variables, geological and geomorphological conditions, biogeochemical parameters, and particle sedimentation. The methods for observation change quickly over time, from the early manual in-situ measurements and sampling to automatic weather stations for continuous data collection, and to aerial and satellite remote sensing. The observation and monitoring cover areas from site-specific to regional and global scales. Combining numerical modeling and big data analysis with super-computers, the full-scale monitoring with modeling greatly improves our understanding of cryospheric processes and their changes under warming climate.

The impact of and adaptation to cryospheric change involve human societies and activities. Their observation and monitoring are quite different from those for natural processes. Social science approaches such as survey and questionnaire are often applied, though more advanced technology such as remote sensing starts to shape up the new development.

1.3.2 *Spatial and Temporal Scales for Cryosphere Development*

The cryosphere is very sensitive to temperature and precipitation. Climatic conditions are critical for the lifecycle of each cryospheric component. Together with terrain factors, the formation process of cryosphere components varies widely.

The cryosphere on earth is a continuous sphere that has a certain thickness of layer along with the atmospheric cryosphere. Considering the effect of latitude and altitude, the lower boundary of the cryosphere near the equator is at the highest elevation, reaching 5000–6000 m, such as glaciers over Kilimanjaro (3° 03′ 39.11″ S, 37° 21′ 35.69″ E) at 5897 m a.s.l. Elevation of the boundary of the cryosphere is lowered with an increase in latitude. At a very high latitude, it can drop to the sea level and even below it. For an example, subsea permafrost may exist at the bottom of Arctic Ocean.

A scale of spatial distributions of the cryosphere can be quite different. The continental cryosphere includes components with large area extent such as frozen ground (permafrost and seasonally frozen ground), Antarctic and Greenland ice sheets, and snow cover. It also includes components with relatively smaller extent such as mountain glaciers, river ice and lake ice. The component with a largest extent in marine cryosphere is sea ice, while subsea permafrost may be the second. Aerial cryosphere is in the shape of an ellipsoid around the Earth and may be the largest on the aspect of an area extent. The life cycle of cryosphere components also varies at different time scales (Fig. 1.6), in diurnal, seasonal, annual and inter-decadal periods.

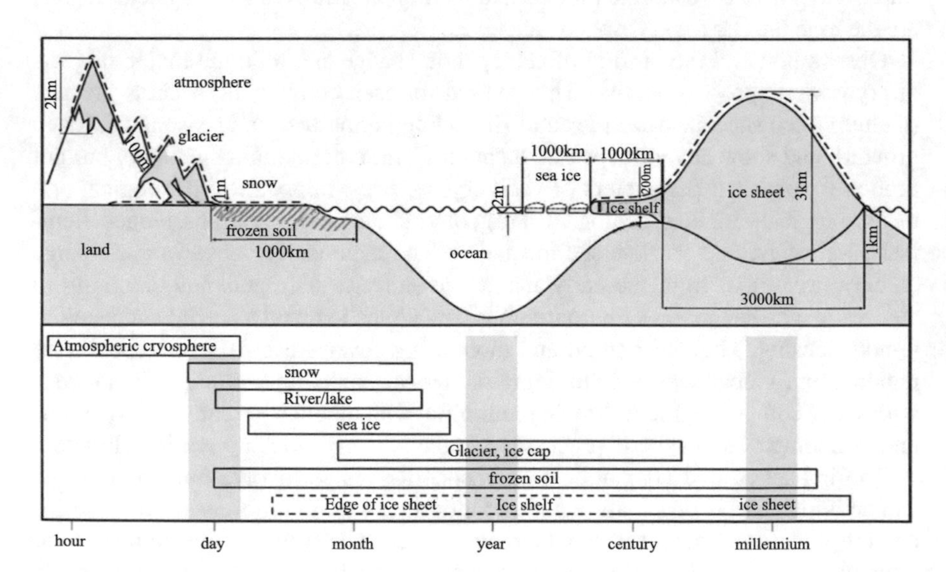

Fig. 1.6 Cryospheric components and their spatiotemporal scales (modified from IPCC AR4 WGI 2007)

For mountain glaciers in the continental sphere, ice flowing from the accumulation area to the terminus, the required time depends on glacial size, physical characteristics, terrain and climatic conditions, in a range from decades to thousands of years. For the Antarctic Ice Sheet and the Greenland Ice Sheet, it may require hundreds of thousands to millions of years. Over the past 1.4 million years, the East Antarctic ice sheet has remained stable. The oldest ice on the East Antarctic ice sheet may be over a million years considering rheological properties of ice body, climatic and terrain conditions. The extent of permafrost is much larger than one of glaciers and ice sheets. With global warming, permafrost is degrading from continuous to discontinuous one or seasonally frozen ground, while permafrost thickness would also decrease from both the upper and the lower boundaries on its vertical profile. River ice and lake ice usually exist during cold seasons and disappear completely when in summer. Snow cover melts every spring and may cause floods in mountain regions in the season.

In the marine cryosphere, the sea ice in the Southern Ocean and Arctic Ocean advances in cold seasons and melts in warm months. In general, the lifecycle of sea ice is around 12 months. The majority of sea ice belongs to the first year sea ice, however, there is a small fraction of areas where multi-year ice exists. Ice shelves exist from a few decades to several thousands of years. Iceberg mainly develops in Southern Ocean and Arctic Ocean with a lifecycle ranging from several months to several hundred years. The lifetime of iceberg depends on atmospheric circulation, sea water temperature, and ocean current, as well as its size, location, and time of the year breaking from ice sheet and ice shelves. The lifecycle of frozen water or ice crystal in aerial cryosphere is usually in days or even in hours.

1.3.3 Interactions Between the Cryosphere and the Other Spheres

Besides the cryosphere, the other spheres in the climate system refer to atmosphere, hydrosphere, biosphere, lithosphere, and anthroposphere. The cryosphere interacts with the other spheres. For example, it directly or indirectly affects sustainable development, geopolitics and national interests in anthroposphere, which also plays crucially important roles in cryospheric change (Fig. 1.5).

Cryospheric research integrates the interactions with impact and adaptation of cryospheric changes as well as sustainable development of human society, while linking natural and social sciences, science and policy making. It in turn enriches the contents of cryospheric science, and enhances its scientific and social values (Fig. 1.5, in the parts crossing spheres). A brief introduction is given here considering the richness in content about the interaction between cryosphere and the other spheres, but will be discussed great details in the following chapters (Fig. 1.7).

Cryosphere and Atmosphere: The cryosphere is a product of climate and it in turn affects weather, climate and climate systems through phase change, snow and

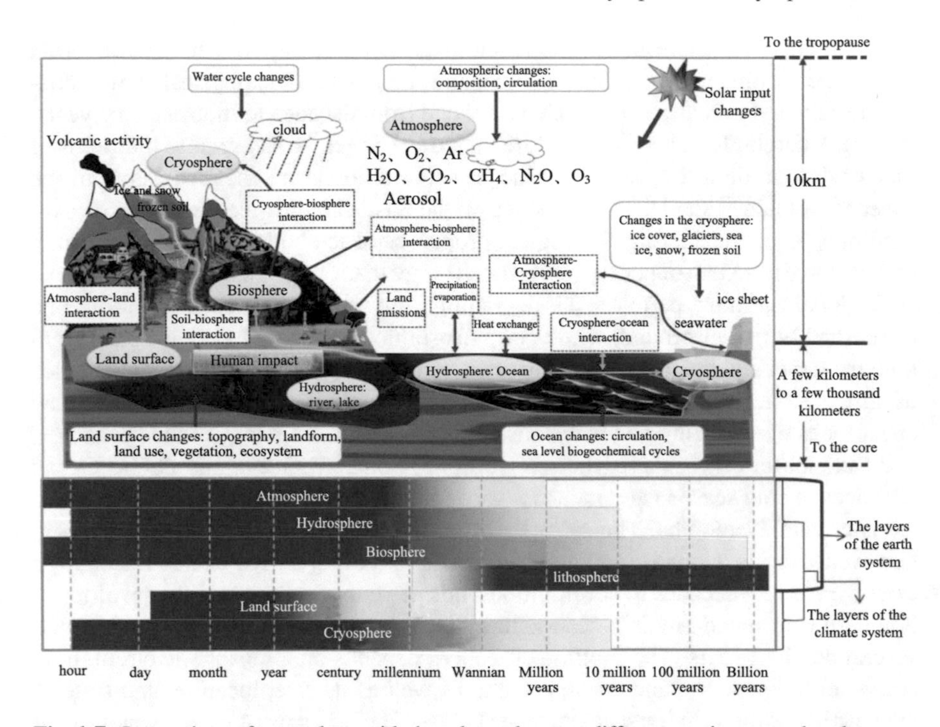

Fig. 1.7 Interactions of cryosphere with the other spheres at different spatiotemporal scales

ice albedo, and spatial changes as well. Interactions between the cryosphere and the atmosphere along with their physical dynamics is of interest to many disciplines.

Cryosphere and Hydrosphere: Ice is naturally "solid water". In the conventional definition of "four spheres" in the climate system, ice and water are in one sphere, or say, the cryosphere belongs to the hydrosphere. However, the perception changed in the 1980s. The cryosphere was separated from the hydrosphere as an independent sphere, considering that the impacts of cryosphere change on surface albedo and latent heat due to phase change. It may change the thermohaline of ocean currents and subsequently affect ocean circulation. Its regional and global implications in the climate system is also the effect on atmospheric circulation. It is understood that the cryosphere and its change can bring in benefits as well as adverse risks to human society. All the implications are different when water exists in a liquid phase in the hydrosphere over global and regional scales.

The cryosphere is a natural solid water reservoir on earth, which contains substantial fresh water. The water in the cryosphere participates in global water cycle, affecting the global and regional sea level, climate, ecosystems, environment, sustainable economic and social development, mostly concerned by scientific communities and broad societies. Cryospheric change causes sea level rise and change the global water cycle, and it hence affects runoff and distribution on regional and basin scales creating a huge impact on allocation and utilization of water resources.

Cryosphere and Biosphere: Biological processes develop over the interface between the atmosphere and the cryosphere, especially in permafrost or frozen ground regions at large. Biogeochemical and carbon cycles in permafrost regions are complex processes, further compounded by snow covers. On one hand, cryosphere change affects soil moisture content, soil thermal regime, and vegetation growth. On another hand, vegetation change significantly affects cryosphere processes, especially on seasonally frozen ground and snow cover. At the same time, cryospheric change affects ocean temperature, salinity, acidity and ocean circulation, further impact on marine ecosystems and associated socio-economy. Ice shelves, sea ice and icebergs affect ocean circulation and marine ecosystems. In addition, the cryosphere is closely related to crop farming, fisheries, forests, grasslands, built and natural ecosystems.

Cryosphere and lithosphere: Erosion and mass transport in the cryosphere can shape up surface terrains and landforms. Frost weathering, nival erosion, frost action, repeated freeze/thaw cycle are the driving forces to transform landscapes. Erosion and accumulation landforms are the results of these dynamic processes that record earth's environment evolution in geological ages. The land affected by the cryosphere in the past is nowadays a good habitat for mankind such as Western Europe and Nordic region where the land is fertile with a favourable climate, high population density and well-developed economy. The loose sediment in cryospheric regions could potentially pose risks of debris flow and flood hazards on downstream.

Cryosphere and anthroposphere: Changes in the cryosphere can generate more pressure on human society. The cryosphere is an important natural heritage to human beings and needs much more respect and protection. With global warming, the regions of cryosphere have increasingly become ecologically fragile. Human beings should revere nature and live in harmony with the cryosphere. It is a bitter lesson that back in the late 1950s, black/grey ashes were spread on the Qilian Mountains glaciers to artificially reduce surface albedo, in order to accelerate the melting of snow and ice for water supply. It is further noticed that local and regional geopolitical problems become clearly more urgent due to cryosphere retreating. One way to deal with the problems is to strengthen investigation of cryosphere for solutions, which supports improved multilateral exchanges and dialogues among countries in relevant regions.

1.3.4 Roles of Cryosphere in Climate System

Cryosphere not only affects atmospheric circulation, surface energy balance, hydrological processes and water resources, but also regulates climate at different temporal and spatial scales. A few examples are here introduced.

1. Snow

Snow is closely related with climate, especially with monsoon and various atmospheric signals. Seasonal variation of snow albedo and latent heat of snowmelt affects surface energy balance. Snowmelt water may penetrate into soils and change soil water content. The latter has a "memory" function, which is one of the main reasons of snow affecting short-term climate. As early as in the late 19th century, scientists found that snow cover had a great influence on the Indian summer monsoon in Himalayas. When larger snow cover occurs in Himalayas, onset of the Indian monsoon comes later which brings less precipitation, and vice verse. The winter snow cover anomaly in the Qinghai-Tibet Plateau also has a great impact on the East Asian and South Asian summer monsoons. Precipitation would increase in eastern and southern part of India and northwest bay of Bengal but decrease in Indo-China Peninsula when snow cover extent increases in the eastern part of the plateau.

Changes in snow cover and depth are also correlated to atmospheric signals such as the North Atlantic Oscillation (NAO) and the Arctic Oscillation (AO). Increase of Eurasian Continent snow depth is linked to water vapor transport path where NAO has been significantly enhanced and cyclonic circulation has been changed significantly since the 1970s. A study of relationship between snow and climate showed that decrease of temperature in autumn and decrease of sea ice extent in Arctic Ocean will lead to increase of sea surface evaporation and tropospheric moisture, which will increase snowfall and snow cover in Eurasian continent.

2. Feedback effect of ice and snow albedo

The surface of clean ice and snow has a significant higher albedo (up to 0.9) to solar radiation than other land surfaces (usually below 0.2). The small changes in surface albedo can significantly alter surface energy balance of the land-atmospheric system, resulting in weather and climatic change. Reduction of surface albedo means that more solar radiation is absorbed by the land surface leading to increase in surface temperature. The long-wave radiation is further enhanced to heat up atmosphere, which accelerates the warming of air. As a consequence, increasing temperature further accelerates snow and ice melting that further reduces the surface albedo, and finally leads to positive feedback effect, known as the so-called "snow and ice albedo feedback".

Global energy distribution is affected by changes in snow and ice coverage, snow particle size, surface humidity, as well as impurities and black carbon content in ice and snow through the albedo effect. Global warming would result in reduction in snow and ice coverage, further impact on surface energy balance and global energy distribution. As a result, the surface would be heated up by absorbing more energy, further increase air temperature, and vice versa.

When snow melts under a warming climate, the size of snow particles will increase and firnification process will accelerate. Observation evidences show that an increase in snow particle size and particle surface humidity would lead to a decrease in snow surface albedo. However, quantitative assessment of the problem is yet sufficient.

Black carbon originates from in combustion of fossil fuel and biomass. Albedo feedback effect of snow and ice will be strengthened when albedo of snow and ice is reduced by black carbon deposited onto their surface. The IPCC AR5 WG1 reported that radiative forcing of black carbon in snow and ice can be as high as 0.04 [0.02–0.09] W/m^2.

3. Cryosphere and ocean

Ocean is the largest component of the hydrosphere. The interaction between the cryosphere and ocean involves sea level change, which has an effect on oceanic conveyor belt and ocean currents, and subsequently global climate.

In winter, the sea surface water of high latitude freezes into sea ice, and salts subsequently precipitate, resulting in increase in the salinity and density of underearth sea water. The more dense sea water slowly sinks to sea bottom forming underlying cold water flow. In contrast, the melting sea ice discharges a large amount of low salinity water into ocean surface in summer. These processes develop into Meridional Overturning Circulation (MOC) due to surface and deep marine water bodies' exchange. The most representative Atlantic Meridional Overturning Circulation (AMOC) is a significant segment of oceanic conveyor belt and a key factor to regulate Earth's climate. Ocean conveyors bring warm ocean current to high latitude area in northern hemisphere and transmit cold water with deep ocean current in the North Atlantic to the Equator as part of ocean circulation and climate interrelation. The key phrase of this activity is the overflow water at Danish Strait where low-density and high-density sea water exchanges through Greenland-Scottish sea ridge. Scientists believed that overflow water at Danish Strait mainly comes from the East Greenland, but Icelandic oceanographers questioned that they found a deep ocean current flowing along the Icelandic continental slope to flow south and named Icelandic current. Data showed that Icelandic ocean current is a key factor contributing to AMOC that really brings overflow water back to Danish Strait.

As a result of global warming, AMOC also shows a sign of gradual slowing-down. Ocean current can change regional climate to make sea ice melting in northern North Atlantic Sea and produce lower salinity water into ocean. In addition, accelerated melting of Greenland ice sheet releases fresh water to prevent from subsidence of sea surface water. As a result, it diminishes the ocean current cycle between low latitude and high latitude, leading to sea surface freeze in this area. Further snow and ice albedo feedback effect causes climate cooling in northern hemisphere, or a sudden change in climate. This scientific speculation is very significant and scientists recognize that AMOC process can look into future climate and circulation relationship.

With the global warming at present, ice volume of the Greenland and Antarctic ice sheets is in negative mass balance. More fresh water is released into ocean, leading to sea level rise, surface water warming and desalting in high latitude ocean with Thermohaline Circulation (THC) affecting global climate.

4. Cryosphere and biogeochemical cycle

Permafrost evolution including the freezing and thawing processes, and biological populations in permafrost regions are all related to global carbon (C) and nitrogen

(N) cycles. The formation and degradation of permafrost directly affect regional and global biogeochemical cycles.

Soil organic C stored in permafrost regions in the northern hemisphere is about 1832 GtC which is 3.3 times of the total human emissions (555 GtC) since 1750. With global warming, permafrost degradation may accelerate the release of carbon stored in permafrost, increasing the concentration of atmospheric greenhouse gases and ultimately accelerating global warming. Changes in permafrost play an important role in C and N cycles between the terrestrial ecosystem and atmosphere. CH_4 release rate continues to increase in permafrost regions across the Arctic. CH_4 has been also released from the Arctic Ocean's subsea permafrost.

1.4 Role of Cryospheric Science in Economic and Social Development

The effect of cryospheric change in a warming climate on social and economic development is important for the continuous efforts in sustainable development, and human benefit.

Announced in December, 2016, the Paris Agreement proposed global average surface temperature would rise less than 2 °C by 2100 in comparison with 1750, with more efforts to reduce emission to keep the rise capped at 1.5 °C. While the eventual outcome depends on the future socio-economic pathway, it would feedback and impact on socio-economic development. Playing a significant role in climate system as well as providing benefit to human, the cryosphere is a critical part in our meeting the target. Cryospheric science should pay attention to the Paris Agreement that is also linked to social and economic transformation.

1.4.1 Water Cycle and Water Resources

Cryosphere is one of the most important freshwater resources. The most of large rivers originate from cryospheric areas and continue providing water supply to a vast land and population. However, all of these would likely be changed as a consequence of glacier retreat, permafrost degradation and snowfall reduction. Currently, many glaciers have been experiencing accelerated ablation leading to an increase in discharge and runoff, but observation and research have indicated that the discharge and runoff would experience a tipping point where they start to decline. By then, food security, human health, eco system and economic and social development would face a great challenge.

In the cryosphere, permafrost contains ground ice. Its change affects surface runoff, groundwater reserve, and exchange between surface and groundwater. Snow can change the yearly runoff distribution of rivers depending on melting seasons.

Snow cover accumulates and its depth generally increases in winter and start to melt in spring. By autumn most of snow melt to supply river runoff with some expecting sublimation into atmosphere. Coupled with terrain factors, spring snow may cause flooding disaster.

Upon the current circumstance in western China, it is necessary to evaluate how the change of cryosphere affects the runoff in an inland river basin, change in water resources in the cryosphere by 2030, 2050 and 2100 when 1.5 and 2 °C global warming. By taking into account the facts in the arid and oasis areas in western China and the threshold of loss in cryospheric service, solutions and pathways for development are optimised by maximising the service.

1.4.2 Cryospheric Disasters

The frequency and intensity of cryospheric disasters are now increasing and the corresponding consequence becomes more severe. It is an important part of cryospheric science to study the reduction of disaster risk, and the development of strategies to minimize the adverse consequences directly or indirectly caused by changing cryosphere.

In the continental cryosphere, the changes in mountain glaciers and subsequently in river runoff lead to triggering hazards such as drought or flooding, affecting agricultural production, ecosystems, human settlements. Landslides and glacial lake outburst flooding (GLOF) can cause fatalities, affect people, damage properties, destroy roads and other infrastructures. Snow and frozen rain disaster may cause freezing injury, road and transmission line icing. Frost heave and thaw settlement can be induced by repeated freezing and thawing processes of the active layer and permafrost degradation, damaging infrastructure and engineering construction in permafrost regions, such as roads, airports, oil pipelines, communication lines etc. (Fig. 1.8). Snowstorm can become a severe hazard that would frequently occur in the area such as northern China during winter and spring. It would bring a very cold front to severely affect agriculture, livestock and transport as well. Moreover, it is found

(a) (b)

Fig. 1.8 Highway and railway damages produced by repeated frost heave and thaw settlement

that the date of spring flooding has shifted one month earlier in the last 30 years in Tianshan Mountains, a northern part of Xinjiang. Also, hundreds of people die due to avalanche each year around the globe.

In marine cryosphere, changes in the extent, thickness and variability of sea ice may cause interruptions to maritime transport, marine operations, offshore facilities and coastal infrastructure. Coastal erosion in permafrost areas impact on infrastructure. At the same time, sea level rise may cause seawater intrusion, land and groundwater salinization. It may also intensify storm surges to threaten coastal cities, in particular, in low-lying areas.

In aerial cryosphere, meteorological hazards are related to low temperature, such as damage caused by frozen-rain, concerned in both cryospheric science and atmospheric science.

The early warning of cryospheric disasters can significantly reduce casualty and property losses. Chinese scientists successfully provided warning on the outburst of Mertzbakher Ice Lake in Kyrgyzstan in 2011 and 2012, which reduced losses to a minimum at downstream areas.

1.4.3 Mineral Resources and Engineering Construction

The demand of mineral resources for economic development has accelerated exploitation in cryospheric regions, where railways, highways, airports, oil fields and number of towns have sharply increased in recent decades. Infrastructure development such as marine drilling, oil and gas exploration, and oil pipeline construction has faced many cryosphere-related technical problems in the regions, for example, Arctic regions in Canada, northern Alaska, Siberia and Nordic land and coastal continental shelf. At the same time, new challenges of environmental protection have also arisen.

For more examples, when Russia constructed a Siberian railway and China Eastern Railway to take into account its geopolitical strategies in the late 19th century, it encountered many problems in association with permafrost. Exploration of oil and gas in the Arctic continental shelf experienced damages of drilling platform crashed by sea ice drifting. It is now understood that construction and operation cost of engineering projects, along with the issues of safety and service life are substantially affected by permafrost, ground ice, sea ice, river ice and lake ice, snow and drifing snow. These have otherwise brought opportunities to develop and improve engineering technologies in cryospheric regions.

In retrospect of the development in China, there has been a huge demand for resource development, transportation and infrastructure construction in cryospheric regions. The development of northeastern region in China in the early 1950s experienced severe engineering geological problems under low temperature conditions such as icing, and engineering structures were damaged by snow, river and lake ice, soil frost heave and thaw settlement. Mining of mineral resources in western China also experienced engineering geological problems in permafrost regions in the 1960s.

The development of the Qinghai-Xizang (Tibetan) Railway and the construction of oil pipeline and communication line from Golmud to Lhasa in permafrost regions in the late 1970s faced similar issues.

By applying the knowledge of cryospheric science for practical purposes, the state-of-the-art "technology and theory of cooling roadbed using crashed rocks" was implemented by Chinese permafrost scientists for the construction of Qinghai-Tibetan Railway, and it successfully solved roadbed stability problems in the 21st century. Recently, advances in the development of Western China and the Belt and Road Initiative proposed by the Chinese government require more contributions from cryospheric sciencc. These include studies of the cryospheric change impact on major projects under different global warming scenarios to ensure their reliability in the future.

1.4.4 Adventure and Tourism in Cryospheric Regions

At high latitude and altitude in cryospheric regions, the environment is generally extreme, such as polar day, polar night, low oxygen, extremely low temperature and severe snowstorms. However, they are also ideal destinations to be explored for tourism, sports and adventure, which would satisfy people's desire to explore as well as to obtain the understanding of the nature, and in turn the awareness of protecting it for its harmonious development.

Antarctic, Arctic, high mountain regions such as Qinghai-Tibetan Plateau and those in all continents are the areas that people are looking for. The early exploration and discovery enhanced the understanding of the cryosphere and laid a foundation for the later scientific investigation as well as resource exploration and regional development. At the same time, more and more publics start to be involved in adventure and tourism in cryospheric regions as their living standard is gradually improved.

There is a wide range of adventure and tourism activities in cryospheric regions which may include, but not limited to, mountaineering, ski, dog sledding, sailing, sightseeing and pilgrimage. In fact, environmental protection and economic development can be combined in association with the activities, benefiting others services such as restaurants, hotels, transportation and so on. In this regard, tourisms in cryospheric regions have been well developed and promoted by many countries. All these play an important role in cultural exchange, employment, and prosperity of regional economy and communities.

Cryosphere in China is mostly located in mid-latitude and low-latitude regions. There are many glaciers and more than 200 skiing fields as abundant tourism resources to be explored or improved. It is envisaged that snow-ice sports and tourism in cryospheric regions may undergo considerable development given the successful bid of the 2022 Winter Olympic Games in Beijing–Zhangjiakou.

Having said that, environmental protection has to be closely watched while developing cryosphere tourism and adventure.

1.4.5 Benefits of Cryosphere to Human Society

Cryosphere can provide a wide range of services on the aspect of climate, ecosystem, resources, infrastructure, tourism, leisure, sports, adventure and culture. Cryospheric services can be generally categorised into provisioning, regulating, supporting and culture. The provisioning service includes water, cold pool, germplasm resources, gas hydrates etc.; the regulating service mainly deal with climate, runoff, water conservation, environment etc. The supporting service can be related to ecosystem and habitats as well as engineering construction. The social and cultural services in cryopsheric regions are those such as snow tourism and leisure, sports, research and education, religious and spiritual demands of indigenous people in the cryosphere.

Playing important roles in the cryosphere, Antarctic and Arctic regions are geopolitically sensitive. It requires visions to strengthen strategic regional collaborations, which includes the enhancement of regional collaboration in scientific research of the cryosphere.

1.4.6 Geopolitics in Cryosphere

The cryospheric geopolitics is a science dealing with issues such as politics, economy, society, culture, history, resources, environment, territory, military and national security, in relation to formation, growth and evolution processes of the cryosphere and its components. Very often, due to thc special geographic location and natural features of the cryosphere, along with the geopolitical development around it, its change would exert threats to national security and human well-being. Meanwhile, normally located in the high latitude and altitude, the cryosphere is often in poor and stricken areas which are paid little attention. With more accessibility to the areas due to global warming as well as technological advances together with the increasing appetite for resources, the cryosphere is of concern, not only to scientific community, but also to policy makers and general public as a whole.

It is noticed that the extent and thickness of Arctic sea ice have significantly declined in the recent decades. The distance of sea transport between East Asia, Europe and North America has been considerably shortened with the Northeast and Northwest channels opening up for navigation. It thus has a dramatic impact on business and environment, considering the subsequent saving of transport time and fuel as well as reduction in energy consumption and carbon emission. However, numerous disputes or political games have emerged for territories and resources, when Arctic the sea ice retreated creating conditions for new mineral resources exploration and development over the continental shelf and Arctic Ocean basin. As reported by the media, the disputes on Arctic continental shelf have been lingering on between the United States and Canada since the 1990s. Eight countries surrounding the Arctic Ocean have claimed more or less the territory and development right for its resources. A Russian expedition team reached seabed of about 4000 m below sea level near the North Pole by using a deep sea submarine and raised a titanium Russian flag on August 2, 2007. Moreover, Russia proposed to the United Nations that they own the right for resource exploration on the seabed of 1.2 million km^2 on August 4, 2015.

Antarctic is one of the Earth's seven continents, also known as the hotspot for scientific research. During the International Geophysical Year 1957–1958, hundreds of thousands scientists from 12 countries stepped on Antarctica for collaborative scientific research, but geopolitics has been plaguing it. The British announced sovereignty over a fan-shaped land and water areas of the Antarctic Peninsula in 1908. After that, Australia, New Zealand, France, Chile, Argentina and Norway also raised a territory requirement. After a broad consultation, the United States, the Soviet Union and other 12 countries signed the Antarctic Treaty in December 1959, which came into force in June 1961. The Antarctic Treaty recognizes that for all mankind's benefit, all States are free to carry out Antarctic scientific expeditions, develop international cooperation, freeze territorial requirements. The Antarctic Treaty ensures that Antarctica is only used for peaceful purposes. It is the most successful international treaty in the 20th century which ensures cooperation between countries and peace in Antarctica. At present, there are more than 150 research stations and bases in Antarctica, belonging to 20 countries respectively. China has established four scientific research stations there.

The Qinghai-Tibet Plateau has been considered as the Third Pole in the world. The change in the cryosphere impacts on rivers originating from the Third Pole, which directly or indirectly supplies water for nearly 2 billion people in the downstream. There is a high risk of international disputes in the region for water resources. Additionally, it has been claimed that the 21st century would be a water century. In this regard, competition for water resources and geopolitical issues may become more prominent. The impact of cryospheric change is not only a regional but also global issue, involving environment, resources, and most importantly, territorial sovereignty.

1.5 Planetary Cryosphere

The Earth is one of the eight planets in the solar system. Although we have had a better understanding of Earth's cryosphere, a key question remains: that is, whether there exists cryosphere on other seven planets. If so, how different are planetary cryosphere from Earth's cryosphere?

Humankind has been improving the knowledge and understanding of planets in the solar system, as a result of more and more spacecrafts launched. It is found that there do exist cryospheres on planets of Mercury and Mars. There are also cryospheres over dwarf planets and moons, such as Pluto, Ceres, Europa, and Titan. The major composition of those planetary cryospheres is water ice. However, their cryospheres have very different physical properties from Earth's cryosphere due to the fact that there are very different temperature and pressure conditions. We use cryosphere on Mars as an example for the following discussion.

1.5.1 Characteristics of Mars' Cryosphere

Mars is a terrestrial planet in the solar system. Its orbital period is about 687 Earth days (1.88 Earth year) or 668.6 Mars days. Mars' rotational speed is very close to Earth's, about 24 h 39 min and 35.244 s, i.e., 1.027 Earth day. Its axial tilting angle is 25.19°, close to Earth's 23.5°. Thus, Mars also has four seasons each year. On September 26, 2013, Mars Curiosity Rover discovered that Mars' soil contains water ranging from 1.5 to 3% by mass. Because Mars is 1.52 times as far from the Sun as Earth, it receives much less solar radiation, or only 43% of the amount received by Earth. Therefore, its surface temperature is much lower. The global mean surface temperature is about −63 °C. The Mars Rover's measurement indicates the Martian surface temperatures are in a wide range from −80 to 0 °C. Mars has a thin atmosphere. Surface air pressure is less than 1% of Earth's surface pressure. The average surface air pressure is about 600 Pa. The major composition of the Martian atmosphere is CO_2, approximately 96%. The other components include N_2 (3%), Ar (1.6%), O_2, and water.

Mars has permanent polar ice caps (Fig. 1.9). The ice caps were believed to be CO_2 dry ice. It is now known that the ice caps at both poles consist primarily of water ice (70%), while CO_2 dry ice is only a tiny thin layer on the surface. The dry ice layer on the north pole is about 1 m thick in the northern winter when CO_2 is frozen into dry ice, and it sublimes in the northern summer. By contrast, dry ice on the southern cap is permanent, which is about 8 m thick. The northern polar ice cap has a diameter of 1100 km and a depth of 2 km. Its volume is about 1.9×10^6 km^3, which is comparable to that of the Greenland ice sheet whose volume is 2.85×10^6 km^3. The southern polar cap has a diameter of 370 km and 3.0 km in thickness. Mars

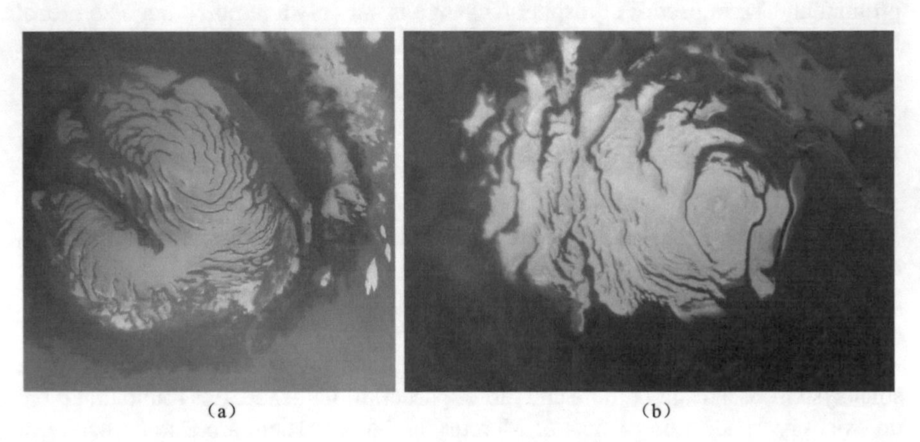

Fig. 1.9 Mars' polar ice caps. **a** North polar ice cap in the early summer of 1999 (from NASA/JPL/MSSS, http://photojournal.jpl.nasa.gov/catalog/PIA02800), and **b** South polar ice cap in the middle summer of 2000 (from NASA/JPL/MSSS, http://photojournal.jpl.nasa.gov/catalog/PIA02393)

has two kinds of cryospheres. One is water ice cryosphere, and the other one is CO_2 dry ice cryosphere.

Both polar ice caps on Mars show spiral troughs. Recent studies suggest that the spiral troughs are caused by descending air flows over the cold ice caps (katabatic winds) due to the Coriolis effect. The spectrometer at Mars Spacecraft Odyssey revealed that a large amount of water is trapped in the permafrost mantle, and that the permafrost area extends from the pole to 60° where the surface soil layer of one meter thick contains water ice by 60%. Mars' landscape is represented with valleys and channels. Some of them seem to be carved or eroded by runoff liquid water in early Mars history.

At present, liquid water cannot exist on Mars surface because of very low surface temperature and pressure. However, many studies suggested that Mars was much warmer in its early history, and that it was warm enough to hold liquid water on the surface. Some scientists even suggested much of the low plains of the Northern Hemisphere on Mars was covered with an ocean of hundreds of meters deep. Given that there exists liquid water on Mars today, it is likely that liquid water can only exist in lakes under the polar ice caps. Water ice at middle and low latitudes can only exist in permafrost soil.

1.5.2 Evidence of Mars Water–Ice

On July 8, 2005, European Space Agency (ESA) released a picture that shows water ice in a crater (Fig. 1.10a). This crater is located at the northern plain (70.5° N, 103° E), with a diameter of 35 km and depth of about 2 km. The water ice in the crater is about 200 m thick. Mars surface demonstrates glacier landforms. The most recent evidence shows that water ice exists on Mars in the form of glacier, which is under dust and rock debris. In March of 2010, radar detection showed that there is water

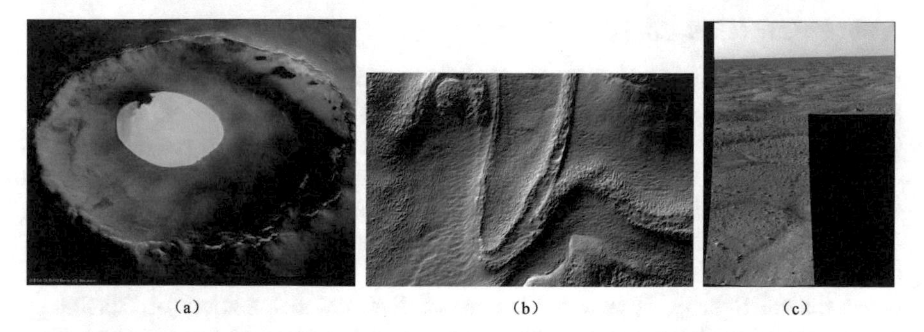

Fig. 1.10 **a** A photo of crater with water ice by Mars Express of European Space Agency (ESA) (from ESA/DLR/FU Berlin); **b** the tongue shaped glacier of Mars (from http://www.msss.com), and **c** a photo of polygonal ground by the phoenix Mars lander near the landing point in the north pole of Mars (from NASA/Jet Propulsion Lab/University of Arizona)

ice several meters under the surface. Dust and rocky debris on ice glacier indicate the direction of glacier movement. Ice sublimation results in rough ice surface with holes, and dust and rock fill in the holes. As the glacier melts and retreats, dust and rock are left behind to form glacier debris. Photos show twisted ridges of glacier debris, which likely form in a late time (Fig. 1.10b).

Planetary scientists have long believed that there are permafrost regions on Mars. Observations indicate that there is water ice under the surface in these regions. Some areas of Mars surface show polygon-shaped features (Fig. 1.10c). On Earth, such a landscape feature can be seen at high latitude permafrost regions, which is caused by repeated frost crack. Thus, the polygon-shaped landforms may also be due to frost crack of Martian soil. On July 31, 2008, NASA announced that Mars Phoenix lander discovered water vapor as Martian soil was baked, confirming the existence of water on Mars.

Questions

1. Explain the following terms according to this course or related literature: cryosphere, cryospheric science, climate system, climate change, greenhouse gas, greenhouse effect, radiative forcing, attribution, impact, adaptation and mitigation of climate change, socio-economic scenario, climate scenario, projection of climate change.
2. What is cryospheric science about?
3. How do you understand the relationship between cryosphere and the other spheres of climate system? Briefly explain the relationship between cryosphere and human economy and society.
4. Traditional geography classifies cryosphere into hydrosphere, while climate system lists it as an independent sphere, why?

Extended Readings

Representatives

1. Shi Yafeng (1919–2011)

Born in Jiangsu Haimen, he was a geographer, glaciologist, and a pioneer of glacier and permafrost research in China. He was elected as a committee member of the Chinese Academy of Sciences in 1980, known as academician. He obtained his master degree in 1944 from History and Geography Institute, Zhejiang University. He was honorary Director of Cold and Arid Regions Environmental and Engineering Research Institute in Chinese Academy of Sciences. He was also Research fellow of Nanjing Institute of Geology and Limnology, Honorary Director of Chinese Geographical Society, Honorary Member of International Society of Glaciology and International Quaternary Association, and member of Royal London Geological Society. Shi Ya-feng took on the research of glaciers and geographical environment for a very long time. As early as in the 1950s, he led teams for many times, carrying out surveys of Qilian Mountains, Tianshan Mountains, Himalayas and Karakorum Mountains glaciers, and subsequently suggested that high Asian glaciers could be divided into oceanic glaciers, subcontinent and polar continental glaciers. He was a lead author for the multiple volumes of Chinese glacier inventory, which investigate the glaciers in a great detail. He proposed the maximum ice age in the Qinghai-Tibet Plateau that occurred 600,000–800,000 years ago, but did not form ice caps. He reconstructed the extent and climate for the last glaciation 20,000 years ago, which suggested prevailing warm and humid climate in the Qinghai-Tibet Plateau and the East Asian continent 30,000–40,000 years ago. He had a strong view that glaciers existed only on a few mountains in eastern China, but not in Lushan, Huangshan, or Beijing Xishan. He pioneered the research on permafrost and debris flow, and made a great contribution to the knowledge of water resources in northwest China, quaternary environmental evolution, and effects of global warming on sea level rise. He published more than 200 papers, edited more than 20 monographs. He was nationally awarded, including the 1st, 2nd and 3rd prize of National Natural Science, the 2nd prize of National Science and Technology Progress, several of the 1st and 2nd prize of Chinese Academy of Sciences on Natural Science. Among many others, he was also awarded with The Holeung Ho Lee Foundation for Science and Technology Progress. He was awarded with Achievement prize from China Geographical Society Geographic Science, great contributor to Science and Technology by Gansu Provincial government as well as by China Quaternary Research Society.

2. Kudlyavtsev B. A. (1911–1982)

Professor B. A. Kudryavtsev was a founding member of University of Moscow permafrost school and made an important contribution in many areas of modern permafrost. As early as in 1950s, B. A. Kudryavtsev inherited and developed geophysical research of permafrost. He advocated combination of geography and mathematical physics in permafrost and applied thermal view to studying permafrost formation and development. He believed that permafrost is a product of heat exchange between crustal surface in geologic environment and atmosphere and changes with each factor change. He proposed the system of permafrost formation classifications, which were recognized and affirmed by the Soviet permafrost community. He summarized predecessor's trends and changes of permafrost development in geological and geographical factors and made these factors become basis of judging evolution of permafrost. Professor B. A. Kudryavtsev not only made a creative contribution to basic theory of permafrost science but also applied the latest theory of permafrost into permafrost engineering. He proposed the method of mapping permafrost and led the compilation of 1: 200,000 permafrost map. He led to establish permafrost-geological survey method and standard which approved and published by former the Soviet Union Department of Geology and Mineral Inspection in permafrost. Professor B. A. Kudryavtsev was awarded the first Lomonosov Prize in 1977.

Classic Work

The global cryosphere: past, present and future

Authors: Roger G. B., Thian Y. G.

Publisher: Cambridge University Press, 2011

Introduction: This is a book that introduces distribution, classification and evolution and future changes of global cryosphere. This book provides a detailed introduction to main elements of cryosphere such as glaciers, ice sheets, snow, river ice, lake ice, permafrost, sea ice and iceberg with emphasis on status, change and future change. This book is professor Barry's experience and summary of his teaching and research work in many years. However, this book did not deal with the adaptation and sustainable development and the other issues associated with the changing cryosphere.

The book has a total of 458 pages, which are divided into four parts, eleven chapters. The four parts are continental cryosphere, marine cryosphere, the past and the future of cryosphere and application of cryosphere, respectively. The book is rich in content, material and information. There are many experts' and scholars' biographies, color pictures, photos, annotations, definitions and explanations etc. It is a more comprehensive reflection of the whole picture of cryosphere related English monographs, read the book is very helpful to understanding cryospheric science especially for high schools and above qualifications in environmental science, geography, geology, geomorphology, climatology, hydrology and water resources, marine and climate change science and other fields of scientific and technological personnel.

Chapter 2
Classification and Geographical Distribution of Cryosphere

Lead Authors: Cunde Xiao, Jiahong Wen

Contributing Authors: Shiyin Liu, Tingjun Zhang, Lijuan Ma, Renhe Zhang, Lin Zhao, Minghu Ding

This chapter describes the classification of the cryospheric components and their geographical distribution in the world. The first part introduces the overall distribution of the continental cryosphere, marine cryosphere and aerial cryosphere on the global scale, followed by the classification and geographical distribution of glaciers and ice sheets, frozen ground, snow cover, sea ice, river and lake ice, and all solid water in the atmosphere. The geographical distribution of a single cryospheric component is introduced at the global scale, then at the hemispheric scale and finally at the national scale.

2.1 Distribution, Composition and Classification of Cryosphere at Global Scale

2.1.1 Zonality of Cryosphere

Climate is the primary factor that determines the formation and development of the cryosphere. Therefore, distribution of the cryosphere is generally consistent with the specific climatic zone and shows certain latitudinal and vertical zonalities.

Latitudinal and vertical zonation of the continental cryosphere. Because the earth is spherical and the solar zenith angle causes the solar radiation to be distributed unevenly along the latitude, the zonal distribution of solar radiation influences various processes of the climate system directly and indirectly. Primarily, it generates all of the atmospheric processes and meteorological factors showing zonation, such as air temperature, pressure, circulation, evaporation, atmospheric humidity, cloud and precipitation. As a synthesis of these factors, the climate system also eventually shows the zonal distribution or the zonal climate differences. The continental cryosphere distribution shows obvious variability of latitude, and cryospheric components, such as glacier, snow cover, sea ice, river ice, lake ice, seasonal frozen ground and permafrost, are mainly distributed in the mid-high latitudes. The Antarctic ice

D. Qin et al. (eds.), *Introduction to Cryospheric Science*, Springer Geography,
https://doi.org/10.1007/978-981-16-6425-0_2

sheet and the Greenland ice sheet are each located on a high-latitude continent in opposite hemispheres, so the continental cryosphere is distributed in an obvious latitudinal zonation.

As stated above, the continental cryosphere also shows the law of vertical zonation. Because temperatures generally decrease with increasing mountain height, rainfall and humidity of the air at certain heights increase with altitude, thus causing the phenomenon of vertical changes to natural environment and its composition, which is known as vertical zonation. Unlike the latitudinal zonality, the temperature of the vertical zone decreases with height is caused by the rapid increase of long-wave radiation with height, which results in the drop of radiation balance and temperature, rather than the change of the incident angle of the sun's rays, which causes the decrease of solar radiation and temperature. Generally, there exits vertical zonal differentiation once there is sufficient altitudinal changes. Significantly affected by temperature, the continental cryosphere shows clear vertical zonation. The snow-ice belt is always distributed in the highest altitude of the vertical zones, which is controlled by latitude and, therefore, shows different altitudes at different latitudes on Earth. This differentiation is especially obvious in mountainous and plateau regions; For instance, the summit of the Mount Kilimanjaro, which is located near the equator, is covered with snow year round. In the Qinghai-Tibetan Plateau, due to the high altitude and low temperature, except for sea ice, there are developed most components of the cryosphere, such as glacier, permafrost, snow, river ice and lake ice. Vertical zonation is the mesoscale regional differentiation and is restricted by large-scale geographical differentiation. The type of vertical zonation spectrum is closely related to the basal belt (controlled by the location of the mountain) and the mountain height. In the Qinghai-Tibetan plateau, the distribution of cryospheric components is significantly controlled by altitude; moreover, it is also influenced by both the latitudinal zonality and the distance to the ocean. For example, in the western region of Mt. Kunlun, the snow line is at the elevation of 5600–6000 m a.s.l., and the lower boundary of permafrost is at approximately 4550 m. Another example is the average snowline of the Xiao Dongkemadi Glacier, which is 5620 m a.s.l., in Mt. Tanggula of the middle Tibetan Plateau, while the lower boundary of the permafrost, 4780 m a.s.l., is in the southern foothills of this region. However, in the eastern Tibet Plateau, the snow line of the Mt. A'nyemaqen ranges from 4950 to 5200 m a.s.l., and the lower boundary of the permafrost is 4000–4050 m a.s.l.

Latitudinal zonation of the marine cryosphere. The marine cryosphere is controlled mainly by latitudinal zonality. Solar radiation varies with latitude; the higher the latitude, the less solar radiant energy. Therefore, the distribution of sea ice has a strong latitudinal zonality. For example, sea ice over the Southern Ocean expands in winter and subsides polarward in summer, showing a significant latitudinal zonality.

2.1.2 Components and Distribution of Cryosphere

The main components of the cryosphere include glaciers, ice sheets, permafrost, snow, ice, ice shelves, icebergs, river ice, lake ice and solid precipitation (Fig. 2.1).

The snow cover in the northern hemisphere reached its maximum in January, with an average of 45.2×10^6 km^2 from 1964 to 2004 and reached its minimum in August with an average of 1.9×10^6 km^2, from November to the following April. Snow covers more than 33% of the northern hemisphere's land area on average, and as much as 49% in January. In the southern hemisphere, snow is mainly distributed in the southern part of South America, on the southern island of New Zealand and in the high mountains of eastern Australia. The role of snow in the climate system differs in different latitudes and seasons, giving strong, positive feedback related to albedo and weak feedback related to water storage, latent heat and the surface heat insulation effect.

Rivers and lakes in high latitudes are generally covered by ice during winter. Although the ice volume and surface area are smaller compared with other cryospheric components, yet they play an important role in freshwater ecosystems, winter traffic, bridge, and pipelines. Therefore, the thickness and duration of ice on these water bodies has an important impact on the natural environment and on human activities. For example, the melting and breaking-up of river ice often forms an "ice jam", which impedes river channel and causes severe ice jam flood.

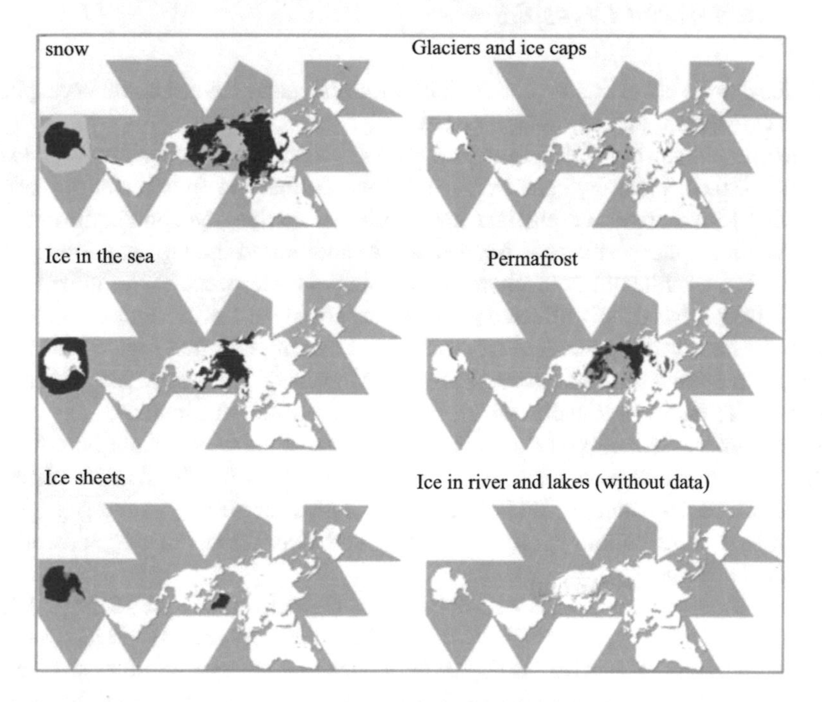

Fig. 2.1 Distribution of major cryospheric components (dark blue) (after UNEP 2007)

The continental cryosphere stores 75% of the world's fresh water. The amount of water stored in the Greenland and Antarctic ice sheets would account for a sea level rise of approximately 7.36 m and 58.3 m, respectively. The change in the amount of continental ice has led to sea level variation over the last century, and glaciers and snow are still important sources of fresh water.

Currently, ice covers 10% of the land surface, and Antarctica and Greenland make up the bulk of it (Table 1.1). On average, 7% of the ocean's surface is covered by ice. In the midwinter, 49% of the land surface area of the northern hemisphere is covered with snow. Frozen ground covers the largest area of all cryospheric components (Fig. 2.1). Changes in different cryospheric components occurred at different time scales based on thermodynamic character.

The maximum extent of Arctic sea ice is 15×10^6 km^2, while in summer, the minimum is 6×10^6 km^2. The Antarctic sea ice extent (SIE) changes by a large seasonal amplitude, with the maximum SIE larger than 18×10^6 km^2 in winter, while the minimum SIE in summer is only 3×10^6 km^2. Most of the sea ice is part of the moving "pack ice" that drifts in the polar ocean with wind and surface currents. The floating ice is extremely uneven in thickness, ice age, snow cover area and open water distribution.

2.1.3 Continental Cryosphere, Marine Cryosphere, and Aerial Cryosphere

The global cryosphere can be divided into three categories: continental cryosphere, marine cryosphere, and aerial cryosphere.

The continental cryosphere refers to the cryosphere that develops and exists on land surface, including glaciers, ice sheets, permafrost (not including subsea permafrost), snow, river ice, and lake ice. Unlike the marine cryosphere, the majority of the continental cryosphere is fresh water, in addition to the lake ice above a few lagoons. The continental cryosphere is an essential component of the global water cycle. Firstly, it has deeply affected sea level variations, as well as ocean circulations due to the cold water generated from ice sheet calving and melting, which is one of the main drivers of global ocean currents. Moreover, continental ice melt runoff is part of the land water cycle, which is often the source of great rivers as critical water resources. Secondly, the seasonal alternation of snow cover not only has the water circulation effect but also exert climatic effect. Thirdly, the sudden release of continental frozen water on land often causes a disaster effect, resulting in glacial lake outburst, mountainous spring meltwater flood, river flood, thawing and collapse of the land surface, etc.

The marine cryosphere has two definitions: Climate and Cryosphere of World Climate Research Program (WCRP/CliC) defines it as only sea ice and its overlying snow; another by IPCC AR5 WGI (2013) defines marine cryosphere as sea ice and its overlying snow, plus ice shelves and icebergs. In addition to the above three elements, the submarine permafrost (subsea permafrost) should also be classified into the marine cryosphere. Therefore, we define the marine cryosphere as including sea ice and the overlying snow cover, ice shelves and icebergs, and subsea permafrost. Sea ice is the main body of the marine cryosphere, and the global sea ice cover extent is 19×10^6 to 27×10^6 km^2. In the northern hemisphere, the southmost boundary of the sea ice can reach Bohai Sea of China (approximately 38° N), while in southern hemisphere, sea ice mainly formed around the Antarctic, and its northmost boundary can reach to 55° S (Fig. 2.2). Approximately 15% of the world's oceans are covered by sea ice that existed in different time lengths over the year.

Due to polar locations and strong seasonality, the marine cryosphere varies with the seasons, which greatly affect global energy balance through the albedo of the marine cryosphere and the salty and cold sinking water that drives the global current

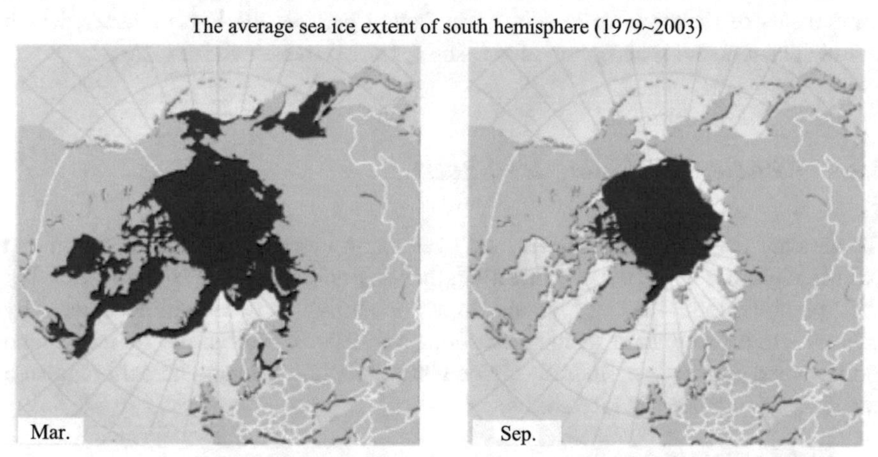

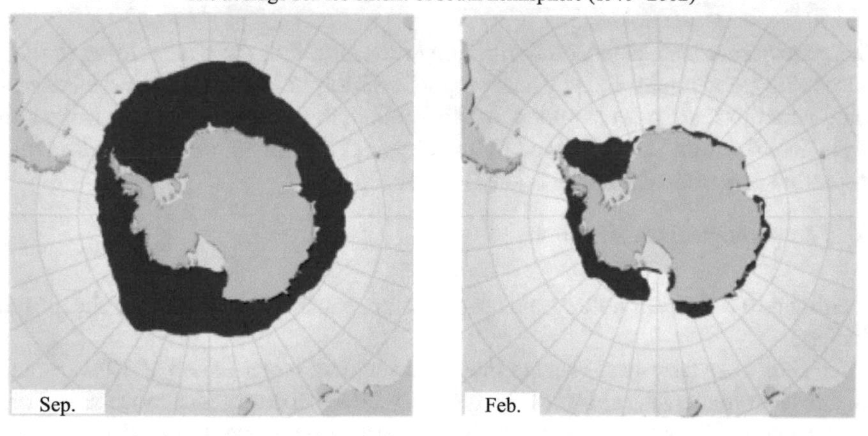

Fig. 2.2 Distribution of the main part of the marine cryosphere in the north and south hemispheres (after UNEP 2007)

when forming sea ice. In addition, the marine cryosphere has a considerable influence on the navigation and habitat of marine biology.

The aerial cryosphere refers to solid water that exists over the troposphere and stratosphere spaces where the temperature drops below the freezing point (<0 °C), and the solid forms of water in this space mainly are in the forms of snow, hail, graupel and ice pellet, etc.

2.2 Classification and Distribution of Continental Cryosphere

This section provides an overview of the classification and distribution of the main components of the cryosphere—the main frozen parts of the Earth's water, which include glaciers, ice sheets, permafrost, snow, frozen rivers and ice lakes.

2.2.1 Glacier (Including Ice Sheet)

Glaciers are ice bodies that are formed over many years by the accumulation and evolution (compression, recrystallization, thawing and refreezing) of snow or other solid precipitation. Glaciers typically move according to deformations caused by internal strain or by sliding along a bottom interface. According to their size and morphology, glaciers are generally divided into mountain glaciers (hereafter referred to simply as *glaciers*) and ice sheets.

1. Classification of glaciers

There are numerous glaciers on Earth. They have different forms according to their morphology, geography and physical properties. Morphological and physical classification methods are most common. In this section, the morphological classification of glaciers is introduced, followed by the physical classification. Finally, the types of glaciers present in China are briefly introduced.

1) Morphological classification

According to their morphology and size, glaciers can be divided into ice sheets and glaciers.

Ice sheets are terrestrial glaciers that cover an area larger than 5×10^4 km^2 and are unconstrained by the terrain. They typically flow in a divergence movement from a central point. Only two ice sheets exist on Earth at the present time: the Antarctic ice sheet and the Greenland ice sheet. Apart from a few areas with mountains that rise above the ice sheets (e.g., nunatak), high-latitude areas with cold weather and sufficient snowfall allow ice bodies hundreds or thousands of metres thick to form, which eventually becomes a dome-shaped ice sheet. An ice sheet is a unique form of glacier, and its classification and distribution will be introduced later in this section.

Glacier: A flowing ice body formed by the accumulation and dynamic metamorphism of snow and other solid precipitation. Compared with ice sheets, glaciers are generally constrained by the mountainous topography and have far smaller sizes and thicknesses. Glaciers are terrestrial and include ice caps, ice fields and mountain glaciers. Glaciers have different shapes and sizes according to the region and topographic conditions.

The main shape types of glaciers are:

Hanging glacier, which hang on to mountainsides and don't extend into surrounding flat areas.

Slope glacier, which are less steep than hanging glaciers but are larger and unrestricted by the terrain.

Cirque glacier, which develop on a slope or valley head with the shape of a saddle.

Hanging cirque glacier, which are hanging glaciers that exceed the range of cirque glaciers.

Cirque valley glacier, which exceed the range of cirque glaciers and extend into valleys.

Valley glacier, which are ice bodies that extend into valleys and are distributed according to the valley topography. They have multiple shapes, including simplex valley glaciers, composite glaciers, arborization valley glaciers, network valley glaciers, and certain other special types of glaciers. The convergence of dry ice flows may cause mutually parallel or mutually overlapped glacier combinations.

Fjord glacier, which are similar to valley glaciers that develop into fjords, but without any glacial firn basin. They are typically thin and long.

Ice cap, which have a similar appearance to ice sheets but are much smaller in size and have a dome shape. The ice body flows outward from the centre and can be further divided into *plateau* and *island* types according to the basal terrain.

2) Physical classification of glaciers is mainly based on the temperature or thermodynamic features of the ice. There are a diversity of forms, as follows.

Le Gary classification. Le Gary divided glaciers into *temperate*, *transition* and *cold* types according to the thermodynamic properties of the constant temperature zone below the active layer (15–20 m depth below the glacier surface, Chap. 4). Temperate types have ice layers below the active layer that reach melting temperature under a certain pressure. Cold types have ice layers below the active layer that always remain below the melting point. Transition types have surface temperatures below 0 °C and near-bottom ice bodies that reach the melting point under pressure.

Yaroslav classification. Yaroslav divided glaciers into five types according to local climatic conditions and glacier temperature. ① Dry polar type: the overall ice temperature is lower than the melting point and is slightly lower than the local annual average temperature. ② Wet polar type: occurs when the air temperature is above 0 °C in summer and the superficial ice layer is close to 0 °C, resulting in some melting and meltwater. ③ Wet-cold type: occurs where the average temperature of the ice is higher than the annual average temperature, but is still below 0 °C. Meltwater can penetrate into the bottom of the active layer. ④ Maritime type: the temperature of the ice body is determined by meltwater. The temperature of the active layer is 0 °C in summer and below 0 °C in winter. The temperature at the upper part of the active layer is lower than that at the lower part. The temperature in the deep layers is 0 °C. ⑤ Continental type: characterized by strong radiation and low precipitation. Ice temperatures at different depths are below 0 °C. The temperature at depths of 5–10 cm can reach 0 °C in summer and the ice temperature in deep layers is always below the freezing point.

Arles classification. A geophysical classification method. Arles identified *temperate glaciers*, *sub-polar glaciers* and *high-polar glaciers* according to the snow physics of the upper part of the glacier and glacier temperature. Temperate glaciers have strong infiltration and recrystallization of meltwater. The temperature of the whole ice body is at melting point at a certain pressure, and the superficial layer is below 0 °C only in winter. Most polar glacier ice bodies have temperatures <0 °C. The accumulation areas of high polar glaciers are composed of thick <0 °C firn that generally does not melt in summer. The accumulation areas of subpolar glaciers are composed of 10–20 m thick firn, and the summer temperature is close to 0 °C when melting occurs.

3) Glacier types in China

In terms of the different international classifications, glaciers in China presently are divided into the continental and maritime types. The former are divided into polar continental glaciers and subcontinental glaciers. The threshold standards of polar continental, subcontinental and maritime glaciers are determined by integrated analyses of their physical indexes, which are used to classify their distributions. Based on these studies, the three types of glaciers in China have the following characteristics and spatial distributions.

Maritime or *temperate glaciers*: These types occur where the annual precipitation at the equilibrium line altitude (ELA) reaches 1000–3000 mm and the annual average temperature is higher than –6 °C. The monthly average temperature from June to August is 1–5 °C and the ice temperature is 1–0 °C. Ice bodies move quickly, and the annual distance moved is greater than 100 m. Strong ice thawing occurs and the annual thawing depth at the lower end of the glacier tongue is 10 m. Maritime or temperate glaciers are mainly distributed in the eastern section of Nyainqentanglha, the Hengduan Mountains, and the eastern Himalayas.

Subcontinental or subpolar glaciers: These types occur with an annual average precipitation (at ELA) of 500–1000 mm and an annual average temperature ranging between 6 and 12 °C. The monthly average temperature from June to August is 0–3 °C, and the ice temperature is 10–1 °C in the superficial layer (<20 m). The ice body moves quickly, with an average speed of 5–100 m per year. The annual thawing depth of the ice surface is 2–8 m. Subcontinental or subpolar glaciers are mainly distributed in the Qinghai-Tibet Plateau except for the peripheral mountains of southeast Tibet, Pamirs, the Tianshan Mountains, and the Altai Mountains.

Continental or polar-type glaciers: The annual average precipitation at ELA is 200–500 mm, and the annual average temperature is lower than 10 °C. The monthly average temperature from June to August is lower than 1 °C. These glaciers move slowly, at 30–50 m per year. The annual melting depth of the ice tongue is 1–2 m. This type of glacier is mainly distributed in the Qinghai-Tibet Plateau and the western Qilian Mountains.

2. Global distribution of glaciers

According to IPCC AR5, excluding ice sheets but including their surrounding ice caps, there are 168,331 glaciers in the world, which cover a total area of 726,258 km^2 and have a mass of 113,915–191,879 Gt. Their potential contribution to the sea level rise is 412 mm. The quantities and distributions of glaciers are shown in Table 2.1. Some distribution characteristics are highlighted as follows.

(1) The Northern Hemisphere has the majority of mountain glaciers. There are 143,450 contemporary glaciers covering an area of 560,915 km^2 with an ice reserve of 82,270–141,762 Gt (water equivalent) and a sea level equivalence by 301.4 mm. Mountain glaciers comprise 85.2% of the worlds' glaciers, 77.23% of their area, 72.2–73.9% of their ice reserves, and 73.2% of their sea level rise potential.
(2) Although there are more glaciers in middle-altitude and low-altitude regions compared to high-altitude ones, the coverage area, ice reserves and sea level equivalence of high-altitude (>50°) glaciers are higher. For example, there are 87,360 glaciers in the middle-altitude and low-altitude regions of the Northern Hemisphere (rows of 2, 11, 12 and 13 in Table 2.1), which cover an area of 137,849 km^2, with ice mass 9,103–12,900 Gt, and a sea level equivalence of 33 mm. There are 56,090 glaciers in high-altitude regions north of 50° N, less than the amount in middle-altitude and low-altitude regions. However, their area, ice reserves and sea level equivalence are 423,065 km^2, 72,567–126,759 Gt and 268.4 mm, respectively. All three parameters are higher than those of middle-altitude and low-altitude regions between 0–50° N. There are 21,607 glaciers in middle and low-altitude regions in the Southern Hemisphere (rows of 16 and 18 in Table 2.1), with an area of 33,076 km^2, with an ice reserve of 4,421–6,345 Gt, and a sea level potential of 14.2 mm. In the high-altitude regions south of 50° S, there are 3,274 glaciers covering an area of 132,267 km^2, with an ice reserve of 27,224–43,772 km^3, and a sea level potential of 96.3 mm. Glaciers area in these region is greater than the glaciers of the middle-altitude and low-altitude regions of the Southern Hemisphere.

3. Distribution of glaciers in China

According to the climatic conditions in which glaciers developed, glaciers in China can be divided into maritime glaciers, continental glaciers and extreme-continental glaciers (Fig. 2.3).

Table 2.1 The 19 regions used in IPCC AR5 and the characteristics of their glaciers

No	Region	Number of glaciers	Area (km^2)	Minimum ice reserves (Gt)	Maximum ice reserves (Gt)	Sea level rise potential (mm)
1	Alaska	32,112	89,267	16,168	28,021	54.7
2	Western Canada and America	15,073	14,503.5	906	1148	2.8
3	Canadian Northern Arctic	3318	103,990.2	22,366	37,555	84.2
4	Canadian Southern Arctic	7342	40,600.7	5510	8845	19.4
5	Greenland	13,880	87,125.9	10,005	17,146	38.9
6	Iceland	290	10,988.6	2390	4640	9.8
7	Svalbard	1615	33,672.9	4821	8700	19.1
8	Scandinavia	1799	2833.7	182	290	0.6
9	Northern Russia	331	51,160.5	11,016	21,315	41.2
10	Northern Asia	4403	3425.6	109	247	0.5
11	Central Europe	3920	2058.1	109	125	0.3
12	Cadence & Cascade	1339	1125.6	61	72	0.2
13	Central Asia	30,200	64,497	4531	8591	16.7
14	Western South Asia	22,822	33,862	2900	3444	9.1
15	Eastern South Asia	14,006	21,803.2	1196	1623	3.9
16	Low-altitude regions	2601	2554.7	109	218	0.5
17	Southern Andes Mountain	15,994	29,361.2	4241	6018	13.5
18	New Zealand	3012	1160.5	71	109	0.2
19	South Pole and Subantarctic	3274	132,267.4	27,224	43,772	96.3
Total		168,331	726,258.3	113,915	191,879	411.9

Data source IPCC AR5 WGI (2013)

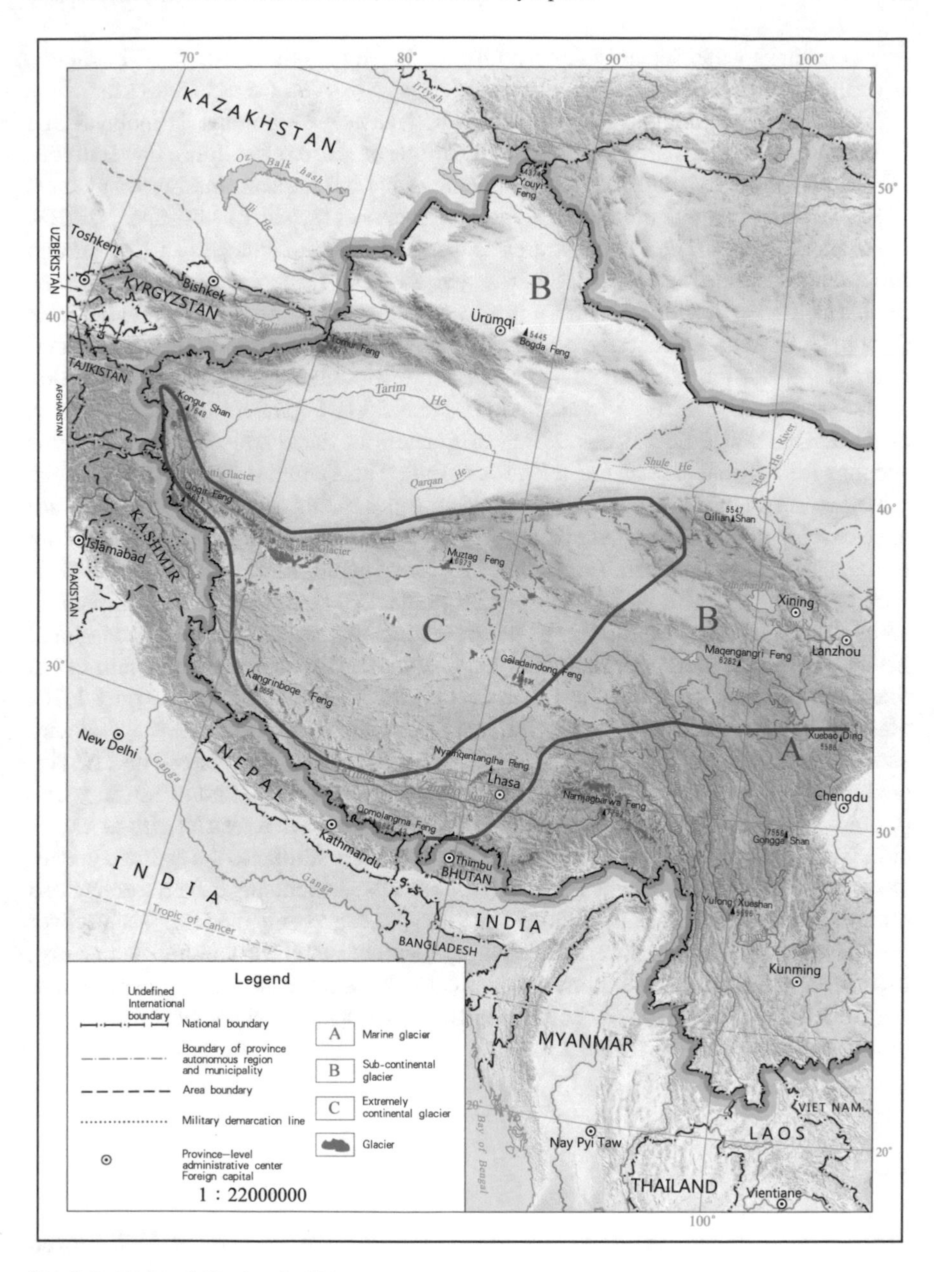

Fig. 2.3 Types of Glaciers in China

According to the Second Chinese Glacier Inventory, there are 48,571 glaciers covering an area of 51,766 km^2 with an ice reserve of 4494.00 $\pm$ 175.93 km^3.

The Kunlun Mountains, Nyainqentanglha, Tianshan Mountains, Himalayas and Karakoram Mountains are the dominant locations of glaciers in China. The statistics (Table 2.2) show that the Kunlun Mountains have the highest number of glaciers (8,922), and the largest glacier area and ice reserves (11,524.13 km^2 and 1,106.34 km^3, respectively). The quantity, area and ice reserves of glaciers in the Kunlun Mountains account for 18.37%, 22.26% and 24.62%, respectively, of the national total. The Tianshan Mountains have the second-largest quantity of glaciers but rank third in terms of area and ice reserves after the Kunlun Mountains and Nyainqentanglha. Moreover, the Himalayas and Karakoram Mountains have more than 5,000 glaciers each. Above five mountains comprise 35,104 glaciers covering an area of 41,073 km^2. The quantity and area of glaciers in these five-mountain systems account for 3/4 and 4/5 of the total for China. The Qangtang Plateau lies in the hinterland of the Qinghai-Tibet Plateau. Some flat mountains (altitude >6,000 m) have developed in the Qangtang Plateau. Many large ice caps and glaciers, including Purogangri, Zangsergangri, Tuzegangri and Jinyanggangri, are centred over these mountain peaks. Additionally, there are large glaciers ($\geq$2 km^2) in these mountain areas. Large glaciers account for 78.64% of the total glacier area in the region. Small glaciers only account for approximately 21%, even though there is a huge number of them, with an average glacier area of only 1.65 km^2. Although there are only 1,612 glaciers in the Pamirs, the total glacier area is 2,159.62 km^2 and the average glacier size is 1.34 km^2, which is next in size only after the Qiangtang Plateau and Nyainqentanglha (1.39 km^2). The Himalayas, where the highest mountain in the world (Qomolangma, 8844 m) is sharp terrain, has narrow ridges that restrict glacier extension. The average glacier area is only 1.12 km^2, which is similar to the average glacier size for the Karakoram Range. Although the Gandise Mountains have a tremendous number of glaciers (3,703), the total glacier area (averaged only 0.35 km^2) is just over 50% of that in the Pamirs, where the mountain with the smallest average glacier size. Musitaoling, the Altai Mountains and the Altun Mountains are the three mountain systems with the lowest glacier number and area. The average glacier area in all three mountain systems is less than 0.75 km^2.

2.2.2 *Frozen Ground*

Frozen ground defines as all kinds of ground temperature at or below 0 °C. Frozen ground is a multi-component and multiphase substance that is composed of mineral particles, ice, unfrozen water and water vapour and organic matter.

Table 2.2 Statistics on the number of glaciers in various mountain systems (plateaus) in western China (Liu et al. 2015)

Mountain (plateau)	Quantity		Area		Ice reserves	
	Quantity	Proportion/%	Area/km^2	Proportion/%	Ice reserves/km^3	Proportion/%
Altai Mountains	273	0.56	178.79	0.35	10.50 ± 0.21	0.23
Mustau ling	12	0.02	8.96	0.02	0.40 ± 0.03	0.01
Tianshan	7934	16.33	7179.77	13.87	707.95 ± 45.05	15.75
Karakoram	5316	10.94	5988.67	11.57	592.86 ± 34.68	13.19
Pamirs	1612	3.32	2159.62	4.17	176.89 ± 4.63	3.94
Kunlun Mountains	8922	18.37	11,524.13	22.26	1106.34 ± 56.60	24.62
Altun Mountains	466	0.96	295.11	0.57	15.36 ± 0.65	0.34
Qilian Mountains	2683	5.52	1597.81	3.09	84.48 ± 3.13	1.88
Tanggula Mountain	1595	3.28	1843.91	3.56	140.34 ± 1.70	3.12
Qiangtang Plateau	1162	2.39	1917.74	3.70	157.29 ± 3.11	3.50
Gangdis	3703	7.62	1296.33	2.50	56.62 ± 3.43	1.26
Himalayas	6072	12.50	6820.98	13.18	533.16 ± 8.71	11.87
Nyainqentanglha Mountain	6860	14.12	9559.20	18.47	835.30 ± 31.30	18.59
Hengduan Mountain	1961	4.04	1395.06	2.69	76.50 ± 2.41	1.70
Total	48,571	100.00	51,766.08	100.00	4494.00 ± 175.93	100.00

1. Classification of frozen ground

The following is a brief introduction to the common classification of permafrost and seasonally frozen ground. Types of frozen ground under perspective of soil genesis are belong to the category of soil science, thus are not described here.

1) Classification based on the duration of existence

According to the duration of soil frozen state, frozen ground can generally be divided into four types: temporal frozen ground (the frozen period is from several hours, several days to half a month), seasonally frozen ground (the frozen period is from half a month to several months), pereletok (the frozen period is more than 1 years but less than 2 years) and permafrost (the continued frozen period is more than 2 years). The temporal frozen ground, normally has a short existence time and frozen depth is generally less than 30 cm, is mainly distributed in the middle and low latitudes, and its distribution is affected by the cold airflow of the synoptic scale. The mean annual area of seasonally frozen ground is approximately 4812×10^4 km^2, which accounts for 50.5% of the land area in the northern hemisphere. Pereletok exists in theory, but it has not been reported. The mean annual area of the short-term frozen ground in the northern hemisphere is 627×10^4 km^2, which accounts for approximately 6.6% of the land area in the northern hemisphere. The distribution of seasonally frozen ground is large, and it occurs mainly in the middle and high latitudes. In particular, the active layer in the permafrost region belongs to seasonally frozen ground, because the active layer is the soil layer that is frozen in winter and thawed in summer. Permafrost is mainly distributed in the northern hemisphere and accounts for approximately 24% of the land area in the hemisphere.

2) Classification based on the spatial continuity

In permafrost regions, frozen ground is divided into continuous permafrost (continuity coefficient is over 90%) and discontinuous permafrost (continuity coefficient is less than 90%) according to spatial continuity. In discontinuous permafrost regions, frozen ground can be subdivided into intermittent permafrost (continuity coefficient is 75–90%), sporadic permafrost (continuity coefficient is 60–75%), island permafrost (continuity coefficient is 30–60%) and sparse island permafrost (continuity coefficient is less than 30%). The continuity coefficient is the ratio of the permafrost area to the region area.

The classification index of spatial continuity from the International Permafrost Association (IPA) is continuous permafrost (continuity coefficient is over 90%), discontinuous permafrost (continuity coefficient is 50–90%), sporadic discontinuous permafrost (continuity coefficient is 10–50%) and sparse island permafrost (continuity coefficient is less than 10%).

Although the division of permafrost regions based on the spatial continuity index is widely applied, the division principle of the zone division is very imprecise, and the problem of how much spatial scope to use when calculating the spatial continuity has not been solved. For example, a region with an area within 1 km^2, of which 90% of the area is permafrost, should belong to continuous permafrost, according to the definition of IPA. However, within 100 km^2, where the permafrost area is only 20%, the region should belong to sporadic discontinuous permafrost.

3) Classification based on the thermal stability

According to the temperature and thickness of permafrost, it can be divided into the following types:

Extremely stable type: the mean annual ground temperature (MAGT) is less than –5 °C and a permafrost thickness is more than 170 m;

Stable type: MAGT is –5 to –3 °C, and the permafrost thickness is 110–170 m;

Metastable type: MAGT is –3.0 to –1.5 °C, and the permafrost thickness is 60–110 m;

Transitory type: MAGT is –1.5 to –0.5 °C, and the permafrost thickness is 30–60 m;

Unstable type: MAGT is –0.5 to 0.5 °C, and the permafrost thickness is 0–30 m;

Extremely unstable type: the remaining type of permafrost has higher temperature and thinner thickness.

The temperature and thickness of permafrost are determined by many factors and are comprehensive indexes that reflect the survival of regional or local permafrost. For the same region or place, the change of permafrost types reflect the change of the local climate or regional conditions, which can be used as the main evidence for the change of permafrost.

4) Classification based on temperature

According to the mean annual ground temperature, permafrost can be divided into the following types:

Low temperature permafrost, with a mean annual ground temperature that is generally below –2.0 °C, which can recover in a relatively short freeze–thaw cycle period (1–2 years) after it has been under a short-term disturbance.

Moderate temperature permafrost is a transitory type, and when the mean annual ground temperature is below –1.0 °C, its corresponding thickness is generally greater than 50 m; these regions are relatively stable, and the mean annual ground temperature fluctuation is generally at ±0.1 °C.

High temperature permafrost, with a mean annual ground temperature of generally –1.0 to ~ –0.5 °C, is sensitive to the effects of surface disturbance, and the change of permafrost caused by the surface disturbance is irreversible.

Extremely high temperature permafrost, with a mean annual ground temperature of generally –0.3 °C, is extremely sensitive to the effects of surface disturbance, and the change of permafrost is irreversible under the influence of human factors.

This division system is widely used in Chinese literature.

5) Classification based on ice content

According to the ice content, permafrost can be divided into dry cryogic ground, ice-poor permafrost, ice-medium permafrost, ice-rich permafrost, ice-saturated permafrost and ice layer with soil inclusions. Dry cryogic ground is the dense rock and soil at or below 0 °C without water or ice. The volumetric ice content in ice-poor permafrost is generally less than 3%; in ice-medium and ice-rich permafrost is 3–20%; in ice-saturated permafrost is 20–40%; and in ice layer with soil inclusions, it is generally greater than 40%.

According to the division based on the total water content of permafrost, for crushed gravel soil, the total water content in ice-poor permafrost is less than 10%; in ice-medium permafrost is 10–18%; in ice-rich permafrost is 18–25%; in ice-saturated permafrost is 25–65%; in ice layer with soil inclusions, it is greater than 65%. For sand and sandy soil, the total water content in ice-poor permafrost is less than 12%; in ice-medium permafrost is 12–21%; in ice-rich permafrost is 21–28%; in ice-saturated permafrost is 28–65%; in ice layer with soil inclusions, it is greater than 65%. For silty and clay soil, it is—generally divided according to the plastic limit and liquid limit of the soil.

The ice content in permafrost is the main factor in determining the properties of permafrost engineering, and it is the main parameter for the classification of permafrost engineering.

6) Classification based on engineering

There is a need to establish a classification of permafrost engineering for engineering application. It considers the relationship to building foundation and can fully reflect the main destructive factors of permafrost to engineering buildings, which is mainly the thaw settlement of permafrost under the thermal effect. Meanwhile, it can also reflect objective differences and integrate the permafrost structure with the physical mechanic index. The classification is not only applicable for permafrost but is also applicable for a seasonally thawed layer on permafrost.

According to the above principles of classification, there are three classification views from different index categories:

(1) Thaw settlement view. Based on the thaw settlement coefficient (A), it is divided into 5 categories: non-thaw settlement soil, weak thaw settlement soil, thaw settlement soil, strong thaw settlement soil and strong thaw collapse soil;
(2) Frost heave view. Based on the frost heave coefficient (η), it is divided into 5 categories: non-frost heave soil, weak frost heave soil, moderate frost heave soil, frost heave soil and strong frost heave soil;
(3) Strength view. Based on relative strength, it is divided into 4 categories: ice-poor permafrost, ice-rich permafrost, ice-saturated permafrost and ice layer with soil inclusions.

The classification based on permafrost engineering is very important for the engineering design, construction and operation in permafrost regions and in deep seasonally frozen ground regions. At present, there is not a standard classification index in the world. However, according to the construction codes of various countries, governments have developed their own corresponding classification of permafrost engineering and building codes based on the permafrost engineering types.

7) Classification of seasonally frozen ground

A relatively systematic classification of seasonally frozen ground was the Soviet Kudriavtsev's classification. He divided seasonally frozen ground into two major categories: seasonally frozen layer and seasonally thawed layer. The seasonally thaw ground (active layer) is on permafrost and connects with the permafrost table, but seasonally frozen ground occurs on non-permafrost. The principle of Kudriavtsev's classification is to consider not only the climate zonation and the continentality but also the regional differences that affect the development of permafrost. Kudriavtsev used four indexes, which are the mean annual ground temperature, the annual surface temperature range, the lithology and the soil water content to classify the seasonally frozen ground systematically. The mean annual ground temperature mainly reflects the climate zonation, the annual surface temperature range mainly reflects the continentality, while the lithology and the soil water content mainly reflect local and regional differences. There are different limit indexes in the four indexes. After every possible combination, seasonally frozen ground is divided into more than 1200 categories by this classification method. The principle of this classification is comprehensive and discusses the distribution rules and categorization of seasonally frozen ground from soil genetic theory and the influence factor on the development of permafrost. The actual operability, however, is poor. On the one hand, the data are lacking; on the other hand, it has not been used widely because of too many frozen ground categories and poor application. According to the frost depth, seasonally frozen ground whose depth is more than 1 m can be defined as deep seasonally frozen ground, whose depth is less than 1 m and can be defined as shallow seasonally frozen ground. The U.S. Army Cold Regions Research and Engineering Laboratory divides the region where the soil frost depth is more than 30 cm in winter into the seasonally frozen ground region. According to the soil frost period, the soil near surface whose frost period is more than 15 days is defined as seasonally frozen ground, and soil whose frost period is less than 15 days is defined as short-term frozen ground.

2. Distribution of frozen ground

As far as the spatial scope is concerned, frozen ground including permafrost and seasonally frozen ground, is the largest component of the cryosphere. The frost and thaw state of soil have a great influence on the soil thermal and physical properties, thus playing a very important role in the hydrothermal exchange between air and ground surface, eco-hydrological process, carbon cycle, and eventually climate system. Therefore, it is very important to study the temporal and spatial distribution of frozen ground.

1) Distribution of permafrost around the world

The global permafrost is mainly distributed in the polar regions of the northern hemisphere and in the alpine regions of North America and Asia. The frozen ground of southern hemisphere is mainly distributed in the Andes and the Antarctic continent bare areas not covered by glaciers. In addition, based on preliminary analysis, there should be permafrost under the alpine continental glacier, but whether there is permafrost under the maritime glacier is still unknown and needs further study. There is also a large amount of permafrost under the Arctic continental shelf, known as offshore permafrost, which is a part of the sea ice circle. This part should belong to the marine cryosphere.

2) Distribution of frozen ground in the northern hemisphere

The permafrost of the northern hemisphere is mainly distributed in Eurasia, North America, the northern Arctic Ocean island (including Greenland and Iceland), Qinghai- Trbetan Plateau and high mountains, as well as on the continental shelf and ocean floor (Fig. 2.4). The area of permafrost in the northern hemisphere is 2279 $\times 10^4$ km^2, which accounts for 23.9% of the land area in the northern hemisphere; the mean annual area of seasonally frozen ground is approximately 4812 $\times 10^4$ km^2, which accounting for 50.5% of the land area in the northern hemisphere; the mean annual area of the short-term frozen ground in the northern hemisphere is 627 $\times 10^4$ km^2, which accounts for approximately 6.6% of the land area in the northern hemisphere. Under extreme conditions, the area of seasonally frozen ground can account for more than 80% of the land area in the northern hemisphere.

The latitudinal distribution of the permafrost in the northern hemisphere is southward to the Himalayas at 26° N from Greenland at 84° N, of which 70% is distributed between 45–67° N. The altitudinal distribution shows that approximately 62% of the permafrost in the northern hemisphere is distributed below 500 m a.s.l., and 10% of the permafrost is distributed above 3000 m a.s.l.

From the ice content point of view, the ice-rich permafrost whose volumetric ice content is more than 20% is mainly distributed in high latitudes in the northern hemisphere, and accounts for approximately 8.57% of the permafrost area. Permafrost whose volumetric ice content is less than 10%, which is mainly distributed in the permafrost regions of high altitude mountain, accounts for approximately 66.5% of the permafrost area.

3) Distribution of frozen ground in China

China is one of the three largest permafrost countries in the world. The permafrost area in China is approximately 2.2 $\times 10^6$ km^2, which accounts for 22.3% of the territory and the third largest in the world. Among them, the high-altitude permafrost area of China is the highest in the world, and the seasonally frozen ground is distributed across most of the territory. Permafrost is mainly distributed in the Great Khingan

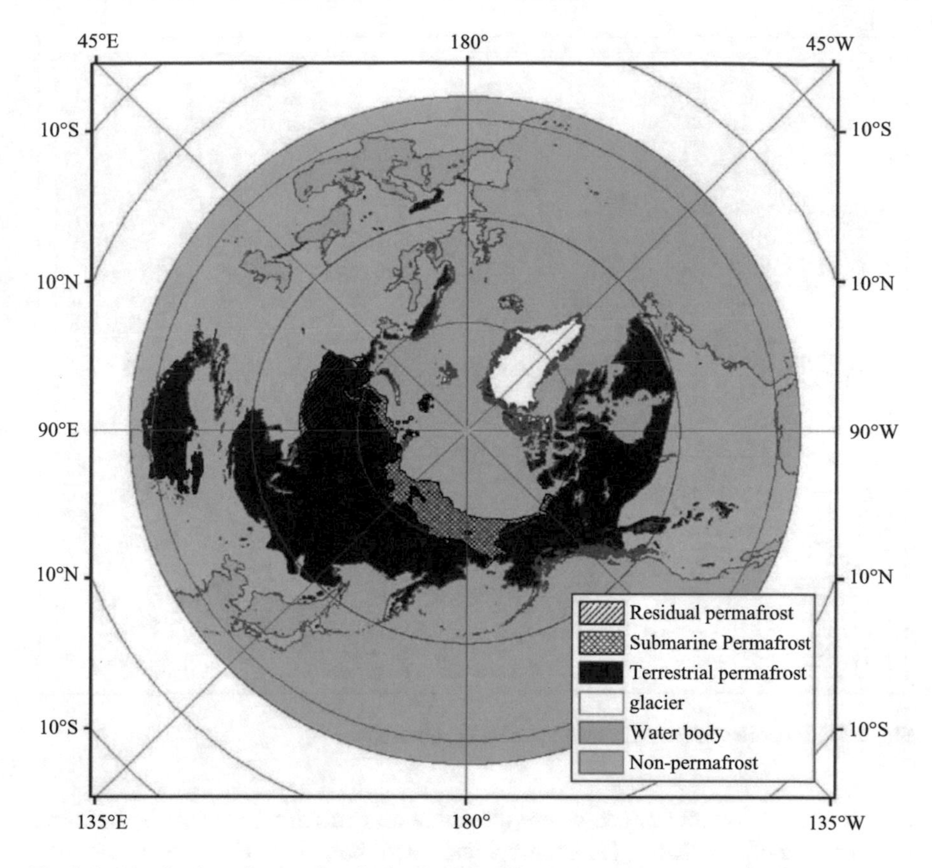

Fig. 2.4 Distribution of permafrost in the Northern Hemisphere (Zhang et al. 2008)

Range, Lesser Khingan Range and the northern part of the Songnen Plain in Northeast China as well as the western mountains and in the Qinghai Tibet Plateau, while some is distributed on some high mountains in the seasonally frozen ground regions (Fig. 2.5). Among them, the northeastern permafrost region is located in the southern margin of the permafrost region of Eurasia, with an area of approximately 39×10^4 km^2, located between 46° 30′ N and 53° 30′ N. Its main distribution characteristics are: the main restriction from the latitudinal zonality, more characteristic northeast permafrost distribution because of the superposition of altitude, harsher permafrost distribution conditions in a low terrain area, sparse island permafrost and sporadic permafrost regions with a north to south width of 200–400 km whose area is much greater than the area of continuous and continuous-island permafrost regions. The permafrost regions in the Chinese western mountains and plateau are mainly distributed in the Altai Mountains, Tianshan, Qilian Mountain and Qinghai Tibet Plateau. Because the distribution of the permafrost is mainly controlled by altitude, it is called high-altitude permafrost, also known as mountain or plateau permafrost. The alpine plateau permafrost only appears above a certain altitude, and the line

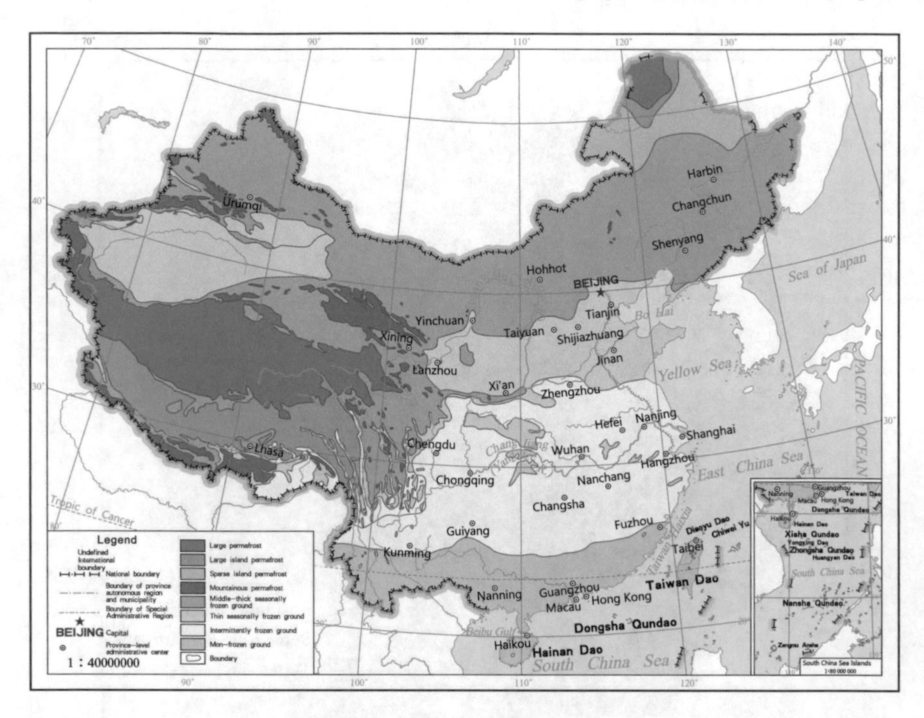

Fig. 2.5 Distribution of frozen ground in China (Zhou 2000)

connecting the lowest altitude where the sporadic permafrost appears is the lower limit of permafrost; that is, the geographical lower boundary. From the lower limit up, the spatial continuity of permafrost increases, from sporadic distribution to large continuous distribution, and finally, to continuous distribution, with a decrease of the permafrost temperature and an increase of the permafrost thickness, which has an obvious vertical zonality. In addition, seasonal freezing and thawing, and the cold process and phenomena also change regularly with the increase of altitude.

Category of permafrost continuity in China used to be different from the one by International Permafrost Association (IPA). Permafrost in China is mainly divided into middle and high latitude permafrost and high-altitude permafrost. Among them, the area of the high-altitude permafrost accounts for 92% of the total permafrost area in China, which is mainly distributed in the Qinghai Tibet Plateau and in the western alpine regions. According to different estimation models, the permafrost area on Qinghai Tibet Plateau is approximately 130×10^4 km^2, which accounts for 13.5% of the territory, 87.2% of the total permafrost area in China, and 94.5% of the high-altitude permafrost area. The middle and high latitude permafrost area is approximately 12×10^4 km^2, which accounts for 7.8% of the total permafrost area in China, which is mainly distributed in the Da Hinggan Mountains, Xiao Hinggan Mountains and in the northern part of the Songnen Plain in Northeast China. The area of seasonally frozen ground, including the active layer in the permafrost regions, accounts for approximately 70% of China's total land area, or 90% when the short-term frozen ground is included. Because of the influence of the monsoon climate, snowfall is small in winter in China, which leads to the well development of seasonally frozen ground. The seasonally frozen ground in China is mainly distributed in the regions north of 25° N (Fig. 2.5). In western China, because of the effect of altitude, the southern boundary of seasonally frozen ground distribution can reach 25° N, but in eastern China, because of the lower elevation, the southern boundary of seasonally frozen ground distribution moves northward, which fits to the 30° N latitude line approximately. The variance of seasonally frozen ground thickness is substantial. In the southern boundary regions, the thickness of seasonally frozen ground is only more than ten centimetres or tens of centimetres, while in northern regions, the thickness of seasonally frozen ground can be higher than 2 m. Under the climate warming conditions, the frost period of seasonally frozen ground is shortening, while the thickness of seasonally frozen ground is thinning.

2.2.3 Snow Cover

Snowpack covering land surface for less than one year is called seasonal snow cover, or snow cover in short.

1. Classification

Snow cover can be classified at either the micro or macro scale. On the micro-scale, grain shape, grain size, liquid water content, snow hardness, and snow temperature are often used to classify snow cover into many different types (Table 2.3). According to the impurity, it can be classified into clean snow, dirty snow, and so on; and there are also many other classifications according to density, hardness, liquid water content, or snow temperature.

On the macro-scale, the classification of snow cover for the whole snow area is mainly base on physical properties, such as snow depth, snow density, thermal conductivity, interaction between snow layers, horizontal variability, changes of snow cover with time, and etc. Defined by a unique ensemble of textural and stratigraphic characteristics, including the sequence of snow layers, their thickness, density, and the crystal morphology and grain characteristics within each layer, global snow is classified into six classes, including tundra snow, taiga snow, alpine snow, maritime snow, prairie snow, and ephemeral snow. They can also be derived using a binary of system of three climate variables: precipitation, wind, and air temperature.

According to the annual snow cover days (SCD), snow covered area can be classified into stable, with annual SCD no less than 60 days, and unstable snow cover areas, with annual SCD less than 60 days. The latter can be further classified into two sub-regions: unstable snow cover areas with periodicity, where annual SCD falls between 10–60 days, and unstable snow cover areas without periodicity, where annual SCD less than 10 days. Based on observed snow depth data from meteorological stations,

Table 2.3 Classification of snow cover

Terms	Symbol	Subclass						Remarks
Grain shape	*F*	Precipitation Particles (PP)	Machine-Made snow (MM)	Decomposing and Fragmented precipitation particles (DF)	Rounded Grains (RG)	Faceted Crystals (FC)	Depth Hoar (DH)	Name (code)
		Surface Hoar (SH)	Melt Forms (MF)	Ice Formations (IF)				
Grain size	*E*	Very fine	Fine	Medium	Coarse	Very coarse	Extreme	
		<0.2	0.2–0.5	0.5–1.0	1.0–2.0	2.0–5.0	>5.0	Unit: mm
Liquid water content	*LWC*	Dry	Moist	Wet	Very wet	Soaked		
		0	0–3	3–8	8–15	>15		Volume fraction in %
		D	M	W	V	S		Code
Snow hardness	*R*	Very soft	Soft	Medium	Hard	Very hard	Ice	Hand test
		Fist	4 fingers	1 finger	Pencil	Knife blade	Ice	
		F	4F	1F	P	K	I	Code
Snow temperature	T_s	$T_s(H)$	$T_s(-H)$	T_{ss}	T_a	T_g		Mark the observation position and value
		Snow temperature at height H in cm above the ground	Snow temperature at depth–H in cm below the surface	Snow surface temperature	Air temperature 1.5 m above snow surface	Ground surface temperature		
Snow density	ρ_s, *G*	Mass per unit volume (kg m^{-3})						
Impurities	*J*	The type of impurity should be fully described and its amount given as mass fraction (%, ppm)						

(continued)

Table 2.3 (continued)

Terms	Symbol	Subclass					Remarks
Microstructure	F	Porosity	Specific surface area	Curvature	Tortuosity	Coordination number	Name and value

Data source Fierz et al. (2009)

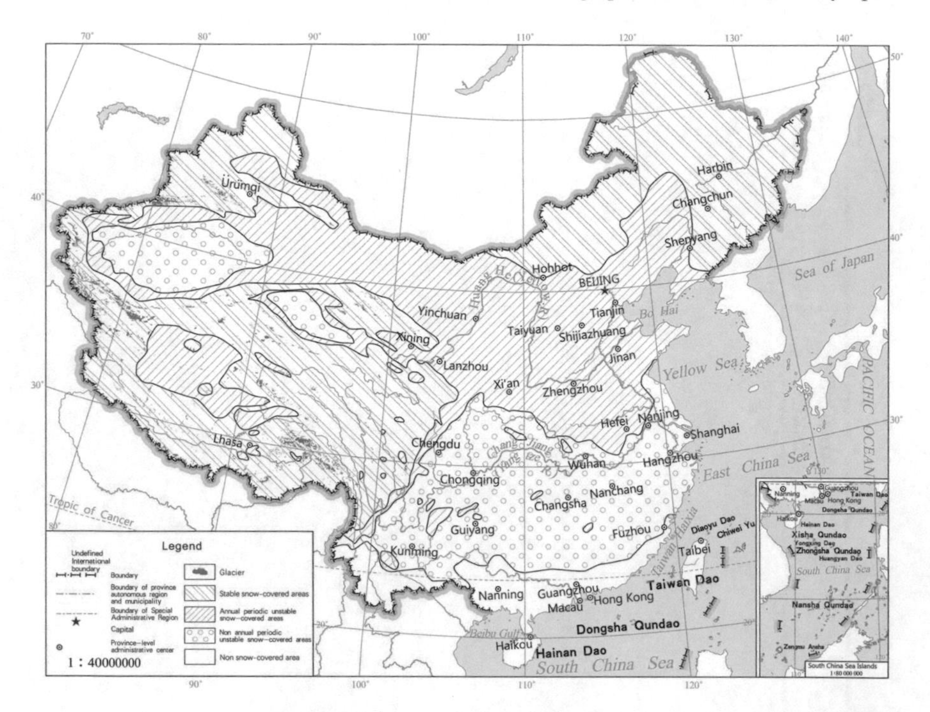

Fig. 2.6 Distribution of snow cover classes according to annual snow cover days in China (**a**) based on observed data from meteorological stations (modified from Li and Mi 1983) and western China (**b**) based on SMMR and SSM/I remote sensing data (He and Li 2011)

a classification graph of snow covered regions in China is shown in Fig. 2.6. Before 1980s, the stable snow cover area mainly locates on the Qinghai-Tibet Plateau (except for the region surrounding the Qaidam Basin and the region eastward with respect to the central part), the northeastern China, and the northwestern China.

Regions with periodic unstable snow cover include the vast area from the Liao River Basin to Qinling and Dabie mountains. Non-periodicity unstable snow cover area lies in the south of Qinling and Dabie mountains, and the Tarim and Qaidam Basins. Regions south of 25° N in China are snow-free areas. Combining ground-based observations and SMMR and SSM/I remote sensing data, snow cover in the western China was reclassified for the period of 1951–2004 (Fig. 2.6b). A big difference is that wide ranges of stable snow cover area over the Qinghai-Tibet Plateau transformed into unstable snow cover area with periodicity, and the others remained almost the same.

Besides annual SCD, an index of continuous snow cover days is also taken as a criterion to classify snow cover. In the northern Eurasia, most regions north of 38° N are classified stable snow cover area, with an annual continuous SCD no less than 30 days, corresponding to annual mean snow depth no less than 7 cm.

In addition, two ratios are also used to classify snow cover. One is the proportion of snowfall in total precipitation, and the other is the proportion of accumulated snow

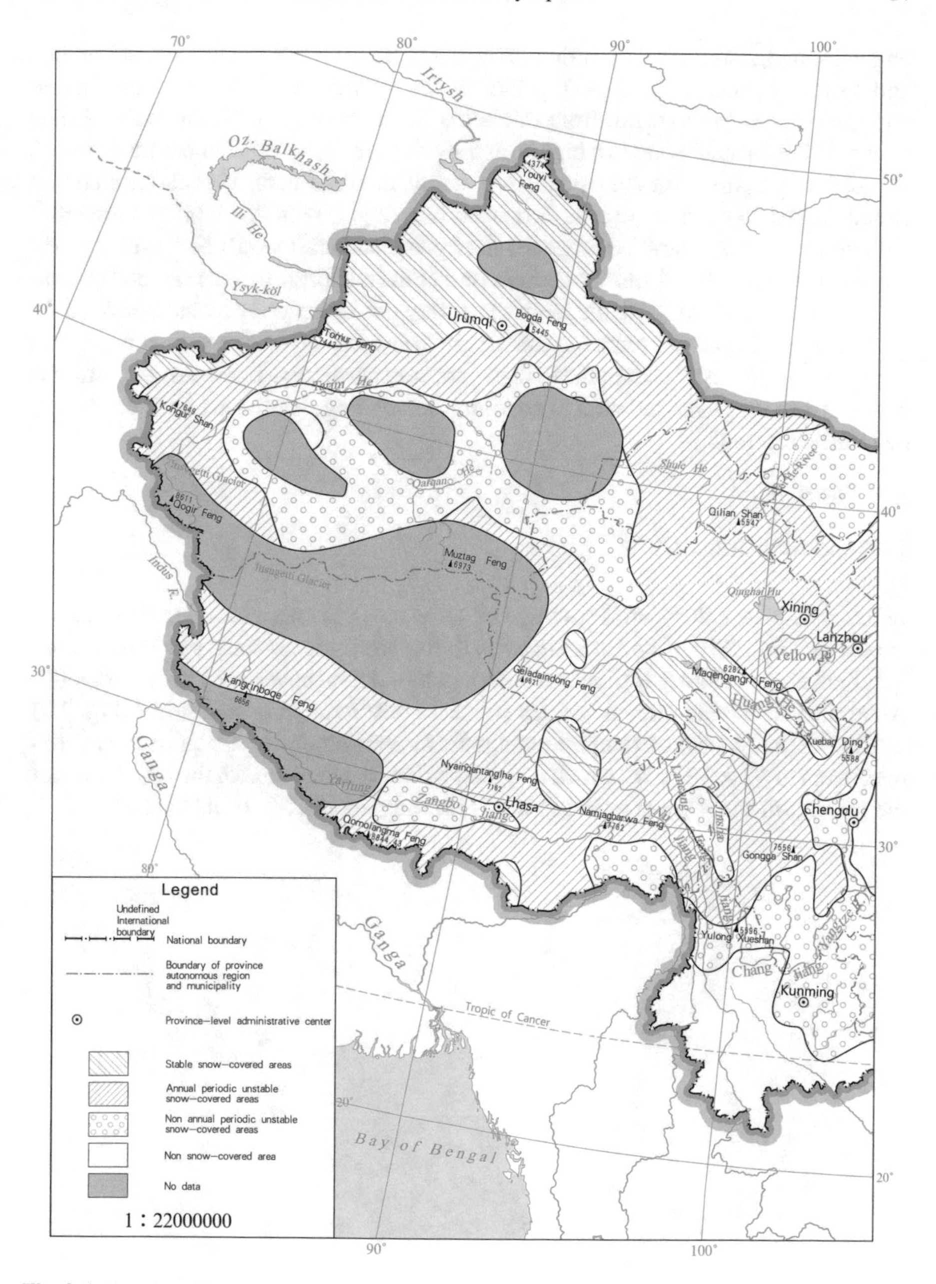

Fig. 2.6 (continued)

on total snowfall amount. With climate warming, the ratio of snowfall to precipitation and the ratio of accumulated snow to snowfall are both reduced. A critical air temperature to distinguish snowfall from rainfall is determined as 0 °C, and snow falling above 0 °C is unstable due to a higher air temperature. When air temperature rises, it is possible that part of the solid precipitation falls in liquid state. This part is therefore called "at-risk" snowfall, and the corresponding region with this kind of snowfall is defined as "at-risk" snowfall areas. Basically, the snowfall won't be "at-risk" if the mean amount reaches 4 mm in autumn or 3 mm in spring. But it may be "at-risk" accumulating snow because the fresh snow may not accumulate to snowpack on the ground due to higher air temperature. The region with "at-risk" accumulating snow is defined as "at-risk" accumulated snow areas accordingly. However, for different snow covered regions, the critical condition for distinguish these two indexes may be totally different.

2. Distribution

1) Global distribution

In the cryosphere, the spatial coverage of snow cover is only second to the seasonal frozen ground, 98% of which distributes in the Northern Hemisphere (NH). In the Southern Hemisphere (SH), there are few patches of land covered by snow, except for Antarctica. The inner-annual variation of snow cover extent is significant (Fig. 2.7). In the NH, snow cover extent may range from a minimum of 1.9 million km^2 to a maximum of 47 × 10^6 km^2, which is almost half of the land area of the NH. However, the maximum snow cover extent in the SH is only about 25% of its land area.

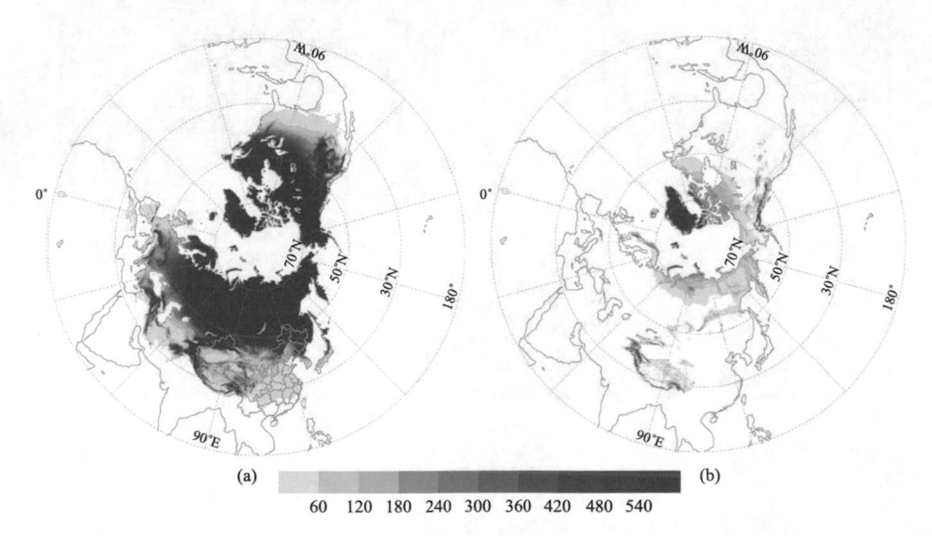

Fig. 2.7 Snow cover distribution in winter (**a**) and summer (**b**) in the Northern Hemisphere on an averaged climate regime (unit: km^2; *Data source* IMS product from NIC, USA)

Table 2.4 Snowline distribution of mountains in the Northern Hemisphere

Latitude/°	80	70	60	50	40	30	20	10
Highest altitude/m	600	1500	2600	3700	5100	6100	5300	4700
Lowest altitude/m	100	300	700	1100	2500	4200	4700	4500

Note The latitude refers to the central value of each latitude zone. E.g., 80° refers to the latitude zone of 75°–85°
Data source Shutov (2009)

Snow cover usually lies north (south) of or above the seasonal snow line in the NH (SH) or mountainous regions. The seasonal snow line is the southernmost/northernmost boundary or the lower limits of snow cover in the NH/SH or in mountain areas, and it moves towards higher latitude or higher altitudes with the snow melting (Table 2.4). When snow cover melts away, the seasonal snow line disappears. Therefore, it varies seasonally.

2) Distribution in China

The geographical distribution of snow cover in China is wide, but extremely uneven. The annual mean snow cover extent in China is approximately 3.4 million km^2 and mainly distributes in the northeastern China, the northern Xinjiang, and the Qinghai-Tibet Plateau. The annual mean snow depth, snow density, snow water equivalent is 0.49 cm, 140 kg/m^3, 0.7 mm, respectively. The region with the most annual SCD is located in the Xing'an Mountains and then followed by the Pamir Plateau, Karakoram Mountain, Himalayas, and Tian Mountain. As a whole, annual SCD is greater in the Qinghai-Tibet Plateau and Xinjiang Uygur Autonomous Region (XUAR) than that in the northeastern China, but the snow cover extent in the northeastern China is much larger (Fig. 2.7). The annual SCD over the Qinghai-Tibet Plateau reaches the most in January, which is approximately 7 days and is the least during June to September, which is less than 1 day. Seasonally, SCD reaches the most in winter, more than 18 days in average and is followed by spring and autumn, more than 14 days and 8 days, respectively.

2.2.4 River Ice and Lake Ice

River ice and lake ice are interwoven into the terrestrial landscape through the major freshwater flow and storage networks and, as a result, have enormous significance for the physical, ecological and socio-economic systems within the high-latitude and altitude cold regions.

1. Classification of river ice

River ice cover types and specific river ice processes are categorized in terms of ① channel types, ② channel sizes, ③ winter severity. There are six ice cover types:

ice shell, suspended ice cover, floating surface ice cover, confined surface ice cover, solid ice cover and no ice.

River ice shells form in most steep channels during the early freeze-up period. They are rigidly attached to cold surfaces, such as emerging rocks, banks, or hanging branches. They are not floating ice features. Ice shells can form even if the water temperature is above 0 °C which explains how they can be observed in relatively warm headwater channels.

Suspended ice cover features form in most steep channels as a result of dynamic ice production once the water temperature has dropped to or slightly below 0 °C. The first cryologic stage leading to the development of a suspended ice cover is the production of anchor ice. The second cryologic stage consists of the emergence of anchor ice accumulations, leading to the development of ice dams. The third cryologic stage consists of the breaching or perforation of ice dams, which release the water stored behind the dams, leaving the ice cover cantilevered, or suspended above the flowing water.

Floating surface ice is the most documented in the river ice literature. Floating ice can be in motion (e.g., frazil slush, frazil pans, or brash ice), or can be immobile ("fast") through frictional forces (e.g., surface ice jam) or by freezing into the banks (a fast ice cover). A complete floating ice cover normally forms in relatively slow flowing channels (low-gradient reaches) by combining ① border ice lateral migration, ② frazil or ice pan congestion, and ③ frontal progression.

Confined surface ice cover. Initially, a surface ice cover that forms in low-gradient reaches is necessarily floating. When the ice cover thickens, then depending on the ice-to-bank links, it can momentarily support the pressure induced by an increasing channel discharge or local water level. A confined ice cover is characterized by its protracted resistance to pressurized flow conditions or submergence.

Solid ice cover forms when the channel discharge is depleted or when the ice cover (surface, confined or suspended) freezes down into the channel bed. This type of ice cover is mostly associated with headwater and the intermediate channels of Arctic regions, independently of their gradient. The possible development of a solid ice cover presents two consequences. First, it presents a serious limitation for water consumption in winter. Second, it can lead to the development of thick in-channel and overbank aufeis accumulations.

Ice free: Some channels never develop a stable ice cover, particularly warm headwater channels. The groundwater temperature feeding a channel largely depends on year-averaged air temperatures. In low-gradient headwater channel segments where no splashing occurs, any surface ice development is ephemeral, and the water remains above freezing during most of the winter period.

2. Classification of lake ice

Black ice is formed as lake water freezes to the underside of the ice cover. Black ice grows from the bottom of the cover, resulting in a sheet of ice composed of columnar crystals elongated in the vertical direction. There is a distinct interface

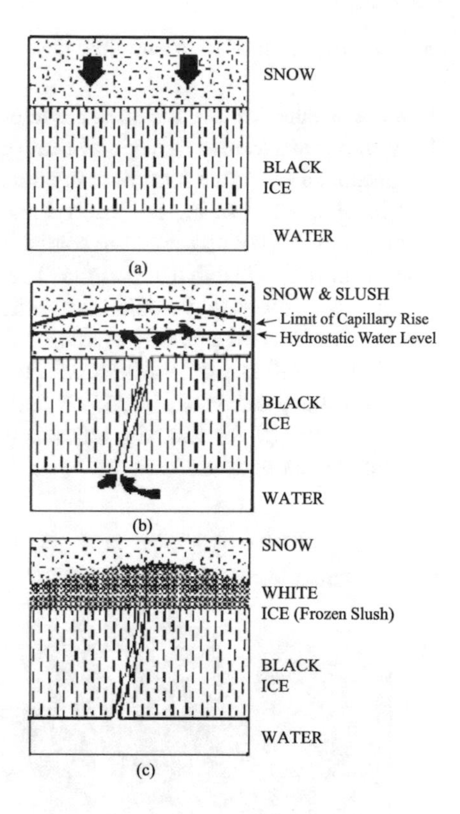

Fig. 2.8 Lake ice types and its formation (Adams and Lasenby 1985)

between white ice and black ice, where it is possible to see through black ice to some extent (Fig. 2.8).

White ice is formed on the top of the ice cover when slush that had developed at the bottom of the lake snow cover freezes. White ice is granular and opaque, with randomly oriented ice crystals. The most common process of white ice production is: the ice is depressed below the hydrostatic water level by its load of snow, lake water enters the snow through cracks in the ice and slush is formed. White ice contains distinct layers of varying bubble density, which presumably indicate different rates of freezing and/or the effects of the stratigraphy of the snowpack, which was slushed to produce the white ice.

The slush freezes from its upper surface downwards to form a layer of white ice. White ice is also formed when meltwater, rain, or river water infiltrates the snow down to the top of the ice sheet and freezes.

Snow ice: There is almost always snow ice occurrence when the positive isostasy that occurs although the start of snow depletion is delayed a week or two.

Snow slush: Water rises through a crack at a time when the surface of the ice cover is below the hydrostatic water level. The snow and ice load exceeds the lifting force, water seeps into the snow, and a slush layer is formed at the bottom of the snow layer on the ice.

3. Distribution of river and lake ice in north hemisphere

River and lake ice are the major components of the cryosphere and are particularly important to many river and lake processes, including extreme events. Its full geographical coverage has never been documented. Recognizing that the freeze-up and breakup of river ice is closely linked to the timing of 0 °C air temperatures, Bannet and Prowse analysed the spatial extent of river networks relative to the location of three 0 °C isotherm periods. These were defined to represent a suite of ice-affected conditions that would be experienced for 6-, 3- or 0·5-month periods, with the briefest interval possibly leading only to a very thin and transient ice cover or simply border/frazil ice formations (Fig. 2.9). The related southern position of the isotherms ranged from 33° N, 35° N and 50° N in central North America to a nearly consistent 27° N for Eurasia, reflecting the influence of the high-elevation in the central plateau region.

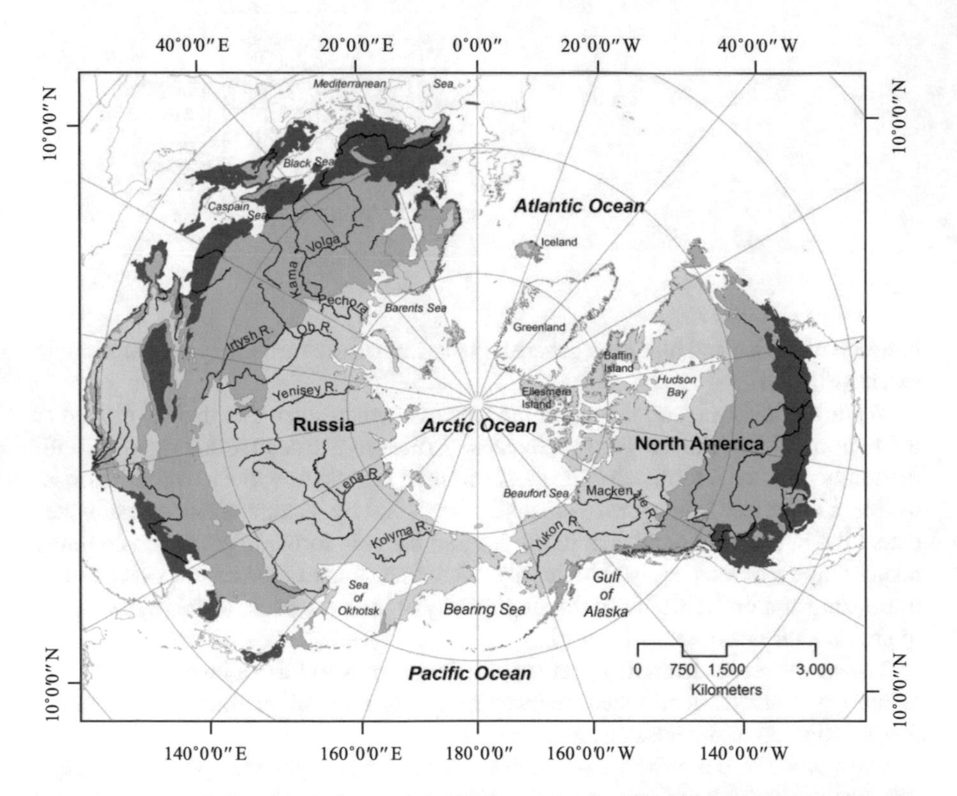

Fig. 2.9 Three isotherms indicated by different frozen time (Bennett and Prowse 2010). Light grey depicts areas within the 0 °C annual isotherm (I6); mid grey, the October–March 0 °C isotherm (I3), and dark grey the 0 °C January isotherm (I0.5). Polar stereographic projection, scale 1: 90 000 000

2.3 Classification and Distribution of Marine Cryosphere

The marine cryosphere consists of ice shelves, icebergs, sea ice with overlying snow, and submarine permafrost.

2.3.1 *Ice Shelf and Iceberg*

Ice outflows from the centre of ice sheets to the periphery, forming ice bodies that extend and float on the ocean surface at their front end, called an ice shelf. Some ice shelves are hundreds of kilometres long. Large or small ice bodies at the edge of ice sheets and ice-shelves or the glacier terminal calve into the ocean, called icebergs.

1. Classification and distribution of ice-shelves

According to the stress, the velocity distribution and the configuration, ice shelves can mainly be divided into three types:

Unconfined ice-shelves: Ice-shelves freely extend outwards; i.e., any of the longitudinal flowlines extend the same length.

Confined ice-shelves: Ice shelves are restrained by parallel walls on both sides and their flow rates are controlled by the depth of water, the speed of front ice and the thickness of the ice shelves.

Fjord ice-shelves: Ice-shelves develop in fjords and the configuration of fjords can hinder their movement. According to the distribution pattern in the fjords, the ice-shelves are further divided into fjord expansion and fjord contraction ice shelves.

Globally, ice shelves are only found in Antarctica, Greenland, Canada and the Russian Arctic. Figure 2.10 shows the distribution of Antarctic ice shelves, and the state of iceberg calving and basal melting.

In the context of global warming, ice shelves continue to collapse or even disappear, which is a rapidly changing component of the cryosphere.

Larger scale ice-shelves have been developed in Canada's high-Arctic Ellesmere archipelago. However, with the remarkable warming of the late 20th century, many ice shelves in the Ellesmere Islands gradually disintegrated. So far, the number and area of ice shelves have greatly shrunk. Vincent et al. (2001) reconstructed the evolution of the Ellesmere ice shelf in the 20th-century (Fig. 2.11).

2. Classification and distribution of icebergs

The Antarctic ice sheet and the Greenland ice sheet are the main sources of icebergs. Icebergs are freshwater ice, and a large number of icebergs into the water can change the ocean's temperature and salinity. Drifting icebergs may greatly threat maritime voyage safety.

After the Ellesmere ice shelf was mapped by Marvin in 1906, the sequence of ice shelf changes was mapped based on previous surveys and remote sensing data

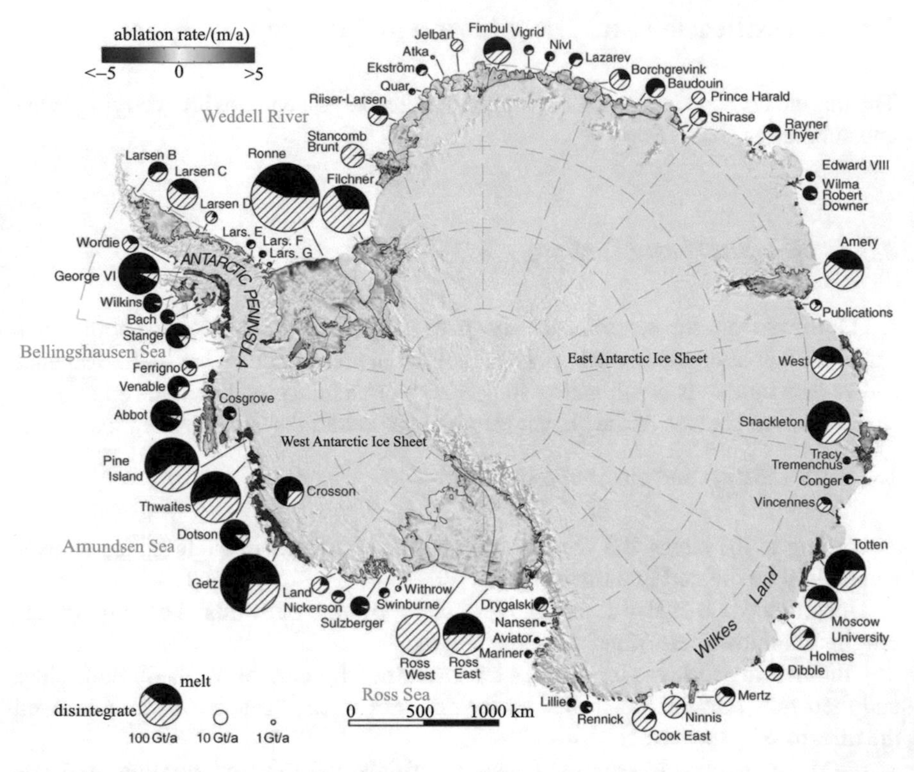

Fig. 2.10 Distribution, estimated melting rates of ice shelves surrounding Antarctic Ice Sheet (Rignot et al. 2013)

(of which data in 1998/1999 was acquired based on RADARSAT-1 remote sensing images). The bottom figure shows the location of the ice shelf.

An iceberg is mainly classified according to the shape and size. Based on the shape and size of icebergs, the World Meteorological Organization divides icebergs into iceberg, small iceberg and growler. Icebergs with freeboard height higher than 5 m can be subdivided into flat top, dome-shaped, spire iceberg, etc. The freeboard height of small icebergs is 1–5 m, and the area is usually 100–300 m^2; the freeboard height of a growler is less than 1 m, and the area is generally approximately 20 m^2. The International Ice Patrol (IIP) has established a classification system based on the size of icebergs. At present, the international classification of the size of icebergs is mainly based on the use of the classification table designed by the International Ice Patrol (Table 2.5).

Most of the earth's icebergs originate from the Antarctic ice sheet. The total number of icebergs in the South Ocean reaches approximately 200,000, which accounts for approximately 93% of the global icebergs and has a total weight of 10^{12} tons, which concentrates on the sea surface in the Southern Ocean and flows from east to the west with the coastal currents. Sometimes, the icebergs can drift into the areas in the South Atlantic near New Zealand and the South Pacific near

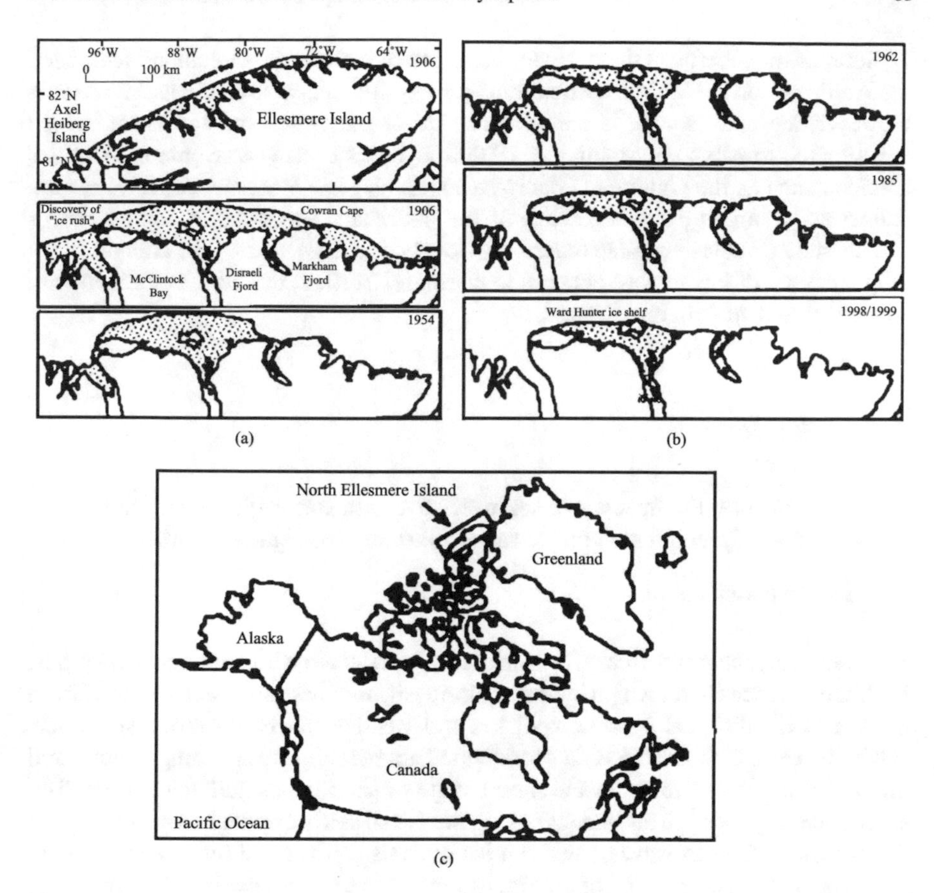

Fig. 2.11 Ice shelves and changes in the northern Ellesmere Island (Vincent et al. 2001)

Table 2.5 Classification based on the size of icebergs

Size	Height/m	Length/m
Growler	<1	<5
Bergy bit	1–5	5–15
Small	5–15	15–60
Medium	15–45	60–120
Large	45–75	120–200
Very large	Over 75	Over 200

the coast of the South America. The National Center for Snow and Ice at Colorado and Brigham Young University have established databases for icebergs in the whole Antarctic region over the past few decades. The large tabular icebergs of 176–2109 km^2 have been continuously tracked and monitored for 15–20 days.

Sources of icebergs in the northern hemisphere include the Greenland Ice Sheet, the Arctic region in Canada, Svalbard in Norway and many parts of Russia's Arctic. However, the main source is the western side of the Greenland Ice Sheet, which is estimated to calve approximately 10,000 icebergs each year. Some of Alaska's glaciers, such as the Columbia Glacier, also disintegrate icebergs. All icebergs are dangerous to shipping but, depending on the size, shape and location, some icebergs can be more troublesome than others. Obviously, icebergs nearest the Atlantic shipping lanes are of the greatest concern to mariners. In 1912, the Titanic collided with an iceberg and sunk in the Atlantic.

2.3.2 Sea Ice

The ice formed by the freezing of sea water over sea surface is called sea ice, and the refreezing of precipitation on the sea ice surface also regard as part of sea ice.

1. Sea ice classification

When sea water begins to freeze, the surface water is mixed with scattered ice crystals, ice frazils and ice slices, which have no definite shape. There are many forms of fresh ice due to the different conditions of sea and weather (such as calm, disturbance, snowfall, etc.). The fresh ice can be divided into frazil, grease, shuga, slush and nilas. The frazil that forms at the initial stage of sea ice is small acicular or disc-shaped ice suspended in the water, making the ocean surface appear soup-like. Under the dynamic effect of waves, the fresh ice crystals can reach a few metres deep in open water. The coalescence of the frazils creates grease ice that is lighter in colour and appears as a blanket on the sea surface. Slush is the sea ice formed by snowfall. Shuga is formed on the water surface with disturbance. The white spongy-like sea ice clusters of several decades of centimetres in size usually formed by grease, slush, or anchor ice that floats to the surface. Under the effect of wind and waves, the shuga easily lines up along the prevailing wind direction to form an ice belt. Nilas is an elastic thin layer of sea ice that is formed by frazil, grease and shuga. It is easy to form finger rafting under the surging waves, and it can be divided into dark Nilas, which are typically 0–5 cm in thickness, and light Nilas, which have a thickness of generally 5–10 cm ice.

Persistent low temperature causes further condensation at the bottom and edges of sea ice, which thickens the sea ice and changes the colour. It is called young ice when sea ice thickness is 10–30 cm. The initial ice can be divided into grey ice and grey-white ice. The thickness of grey ice is between 10 and 15 cm, and it is not as elastic as Nilas. Under the surging waves, the grey ice is easy to break and becomes rafted. The thickness of grey-white ice is generally 15–30 cm; it is more prone to be ridged under the effect of wind pressure, rather than rafted.

Sea ice that only experiences one winter formation season is called the first-year ice. It develops from the initial ice. Without deformation, the thickness of one-year ice is 30–200 cm, while the first-year ice that experiences dynamic deformation can reach to more than 2 m. Sea ice formed by thermodynamic processes is rarely more than 2 m in Antarctica due to the effect of ocean heat flux at the bottom of the sea ice.

Sea ice that experiences at least one melting season is called old ice. Salt composition of old ice is usually lower than that of the first-year ice, and the surface has been subjected to more weathering. It can be further divided into second-year ice; that is, it experiences one thawing season. Multi-year ice refers to sea ice that does not melt for at least two summers.

The types of sea ice described above belong to various stages of thermodynamic development. In the course of sea ice development, there is also a kind of sea ice that is formed by dynamic processes, such as the pancake, which is also known as an ice cake. It refers to disc-shaped sea ice with a diameter of 30 cm–3 m and a thickness of approximately 10 cm. Ice cakes have bulged edges due to collisions with other ice cakes. Under light waves, the ice cake is formed from grease, shuga or ice rind and nilas through collision with each other. It can also be formed by the grey ice in rough waves.

According to its dynamic characteristics, sea ice can be divided into fast ice and drift ice. The former does not move with ocean currents and wind force, whereas the latter is driven by ocean currents and sea surface winds. Fast ice refers to the sea ice developed and adhered to on the coast, the ice wall, the front of the glacier, and between two shoals or stranded icebergs. Fast ice can be formed in situ by sea water, or by ice floe of different ice ages, which is fastened to the shore. Fast ice extends a few metres to hundreds of kilometres from the shore to the sea. If the ice age of fast ice is more than one year, it is old ice; that is, second-year ice or multi-year ice. Divided according to morphology, fast ice that attaches to the shore is ice foot; attaches to the shoals is shore ice; and fast ice that is frozen to the seafloor in shallow waters is anchor ice.

Floe refers to the broken pieces of different sizes that fragment under the effect of wind, sea water, current and tide after the formation of sea ice. It can be divided into brash, pancake, ice cake, small floe, medium floe, big floe and vast ice according to its size (Table 2.6).

Table 2.6 Sizes of different types of floe

Type	Size/m	Reference object
Brash	<2	
Pancake	0.3–3	Billiards table
Ice cake	$\leq$20	Volleyball court
Small floe	20–100	Warehouse
Medium floe	100–500	A block of a city
Big floe	500–2000	Golf course
Vast ice	$\geq$2000	Townlet

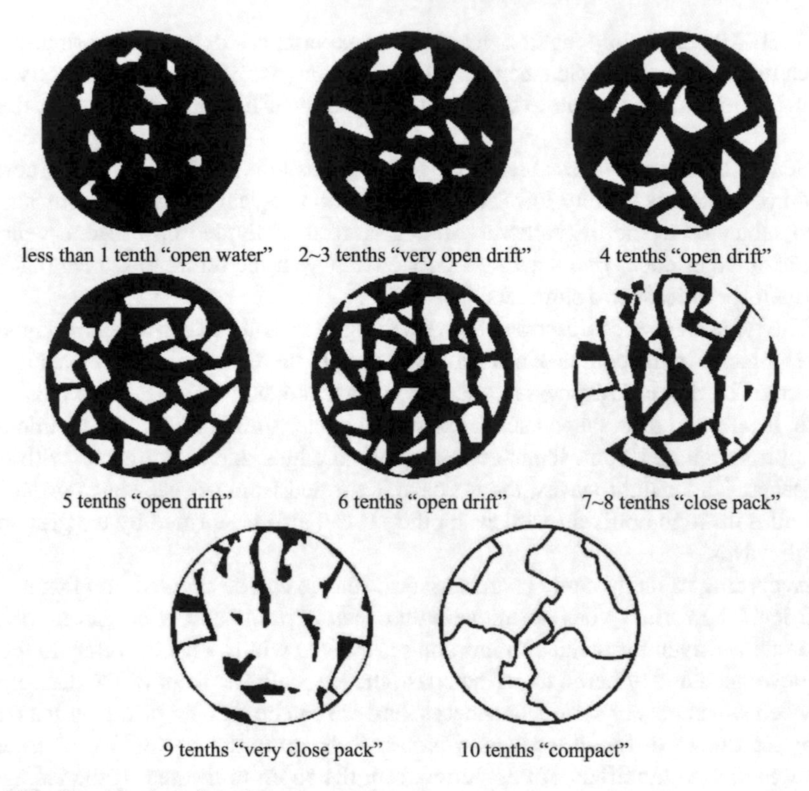

Fig. 2.12 Classification of sea ice concentrations

Sea ice concentration refers to the ratio of the area of ice features to the total area of a sea part (zone), expressed in tenths (Fig. 2.12).

(1) Open water: A large area of freely navigable water in which sea ice is present in concentrations of less than 1/10 and ice of land origin is absent;
(2) Very open pack ice: The concentration of floating ice is 1/10 to 3/10 and water preponderates over ice;
(3) Open pack ice: The ice concentration is 4/10 to 6/10, with many fractures and floes that are generally not in contact with one another;
(4) Close pack ice: The ice concentration is 7/10 to 8/10 and is composed of floes mostly in contact;
(5) Very close pack ice: The ice concentration is between 9/10 and 10/10;
(6) Compact pack ice: The ice concentration is 10/10, and no water is visible.

According to the characteristics of sea ice, it can be divided into level ice and deformed ice, bare ice and snow-covered ice, dirty ice, etc. Level ice is a kind of sea ice not affected by deformation. Deformed ice is a general term for sea ice that is squeezed and fractured along with the convergence of surface and underwater sea ice. It can be subdivided into rafted ice, ridged ice and rough ice, etc. The convergence of the thin ice usually overlaps, forming overlapping ice; and the thick ice is usually compressed, forming an ice ridge. The phenomenon of rafting (overlapping) usually appears in the early stage of sea ice growth, and its effect causes the thickness of sea ice to increase rapidly to 0.4–0.6 m. When the thickness of ice floes is more than 0.4 m, it is prone to be ridged due to the convergence of ice floes, and the rafting and ridging play an important role in sea ice growth and thickness distribution. Overlapped and ridged ice will significantly increase the thickness of sea ice. The degree of deformation of the ridged ice beneath the water level is usually greater than that above the water surface. Ridges with greater deformation may have a surface height up to 12 m and an ice draft of 45 m at the bottom.

Bare ice refers to the sea ice without snow cover. Dirty ice contains minerals or organics on the surface or inside the ice that come from nature or human origin. Due to the high concentration of algae in the ice, the sea ice is brownish yellow, called brown ice, and the brown ice may appear in all layers of sea ice.

Waters surrounded by ice include fracture, lead, polynya and tide crack, etc.

Fracture is a phenomenon of permanent deformation and rupture of sea ice under pressure. Break or rupture occurs through close ice, compact ice, consolidated ice, fast ice, or a single floe resulting from shears and a deformation process. The crack length varies from a few metres to thousands of metres. The development of cracks form leads, and leads are prone to form fresh ice. Leads and cracks appear in linear form. Open water that is larger than lead is called polynya. The lead between a close pack of ice and the shore is called shore lead. A passage-way between a close pack of ice and fast ice is called a flaw lead. A crack is much narrower than a lead, and it does not help ship navigation. However, lead is navigable to the voyage of vessels. The occurrence of cracks and leads enhance the heat exchange between the ocean and atmosphere, leading to the appearance of water vapor, sea fog or frost smoke. Cracks and lead also provide seals and penguins with access to the ocean and provide whales with breathing holes.

A polynya is an area of non-linear open water that is surrounded by sea ice. Polynyas may be covered with fresh ice, nilas or young ice. The submarines treat the ice lake as a skylight. Polynyas can be divided into:

(1) Shore polynya: A polynya is restricted by the shore from one side.
(2) Flaw polynya: A polynya which lies between pack ice and fast ice.
(3) Recurring polynya: A polynya recurs in the same position every year.

A polynya's area ranges greatly. The biggest polynya was observed in the Weddell Sea during 1975–1977 and covered an area of 20×10^4 km^2.

Polynyas can be divided into latent heat polynyas and sensible heat polynyas according to their formation mechanism. Latent heat polynyas form under the effect of frequent katabatic wind. The newly formed sea ice drifts northward, forced by the wind, which leads to the occurrence of open water. Then, it will result in the formation of new ice. Sensible heat polynyas are known as the factory of sea ice. Most shore polynyas are latent heat polynyas. They form under the effect of the upwelling warm current. Shore polynyas do not form a large amount of new ice in sensible heat polynyas. Nevertheless, there are also some polynyas that formed under the combined effect of latent heat and sensible heat.

Due to the tidal effect, sea surface rises and falls, and tide flaws occur in fast ice. Tide flaws provide access for penguins and seals to get in and out.

The formation of sea ice can begin in any layer of seawater, even on the ocean floor. The ice that is formed under the surface of the water is called frazil. It is formed by the freezing of supercooled water. The ice that adheres to the ocean floor is termed anchor ice. After the formation of anchor ice, because the ice density is lower than that of seawater, the ice will gradually rise and combine with the sea ice formed on the sea surface. The process will make the sea ice on the sea surface become thicker.

2. Distribution of sea ice

1) Global distribution of sea ice

Sea ice covers approximately 7% of the Earth's surface and 12% of the world's oceans. Sea ice exits all year in the multi-year sea ice area, including the centre of the Arctic Ocean, as well as in parts of Antarctica, mainly in the west Weddell Sea. Those regions where sea ice only occurs in winter are called seasonal sea ice regions, which can extend to locations where average latitude is approximately 60°. Most of the world's sea ice is concentrated in the Earth's polar oceans. In the Southern Hemisphere, sea ice is found mainly in the Southern Ocean around the Antarctic continent. The sea ice cover in the Southern Ocean is actually ring-shaped with its length of approximately 2×10^4 km. In summer, its width is almost zero, while in winter, it can reach up to 1000 km surrounding the Antarctic continent, which is concentrated between 60° S and 70° S. In the Northern Hemisphere, sea ice is found mostly in the Arctic Ocean and nearby areas, as well as in other cold waters and bays, such as the Sea of Okhotsk, the Bering Sea, the Baffin Bay, the Hudson Bay, the Greenland Sea, the Labrador Sea, the Baltic Sea and the Bohai Sea, etc. The sea ice with the lowest latitude is found in the Yellow Sea and in the Bohai Sea of China and ranges from 37° N to 41° N.

Since the early 1970s, passive microwave remote sensing data has provided the most complete records of sea ice extent. Before this, sea ice observation on the shore can only be conducted at some locations and at a specific time.

Sea ice varies significantly at seasonal and annual scales. The extent of sea ice in the Northern Hemisphere reaches its maximum in February or March and minimum in September. Arctic sea ice cover with an average ice extent varies between approximately 6×10^6 km^2 in summer and approximately 15×10^6 km^2 in winter. The minimum ice extent of the Arctic Ocean in 2012 was only 3.44×10^6 km^2, which is the smallest size recorded since the advent of satellite observation in 1979 (see Fig. 2.2). The seasonal extent of the Antarctic sea ice cover varies more significantly. It reaches a maximum extent of over 18×10^6 km^2 in September whereas its minimum of approximately 3×10^6 km^2 occurs in February.

2) Distribution of sea ice in China

The Bohai Sea and the Yellow Sea are located in the middle latitudes of the East Asian monsoon climate zone, which is one of the lowest latitudes where sea ice

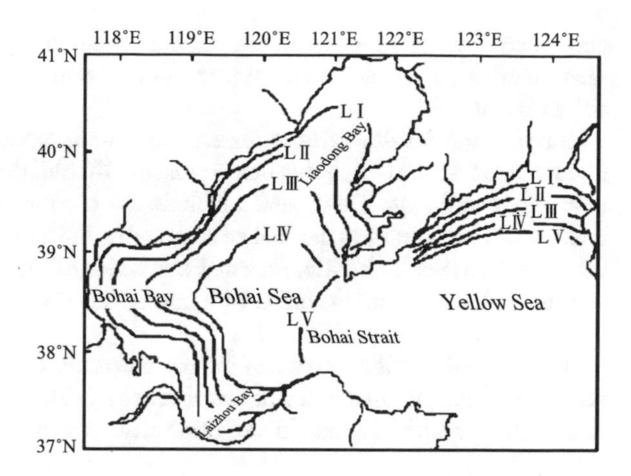

Fig. 2.13 Classification of sea ice severity in the Bohai Sea and the north Yellow Sea, China (Bai et al. 2001)

forms in winter. The sea ice in the Bohai Sea and the north Yellow Sea vary greatly with winter climate each year. In warm winters, the sea ice covers less than 15% of the waters, while in cold winters, it can cover more than 80%. In the 20th century, the sea ice was three times greater, so that the Bohai Sea was almost fully covered by ice from January to February in 1936, January to February in 1947, and February to March in 1969.

According to observations and historical records, the severity of sea ice in the Bohai Sea and the Yellow Sea can be divided into 5 levels based on the ice thicknesses and extents. Figure 2.13 indicates the classification of sea ice severity, with the distribution of the ice marginal line and the corresponding levels.

As the climate gets warmer, the sea ice in the Bohai Sea and in the Yellow Sea has been continuously decreasing since 1980s, because the first-ice date is delayed and the end-ice date shifted earlier. The number of ice days currently is approximately 30 days less than that of the 1960s.

2.3.3 Subsea Permafrost

Sub-sea permafrost, alternatively known as submarine permafrost and offshore permafrost, is defined as permafrost occurring beneath the seabed. It exists in continental shelves in the polar regions. During the Ice Ages as well as during the Last Glacial Maximum, the sea level may have been 100 m lower than it is today. As a result, the shallow continental shelf in the polar regions that were not covered by ice sheets were exposed to low air temperature. Permafrost aggraded in these shelves from the exposed ground surface downwards. With the disappearance of ancient ice sheets and the rising of sea levels, permafrost degrades in the submerged shelves under relatively warm and salty boundary conditions. Sub-sea permafrost differs from other permafrost in that it is relic, warm, and generally degrading. Sub-sea permafrost derives its economic importance from current interests in the development of offshore petroleum and other natural resources in the continental shelves of polar regions.

Based on the transition from sub-aerial to sub-sea conditions, sub-sea permafrost is divided into 5 zones (Fig. 2.14) that represent different thermal and chemical surface boundary conditions. These zones include the onshore and the beach regions, the areas where ice freezes to the seabed seasonally, the areas where restricted under-ice circulation causes higher-than-normal sea water salinities and lower temperatures over the sediments, and the areas with normal sea water over the seabed throughout the year.

The detailed distribution of sub-sea permafrost has not been fully investigated. In particular, it is still unknown whether there exists sub-sea permafrost in Antarctica. Around the Arctic is the main distribution area, of which the Eurasian side is the key area.

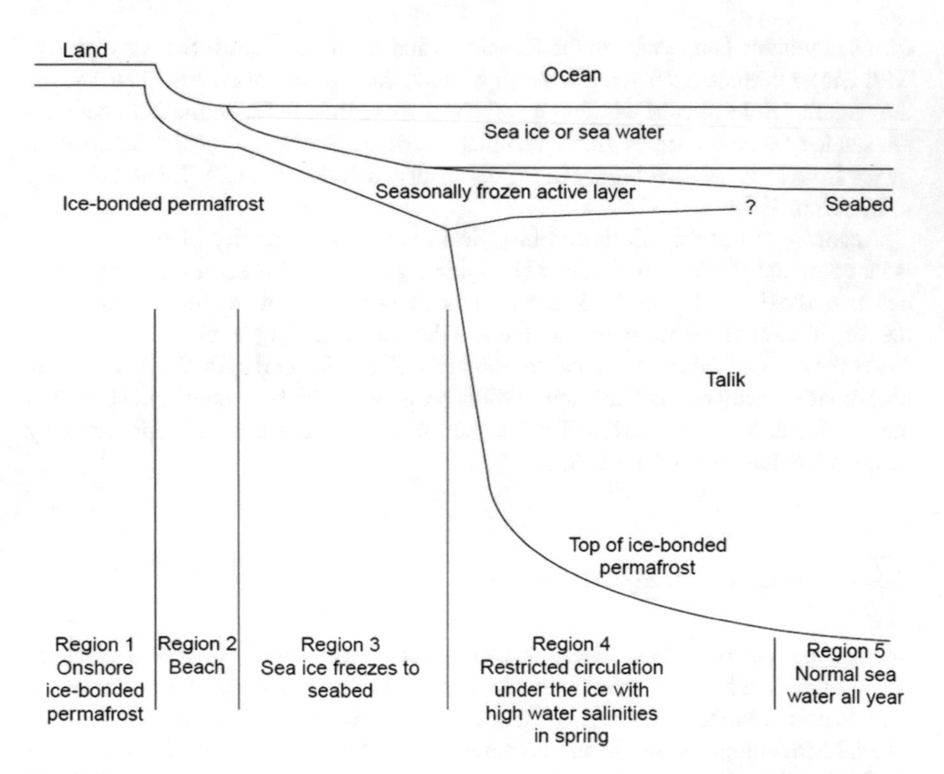

Fig. 2.14 Schematic illustration of the transition of permafrost from sub-aerial to sub-sea conditions. There are five potential regions with differing thermal and chemical seabed boundary conditions (after Osterkamp 2001)

2.4 Classification and Distribution of the Aerial Cryosphere

The space within the troposphere and stratosphere with temperatures below the freezing point (0 °C) is the scope of the aerial cryosphere, such as ice clouds and solid precipitation. The main form of the aerial cryosphere is solid precipitation existing in the atmosphere in various forms before descending to the ground. The snowflake, graupel, hail, and other ice crystals before descending to the ground are components of the aerial cryosphere, which are completely distinguished from fresh snow on the ground. Solid precipitation after dropping on the land is part of the terrestrial cryosphere, such as snow cover on the ground, but becomes part of the marine cryosphere if it drops onto sea ice, such as snow cover on sea ice.

2.4.1 Classification

Due to the difference of meteorological conditions and the developing environment, various kinds of solid precipitation exist in the atmosphere and are quite different. In 1951, the International Commission on Snow and Ice proposed a system of classification for solid precipitation, which quickly became the standard system used in the field. The commission identified a total of ten types of solid precipitation—seven types of snowflakes, plus hail, ice pellets, and graupel (Fig. 2.15). The seven types of snowflakes are as follows: plate, stellar crystal, column, needle, spatial dendrite, capped column, and irregular forms. These 7 kinds are solid precipitation formed through the condensation of water vapor in the atmosphere. Ice pellets, graupel, and hail are types of solid precipitation that water vapor transforms first into liquid water, and then into ice crystals. Ice pellets are often referred to as "sleet." Graupel, also called "soft hail", is a combination of hail and snow where a snowflake forms the nucleus around which supercooled water solidifies as a hail stone.

In 1966, Magono and Lee classified the naturally occurring snow crystals in more detail, with a total of 80 species. This classification was used widely in snow and ice crystal studies until 2013, when 41 new kinds were added after observation of snow crystals from Japan to the polar regions. This updated classification system is considered to be appropriate for solid precipitation classification globally. This global classification consists of three levels—general, intermediate, and elementary—which are composed of 8, 39, and 121 categories, respectively.

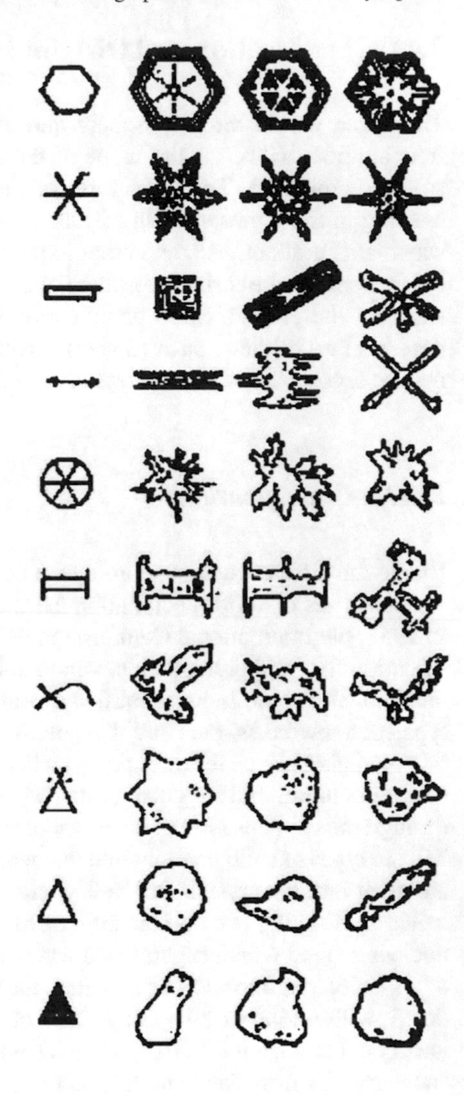

Fig. 2.15 Ten forms of solid precipitation. Top to bottom: plate, stellar crystal, column, needle, spatial dendrite, capped column, irregular forms, graupel, ice pellet, and hail

2.4.2 Distribution

1. Global distribution

Whenever snow crystals grow to a size where the gravitational pull exceeds the buoyancy effect of air, snowfall occurs. Snowfall is the most widespread, common and primary form of atmospheric solid precipitation. Snowfall is a random process that occurs in many parts of the world, and its distribution is affected by regional climatic and geographical locations (latitude, longitude and elevation).

The global distribution of solid precipitation can be roughly seen as the scope of global snow cover. However, because the snow falls around the southern boundary in NH or northern boundary in SH, solid precipitation may rapidly melt without accumulation; in general, the distribution of solid precipitation should be greater than that of snow cover.

This section will only present some examples of solid precipitation distribution in the United States and China as examples.

In the United States, according to the average annual frequency of events, a snowstorm event occurs when the amount of snowfall accumulates to over 15.2 cm in 1–2 days. The snowstorm has a large spatial variability; in the eastern United States, snowstorm frequency is distributed latitudinally. In the southern hinterland, the snowstorm frequency is about once per decade and increases northward to twice per decade along the US and Canadian border. The frequency of the leeward snowfalls of the Great Lakes and the Appalachian Mountains are relatively high. Snow falls at lower-elevation sites with an average of 0.1–2 storms per year in the western region of the United States, but the annual minima are 1 or more storms in the high-elevation areas of the West and Northeast. Snowstorms first occur in September in the Rockies, in October in the high plains, in November across most of the United States, and in December in the Deep South. December is the month when the season's last storms occur in the South and then shifts northward, with April being the last month of snowstorms across most of the United States.

2. Distribution of the aerial cryosphere in China

Solid precipitation is an important form of precipitation in western China, where plateau and high mountains extend throughout the region. However, the coverage for solid precipitation monitoring is far from enough due to the harsh natural conditions.

1) Snowfall

The snowfall in China is characterized by distribution mainly in regions with high latitudes or high altitudes and is concentrated in southern China as well. There are four regions with high snowfall amounts: the northern and eastern parts of Northeast China and Changbai mountains, the northern XUAR and the western Pamirs, the Qilian mountains and the eastern and southern parts of Tibetan Plateau, and the middle and lower reaches of the Yangtze River.

Northern XUAR and the northern and eastern parts of northeastern China are located at high latitudes, which are affected by the cold air mass from the north, with great intensity, high frequency, and long duration. In northern XUAR, the snowfall usually begins in September and ends in June of the following year, and in northeastern China, the snowfall lasts until the following May. In the middle and lower reaches of the Yangtze River, the water content of snow is larger, since they are located in the subtropical monsoon climate zone. Once the cold air moves southward, it is easy to develop a weather process of combined rain and snow. Due to the block of the Qinling Mountains, the cold air is difficultly entering into southwest China, which is centred on the Sichuan Basin, resulting in less snow there except for in the western edge of the basin that borders the Tibetan Plateau. The North China

Plain and the Northeast China Plain, located in the temperate continental monsoon climate, are semi-arid and semi-humid regions in China. In winter, these regions are affected mainly by the cold air from the north and the westerlies systems, and the annual snowfall amount is approximately 30 mm. The dry desert areas in inland China, including western Inner Mongolia, are areas with less snow, where the annual snowfall amount is below 10 mm.

The snowfall event is controlled by the air temperature and water vapour content. Consequently, the geographical distribution of snow in China is closely related to the area affected by the cold wave activities. The zonal distribution of snowfall is mainly affected by air temperature, and the meridional distribution is mainly controlled by water vapor conditions. The snowfall event in high latitudes is affected by both the air temperature and the local water vapour content, where snowfall, especially heavy snow, occurs more frequently.

2) Hail

Hail refers to solid, conical or irregular solid precipitation and is common in summer or at the end of Spring and in the beginning of Summer. Some are as small as mung beans and yellow beans, and some are as large as chestnuts or eggs. Hail is rare in the Guangdong, Hunan, Hubei, Fujian, and Jiangxi provinces of South China, but the damages caused by hail hit the other provinces each year, to different extents. In the northern mountains and hilly areas with complicated terrain and changeable weather, hail occurs in high frequency and is very harmful to agriculture.

From the regional distribution of hail in China, more hailstone occurs in one zone and two belts. The one zone refers to the Tibetan Plateau, and the two belts refer to the southern and the northern parts of China. The southern belt centres on the Yunnan-Guizhou Plateau mainly at altitudes of 1000–2000 m and extents eastward to the border of the western Hunan, Sichuan and Hubei provinces. The northern belt extends from the Qilian and Liupan Mountains to the Inner Mongolian Plateau via the Loess Plateau.

3) Graupel and ice pellets

Graupel and ice pellets are not as common as snow and hail.

Graupel refers to the solid precipitation composed of white opaque quasi-spherical (sometimes conical) ice particles with snow-like structures, also known as snow pellets or soft hail. The diameter of graupel is usually 2–5 mm. It is crisp and fragile and often bounces back when falling onto the hard ground. Graupel usually occurs when the temperature is not so cold, usually before or with snow.

Graupel is not a category of snow. The structure of graupel is more dense than ordinary snow and particles, which is caused by overlying frost. The weight and low viscosity of the graupel make it unstable on the slope; the snowpack layer of 20–30 cm graupel may consequently lead to a severe avalanche. Due to the temperature effect and the characteristics of graupel, graupel becomes more stable in one or two days after an avalanche.

Ice pellets are transparent globular or irregular solid precipitation, which are hard and usually rebound on the ground, usually with a diameter less than 5 mm. Sometimes, there is unfrozen water inside, and when it is broken, only the broken shell is left.

Ice pellets and hail are translucent pellets that are formed by freezing layers of larger drops of water around the condensation nuclei. In meteorology, the forms with a particle size of no more than 5 mm are called ice pellets, and those larger than 5 mm are called hail. In summer, ice and hail are often seen in the northern plains of China. The main difference between ice pellets and snowflakes is that snowflakes are formed at a lower temperature and generally occur in a large range. However, ice pellets tend to appear during convective weather, and hence, often occur locally.

Questions

1. What are the zonal distribution characteristics of the global cryosphere?
2. What is the main basis for the classification of cryosphere components?
3. What are the seasonal changes for the upper and lower boundaries of the aerial cryosphere?

Extended Readings

Classic Works

1. *The outline of China's glaciers*

Author: Shi Y. F.

Publisher: Science Press, 1988

The Outline of China's Glaciers, edited by Shi Yafeng, was published by Science Press in 1988.

For 30 years, from 1958 to 1988, Chinese glacier climbers worked the major mountain ranges and boarded many representative glaciers to identify the basic situation of Chinese glaciers.

The introduction of the Chinese Glacier is the crystallization of this kind of hard work.

The book is based on 30 years of field investigation, indoor statistics compilation and cataloguing work.

The whole book is divided into 12 chapters, 36 thousand words, 40 articles of reference, more than 60 pictures, and the same pictures and letters.

This book systematically expounds the developmental conditions, heat balance, ice formation, material balance, glacier movement, glacier temperature, glacier geochemistry, glacier type distribution, glacier change, glacier hydrology and glacial calamity.

There are unique research or theoretical advances in many aspects of the glacier balance line with temperature and precipitation, glacier movement and the nature of glaciers and glacier ice formation.

In particular, the research data on the changes of China's glaciers are of great value in understanding the laws of climate change and for predicting global changes in the future.

The last two chapters of this book give a detailed introduction to glacier hydrology and glacial calamity in China, which is undoubtedly of great significance for the development and construction of western China.

2. ***The permafrost of China***

Author: Zhou Y. W., Guo D. X., Qiu G. Q., Cheng G. D., Li S. D.

Publisher: Science Press, 2000

The Permafrost of China summarizes the research results of China's permafrost since the late 1950s and shows to the world the remarkable achievements in permafrost and seasonal frozen soil research.

The introduction briefly describes the basic definitions and terms of permafrost. The term used in the book is in accordance with the terms used by the international permafrost community.

The first, second and third chapters of the book are "Frozen Soil Forming Conditions and their Main Characteristics", "Frozen Soil Regionalization and Frozen Soil Characteristics in Frozen Soil Regions", and "Frozen Soil History Evolution and Frozen Soil Area Development" in China.

A book on thc study of frozen soil and the freezing phenomenon, the formation and development of permafrost and seasonal freezing and thawing provides a scientific principle, especially provides the principles and methods for zoning and classification of seasonal frozen soil and permafrost, describes in detail the frozen soil characteristics of China in the permafrost region and in each sub region, and provides a map of the new China permafrost distribution.

On the basis of the background of global change and the particularity of various regions, the author clarifies the historical changes of permafrost and the possible changes in the future.

Based on 40 years of experience and lessons, the book also outlines the principles of engineering geology and environmental protection for the design and construction of engineering structures in permafrost regions. It is the first comprehensive and systematic summary of Chinese scholars' latest works and unique viewpoints on frozen soil research.

3. ***The international classification for snow cover on the ground***

Author: Fierz C., Armstrong R. L., Durand Y., et al.

Publisher: IHP-VII Technical Documents in Hydrology N°83/IACS Contribution N°1

UNESCO Working Series SC-2009/WS/1

Snow research is an interdisciplinary study, which involves a wide range of fields. It is very important to dictate a description of snow normalization and a general method of measurement. The Snow/Ice Committee of the International Association of Hydrological Sciences (IASH) realized this requirement, and in 1948, appointed a special committee, then in 1954 published "The International Classification for Snow Cover on the Ground". As time goes on, people's understanding of the snow process is expanding, and there are greater differences among the observation methods used by different countries. In 1985, the International Committee of Snow and Ice (ICSI) established a new committee on the classification of snow. Five years later, a comprehensive revision and updating of ***The International Classification of Seasonal Snow***

Cover on the Ground was issued. This work has been widely used as the standard for describing the features of seasonal snow cover on the ground.

By 2003, the original classification standard of snow cover (Colbeck et al. 1990) was considered to be updated. Following the principles of the previous version, the snow classification working group worked out a concise document to facilitate the use of scientists in snow research, as well as in other fields and nonprofessional groups. The improved version provides more comprehensive knowledge, and the measurement and observation methods are more advanced.

The updated classification of snow cover adds a new class (machine made, MM). The abbreviation code is no longer alphabetic or numeric, and the structure of the classification tree does not indicate fine changes in snow metamorphism. The new code helps avoid misunderstandings and increases the flexibility of the classification.

Chapter 3
Formation and Development of the Cryosphere

Lead Authors: Lin Zhao, Zhongqin Li, Jinping Zhao

Contributing Authors: Renhe Zhang, Tao Che, Lijuan Ma, Cunde Xiao

The cryosphere is a sphere on the earth's surface with a certain thickness, where temperature is continuously below the freezing point. All the processes related to liquid water freezing and state change of frozen water taken place on the Earth surface naturally are attributed as the formation processes of the cryosphere. It mainly includes the accumulation and transformation of solid precipitation, which forms snow cover and glaciers; the freezing of surface water, which attributed to the formation of sea ice, lake ice and river ice; and the formation and transformation of ground ice, which are mainly related to the formation of frozen ground. The conversion among solid water, liquid water, and vapor meet the law of mass balance, along with the coupling and transformation of energy. Thus, the coupled mass-energy processes provide the physical basis for the formation and changes of the cryosphere. This chapter synthesizes the conditions and physical basis of the formation and development of the cryosphere and its components.

3.1 Conditions of Formation and Development of Cryosphere

The existence of frozen water is an essential feature of the cryosphere. Because the primary condition for the formation and development of solid water is the temperature below freezing point. Cold climate is the crucial factor for forming and developing of the cryosphere. Environmental factors, including the geological, geomorphic, and geographical backgrounds of the continental cryosphere and the surface features and ocean currents of the marine cryosphere, primarily affect the formation and development of cryosphere.

3.1.1 Glaciers

Solid precipitation and low air temperature are two main factors for glacier formation. Regional and local geomorphological conditions have their influences on the degree

D. Qin et al. (eds.), *Introduction to Cryospheric Science*, Springer Geography,
https://doi.org/10.1007/978-981-16-6425-0_3

of air temperature and the amount of solid precipitation accumulation through its impacts on surface energy and water budget.

Low air temperature is an essential requirement for the formation and development of glaciers. The lowest air temperatures commonly exist in the Polar Regions due to air temperature decreases with increasing latitude. Snow and ice in Polar Regions are hardly melted despite the scarcity of precipitation, and the long-term accumulation of snow eventually evolves into ice sheets. Another reason for the Antarctic and Arctic ice sheets is the high albedo that can reflect 70–90% of solar radiation to the atmosphere. In alpine regions, air temperature decreases with increasing altitude, which leads to relatively low temperatures. A large number of glaciers prevalently develop in the valleys and basins of high mountains, while their sizes are much smaller than those in polar regions.

Solid precipitation is a material basis for glacier development. Precipitation is mainly affected by the distance to vapor source. In the mountainous areas, the effect of valley wind might result in higher precipitation in higher elevations. This made the large-scale mountain glaciers to concentrate in high mountains with abundant water vapor.

The seasonal distribution pattern of precipitation also affect the accumulation and ablation characteristics of glacier. Precipitation during winter and spring can effectively enhance the accumulation of glacier. Summer precipitation, to a certain extent, can decrease local air temperature; of which the solid part can increase the surface albedo and thus inhibit the ablation of glaciers, while the liquid part may strengthen glacier ablation. In the context of climate warming and increasing rainfall, the ablation of summer-accumulation glaciers will be severely enhanced.

Another important factor affecting glacier development is the geographic conditions. Snow line is the lower topographic limit of snowpack, which is named as firn line on glacier. Equilibrium line refers to the boundary between the accumulation zone and ablation zone where the mass balance is zero, which is slightly lower than the snow line. The zone between equilibrium line and snow line is known as a superimposed-ice zone, caused by the partial refreezing of meltwater on the glacier surface. In mountain regions, the higher the altitude, the lower the temperature, the more the water vapor interception, and thus the larger the glacier accumulation areas. When topographic conditions meet the requirements for glacier development, the valley glacier may be formed if the altitude of ridges is higher than snow line; otherwise, the cirque glacier or hanging glacier are generally formed. The tabular glacier and ice cap may be formed when the flat summit is above the snow line.

In addition, the relationships between orientation of atmospheric circulation and the trend of mountain ranges and valleys can also impact glacier development. A positive relationship is beneficial to water vapor transport and glacier growth. Furthermore, the aspect of mountain ranges is also a key factor in affecting the distribution of glaciers. The shady slope is favorable for the growth of glacier; however, the sunny slope is detrimental to glacier growth due to stronger solar radiation.

3.1.2 Permafrost

Permafrost is the product of the exchange of energy and water between surface lithosphere and atmosphere under cold climate conditions (Cheng et al. 1992; Brown 1965). As the air temperature is lower enough, heat would be flow from surface lithosphere to the atmosphere. Surface lithosphere would be frozen, and become permafrost as such a frozen state existed more than two consecutive years. So the lower the air temperature, and the colder the permafrost, and thinner the permafrost thickness (Brown 1966b, 1973; Bryan 1992; Zhao et al. 2000, 2019, 2020).

The relationship between permafrost and precipitation is complex because all precipitation parameters, such as the forms, duration, frequency, and intensity of precipitation, have a great impact on surface energy balance. For the same region, long-term increase in precipitation may result in increase in surface evapotranspiration and decrease in surface temperature. This process not only changes surface sensible heat and latent heat but also alters hydrothermal processes along soil profiles, leading to large changes in surface heat flux and thus the permafrost development.

Cloudiness and sunshine can affect surface temperature through surface radiation balance. Due to its high albedo, snow cover can reduce the surface temperature, and even ground temperature. Moreover, snow cover hinders the energy exchange between land and atmosphere due to its low thermal conductivity. Because snow cover absorbs most of the solar radiation by latent heat during melting processes, it can prevent the surface temperature from rising. Thus, the duration, structure, density, and thickness of snow cover can affect the development of permafrost.

Better vegetation conditions, such as higher coverage and height can result in lower surface and ground temperatures through its influence to albedo, shielding, and transportation, then is favorite to permafrost development (Brown 1966a; Bryant et al. 1970). Different soils have different grain-size composition, organic carbon content, and compactions. These all affect surface heat and water balance through its thermal and hydrologic effects.

The formation and development of permafrost are also related to regional geological conditions (Cheng 1982, 2003, 2004; Brown 1973). The annual changes of surface solar radiation can affect the ground temperature (GT) to the depth of 10–20 m; the decadal changes of surface solar radiation can influence GT to the depth of several tens of meters; and the temperature of a few hundreds to thousands of meters deep is mainly controlled by the interaction of the energy balance between surface and internal Earth for a few thousands to tens of thousands of years. Due to the effects of the earthquake, volcano, and tectonic movement, the geothermal flow varies considerably in different regions. In general, the higher geothermal flow is unfavorable for permafrost development.

3.1.3 Snow Cover

Snow cover begins to form when snow accumulates on the ground surface which can be visible and measured by instruments. The formation and development of snow cover are related to surface temperature and morphology, snowfall, wind field, etc. Snow cover can be formed only when the accumulation amount of snowfall and snowdrift is greater than the loss amount. Therefore, the thicker snow depth may occur in the regions with lower surface temperature, more snowfall, and lower wind speed. Due to the latitudinal and altitudinal zonality of climate, the frequency and duration of snow increase with increasing latitude and altitude.

The topographical condition also strongly affects the formation and development of snow cover. The characteristics of snow cover, such as depth and duration, are greatly different due to the differences in solar radiation and wind speeds on different slopes. Flatland and gentle slopes are favorable for accumulating and preserving snow. Slopes facing to the prevailing wind directions have less snow cover in the cold season.

Surface wind speed has a considerable effect on the formation and development of snow cover. Strong wind not only leads the snow to accumulate on the low-lying areas but also increases the sublimation of snow, and thus it is not beneficial to form snow cover. For instance, the snowfall occurs at all seasons on the Tibetan Plateau, whereas snow cover is generally unstable and short duration due to the strong solar radiation and wind.

3.1.4 Lake Ice and River Ice

Lake ice and river ice varies significantly seasonally, which generally freeze in autumn and winter and thaws in spring and summer in the Northern Hemisphere. The duration of lake ice and river ice are related to air temperature and is also affected by wind, solar radiation, precipitation, etc. The formation and ablation of lake ice are mainly controlled by local surface energy balance and the extinction coefficient and albedo of ice that affecting the heat exchange between ice and air. In addition to meteorological conditions, the geometry of rivers and hydraulic dynamics can influence the formation and ablation processes of river ice.

3.1.5 Sea Ice, Ice Shelf, and Iceberg

Sea ice begins to form on the sea surface when the sea surface temperature decreases to the freezing point of seawater accompanied by air temperature decreasing in cold season. Warm ocean current might bring great amount of heat, thus much colder climate is necessary for sea ice formation. For example, there is almost no sea ice in the Barents Sea and the seas in the northern Europe due to the strong warm current, though they are within the Arctic Circle (Lemke et al. 2007; Stroeve et al. 2007; Ukita et al. 2007; Zhang et al. 2008; Comiso et al. 2008; Parkinson and Cavalieri 2008; Screen and Simmonds 2010a; Screen 2017).

Ice shelf is an extension part of the ice sheet into ocean. As the ice shelf breaks off from a glacier and floats freely in ocean, it becomes iceberg. The formation of iceberg is much more related to the characteristics of the glaciers which come from. Its melting speed is mainly controlled by sea water temperature and its size (Morgan and Budd 1978).

3.2 Physical Basis for Formation and Development of the Cryosphere

Energy and water balance processes of ice-water–vapor conversion in each cryospheric component are the physical basis to cryosphere development, including energy-water balance on the cryosphere surface and energy-water transfer processes within each cryospheric components. There are considerable differences in the physical mechanism of the formation and development of different cryospheric components. For example, the physical basis for formation and development of glaciers includes mass balance and dynamic process of glaciers, and the mass balance processes of river ice, lake ice, and sea ice are related to the dynamic processes of rivers, lakes, and oceans.

3.2.1 Surface Energy Balance

The surface energy balance of the cryosphere is the balance between surface net radiation flux and the energy consumption or compensation when it converses to other energy (Harris, 2010, 2013, 2017). For the land, the equation of energy balance is as follows:

$$R = \lambda E + H + G \quad (3.1)$$

where R is the net radiation flux; H is the sensible heat flux; λE is latent heat flux for evaporation; G is the downward heat flux from the surface.

The net radiation flux (R) is the surface total radiation flux budget of the cryosphere, and its equation can be expressed as:

$$R = Q\,(1-\alpha) + R_{\mathrm{L}} - U \quad (3.2)$$

where Q is the total radiation reaching the surface of the cryosphere, including direct and diffuse solar radiation; α is the surface albedo influenced by underlying conditions, color, vegetation, and soil properties, and surface roughness; R_L is the downward atmospheric longwave radiation; U is the longwave radiation released from the surface.

3.2.2 Surface Water Balance

Surface water balance of cryosphere is defined by the budget of inputs and outputs of water, which can be described by the following equation:

$$\Delta W_l = P_l - E_l - E_c - R_l - K + M_l \tag{3.3}$$

where ΔW_l is the change of water storage of various media on the surface of the cryosphere (e.g., permafrost: soil/rock layers; glaciers, and sea ice, river ice, and lake ice: snow and ice) during the study period; P_l is precipitation; E_l and E_c are the direct evaporation and the vegetation transpiration, respectively; R_l is the surface runoff or the lateral runoff of ice; K is the surface infiltration; M_l is the melting amount of surface snow cover or ice.

3.2.3 Heat Transfer in Cryospheric Material

There are great differences in the energy transfer of different cryosphere components. Glacier is moving solid media, and its movement is accompanied by energy transfer. The heat transfer during glacier movement follows the following equation:

$$\frac{\partial T}{\partial t} = k\frac{\partial^2 T}{\partial x_i^2} + v_i\frac{\partial T}{\partial x_i} + \frac{Q}{\rho C} + \frac{\left(\frac{\partial \lambda}{\partial x_i}\frac{\partial T}{\partial x_i}\right)}{\rho C} \tag{3.4}$$

where v_i is the momentum vector along x_i; Q is the velocity per unit volume; T is the temperature at x_i; t is the time; λ is the thermal diffusivity; k is the thermal conductivity; ρ is the density; C is the specific heat capacity.

For those cryosphere components such as permafrost, snow cover, river ice, lake ice, and sea ice, the movements and phase transitions of water and heat are coupled. The heat transfer process in the vertical direction meets the following equations:

$$\frac{\partial (CT)}{\partial t} - L_f\rho\frac{\partial \theta_i}{\partial t} = \frac{\partial}{\partial z}\left(k\frac{\partial T}{\partial z}\right) + C_u T\frac{\partial q_u}{\partial z} + L_v\frac{\partial q_v}{\partial z} - S_h \tag{3.5}$$

$$k = \frac{\lambda}{C} \tag{3.6}$$

$$\lambda \approx \lambda_u^{\theta_u} \cdot \lambda_i^{\theta-\theta_u} \cdot \lambda_m^{1-\theta} \tag{3.7}$$

$$C = C_f + L \cdot \rho \cdot \frac{\partial \theta_u}{\partial T} \tag{3.8}$$

$$C_f = C_{sf} + (\theta - \theta_u)C_i + \theta_u \cdot C_u \tag{3.9}$$

where θ, θ_u, and θ_i is the total volumetric water content, volumetric unfrozen water content, and volumetric ice content, respectively; z is the depth; λ, λ_u, λ_i, and λ_m is the thermal diffusivity of frozen soil, unfrozen (liquid) water, ice, and soil minerals, respectively; C, C_f, C_{sf}, C_i, and C_u is the specific heat capacity of soil, frozen soil, soil minerals, ice, and unfrozen water, respectively; L, L_f, and L_v is the latent heat of phase transition, freeze, and evaporation, respectively; q_u and q_v is convective fluxes of liquid and vaporous water, respectively; S_h is the source and sink items of energy balance.

3.2.4 *Mass Balance*

1. Mass balance of snow cover

The mass balance of snow cover can be expressed as the following equation:

$$\Delta M = P - S + F - E + C + B - R \tag{3.10}$$

where ΔM is the changing amount of total mass of snow cover, including ice crystals and liquid water in snowpack; P is precipitation; S and F represents sublimation and refreezing, respectively; E and C represents evaporation and condensation, respectively; B is the migrating amount of snowdrift; R is the amount of snowmelt. It should be noted that R represents the amount of snowmelt that flows out from snowpack. If only phase transition occurs, and the liquid water persists in the snowpack, this process increases liquid water contents and thus leads to unaltered mass in the snowpack.

2. Mass balance of glacier and ice sheet

Mass balance refers to the budget between total mass inputs and outputs of glacier (or ice sheet) at a time unit (Fig. 3.1). Glacier mass accumulation mainly consists of solid precipitation, including snowfall, sublimation, refreezing, and snowdrift, etc. Glaciers lose mass through the ablation process that is the total outputs, including runoff, evaporation, sublimation, and ice collapse, etc.

The rate of mass gain and loss of glacier per time unit is defined as glacier accumulation rate ($\dot{c}$) and ablation rate ($\dot{a}$), respectively. The mass balance year is a time

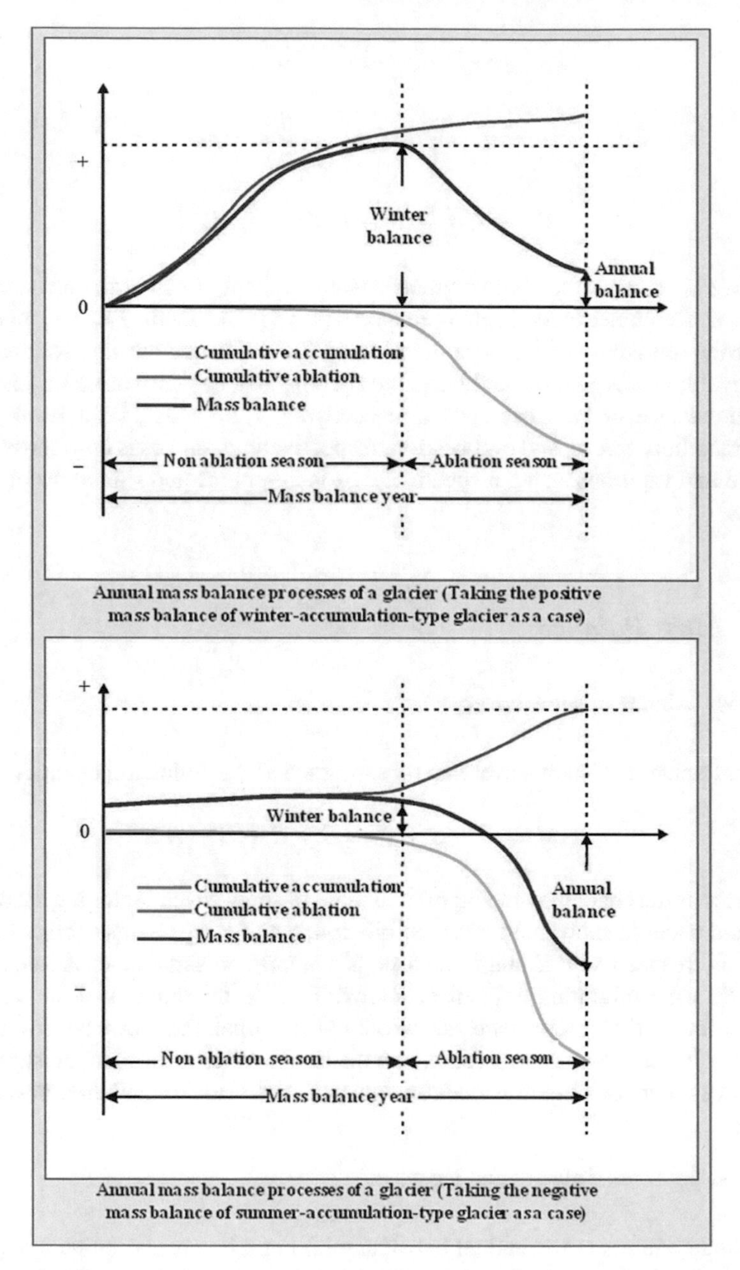

Fig. 3.1 Annual mass balance of glacier (winter-accumulation-type was modified from Cuffry and Paterson (2010), and summer-accumulation-type was observed on the Glacier No. 1 at the headwaters of Urumqi River, Tianshan Mountains) (Li et al. 2003; Cuffey and Paterson 2010)

period from the end of the ablation season to the next. Integral of $\dot{c}$ and $\dot{a}$ over the mass-balance year is total annual accumulation (c_a) and ablation (a_a), respectively, and the annual mass balance (b_a) is the difference in annual accumulation and ablation:

$$b_a = c_a - a_a \tag{3.11}$$

In terms of any study periods, it is called glacier accumulation (c_t), ablation (a_t), and mass balance (b_t) at a specific period. Therefore, the mass accumulation and ablation of winter and summer seasons can be distinguished. For a specific glacier, transient equilibrium-line altitude (ELAt) and annual equilibrium-line altitude (ELA) represents $b_t = 0$ and $b_a = 0$, respectively. The zone above ELA is accumulation area and below ELA is ablation area.

If the horizontal projection area of a glacier surface is *s*, and the integrals of c_a, a_a, and b_a over the area is the total accumulation (C_a), ablation (A_a), and annual mass balance (B_n; Eq. 3.12).

$$C_a = \iint_S c_a d_x d_y,\ A_a = \iint_S a_a d_x d_y,\ B_n = \iint_S b_a d_x d_y \tag{3.12}$$

Actually, the specific mass balance per unit area is commonly used in practical applications, such as the net balance (b_n; Eq. 3.13):

$$b_n = B_n/S \tag{3.13}$$

where *S* denotes the total area of a glacier. Moreover, the mean net accumulation ($\overline{c}$) and ablation ($\overline{a}$), and the mean total accumulation ($\overline{c}_a$) and ablation ($\overline{a}_a$) can be expressed as:

$$\overline{c} = B_c/S, \overline{a} = B_a/S, \overline{c}_a = C_a/S, \overline{a}_a = A_a/S \tag{3.14}$$

The units of above-mentioned variables are millimetre water equivalent (mm w.e.).

3. Mass balance of river ice, lake ice, and sea ice

The mass balance of river ice, lake ice, and sea ice is the hydrothermal coupling process among the surface of the water (ice), ice layer and its underlying waterbody, on the basis of the equations of surface energy balance, internal heat transfer of ice and water, and hydrothermal coupling balance.

As a matter of fact, the mass balance of sea ice is the seasonal changes in the mass of sea ice. In general, polar sea ice can be monitored by ice mass-balance buoys that observe the freezing and thawing processes of the upper and lower of sea ice surface. For the Arctic, the mass balance of sea ice mainly refers to its inputs and outputs. Long-term change in sea ice directly is in response to climate change, and they are also affected by heat transfer of river and wind field.

3.2.5 Soil Moisture Migration

The theories of soil moisture migration in the context of freezing–thawing cycles mainly include the theories of capillary action and film moisture adsorption (Xu and Deng 1991).

1. Capillary water migration theory

Capillary water migration in soils refers to the capillary water upward caused by the micropores between mineral particles in thawed soils and between soil particles and ice in a frozen state. This theory owes the driving force of water movement in the vadose zone to capillary force. Due to the capillary force, water migrates to freezing front along capillaries formed by soil cracks and pores in frozen ground (Yuan 2006).

2. Film water adsorption theory

The water migration theory of film water adsorption considers that soil particles can adsorb water molecules (that is soil water potential), which is related to minerals and grain size distribution of soil and the distance from the surface of soil particles (Yershov 1979). Due to the effects of temperature and soil water potential, the unfrozen film water around soil particles is asymmetric, with thinner films in the colder side and thicker films in the warmer side. As the soil temperature decreased below the freezing point of pore water, the pore water outside the film water around soil particles begin to freeze, and the thickness of this unfrozen film water is reduced near the freezing front, breaking the original balance of the unfrozen water–ice-soil particles system. To maintain a new balance, the unfrozen film water moves from high to low temperature (Mackay 1983, 1984; Anderson et al. 1984; Cook 1955; Li et al. 1995; Kozlowski 2004; Hu et al. 2014, 2020). Therefore, when it exists under negative temperatures and temperature gradient in soils, the unfrozen water migrates from higher unfrozen water content area to lower area. This theory has been widely accepted and used to explain the water migration in fine-grained soils during freeze–thaw processes and the formation of ice lens and thick ground ice.

Under natural conditions, water migration depends on the combination of dynamic, physical, and chemical factors. The kinetic energy of soil pore water is generally low due to its relatively slow movement (less than 0.1 m/h). Thus, soil water potential represents the potential energy of soil water. For the freeze–thaw soil system, the difference between soil water potential at any two points is the driving force of water movement between the two points. Soil water potential theory not only figures out the mechanism of soil water migration (soil water migrates from higher soil water potential area to lower area, and soil water potential gradient is the force to driving water movement in soils) but also possibly allows scientists to quantitatively study spatiotemporal distribution and movement of soil water using physical and mathematical equations.

3.3 Continental Cryosphere

3.3.1 Glacier

1. Ice formation

Falling on the ground (snow or ice surface), snow would gradually evolves into spherical particles without crystal characteristics as it was not be melted. This kind of granular snow is called firn. During the transformation from snow to firn, the hardness and compactness of snow increase significantly and the porosity, brightness, and transparency of snow gradually reduce. Eventually, snow becomes glacier ice. This evolution process from snow to ice is known as ice formation.

The fresh snow density averages 0.13–0.21 g/cm^3, and firn density increases to about 0.55 g/cm^3 due to the rounding-subsidence effects of firn, and it further increases along with the subsequent recrystallization (Shi and Xie 1964). When the pores within snow or firn are completely sealed into bubbles, it turns into ice, with a density of around 0.83 g/cm^3. Ice density can further increase to 0.917 g/cm^3 due to compression of bubbles after ice formation. If the overlying ice is thick enough (usually larger than 800 m), the tremendous pressure can change the air in bubbles into hydrate (Li et al. 2007a). The ice formation process without melting is called dynamic ice formation process, and the ice is called dynamically metamorphic ice.

The evolution of snowpack may be accompanied by the inputs of precipitation, condensation, and snowdrift and the outputs of melt and sublimation of snow. The depth hoar can be formed due to sublimation and condensation and ice lens induced by frequent freeze–thaw cycles. When spring temperature fluctuates around 0 °C, it can form a lot of ice layers (You et al. 2005). In mid-October, the deep frost layer develops if the temperature gradient of snowpack reaches 13.0 °C/m, and it converts to coarse firn due to the influence of melting water in the next June (Wang et al. 2007). In addition, the dust sedimentation can form dust layers (Li et al. 2011). The formation of dust layers in snow is mainly derived from the impurities of melting water in summer season.

Glaciers have vertical zonality due to the large difference in relative height (up to several kilometers). The hydrothermal conditions at different altitudes vary greatly, and ice formation processes are different. Currently, there are two classical theories for dividing glacier band. In 1964, P.A. Shumskii, a glaciologist from the former Soviet Union, divided a glacier into seven glacier bands that include recrystallization zone, refreezing-recrystallization zone, cold infiltration- recrystallization zone, warm infiltration-crystallization zone, infiltration zone, infiltration- freezing zone, and ablation zone. Paterson raised the second division theory in 1969, a Canadian glaciologist, who divided a glacier into five glacier bands consisting of the dry-snow zone, percolation zone, wet-snow zone, superimposed ice zone, and ablation zone (Fig. 3.2).

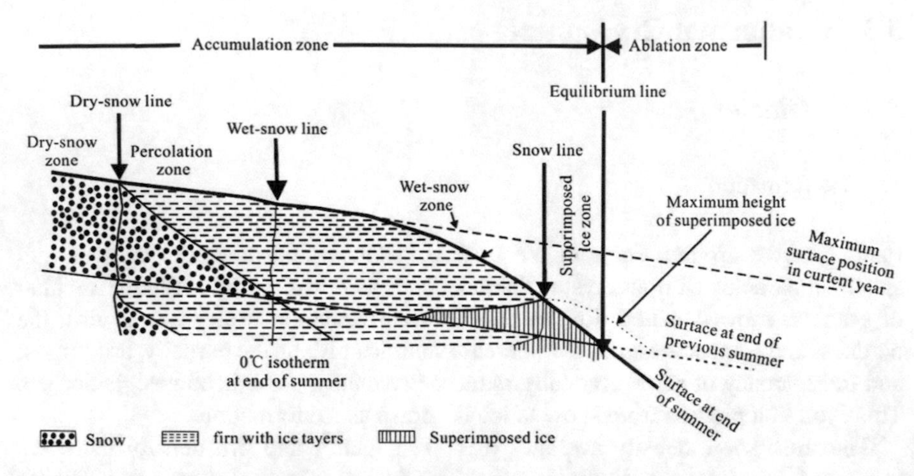

Fig. 3.2 Division of glacier zones (modified from Cuffry and Paterson 2010)

Glacier zones are sensitive to climate changes (Xie et al. 2009). For example, the glacier zones of Glacier No. 1 at the headwaters of Urumqi River in Tianshan Mountains have changed apparently under global warming, leading to disappearances in cold-type ice formation and its corresponding recrystallization zone or dry-snow zone (Fig. 3.3) (Li et al. 2003, 2007a).

The way and duration for transforming snow to ice depends on the hydrothermal conditions. Under the existence of high temperatures and melting water, the melting-refreezing process occurs in snow layers, thus forming the warm-type ice formation (Li et al. 2006, 2011). In regions with low temperatures, such as central Antarctica, glacier formation entirely relies on gravity and the compactly metamorphic effect of firn crystals (that is cold-type ice formation).

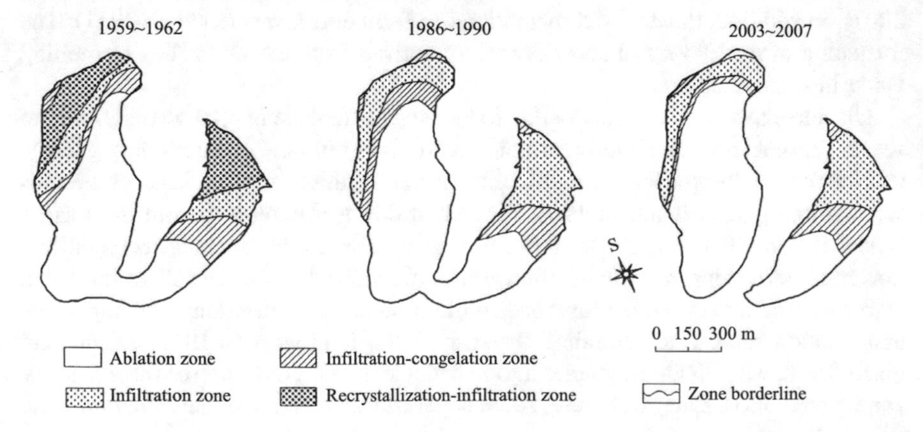

Fig. 3.3 Ice formation zones during different periods on the Glacier No.1 at the headwaters of Urumqi River, Tianshan Mountains

2. Formation of ice sheet

The ice sheet is a type of glacier, and its central part is accumulation area, and the margin part is the ablation area. The ice sheet is almost not constrained by underlying geomorphological conditions. As ice sheet flows down to a coastline and onto the ocean surface, this floating ice is called ice shelf, and then iceberg is formed when ice shelf is broken and floats freely in open water. There are only two ice sheets on the Earth: Antarctic and Greenland ice sheets. Due to the dry-cold climate, extremely low evaporation, and surface condensation phenomena in Antarctic, long-term snowfall accumulation eventually evolves into the Antarctic ice sheet. The significant difference in temperature between Greenland and the warm Atlantic Ocean facilitates the emergence of cyclones, which brings about abundant solid precipitation to Greenland. This has laid a material foundation for the development of Greenland ice sheet (Crowley and Kim 1995; Zachos et al. 2005).

3.3.2 *Frozen Ground*

1. The freeze–thaw process of seasonal frozen ground

The surface soil layer that freezes in cold seasons and thaws in warm seasons is called seasonal frozen ground, which is known as an active layer in permafrost regions. In seasonal frozen ground regions, the seasonal freezing–thawing process of soil layers is abbreviated as a seasonal freezing process. When air temperature decreases to below 0 °C in the cold season, the surface soil begins to freeze. As air temperature continues to decrease, the freezing interface moves downward slowly, and it reaches the maximum depth during the lowest temperature period. Then, when the air temperature is higher than the ground temperature of frozen layers, the frozen layers start to thaw from the bottom up; when the air temperature is higher than 0 °C, the thawing process of frozen layers is bidirectional (Fig. 3.4, left). However, in permafrost regions, the seasonal freezing–thawing process of soil layers is called seasonal thawing process. The dynamic process of freezing and thawing fronts is opposite compared to those in seasonally frozen ground regions (Fig. 3.4, right).

2. Formation of permafrost

When the air temperature drops below a certain degree, the frozen ground layer formed in the previous cold season cannot thaw entirely in the following warm season, leading to the formation of pereletok. If pereletok continues to exist for two consecutive warm seasons or more, it is called permafrost. There are three types of permafrost formation: epigenesis, symbiosis, and the mixed genesis. The epigenetic permafrost is a product of continuous cooling of the climate, which refers to permafrost that formed through lowering of the permafrost base in sediment or bedrock.

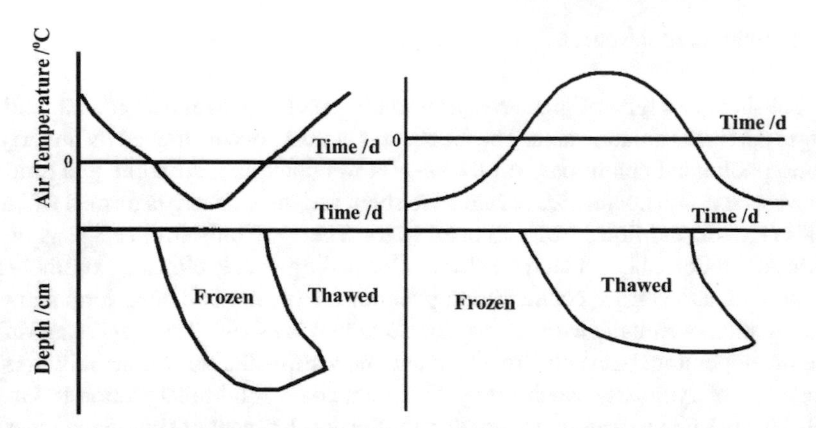

Fig. 3.4 The freezing process of seasonally frozen ground (left) and thawing process of the active layer (right)

The syngenetic permafrost refers to the permafrost developed due to the permafrost table moving upwards accompanied with new sediments deposited on the original ground surface under cold climate, leading to increases in permafrost thickness (that is, the freezing and sedimentation processes occur simultaneously in soils or rocks). The essential difference between syngenetic and epigenetic permafrost is that the former results from the increase in permafrost table with soil deposition, whereas the latter results from the decrease in permafrost base. The formation of permafrost in most areas is caused by mixed genesis that is the interaction between epigenesis and symbiosis.

3. Ground ice and permafrost fabrics

Ground ice is a general term for all types of ice in freezing soils and frozen ground, which may be epigenetic, syngenetic, mixed genesis, residual, perennial or seasonal (Zhao et al. 2010b). Ground ice develops in soil, or rock cracks, caves or other open spaces, and often exists in the forms of the lens, ice wedge, vein, layer, irregular blocks, single crystal or cap. The perennial ground ice only occurs in the interior of permafrost.

Segregation ice is the most commonly developed ground ice, most of which is driven by the effect of repeated ice-segregation. Ice segregation is a process that pore water migrates, freezes, and accumulates to specific parts of frozen mineral or organic matter, which is caused by the migration of pore and film water to freezing front due to the effect of soil water potential (Tsytovich 1975; Cheng 1983; Harry et al. 1988). If there is a temperature gradient in permafrost, it can generate free energy gradient and thus cause moisture to migrate to the area with lower temperature. When water migrates to the vicinity of freezing front, its Gibbs free energy increases, breaking up the original balance between ice and unfrozen water and then resulting in the formation of new ice to reach a new balance. When permafrost develops, the ice formation in freezing soils could result in frost heave. Ice segregation is the main reason for causing frost heave of soils and rock fracture (Taber 1929, 1930a, b).

The ice formed by ice segregation is called segregation ice. Under certain conditions, the volume of segregation ice dramatically exceeds the volume of soil pores before freezing. The thickness of segregation ice can be from a few centimeters to dozens of meters. It is worthy noted that injection ice does not always have a distinct difference from segregation ice, and sometimes they can be intertwined.

During the freezing process, when soil layer is easy to get external water, the growth of segregation ice is parallel to freezing front, and it can grow rapidly until the heat released by a water–ice conversion that warms ice lens boundary, reduces the temperature gradient in soils, and inhibits ice segregation rate. Further cooling can further promote the formation of segregation ice. When available water is limited, the segregation ice grows slowly. Because the warming effect of heat release during the freezing process on the ice lens boundary is insufficient, the new segregation ice will be formed underlying the initial segregation ice layer. In addition, due to the formation of segregation ice, the volume of frozen ground is far greater than the 9% frost-heave amount of the original soil water content.

Thick ground ice is often observed near permafrost table, and it is widely distributed in fine-grained soils of deposited terrains with low ground temperature. Because of the shallow distribution and large thickness, ground ice has a significant impact on the formation of cryogenic phenomena and the stability of engineering constructions in permafrost regions (Mackay 1971b). This special cryostructure is called variegated cryostructure or suspended cryostructure, with a volumetric ice content of more than 50%. The formation of thick ground ice resulting from the repeated ice-segregation near freezing front is known as a repeated ice-segregation mechanism, which also called “Cheng’s hypothesis” (Cheng 1983).

The repeated ice-segregation occurs near the permafrost table. The thick ground ice induced by this mechanism can also be buried below permafrost table along with surface deposition, and the formation conditions are that soil layers near permafrost table are fine-grained soils or mixed deposits containing many fine particles. The modern segregation ice is lenticular or layered, and its upper surface roughly coincides with permafrost table, with a thickness from tens of centimeters to several meters. The main processes of ice formation include ① water migration and ice formation from bottom to top in the active layer during the freeze–thaw cycle; ② non-equalized migration of unfrozen water; ③ self-purification of ice; ④ syngenetic growth of ground ice caused by the sedimentation of surface soils; and ⑤ a long-term repeat of above-mentioned ice formation.

3.3.3 Snow Cover

After falling onto the ground, the initial structure of fresh snow varies with changing external conditions (e.g., temperature, pressure, temperature gradient, etc.), leading to the cohesion and aggregation among snow particles and thus altering physical characteristics of snow cover. This process is called snow metamorphism. According to the feature of snow, it can be divided into dynamic metamorphism of dry snow and thermal metamorphism of wet snow. The former is characterized that the density of fresh snow gradually increases due to gravity, and the density changes of the latter are accompanied by infiltration of melting water. Finally, fresh snow transforms into a different type of snow through different metamorphisms, such as fine particle snow, coarse particle snow, deep frost, ice layer, etc.

Low thermal conductivity is one of the main characteristics of snow cover. For the ground with thick snow cover, the ground temperature is relatively high or near freezing point even if the air temperature is extremely low, resulting in the existence of liquid water in the snowpack. Consequently, the lower part of snowpack remains relatively warm during cold seasons, compared with the upper part of snowpack, leading to a higher temperature gradient from bottom to top. When water vapor transfer from the lower part to the upper part, it may be frozen and thus form the deep frost. Generally, the development of deep frost layer requires thin snow cover, sunny weather, or low temperatures. When snow temperature is between –15 and –2 °C, the deep frost layer develops the best.

With temperature rises in spring, surface snow melts rapidly due to exposure to daytime solar radiation, and the percolation of melting water triggers thermal metamorphism of snow. At night, the liquid water in snowpack is frozen again, forming an ice layer. If fresh snowfalls during this period, the ice layers can be retained in snowpack and inhibit the infiltration of upper melting water. Obviously, the internal structure of snow cover, to a larger extent, depends on local climatic and geographic conditions.

3.3.4 River Ice and Lake Ice

1. River ice

The evolution of river ice can be divided into occurrence, development, and ablation processes, depending on climate conditions, water-flow process, river morphology.

With the decreasing temperature in late autumn and early winter, heat release from water is greater than heat absorption, cooling the water. When water temperature is below 0 °C, fine ice crystals are formed in supercooled water, which are origins of various types of river ice. In the areas with slow flow and weak turbulence, ice crystals float up and gather on the surface of the water to form a thin ice layer, then forming shore ice as water continues to release heat. However, in the region with fast flow and intense turbulence, ice crystals can be formed throughout the depths where the temperature is below 0 °C due to the combined effect of turbulence. The fusion between them forms larger ice units. The density of river ice increases gradually because of continuous heat dissipation and viscous effect of ice, which can form floating ice, ice jam, or ice dam under the favorable hydraulic conditions. As air temperature further decreases, river ice thickness increases due to the continuous freezing on the water–ice interface and the transformation of water to ice.

With the rising temperatures in late winter and early spring, river ice thaws and forms floating ice in rivers under the thermal and mechanical effects. When floating ice moves to the ice covered river and accumulates, leading to the formation of ice jam or ice dam. Given by hydraulic and meteorological conditions, the thawing processes of river ice include tranquil break and violent break. The tranquil break is that river ice breaks apart in situ under the conditions of lower river discharge, warm climate, and sparse rainfall. The violent breakup is that river ice is broken by strong water flow induced by the combined effects of rapid melting and rainstorm. This process can cause much ice to accumulate in curved narrow channels and form ice dams, increasing water level and thus leading to severe river ice flooding.

2. Lake ice

The formation and ablation processes of lake ice are extremely complex, involving meteorological conditions, location and morphology of lakes. Compared to rive ice, the formation and ablation processes of lake ice are less influenced by dynamic effect but, to a larger extent, by heat and luminous fluxes on the air-ice and ice-water interface. Lake ice is usually frozen in autumn and winter and is melted in the following spring and summer. Because of the difference in heat capacity between water and land, the freezing and ablation processes of lake ice firstly occur in the coastal area. In autumn and winter with relatively low solar radiation and temperature, the lakes losing heat decrease water temperature, leading to the formation of ice crystals when the water temperature drops below 0 °C. As lake ice is formed on the lake surface, the solar radiation entering into the lake can be further reduced because ice albedo (about 70%) is far higher than water albedo (~8%). Thus, the decreases in heat flux of water and shortwave radiation can promote the development of lake ice. On the contrary, lake ice is melted in spring and summer because of the rising temperature and solar radiation. Moreover, lake ice shows a significant seasonal variation.

Considering the energy balance, the extinction coefficient and albedo of ice have an important influence on the formation and ablation processes of lake ice. Snow cover of ice surface is another key factor in influencing the energy balance of ice, which can influence the formation and ablation processes of lake ice by reducing the heat exchange between ice and air. Ice thickness is the most comprehensive index to reflecting the formation and ablation processes of lake ice, and its changes originate from the ice growth and ablation at the ice bottom, depending on the effect of luminous flux on the heat flux at the bottom of lake. In autumn and winter, the decrease of solar radiation and the increase of snow thickness enhance the optical thickness of ice layer, resulting in relatively low luminous flux and heat flux and thus increasing the ice thickness. With the elevating solar radiation in spring and summer and the melting of snow cover, the luminous flux at the ice bottom raises rapidly, this leads to the increases in water temperature underlying the ice and heat flux at the ice bottom. In addition, a wet-snow layer is easy to be formed under the pressure of snow cover on the ice surface. Because the increase of temperatures and the melting of snow ice have a direct impact on the lake surface during the ice melting period, the ablation process is more intense at the ice surface than that at the ice bottom.

3.4 Marine Cryosphere

The formation, development, and evolution of various oceanic cryosphere components vary substantially, which can be divided into oceanic components (e.g., sea ice) and the constituents that are formed on the land but developed and perished in the ocean.

3.4.1 Sea Ice

1. Formation of sea ice

Based on the development stages, sea ice is divided into new ice, nilas, pancake ice, young ice, etc. (Petrich and Zicken 2010).

New ice: When the air temperature drops to the freezing point, or snow falls to the sea surface with low temperatures, the new ice starts to form. At this time, sea ice composes of acicular or flaky fine ice crystals; thus, the condensation of considerable ice crystals accumulates to the slimy or spongy sea ice. The surface of the sea is usually dark grey and lackluster, which is not affected by the gentle breeze (Fig. 3.5).

Ice rind: Ice rind is an interim stage from new ice to nilas, which is formed by the growth of new ice or the freezing of calm sea surface. The surface of ice rind is

Fig. 3.5 New ice, including frazil ice, grease ice, slush ice, and shuga ice

Fig. 3.6 Nilas and ice rind

smooth and wet with dark grey, with a thickness of about 5 cm. Ice rind is usually fragile to be broken by wind and ocean current, and it is also called pancake ice.

Nilas: When the new ice grows to about 10 cm in thickness, the ice becomes stretch and flexible to produce finger overlap, and this ice is called nilas, which includes dark nilas, light nilas, and ice rind (Fig. 3.6). Nilas is the earliest formed sea ice with loading capacity and spreadability, and new ice is in fact, an early stage of sea ice formation.

Young ice: Young ice is sea ice in a transition stage between nilas and first-year ice, formed directly by the freezing of nilas or ice rind, ranging from 10 to 30 cm in thickness. Young ice usually includes grey ice and white–grey ice.

First-year ice: The continuous growth of young ice becomes first-year ice, which exists only one winter and melts in warm seasons. The thickness of this ice ranges from 70 cm to 2 m. First-year ice can be subdivided into thin, medium, and thick first-year ice.

During the formation of sea ice, the crystal nucleus can be produced and then form fine ice crystals due to the supercooling sea surface. In fact, ice crystal growth is the crystallization of liquid water. Generally, the crystal nucleus is formed by precipitation or atmospheric ice (Bennington 1963). In addition, during the development of sea ice, the sea ice structure varies from granular ice to columnar ice, and the columnar ice is dominated when sea ice grows to a certain thickness (Kawamura et al. 2001).

2. Structures and changes of sea ice

Salt and air bubbles in ice: During the freezing process of sea water, about 3% external components can be dissociated from ice crystals, forming the brine with high salinity. The accumulation space of brine is called a salt bubble (Schwerdtfecer 1963). The brine accumulates in deep layers due to gravity effect, and this process can break the chemical bonds of sea ice crystals and form narrow and long salt bubbles in the ice. These bubbles are sealed in ice when new ice is formed. In addition to salt bubbles, there are many air bubbles during ice frozen, making crystal structure unable to form.

Sea water enters into salt and air bubbles in summer seasons due to elevated temperature in seawater. The heat absorption of sea water is stronger compared to that of sea ice. Thus, the short wave radiation entering sea ice can be absorbed by the sea water in salt bubbles, enhancing water temperature in bubbles and causing the internal melting of sea ice.

Sea ice ridging process: Upon the wind force, the relative movement occurs between ice floes. Once the ice floes are separated from each other, the leads and polynyas can be formed; however, if ice floes move towards each other, leading to the formation of ice ridges (Fig. 3.7) (Timco and Burden 1997). As ice ridges are caused by the crash of two large ice floes, the ridges generally parallel to each other. Moreover, the formation of ice ridges is tightly related to the crash momentum among ice pieces. The storms may bring strong convergence to generate huge ice ridges, and weak winds form small ice ridges. In general, the keel depth of ice ridges is much higher than their sail height. It has been reported that the maximum keel depth reaches 40 m for a multi-year ice ridge.

Leads and polynyas: Leads are narrow and open water channels between sea ice blocks, while polynyas are open waters with a wider and larger area. Both leads and polynyas designate the open waters in freezing seasons (such as late autumn, winter, and early spring), and there are no essential differences between them (Zwally et al. 2013). In summer, the open water appears caused by ice broken, and the area is mixing with open water and sea ice is called the marginal ice zone.

Leads are formed due mainly to the different movement of ice blocks. The open waters formed by the discontinuous movement between land fast ice and pack ice are called circumpolar flaw leads. There are two mechanisms for polynyas formation:

Fig. 3.7 Pressure ridges of sea ice

Fig. 3.8 A melt pond on sea ice

latent-heat polynyas, driven by persistent winds from the coastline; and sensible-heat polynyas, resulting from convection and exchange of heat in inner polynyas.

Melt ponds: As snow and ice melt in spring, the liquid water collects in depressions and low-lying areas on the ice surface, forming the melt ponds (Fig. 3.8). The geometric topography of ice surface determines the shape of melt pond, and their water basin dominates the depth of melt ponds. Thus, there are significant differences in the shape and area of melt ponds (Fetterer and Untersteiner 1998). In general, the ice ridges lined up parallel and melt ponds are interconnected in a specific region. The formation of melt ponds is related to the roughness of ice surface. The huge ice ridges benefit to form deeper melt ponds, while no visible melt ponds appear on the ridge-free ice. In recent years, the coverage of melt ponds in summer can reach up to 56%.

3. Sea ice melting

The melting of sea ice has obvious seasonal characteristics. First-year ice totally melts in the next summer season; if the ice survives one melting season, forming the second-year ice and even the multi-year ice. Although sea ice does not totally melt during summer seasons, its thickness greatly reduces. It has been reported that the thickness of multi-year ice decreases from 4 m in winter to 1.2 m in summer (Maslanik et al. 2007, 2011). The sea ice melting includes surface, bottom, lateral, and internal melting processes.

Surface melting: The surface melting of sea ice is driven by direct heating of solar radiation. Only the light with shortwave length (400–550 nm) can penetrate the ice surface and enter the interior of ice, increasing the temperatures of sea ice; however, the light with longwave length can be absorbed by the thin ice surface, and the energy enters internal ice through thermal conductivity. When solar radiation intensity exceeds the heat transfer flux of sea ice, the residual energy can increase the surface temperature of sea ice, thus inducing the surface melting of sea ice. The water produced by melting of sea ice may accumulate in melt ponds, or drainage directly into the ocean. The water has smaller albedo and larger heat absorption rate compared to the ice, promoting the absorption of solar radiation and accelerating surface ice melting process (van den Broeke 2005; Duffie and Beckman 2006; Screen and Simmonds 2010; Serreze et al. 2017).

Bottom melting: Warmer sea water provides heat to sea ice through upward oceanic heat flux. Most heat arriving at the bottom of sea ice accumulates to heat and melt the ice. The maximum oceanic heat flux was estimated to be 500 W/m^2, and the melting rate of ice on the bottom can be greater than that on the surface. The bottom melting of sea ice is dependent on the available heat in ocean obtained either from solar heating through polynyas or from the advection of warm water. The warm water may bring huge heat obtained by solar heating in the ice-free area, and its energy is greater than that by local heating. The warming water can directly heat sea ice, forming oceanic heat flux and rapidly melting bottom sea ice (Zhao and Li 2009; Zhao and Cao 2011).

Lateral melting: In summer, sea ice flaws to many blocks of different sizes. For the same area of sea ice, the more ice blocks to be divided, the more lateral area to connect sea water. The sea ice contacting seawater can be melted, which is called lateral melting of sea ice. It includes directly lateral melting, the denudation of sea ice, and lateral crash induced by colliding with ice floes. The lateral melting speed is related to the heat capacity of seawater; the more heat capacity, the faster lateral melting. Owing to Arctic warming and sea ice retreat, lateral melting will increasingly contribute to sea ice intensity.

Internal melting: Sea water can enter salt and air bubbles in sea ice in summer seasons, making sea ice become a water-contained body. Water can absorb more heat than air in the bubbles, increasing water temperatures and enlarging salt bubbles, which is the internal melting of sea ice. The internal melting process does not alter sea ice concentration and thickness but alters the porosity of ice, losing the ice structure and accelerating the melting of sea ice. The most significant internal melting occurs at the bottom of melt ponds. The relatively high water head of melt ponds can penetrate salt bubbles into the ocean and forms freshwater pool, and this process is called flushing effect. Owing to the high temperature of melt pond water, the heat exchange of melt pond water with sea ice can enlarge the salt bubbles. The ice cores in melt ponds have a higher porosity compared to that in sea ice.

3.4.2 Ice Shelf and Iceberg

Differing from sea ice, ice shelf and iceberg mainly come from land. An ice shelf is thick floating ice that forms where glaciers or ice sheets flow down to the coastline and onto the ocean surface. Ice shelves are generally developed around Antarctica, including the Ross Ice Shelf, Filchner-Ronne ice shelf, and Amery ice shelf. Moreover, ice shelves can move to the ocean with a velocity ranging from several meters to several kilometers.

The snowfall accumulated on the surface of ice shelves can evolve into new ice body through compression and metamorphism, increasing the thickness of ice shelves. In recent years, Australian glaciologists have observed that the sea water underlying the Amery ice shelf froze continuously and increased its thickness. During the warm seasons, the flushing effect of seawater can melt parts of ice body around ice shelves, leading to certain seasonal variations in ice shelves.

With the movement of ice shelves into the ocean, many crevices are produced under the combined effects of gravity and sea waves, thus causing large floes to move towards the ocean and forming the iceberg. Icebergs are produced mainly in warm seasons because the warm climate can accelerate to split off the marginal glacier and ice sheet. Ice density is about 90% of seawater density, and the volume of the iceberg on the sea surface only accounts for 10% of the total volume of the iceberg, which is also the origin of "the tip of the iceberg". The iceberg can be faded away ultimately due to the long-term flushing effect of seawater.

3.4.3 Subsea Permafrost

Subsea permafrost of seafloor was found in the coastal regions of the Arctic Ocean and is primarily distributed in the continental shelves of the northern Siberia and northern Alaska, resulting from climatic evolution. The main formation and development processes were considered as.

During the last glacial period, the regions with modern subsea permafrost were higher than the sea surface, and there was no glaciers developed then. Permafrost was developed very well due to the severe cold climate and low sea levels at that time. After then, with the climate warming and elevated sea levels, these permafrost-developed areas were buried and preserved under the sea surface, forming the subsea permafrost.

Nowadays, the coastal zones of circumarctic regions have been eroded by sea water, with the retreat rates ranging from several meters to hundreds of meters per year. This process leads to the loss of ground ice in permafrost and causes new permafrost in coastal zones to bury in the seabed, thus developing new subsea permafrost.

During the cold seasons, the sea surface temperature is lower than that of the deep sea, and the sea water density at lower temperatures is greater compared to that at higher temperatures, resulting in the downward movement of surface sea water and upward movement of deep sea water. By contrast, the sea surface temperature is higher compared to that of the deep sea, causing the deep sea to maintain a lower temperature. In addition, the sea-floor permafrost contains lots of ground ice. Therefore, the thawing rate of subsea permafrost is very low, which is also the reason that permafrost developed before tens of thousands of years has so far been preserved. Overall, climate change leads to the degradation of subsea permafrost.

3.5 Aerial Cryosphere

3.5.1 Snowflake

Snowflakes are aggregates of ice crystals that form in the atmosphere and fall to the ground. The morphology and volume of ice crystals change accordingly when falling. If the water vapor in the air is saturated, it then condenses on the surface of ice crystals and makes the ice crystals grow up gradually. In addition, ice crystals may also merge, forming a variety of snowflakes in shapes of a star, column, flake, etc., while the major pattern is hexagonal. When the temperature of cloud base is lower than 0 °C, snowflakes may keep its state, forming snowfall; by contrast, if the temperature of cloud base is higher than 0 °C, they may melt into or mix with liquid water, forming rainfall or sleet.

When falling, ice crystals grow very slowly if the surrounding air is not fully over-saturated, and will maintain their original columnar, needle-like and flaky shapes. If the surrounding air is over-saturated, the ice crystals will be solidified by water vapor molecules on any angles and projections, and grow to branch or star snowflakes when falling. Naturally, the water vapor content exposed to each branch of ice crystals is different due to the atmospheric motion. As a result, different growth rates form various ice crystal branches. When they fall into the cloud, their shapes vary substantially due to totally different circumstances, forming complex shapes of snowflakes. Snowflakes are easily bonded to each other and become larger snowflakes.

3.5.2 Hail, Snow Pellet, and Sleet

Hail, snow pellet, and sleet are three forms of precipitation. Hail is precipitation in the form of balls or irregular lumps of ice and is always produced by convective clouds, nearly always cumulonimbus. However, precipitation consisting of white, opaque, approximately round (sometimes conical) ice particles having a snow-like structure is called snow pellets. By convention, hail has a diameter of 5 mm or more, while smaller particles of similar origin, formerly called small hail, may be classed as either ice pellets or snow pellets. Sleet refers to precipitation in the form of a mixture of rain and snow.

An individual unit of hail is called a hailstone. Cumulonimbus (also called hail cloud) being characterized by strong updrafts, large liquid water contents, large cloud-drop sizes, and great vertical height, are favorable to hail formation. There exist violent upward and downward airflow in hail cloud. Usually, hail nucleus drops rise with strong upward airflow, and grow up by adsorbing surrounding small ice particles and water droplets. When their weights cannot be held by the updrafts, they begin to fall. When falling, if they suffer a high temperature, the hail drops surface will melt into water and adsorb surrounding small water droplets. Whenever suffering a strong enough updraft, they rise again, and their surface condenses into ice again. Repeatedly, the volume of hail nucleus enlarges continuously, and fall until their weight is greater than air buoyancy. If it does not dissolve into the water when landing to the ground, it is called hail; otherwise, it becomes raindrops.

Snow pellets are crisp and easily crushed, differing in this respect from snow grains. They rebound when they fall on a hard surface and often break up. In most cases, snow pellets fall in shower form, often before or together with snow, and chiefly on occasions when the surface temperature is at or slightly below 0 °C. It is formed as a result of accretion of super-cooled droplets collected on what is initially a falling ice crystal (probably of the spatial aggregate type).

When a lot of over-cooled cloud droplets condenses snow crystals, they become frost. If this process continues and the crystalline form disappears finally, they transform into sleet.

Questions

1. What are the conditions of formation and development of various cryospheric components?
2. What are the snow line and equilibrium line?
3. What are the hydrothermal migration characteristics and their drivers in active layers during freezing–thawing processes?

Extended Readings

1. ***Glaciers***

Authors: M.J. Hambrey and J. Alean

Pressed by: Cambridge University Press, 2004.

This monograph concisely introduced the formation, development, and evolution processes of glaciers, which is one of the most popular books on glaciers.

2. ***General Geocryology***

Authors: Permafrost Institute

Pressed by: Siberian Branch of USSR Academy of Sciences, 1974 (Russian version)

Translated by: Guo Dongxin, Liu Tieliang, Zhang Weixin, et al.

Pressed by: Science Press, 1988 (Chinese version).

This book is a great classic in permafrost, which systematically introduced the basic concepts and definitions of cryopedology, seasonal freeze–thaw processes in soils or rocks, thermal dynamic characteristics of frozen ground, ground ice and cryogenics, the formation, development history, and distribution of permafrost, the components, properties, and fabrics of permafrost layers, and the temperature regimes and thickness of permafrost zones.

Chapter 4
Physical Properties of the Cryosphere

Lead Authors: Jiawen Ren, Qingbai Wu, Zhijun Li

Contributing Author: Tao Che

Physical properties of the cryosphere are the basis to understanding all processes and mechanisms in the cryosphere. Although ice is the core material in all the cryospheric components, there are enormous physical differences between various cryospheric components due to different proportions and properties of other substances in different cryospheric components. Even in an individual cryospheric component, substance composition is not uniform spatially. Therefore, numerous research achievements on physics of individual cryospheric components have been sprung up and some books have been published not only on specific topics but comprehensively on individual cryospheric components, such as *The physics of glaciers and The mechanics of frozen soils*. As a chapter of the conduction of cryospheric science, it introduces only basic concepts of mechanical and thermal properties of pure ice and the main cryospheric components. A few paragraphs are related to optical and electrical properties in consideration of remote sensing and detection technique applications in the cryospheric research.

4.1 Basic Structure and Physical Properties of Pure Ice

4.1.1 Crystalline Structure

1. Crystallographic structure

Ice is the solid phase of water through liquid water freeze or water vapour sublimation and thus has the same material composition as other phases of water. Ice occurs in crystalline form with different structures under varied temperature and pressure conditions. About ten crystalline structures of ice have been identified up to date, but only one structure forms commonly in nature with a hexagonal crystalline structure and the others are unstable without specific laboratorial conditions. Virtually all the ice on the Earth's surface and in its atmosphere is the hexagonal ice, denoted as ice I_h.

The crystalline lattice of ice I_h is a triangular prism tetrahedron. Each oxygen atom at each angle and on each ridge is shared by the adjacent unit cells, and only

D. Qin et al. (eds.), *Introduction to Cryospheric Science*, Springer Geography,
https://doi.org/10.1007/978-981-16-6425-0_4

the central one is independent (Fig. 4.1). This tetrahedron structure is relatively loose because the tetrahedron space cannot be occupied completely by five water molecules and it is a structure relying on the hydrogen bond linkage (Fig. 4.2). As

Fig. 4.1 The lattice structure of a single crystal of ice

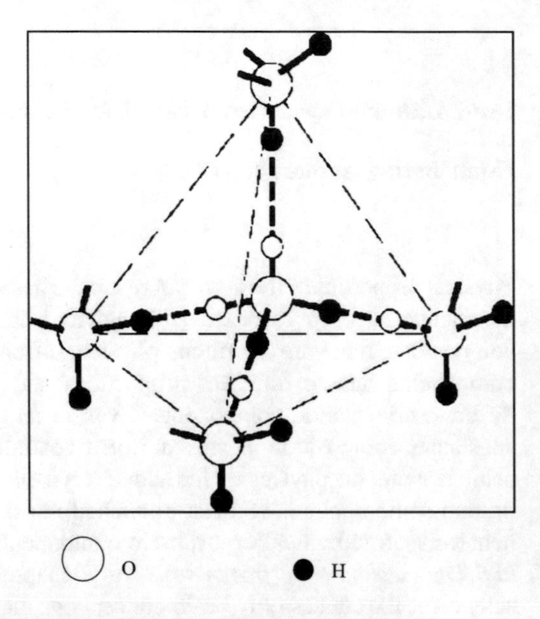

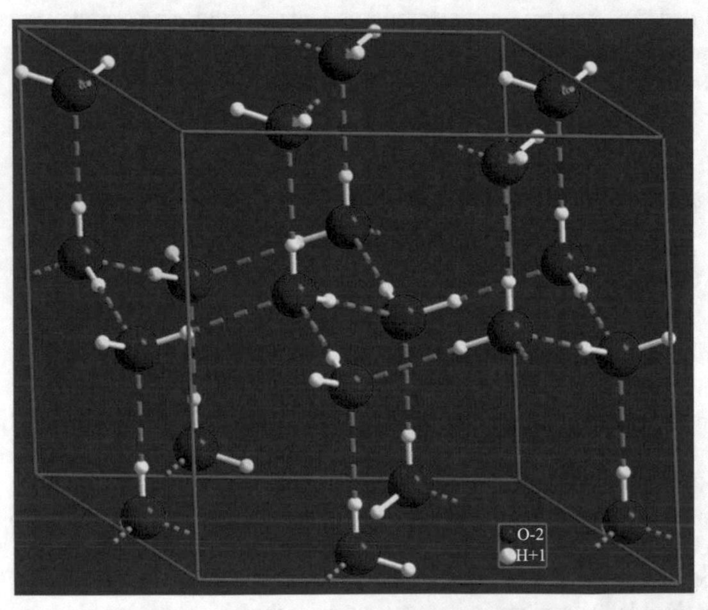

Fig. 4.2 The hexagonal crystal structure of ice (from Wikipedia). The grey dashed lines show hydrogen bonds

a result, therefore, ice is less dense than liquid water. At one atmospheric pressure, liquid water is the densest at about 4 °C, with density of 1000 g/m^3, and becomes less dense as the water molecules begin to form the hexagonal crystals of ice as the freezing point is reached. At 0 °C, water density is 999.87 kg/m^3 while ice density is 917 kg/m^3.

2. Crystal fabrics

Individual ice crystal has four axes, among which three axes intercross at 120° angle on a plane, called the basal plane, and the remaining one perpendicular to the basal plane is called the principle axis, which is also called the optic axis or *c*-axis. The reason why called optic axis is that along this axis light ray propagates the crystal without refraction. A single crystal has only one *c*-axis orientation. If a block of ice is consists of multi crystals and these crystals have a same orientation, the ice is called a single crystal ice or the monocrystalline ice. This kind of ice is anisotropic due to difference in physical properties between the basal plane and the *c*-axis directions. In polycrystalline ice, crystals have different orientations, and if their orientations are extremely random, the ice can be regarded approximately isotropic. Polycrystalline ice may also have several orientations. The c-axis orientation pattern is called the crystal-orientation fabrics or briefly the ice fabrics, and is important to study of ice mechanical and other properties since ice deformation is different for different fabric patterns under a same stress.

Ice in nature occurs mainly in polycrystalline structure. Under stress effects, the original pattern of c-axis orientations can be changed, because in strain process, rotation of crystal grains can produces a preferred *c*-axis orientation pattern corresponding to a certain stress condition. From laboratorial experiments, it is thought that, in general, uniaxial compression easily produces a circle-girdle fabric pattern and simple shear beneficially leads to a single maximum pattern. These two fabric patterns are commonly seen in glaciers and ice sheets.

3. Crystal grain

The smallest unit of ice crystal is crystal cell with a lattice in hexagonal prism shape. In natural environment, however, an individual ice crystalline unit is commonly the aggregate of multi crystal cells, called the ice crystal grain or in brief the ice grain (for the crystalline grain, there is a strict definition in crystallography, but only its generic meaning is taken here). Ice crystal grains formed originally in saturation water vapor may change in shape and volume with changes in temperature, pressure and other factors such as wind and humidity. For example, ice crystal grains form when tiny supercool water droplets freeze in cloud and then grow by condensation of water vapor onto their surface or and bonding between droplets when droplets collide each other. Since differences in temperature, pressure, humidity and wind, the grains change continuously in shape and volume during their dropping down to the land surface, so that multifarious forms of ice grain occur on the earth surface. Particularly it is hard to see the complete same snowflakes in shape, size and volume, although hexagonal shape is basically, and even some snowflakes do not show hexagonal.

After dropped on land surface, snowflakes change into spheric particles rapidly under automatic rounding effect. In snow cover and surface snow layer on a glacier and other kinds of ice such as ice sheet, ice shelf, sea ice, lake ice and river ice, ice grains may experience sintering, extrusion, meltwater infiltration and refreezing, recrystallization and so on. These would result in irregular ice grains. Ice formed in water, such as in sea, river and lake, mainly consists of columnar and granular crystal grains. Columnar (and needle) grain forms easily in calm water, while granular grain in flowing water.

4.1.2 Mechanical Properties

Elastic and plastic deformations are the two main deformation processes of a continuum solid medium under a stress action. As a solid material, however, ice is usually at the temperature near its melting point, and thus plastic deformation occupies the major part of its deformation and viscosity is also important somewhat. Therefore, elastic properties of ice are introduced briefly and the emphasis is put on its plastic properties, especially on ice creep law here.

1. Elastic properties

Mechanical properties of ice are very complicated since it has plasticity, viscosity and brittleness simultaneously. To understand individual mechanical properties of ice, numerous experimental investigations must be carried out at various temperature and stress conditions.

Previous studies have shown that elastic deformation is only transient under a stress for ice. In study of material elasticity, a basic assumption is that the material is isotropic. This is only an approximation for ice, because the single crystal ice is anisotropic and so does the polycrystalline ice with a preferred *c*-axis orientation pattern. Various researches have obtained different results for elastic parameters of ice. The reason is mainly due to differences in experimental technique and conditions as well as ice sample features such as sample sizes, ice structure and impurity. Based on many research results (Hobbs 1974), elastic parameters of pure ice are approximately as following:

Young modulus is 8.3×10^7 hPa to 9.9×10^7 hPa, stiffness modulus (or shear modulus) 3.4×10^7 hPa to 3.8×10^7 hPa, Poisson ratio 0.31–0.37 and bulk modulus 8.7×10^7 hPa to 11.3×10^7 hPa.

2. Plastic deformation

Plastic deformation is the most important in ice mechanics. Numerous experiments show that different relationships between strain and stress appear at different stages with time during the ice deformation.

(1) Elastic strain: It occurs only transiently at initial under stress action, so called transient elastic strain. In this stage, strain varies with stress at the Hooke's law, i.e. strain is proportional to stress and the coefficient is reciprocal of elastic modulus.
(2) Anelastic strain: After unload stress, basic deformation can recover slowly but some degree creep exists, so called primary or transient creep.
(3) Second creep: After the primary creep, deformation increment decreases with time, i.e. strain rate decreases continuously until a minimum value.
(4) Tertiary creep: It is a stage during which strain rate increases again after the minimum strain rate reaches. If test lasts enough long, maybe a constant value of strain rate reach eventually.

3. Ice flow law

Since the major part of ice deformation is creep, the relationship between stress and strain rate is called the ice creep law usually. In the strict sense, the ice creep law is a multinomial covering all individual creep stages. Since the second creep is the most important, however, the creep law can be simply expressed in term of the relation of strain rate versus stress during the second creep:

$$\dot{\varepsilon} = A_0 \exp[-Q/(KT)]\tau^n \tag{4.1}$$

where $\dot{\varepsilon}$ is the effective shear strain rate, τ is the effective shear stress, Q is the activation energy for creep, K is universal gas constant (Boltzmann constant), A_0 depends on crystal orientation, impurity content and perhaps other factors but independent of temperature, and n is a constant. So it is simply expressed by

$$\dot{\varepsilon} = A\tau^n \tag{4.2}$$

where A can be regarded as a constant mainly dependent on temperature for a certain ice sample. From most experiments, n is taken 3 in general. The formula (4.2) is commonly called Glen's law.

Ice creep has some characteristics of viscosity, and glacier movement caused by ice deformation is similar to viscous fluid motion to some extent. The creep law of ice, therefore, is also called ice flow law. However, typical viscous flow, i.e. Newtonian viscous flow, is linear. Another similar deformation to ice is the perfect plastic deformation. Figure 4.3 shows comparison of ice flow law with these two deformations.

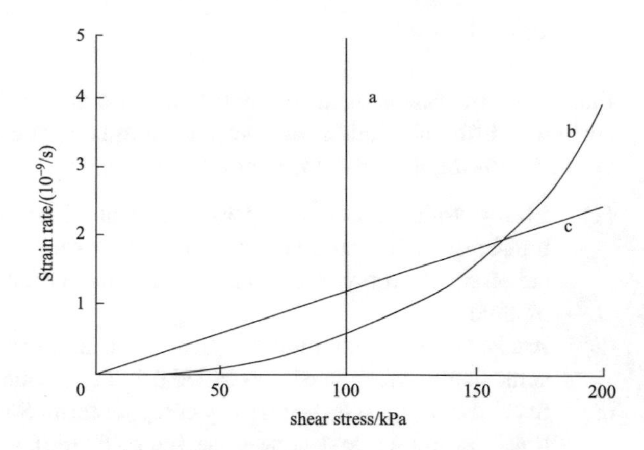

Fig. 4.3 Different types of flow relation (from Paterson 1994): **a** Perfect plasticity with yield stress of 100 kPa, **b** power law of ice with $n = 3, A = 5 \times 10^{-15}\ s^{-1}\ (kPa)^{-3}$, **c** Newtonian viscous flow with viscosity 8×10^{13} Pa s

4. Ice strengths

Strength is the material resistance to deformation and fracture under an external force. According to the stress states, it may refer to compressive strength, tensile strength, flexure strength, shear strength and so on. For ice, the fracture strength is most important to ice engineering.

The fracture strength of ice depends on sample feature and stress condition, mainly including ice type (fresh water ice, salt water ice, glacier ice, etc.), sample shape and size, stress kinds (tensile, compression, shear, impact force) and loading mode (sudden loading or gradual loading, etc.). Study of ice strength, therefore, needs to numerous laboratorial tests for individual types of ice and specific application purposes.

For single crystal ice, some tests give an average range of fracture strength 1.2–3.2 MPa under –90 to –50 °C temperature condition (Hobbs 1974).

4.1.3 Thermal Properties

1. Melting point and latent heat

Thermal properties of a kind of material are generally expressed by a set of parameters, mainly including the melting point (the temperature at which it melts), specific heat (or heat capacity), latent heat of phase change, thermal conductivity (or heat conductivity coefficient), thermal diffusivity (or heat diffusivity coefficient). Individual thermal parameters of ice usually vary with temperature and pressure, and some are related to density.

For most substances, melting and freezing points are approximately equal. The melting point of ice is basically equal to the freezing point of water at one atmosphere of pressure, very close to 0 °C. In the absence of nucleating substances, however, the freezing point of water is not always the same as the melting point of ice. For example, water can exist as a supercooling liquid down to much lower than 0 °C before freezing in the absence of condensation nucleus. The melting point of ice decreases with pressure increasing at a rate of about 0.0075 K per 100 kPa.

In normal condition, i.e. at one atmosphere of pressure, latent heats of ice phase change are constant. The latent heat of ice melting is 333 kJ/kg (about 80 cal/g), equal to the latent heat of water freezing at 0 °C, and the latent heat of ice sublimation is about 2837 kJ/kg (about 676 cal/g).

2. Specific heat and thermal conductivity

Since specific heat refers to mass unit of a material, it is independent of density. And if ice density is approximately regarded constant, the density effect cannot be considered for thermal conductivity of ice. In normal conditions, pressure is usually taken as one atmosphere and so temperature dependency is considered only. Numerous investigations show that the specific heat of ice is less than water at 0 °C and decrease with temperature decreasing, while the contrary is the case for the thermal conductivity of ice, i.e. it larger than that of water at 0 °C and increases with temperature decreasing. Figure 4.4 presents an example of research results on temperature dependency of the specific heat and the thermal conductivity of ice. It can be seen that the specific heat varies with temperature linearly but the thermal conductivity nonlinearly. From various experimental investigations, it can be concluded that typical values of them are as following:

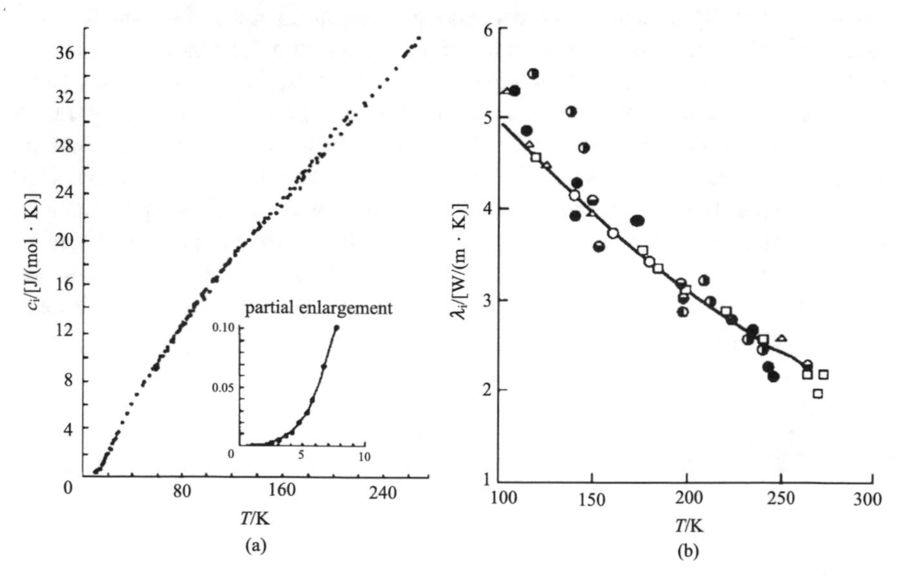

Fig. 4.4 Relations of specific heat (c_i) and thermal conductivity (λ_i) of ice with temperature (T) in Kelvin. Symbols of circle, triangle, square, etc. refer to different experiments (from Yen et al. 1991, 1992)

At 0 °C, the specific heats of ice and water are 2.097 kJ/(kg K) and 4.187 kJ/(kg K), respectively. And the thermal conductivities of ice and water are 2.1 W/(m K) and 0.598 W/(m K), respectively. At –50 °C, the specific heat of ice is 1.741 kJ/(kg K) and the thermal conductivity is 2.76 W/(m K).

3. Thermal diffusivity

Thermal diffusivity is dependent on thermal conductivity, specific heat and density with a relationship of $k = \lambda/(\rho c)$, where k is thermal diffusivity, λ is thermal conductivity, ρ is density and c is specific heat. Therefore, the thermal diffusivity of ice can be calculated when the other relevant parameters are determined. An alternative method to obtain the thermal diffusivity of ice is that after imposing sinusoidal temperature variation to one end of an ice specimen, the attenuation of temperature wave is measured at several points along the specimen and then the thermal diffusivity can be calculated from the attenuation. This method also can be used in field investigation, for instance, measuring the propagation and attenuation of surface temperature wave in a glacier or in other ice. From some research results, if taking the constant density of ice as 917 kg/m, the thermal diffusivity of ice is about 1.09 $\times 10^6$ m^2/s at 0 °C and 1.73 $\times 10^6$ m^2/s at –50 °C, respectively.

4. Other properties related to temperature

Under different pressure and temperature conditions, other properties of ice, such as density, expansion and compressibility, are also concerned for some purposes. In general, ice density increases with decreasing temperature at a certain pressure and does also with increasing pressure at a certain temperature. The thermal expansion is usually expressed in term of its expansion coefficients. All measurements show that the coefficient of linear expansion of ice decreases with decreasing temperature but negative relation below about –220 °C observed in some measurements, and the coefficient of cubic expansion of ice has a similar trend though lesser data below –220 °C. Usually ice is regarded incompressible in discussion on relationship between its strain rate and stress. However, ice is of compressibility to some extent in reality, especially for glacier ice and other ice containing air bubbles and impurities. Ice compressibility is very complicated due to difference in content of air bubbles and impurities and their composition.

4.1.4 Optical and Electrical Properties

1. Optical properties

For an individual ice grain, refraction does not take place in the direction of *c*-axis but does in all the other directions. Based on this character of the refraction of ice crystals, the *c*-axis orientation of each crystal grain in an ice slice can be measured by observing the transmission of a polarized light beam though individual crystal grain. Pure ice has high transparency, but if ice thickness is large, it appears to be blue or dark green somewhat. The reason is that lights with a relatively short wavelength such as blue light are easily absorbed and scattered, similar to sky and ocean. Most ice in nature contains air bubbles and impurities and so transparency decreases with increasing content of bubbles and impurities. The reflectivity of ice is dependent on its cleanliness. The reflectivity of pure ice is related to crystalline structure, temperature and light wavelength.

For ice in nature, such as glacier, snow cover and sea ice, great interest is in its albedo. Clean ice surface has a high albedo up to 0.6 and the albedo of fresh snow is as high as 0.9 or higher.

2. Electrical properties

For electrical properties of ice, much attention is paid to dielectric and conduction. They are usually expressed in terms of the dielectric constant (or the permittivity) and the conductivity, respectively. According to large difference in the dielectric constant for ice and other materials, radio-echo sounding technique is developed for use in detection of glacier and other cryospheric components. Both the high frequency and the static dielectric constants increase with decreasing temperature, but it is very hard to determine their variability due to effect of many other factors, such as crystalline structure, density, the angle between the electric field and c-axis as well as impurities. Some measurements on laboratorial frozen ice and glacier ice samples show that the high frequency dielectric constant is basically 3.1–3.2, and the static dielectric constant mostly 90–110 at temperature between –40 and 0 °C.

The electrical conductivity of ice is particularly sensitive to the impurity composition, besides temperature, ice structure and the electric field. Therefore, the impurity species can be identified from measurement of the electrical conductivity of ice samples in ice core study. Some experiment dada give the direct current conductivity of ice between $10^9/(\Omega\ \text{m})$ and $10^6/(\Omega\ \text{m})$, but most at the order of magnitude of $10^7/(\Omega\ \text{m})$ (or 10^5 S/cm).

4.2 Mechanical and Dynamic Characteristics of Cryospheric Components

4.2.1 *Glacier Motion and Dynamics*

1. Distribution of ice velocity

Glacier movement downstream under the effect of gravity is the basic physical characteristics. Therefore, the longitudinal velocity is the main component of glacier velocity. The transversal velocity is generally small, but it will increase when the longitudinal flow is blocked by the local topography. The vertical velocity depends on glacier surface accumulation and ablation. In the accumulation area, since there is an annual net mass accumulation, glacier velocity has a downward vertical component under pressure of continuous increasing snow/ice. In the ablation area, old ice melt is compensated by upstream ice flow so that glacier velocity has an upward vertical component. Therefore, the accumulation area is the sinking flow zone, whilst the ablation area is the upwelling zone. Around the equilibrium line altitude, ice velocity is parallel to glacier surface with the maximum value and then decreased continuously to upstream and downstream. In the vertical direction, surface velocity is largest. In the transversal direction, glacier flow velocity is largest in the middle, but if glacier makes a bend, the largest velocity deviated to the outside.

2. Flow velocity caused by ice deformation

Assuming the glacier thickness, width, and slope constant in the spatial range discussed, width and length is much larger than thickness, and the ice deformation is caused only by shear stress, the flow vectors (flow-lines) are therefore parallel to the surface and the ice velocity changed only along with depth. Glaciologists call this laminar flow. In this simplest circumstance, the shear stress reached the largest at the glacier bottom. It can be expressed as following:

$$\tau_b = \rho g h \sin a \tag{4.3}$$

where τ_b is shear stress at the base, ρ is density, g is gravitational acceleration,h is ice thickness, and α is slope of bedrock which is equal to surface slope.

According to laminar flow assumption, glacier flow along with the direction to the downstream, and there is only one shear stress as non-zero stress component. From the relation between glacier flow velocity and strain rate and ice flow's law, therefore, the glacier flow velocity can be expressed as:

$$u = u_s - 2A(\rho g \mathrm{sin} a)^n y^{n+1}/(n+1) \tag{4.4}$$

where, u is flow velocity, u_s is glacier surface flow velocity, A and n is constant in the ice flow law. The value of n is usually deemed as 3 and A is determined based on temperature.

Several important conclusions can be obtained based on these two simple formulas (4.3) and (4.4). τ_b can be calculated according to ice thickness and surface slope. If ice was regarded as perfect plasticity, it can be expressed as $h = \tau_0/(\rho g \ \mathrm{sin}\alpha)$. Due to $h \sin \alpha$ are constant, glacier surface slope is inversely proportional to ice thickness.

Obviously, it is only approximate in a very short distance that slopes of glacier surface and bedrock are the same, the slopes and ice thickness keep constant and ice was only acted by the longitudinal shear stress. If slope and thickness changed, ice would be extended or compressed, which can be deduced from velocity variation. Between the two adjacent sections, if ice is compressed, velocity in upper stream is relatively larger, and conversely, ice is extended. In the compressing flow area, ice thickness tends to increase, while in the extending flow area, fracture and crevasses are likely to appear.

The assumption that glacier width is far larger than ice thickness is often deviated from the actual glaciers, and so the influence of the two side walls of glacier valley cannot be ignored. In order to make improvement to some extent, the shape factor of glacier cross-section can be added to the shear stress. The factor is determined by section configuration and the ratio between width and thickness. Generally, the value is between 1 and 0.5.

3. Glacier basal sliding

The glacier sliding along its bedrock is an important component of glacier velocity. When the sliding occurs, the sliding velocity is often exceeding the velocity component of ice formation. Generally, the prerequisite of glacier sliding is the basal temperature reaching or near to the melting point. When the temperature is at the melting point and liquid water exists between the basal ice and the bedrock, the ice is easy to slide along the basement. If the temperature approaches to melting point but without liquid water, regelation can happen (ice can be melting instantaneously and then refreezing) and increase of local stress enhances plastic deformation. In the case of lack liquid water, therefore, ice can be still sliding, although the sliding velocity is smaller than the situation of liquid water existing.

Based on the basic theory model of glacier sliding, sliding velocity is mainly determined by the bottom shear stress and the bedrock roughness, but their relations are nonlinear. Classically, if the exponent in the ice flow law is taken as 3, the sliding velocity is positively proportional to the square of the bottom shear stress, and inversely to the fourth power of the bedrock roughness.

4. Deformation of the subglacial debris layer

From the glacier drilling down to its bedrock, artificial ice cave observation and evidence of glacial geology and geomorphology, there exists a debris layer at the bottom for most glaciers. Usually this debris layer is called the subglacial till. The observed till thickness is about dozens of centimeters to more than one meter. Generally, rock debris has an effect to decrease glacier sliding. If the till becomes loose rocks under the meltwater wash, the rocks can slide or roll on the rigid bedrock surface. In addition, there may not be clear boundary between the ice with less debris proportion and the ice with more debris proportion. According to the artificial ice cave studies of Urumqi Glacier No. 1 (Echelmeyer and Wang 1987), there is a debris layer with the thickness less than 1 m and 30% of the ice content, and the movement of this debris layer accounts for 60–80% of the glacier surface velocity. The movement of the debris layer consists of two parts: continuous deformation of the debris layer and sliding the debris sliding along the shearing surface or shear zone. The former has a contribution of approximately 60% to the glacier velocity and the latter accounts for within 20%. The glacier also can slide when the bottom temperature is less than the melting point, although the magnitude is less.

5. Ice sheet movement

In an ice sheet, ice moves radially from center to the margin. When the bedrock is flat, the typical idealized cross-section of ice sheet and ice cap is parabola. If assuming the perfect plasticity, the cross-section profile of ice sheet can be described as $(h/H)^2 + (x/L) = 1$, where h and x are ice thickness and the horizontal distance to the center, respectively, H is the thickness at the center, L is the distance from center to margin. If the ice sheet is in steady state and the whole surface is the accumulation area, the horizontal velocity is $(b/h)\,x$ due to that the ice flux at x is equal to the accumulation over the upstream area. Here b is the surface average accumulation rate. Therefore, the horizontal velocity at the center of ice sheet is zero and increases with increasing distance towards margin.

The slope and thickness of ice sheet change not much in a large distance from center toward margin, and thus the horizontal velocity increases slowly. In the steady state, if the basal sliding does not exist, the vertical velocity at the surface is equal to the accumulation rate and is zero at the bottom. If assuming the vertical velocity changes linearly with the depth, the required time of the ice movement from surface to a depth is $(h/b)\ \ln(h/y)$, where y is the distance from the bedrock upward to the depth.

Practically, the bedrock undulation of ice sheet has an important impact on ice movement and the basal sliding exists in many places. The rapid ice stream can be formed in the areas with the larger bedrock slope and valley topography. In most areas of West Antarctic Ice Sheet, the bedrock is below sea level and so there are many rapid ice streams. The Lambert Glacier basin in East Antarctic Ice Sheet also belongs to rapid ice stream area.

6. Ice shelf movement

The main feature of ice shelf movement is that the horizontal velocity keeps constant from surface to bottom because there is no shear stress at the bottom. The velocity of the unconfined ice shelf changes very little in the horizontal direction except the areas near the grounding line. The ice expands when the surface is accumulated and keeps the ice thickness unchanged. The ice thickness will decrease when the bottom melting exists. The velocity along the center line of a confined ice shelf is faster than the two lateral sides, which is because of dragging from the land in the sides. Ice shelf movement is also restricted by islands and grounding on shoals. The velocity of the confined ice shelf is faster than that of the unconfined ice shelf at most cases because the ice from upper stream must pass through the confined narrow channel such as valley and bay. After the front part of ice shelf calved to be iceberg, velocity of the successive ice will increase because the previous obstacle disappeared. If massive iceberg calving happens or ice shelf velocity increases rapidly, the resistance to the grounded ice will decrease and thus velocity of ice sheet will increase.

7. Glacier surging

Surging glacier is the glacier with the characteristic of intermittent rapid movement. This kind glacier moves suddenly at a speed many times faster than normal and its terminus advances a long distance within a short time, usually several days to months, and then its movement slows down to normal and terminus retreats continuously until its size recovers to the approximate extent before the rapid movement. After a period of time, it may move rapidly and extend again. Since difficulty to observe the glacier surging directly, it is speculated that the surging mechanism is probably related to structural factor, creep instability, hydrothermal instability, etc. The structural factors include the glacier developing at the active fault zones, geothermal anomaly activity, etc. Creep instability is that if the stress and strain increase, ice temperature will rise and ice deformation will enhance in turn. This positive feedback could cause the basal frozen ice on its bed to reach the melting point and then result in basal sliding. On the other hand, if there is warm ice layer at the deep of the glacier or near the bottom, when the shear stress at the interface between warm ice layer and upper cold ice layer reach to the critical value, the cold ice will slide and the sliding may extend rapidly. The hydrothermal instability is that when the melting water at the bottom gathers to an extent, it will infiltrate the obstacles at the bottom and decrease the resistance to ice sliding and on the other hand, the increasing water pressure will make the sliding much faster.

4.2.2 Mechanical Properties of Frozen Ground

1. Strength of frozen soil

Strength of frozen soil is the capacity of frozen soil resisting to external damage and its value is the maximum withstanding stress under a certain stress condition. Strength of frozen soil can be divided into instantaneous, short-term and long-term strengths according to the time of load, basic, standard, design, critical, ultimate, yield and failure strengths according to the different stress stage, static, dynamic and fatigue strengths according to the reaction of frozen soils and shear, compressive, tensile and cutting strengths according to the stress state.

1) Strength and failure of frozen soil

Frozen soil is a special soil containing ice. Since the strain of ice is different under different temperature, pressure and action time, the mechanical properties of frozen soil is unstable. The ice in frozen soil has not only the role of cementing soil particles, but also the encased ice body formed due to the water condensation and migration in the freezing process has the particularity of resisting the action of external forces. When the ice content is less, ice cannot cement all of the mineral particles into a hard solid mass and so the strength of frozen soil is slightly higher than that of unfrozen soil. The strength of ice-rich frozen soil is much higher due to the increasing cementation of ice. The strength of saturated-ice frozen soil and ice layer in soil is obviously reduced on the contrary and gradually close to the pure ice with increasing ice content.

The stress–strain experiments of frozen soil show that the failure of frozen soil can be divided into two forms: plastic failure and brittle failure. The plastic failure means that the stress–strain curve has no obvious turning point, while brittle failure has an obvious peak. The main factors affecting the failure form of frozen soil are:

Soil particle composition: Generally, the coarse granular frozen soil shows a brittle failure and the viscous frozen soil shows plastic failure. Under the same conditions, the peak strength of brittle failure is higher.

Soil temperature: Frozen soil tends to a brittle failure under a lower temperature and a plastic failure under higher temperature. The lower the soil temperature is, the higher the frozen soil strength.

Water content: Frozen soil strength increases with the increasing water content. The failure of frozen soil will be usually transited from brittle to plastic failure. However, the failure of frozen soil will be transited from plastic to brittle failure when the moisture content increased because the ice shows a brittle failure.

Strain rate: Frozen soils with a large strain rate show a brittle failure, conversely, plastic damage.

2) Elastic modulus and compression modulus of frozen soil

The deformation of frozen soil can be divided into instantaneous, long-term and failure deformation stages. In the instantaneous stage, elastic deformation is dominating and has important influence on the state of frozen soil under dynamic load (impact, explosion, seismic wave, vibration, etc.).

Some experiments show that the elastic deformation of frozen soil only accounts for 10–25% of the total deformation under the temperature above –5 °C and can account for up to 50–60% of the total deformation under the lower temperature. The water content of frozen soil has a great influence on elastic deformation. For fine grained soil, the proportion of elastic deformation increases with increasing moisture when the moisture is less than the plastic water content. The experiment results show that elastic modulus is the maximum for frozen sand, the minimum for frozen clay and medium for frozen silty clay. Elastic modulus is not only related to soil property, temperature and moisture, but also to the stress value (Fig. 4.5). The lower external pressure, the larger the influence of soil temperature on the elastic modulus of frozen soils is. Temperature and external pressure have opposite effects on the elastic modulus of frozen soil.

Temperature is an important factor affecting the compression deformation of frozen soil. As the temperature decreases, the unfrozen water content decreases and the cementing force of the solid particles increases. At –5 °C, frozen clayey soil deformation mainly is compaction and the relationship between the deformation and stress approaches to be linear. At temperature below –10 °C, deformation is mainly composed of creep deformation. Figure 4.6 shows that clay compression modulus

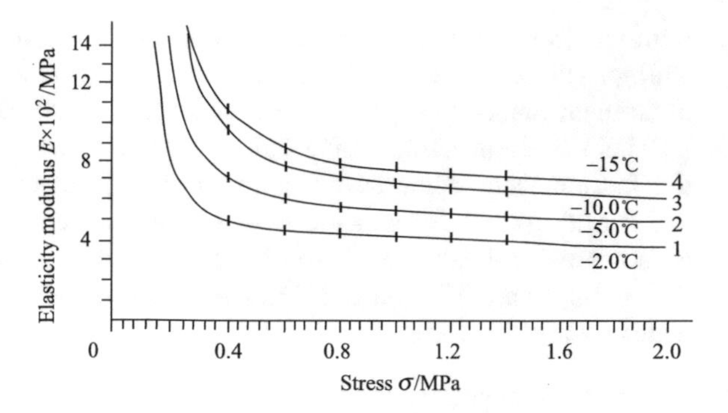

Fig. 4.5 Elastic modulus versus stress under different soil temperatures (Wu and Ma 1994)

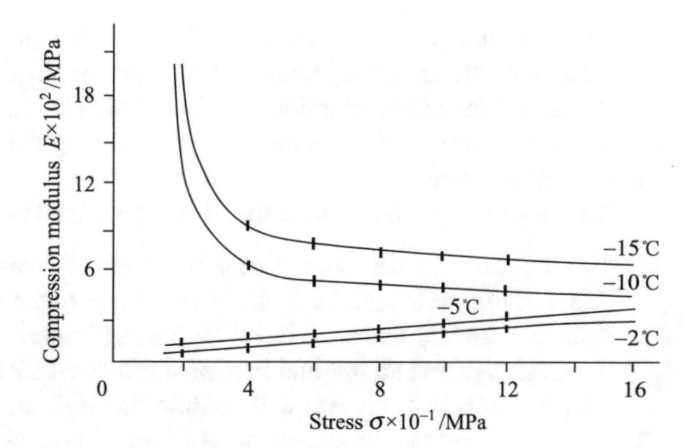

Fig. 4.6 Compression module versus stress for Huainan clay (Wu and Ma 1994)

increases with stress when the temperature is at –2 and –5 °C and it decreases with stress at temperature below –10 °C. For frozen sand soil, however, the compression modulus increases with increasing stress only at temperature around –2 °C and decreases with stress at temperature lower to –5 °C.

3) Instantaneous and long-term strengths

The instantaneous strength is usually expressed by ultimate strength or short-term strength. The long-term strength is that deformation has the attenuation characteristic under this resistance, but it has not yet transitioned to gradual damage. The ultimate compressive strength of the frozen soil is extremely high, even if it is not under the maximum loading speed, up to several to dozens of MPa. Data shows that the compressive strength is above 15.4 MPa for frozen sandy soil and up to 15 MPa for frozen clay under the loading speed of 50–90 MPa/minutes and the temperature of –40 °C. Therefore, frozen soil has a strong capacity of resisting to short-term loading.

The temperature is the main factor controlling the compressive strength of the frozen soil. The compressive strength of both coarse and fine grain frozen soils increases with decreasing temperature. In the extensive phase change temperature range (about 0 to –1 °C for sandy soil and –0.5 to –5 °C for clay), the increase of frozen soil compressive strength is most intensive, and freezing of pore water is fastest with the decreasing of temperature. Under the lower temperature, the compressive strength still increases, but the increase rate changes more complex and the increase of strength cannot be explained by the increase of ice content.

The long-term compressive strength of the frozen soil is much lower than the instantaneous compressive strength. The instantaneous compressive strength is 7.5 MPa for the frozen sand with a water content of 19.3%, but the long-term compressive strength is only 0.65 MPa. The instantaneous and long-term compressive strengths are 3.5 MPa and 0.36 MPa, respectively, for the frozen silty clay with water content of 31.8%.

4) Shear strength

The shear strength of the frozen soil reflects the conglutination force of the frozen soil, especially the cementing force of the ice in the soil. The experimental data shows that the limit (damage) strength of the frozen soil is related to the normal pressure under the plane shear, which is restricted by not only the cohesion force but also the internal friction force.

There are three main factors influencing the shear strength of frozen soil:

(1) Soil grain composition: the shear strength of coarse grain soil is higher than that of fine grain soil. In the same soil temperature (–8.0 to –9.0 °C), cohesion force of frozen fine sand is 1.57 MPa, and internal friction angle is 24°, but cohesion force and internal friction angle are 1.27 MPa and 22°, respectively, for the frozen clayey soil with middle liquid limit.
(2) Temperature: the shear strength of frozen fine sands increases with decreasing soil temperature, because the cohesion and internal friction angle increase with the decrease of soil temperature. When soil temperature is approaching to 0 °C, the internal friction angle of frozen soil is close to that of unfrozen soils, but cohesion force of frozen soil is much larger than that of unfrozen soil.
(3) Load time: shear strength of the frozen soil decreases greatly under the long-term effects of load. When temperature is at –2.0 °C, instantaneous shear strength is 1.3 MPa, but the long-term shear strength is only 0.11 MPa for frozen clayey soil with reticular structure and the water content of 33%. The decrease of shear strength is mainly caused by the decrease of cohesion force. The cohesion force attenuates rapidly within 4 h of the loading, but cohesion force attenuates slowly after 24 h of the loading.

Figure 4.7a is the experimental result for the same kind frozen soil with water content of 33% in the temperature of –1.0 °C. The line 1 represents the shear strength under the fast loading of different normal pressure (P), and line 2 is the limit shear strength under a long-term load. Figure 4.7b indicates that the cohesion force of clay slacks with time. The internal friction angle of frozen clay decreases from 14° (fast shear) to 4° (long-term shear), and the cohesion force decrease from 0.52 MPa (fast shear) to 0.09 MPa (long-term shear).

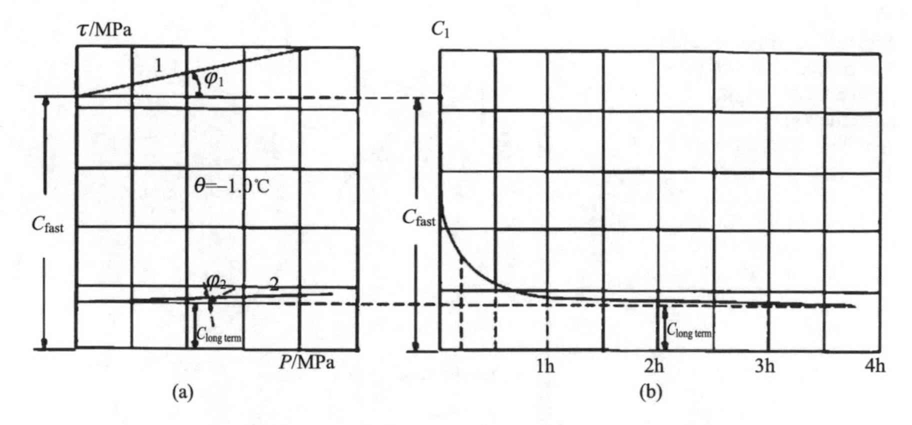

Fig. 4.7 Shear strength of frozen soil versus load action time (Tsitovich 1995)

2. Deformation and creep characteristics of frozen soil

1) Deformation of frozen soils

Frozen soil is compressible to some extent, but it can be considered approximately to be incompressible under a much lower temperature. Under the external load, the compressive deformation of frozen soil will be changed with the external load and the action time. Even under the small load, the frozen soil can have compression deformation. The load makes the unfrozen water migrate, the ice between the mineral grains thaw and the pores decrease. The compacting deformation is no more than a third of the total deformation, and the other deformation is damping deformation controlled by the irreversible shear deformation caused by inter-moving of soil particles from high stress area to low stress area under the action of pressure. At the initial stage, the deformation of frozen soil is very fast, and the deformation gradually slows down and eventually becomes relatively stable with the extension of stress action time.

Generally, compression curve of frozen soil under a constant negative temperature can be divided into three basic sections (Fig. 4.8). The line aa_1 means an elastic deformation and structure reversible deformation when frozen soil is compressed and its deformation rate is very large but is instantaneous. The corresponding pressure at a_1 is close to the structural strength of the frozen soil and the frozen soil starts to be compacted when pressure exceeds this value. Under a stress within the structure strength (about 50–100 kPa), the total deformation is reversible. The line a_1a_2 means the irreversible deformation of the structure, approximately 70–90% of the total deformation, which is caused by the irreversible shear of the soil aggregate. The line a_2a_3 represents strengthening of frozen soils, which is mainly caused by the enhancement of intergranular molecular bonding due to the distance shortening among grains.

Therefore, the deformation of frozen soil is composed of three parts within the range of long-term limit strength, namely instantaneous deformation, unstable deformation and attenuation deformation. The instantaneous deformation is generally very

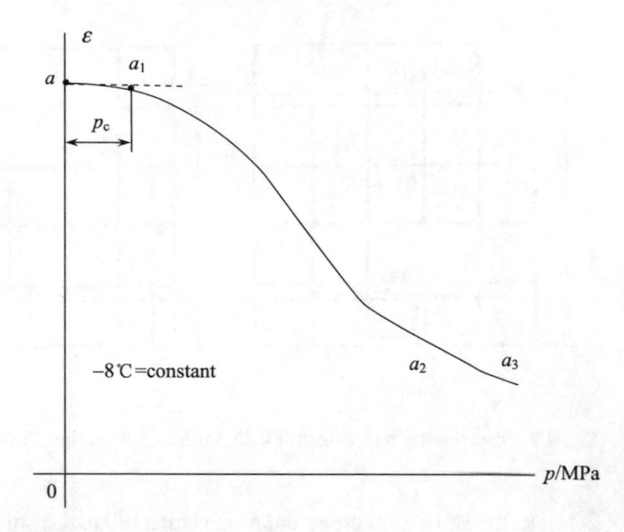

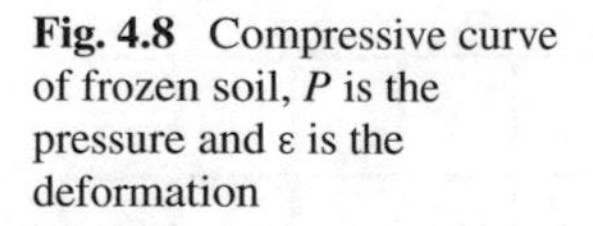
Fig. 4.8 Compressive curve of frozen soil, P is the pressure and ε is the deformation

short, which is negligible compared with the total deformation. The ratio of unstable to attenuated deformation decreases with stress increasing. Within the limit of long-term strength, creep deformation and creep stabilization time are restricted by soil temperature and moisture.

2) Creep and creep strength of frozen soil

Because of ice inclusion within the frozen soil, the ice cementation action restricts the strength and deformation of frozen soil. Any load, therefore, will lead to the plastic flow of ice and the redirection of ice crystals and the reworking action of irreversible structure will occur so that stress relaxation and creep deformation will occur under a very small load. When the stress is less than the long-term limit value, the deformation of frozen soil shows an attenuated creep with time.

When the whole process of frozen soil deformation is described, the creep equation must consider the stress, the soil temperature, the stress duration and frozen soil structure, especially the ice content. Normally, the creep of frozen soil can be expressed as:

$$\varepsilon = \varepsilon_0 t^{\alpha} \text{ or } \frac{d\varepsilon}{dt} = \varepsilon_0 \alpha t^{\alpha-1} \tag{4.5}$$

where, ε is the creep deformation quantity; ε_0 is the initial deformation; t is creep time; α is the test coefficient, $\alpha < 1$; $d\varepsilon/dt$ is the deformation rate.

4.2.3 Dynamical Characteristics of Snow Cover

The dynamic characteristics of snow cover are mainly determined by snow layer structure and temperature condition. If the structure of snow layer is uniform, e.g., snow grain size and snow density are homogeneous, and snow temperature is below the melting point, the viscosity would be low and snow layer is likely to be influenced by external force, e.g., snow layer surface would be easily eroded by wind. Once snow grains move along with the wind, the snow drift is formed, which is also known as the blowing snow. The blowing snow is mainly controlled by the wind strength and the viscosity of snow layer, and the viscosity is further relevant to snow density, grain size, temperature and moisture, etc. Generally, the blowing snow belongs to a solid–gas two phase flow, but if the humidity is high, liquid-phase water would also be involved to some extent. In addition, the movement and properties of blowing snow would be more complex than that of wind-drift sand if snow particles get clustered or collapse along the collision process of the movement.

Snow layer collapses sometimes occur when snow is accumulated on the hillside, and this phenomenon is known as avalanche. Once the weight of snow layer exceeds the critical cohesion of snow layer, snow particles would move downhill along slope. If there is an abrupt change in density or grain size at a depth in snow cover, a sliding interface at the depth would form for overlying snow layer and avalanches are more likely to occur. The critical condition of an avalanche occurrence can be roughly estimated based on a set of parameters such as the slope, snow density, snow depth, snow grain size and temperature, etc.

4.2.4 Dynamical Characteristics of River and Lake Ice

The research on the mechanics of river ice is mainly divided into static ice mechanics and motion ice mechanics. The former mainly studies the effect of thermal expansion of stationary ice on river embankment and buildings, and the carrying capacity of river ice. The latter is mainly concerned with the impact and destruction of ice bodies on river banks and buildings with the movement of river ice.

The thermal expansion of the static river ice is mainly characterized by the expansion coefficient of the ice. The linear expansion coefficient of pure ice at 0 °C is about 51×10^{-6}/K, but river ice sometimes contains dust and other impurities, thus, ice samples of specific locations need to be tested in some cases. The frozen river, if used as temporary bridge, is needed to be tested the bending strength of cantilever beam in order to calculate the bearing capacity.

The impact force of river ice is related to ice speed, temperature, and ice structure. Therefore, it is also need to tested and simulated based on the ice, temperature and water flow based in the specific locations.

The formation, development, and movement of river ice are closely related to river dynamics and thermodynamics. From the perspective of hydraulics, once ice formed in a river, the ice must change the river hydraulic condition. Therefore, laboratory experiments to reveal the hydrodynamic characteristics and the river ice movement under various river and ice conditions are one of the main topics of the research on river ice dynamics.

There are few studies on the mechanical and dynamic characteristics of lake ice. In winter, however, large frozen reservoirs usually have ice monitoring and survey of ice physical characteristics. The results of the observations of the reservoir ice can provide some references for the understanding of the characteristics of the ice in the small freshwater lake.

4.2.5 Dynamical Characteristics of Sea Ice

Sea ice is frozen by salty water. There are salt and brine in the grain gaps, and most of sea ice contains bubbles. Sea ice density is very different but mostly lower than that of fresh water ice.

The research on the mechanical properties of sea ice is mainly focused on the strength experiment due to the need of ice engineering, such as ice breaking and the damage of sea ice to the bank and the buildings. Because of the difference in the climate and marine environment in different sea areas, the sea ice structure and the impurities are different, and the sea ice strength is not consistent. For example, the compressive strength of sea ice in North American sea area is about 2.8–4.2 MPa, and that of Bohai sea ice in China is 0.5–1.9 MPa. Generally speaking, the main factors affecting the mechanical properties of sea ice are impurities (composition and content), structure (crystal structure, grain size, and density), temperature, and load mode.

Generally, most of the impurities in sea ice are salt, which can be characterized by the volume of brine or salinity. The experiments show that the sea ice strength decreases with the increase of the volume of the brine, and has the negative and positive correlations with grain size density, respectively. The relationship between sea ice strength and temperature is more complex. It seems that the strength decreases with the increase of temperature. However, under high strain rate, the strength decreases with decreasing temperature. The strength changes with different loading direction and loading rate.

The most important effect in the static failure of sea ice on bank and building is the change of sea ice volume with, that is, the thermal expansion force. Because the thermal expansion coefficient of salt and water is smaller than that of ice, the coefficient of thermal expansion of sea ice is lower than that of pure ice. However, the actual number of ice samples needs to be tested, or estimated based on the composition, the content of impurities, and temperature range.

Sea ice is always in the state of movement and change since it forms. Ocean currents are the main factor to control the movement of sea ice. The key to the study of sea ice dynamics is to accurately describe three elements. The first is the action of ocean currents and wind, the second is the thickness of ice and the spatial variation in ice structure, and the last is the thermodynamic process of interaction between sea water and ice body.

4.3 Thermal Properties of Main Cryospheric Components

4.3.1 Temperature in Glacier

1. Temperature in the surface layer

General equations for the surface heat budget are valid for all components of cryosphere (see Chap. 3). Since proportions of individual components of heat budget are different for various cryospheric components and surface characteristics under different climate conditions, heat budget is usually estimated from in-situ observations or simplified models for a specific surface.

Heat transfer from surface to interior determines temperature distribution in an object and thus its temperature distribution can be expressed by the heat transfer equation. As like as other land surfaces, temperature increases with increasing depth from surface in a glacier or an ice sheet. If a glacier or ice sheet is regarded as uniform continuum and only vertical heat transfer is considered, the one-dimensional equation of heat conductivity can be used to describe the vertical temperature distribution:

$$\frac{\partial T}{\partial t} = k \frac{\partial^2 T}{\partial y^2} \tag{4.6}$$

If the boundary condition, i.e. the surface temperature, is assumed to be

$$T(0, t) = T_0(0) + A \sin(\omega t) \tag{4.7}$$

The solution is

$$T(y, t) = T_0 + A \exp\left\{-y[\omega/(2k)]^{1/2}\right\} \sin\left\{\omega t - y[\omega/(2k)]^{1/2}\right\} \tag{4.8}$$

where T is temperature, t time, y depth down from surface, A the amplitude of temperature variation at surface, ω the angular frequency of temperature variation, T_0 the equilibrium temperature and k the thermal diffusivity.

The formula (4.8) shows that the higher the frequency (the shorter period), the more rapid the amplitude attenuation with depth. If taking k the value of pure ice and a year the period, for example, the amplitude at 10 m depth is 5% of that at surface, 1.1% at 15 m and 0.24% at 20 m. So the depth between 10 and 20 m is usually regarded as the maximum depth of seasonally variant temperature or the surface layer bottom.

Besides conductivity, some other factors may have important influence on temperature distribution in the surface layer of a glacier, such as the percolation of surface meltwater and snow/ice movement. When surface melting is intensive and medium is snow rather than ice, heat transfer from meltwater percolation and refreezing latent heat can be dominant instead of heat conductivity. In this case, temperature in surface layer may be higher in the accumulation area than in the ablation area since ice is impermeable and meltwater runs off. The vertical movement caused by mass gain or loss also influences on the propagation of surface temperature wave to some extent, and mass gain has an attenuation effect and mass loss reversely. The horizontal movement effect is negligible in the surface layer since much less difference in velocity within a short vertical distance and horizontal temperature gradient.

2. Temperature in deep ice

Beneath the surface layer, if let the origin of coordinates at the glacier bottom, the y-axis vertical and positive upwards, the x-axis horizontal and in the direction of glacier flow and the z-axis horizontal and in the transverse direction, heat transfer along the z-axis can be neglected since both ice velocity and temperature gradient are negligible in this direction. When thermal parameters are taken as constant, we can obtain the heat transfer equation as following:

$$k\frac{\partial^2 T}{\partial y^2}+u\frac{\partial T}{\partial x}+v\frac{\partial T}{\partial y}+\frac{Q}{\rho c}=\frac{\partial T}{\partial t} \tag{4.9}$$

where u and v are the ice movement velocities in the directions of x-axes and y-axes, respectively, and Q is the rate of internal heat production per unit volume.

In the specific case, for instance, under the steady state assumption and taking both the horizontal movement effect and the internal heat as constant and the linear variation in the vertical velocity with depth, the simplest solution can be obtained as the following:

$$T-T_s=\frac{1}{2}\pi^{1/2}l(\partial T\partial y)_b[\mathrm{erf}(y/l)-\mathrm{erf}(h/l)] \tag{4.10}$$

where $\mathrm{erf}(z)=2\pi-\frac{1}{2}\int_0^z\exp(-y^2)\mathrm{d}y$, the error function, $l^2=2kh/B$, h is ice thickness, B the surface mass balance, $(\partial T\partial y)_b$ the temperature gradient at the glacier bottom and T_s the surface temperature (taking as the temperature at the surface layer bottom).

The conditions to obtain the formula (4.10) are rigorous and so may be approached in the central area of ice sheet or ice cap with the flat bed. For mountain glaciers and the areas of ice sheets except of the flat-bed central area, effects caused by horizontal movement and ice strain in deep cannot be neglected and distributions of mass balance and ice velocities are complex. Furthermore, the steady state assumption is not satisfied often, and thermal parameters may not be constant. Therefore, the model of temperature distribution of glaciers should be determined and simplified according to accuracy requirement of research results and data set.

3. Temperature in ice shelf

Researches on temperature in ice shelves are much less relatively to that in ice sheets, possibly due to simplicity of temperature distribution in a floating ice. The basal boundary condition is that the temperature is fixed at the freezing point of sea water. Whether the basal ice absorbs or releases heat depends mainly on the temperature, salinity and circulation of the sea water. If the ice shelf is in steady state, the surface mass balance and the vertical velocity must be balanced by the basal melting or accretion so that the ice thickness keeps constant. At different sites, the vertical profiles of ice temperature are various due to differences in land ice influence, ice velocity, surface mass balance, ice thickness and basal situation. The closer to the land, the larger the land ice influence, and the farther into sea, the temperature profile is close to linear and the gradient is dependent on temperature difference between the surface and the base and ice thickness.

4.3.2 Water Migration and Heat Transfer in Frozen Ground

1. Water migration

1) Unfrozen water

A large number of experiments show that not all water in the fine grain soil will freeze into ice under temperature below 0 °C. The freezing point of bound water and capillary water in soil will be depressed due to the molecule absorption of soil particle surface. The strong bound water is not still frozen even at –78 °C, the weak bound water is completely frozen at –20 to –30 °C, and the freezing point of capillary water is slightly lower than 0 °C. The unfrozen water content in frozen soil is mainly affected by the temperature, the specific surface of mineral particles, mineral sort, pore volume distribution, solute content of pore water and exchangeable ions, etc.

In addition, the freeze–thaw process and external load have some influence on the unfrozen water content. Studies show that the unfrozen water content in the freezing process is always greater than that of the thawing process, and the unfrozen water content curve in the thawing process has a significant hysteresis. Under the same temperature, unfrozen water content increases with increasing pressure and the main reason is that the pressure influences the soil freezing temperature. Freezing temperature linearly decreases with increasing stress at a decline rate close to 0.075 °C/MPa.

2) Water migration in freezing process

As soil is freezing, when phase equilibrium of soil is destroyed and the external action is changed (such as gradient of temperature, pressure, water content, surface energy of mineral grains, molecular activity in water film), water will migrate to the freezing front. This physical and chemical process is called as water migration of freezing soil.

Because the water in the soil will contain soluble salt, water migration will carry some soluble salt to migrate together, and produce desalting when water phases into ice. Therefore, a high concentration of soluble salt will form on both sides the ice lens. When the water changes into ice, the volume is changed and the mineral grains also displace, resulting in a series of phenomena, such as frost heaving, thaw settlement, salinity heaving and secondary salinization of the ground surface.

2. Thermal properties of frozen soil

1) Thermal conductivity of frozen soil

As a kind of multi-material mixture, the thermal parameters of frozen soil change with temperature, soil sort, water content, saturation and soil density. Since the frozen soil is composed of mineral composition, ice, water and gas, change of the proportion of these components will cause change of the thermal conductivity of frozen soil. Therefore, the thermal conductivity of frozen soil varies with dry density, water content, salt content and adsorption cation composition. The thermal conductivity of both the thawed and frozen soils has a logarithmic or exponential increasing with the increase of the dry density, but it can be approximately regarded as a linear relationship in the measurement range. The thermal conductivity changes with water content for the same soil under two different states of freezing and thawing, which can be divided into three stages: the ratio of thermal conductivities decreases with increasing water content at the first stage, rapidly increase at the second stage, and slowly increase at the third stage.

The thermal conductivity of coarse grained soil is larger than that of fine grain soil under condition of the same dry density and water content. The reason is that the total porosity of coarse grain soil is smaller than that of fine grain soil. The difference of mineral composition and dispersion degree of homogeneous soil can make the mean variance of thermal conductivity up to plus or minus 5–11%. Mean variance of thermal conductivity is up to ±5–11% due to the difference in mineral composition and dispersion degree of the same type soil.

The thermal conductivity of frozen soil increases slightly with temperature. Soil temperature changes by 1.0 °C, the change of thermal conductivity is less than 5%.

2) Specific heat of frozen soil

The heat capacity of frozen soil is described by the volumetric heat capacity, which is defined as the heat amount required by a unit volume of soil to change a temperature unit, namely, the product of specific heat and density.

Frozen soil usually is a multi-phase fine media composed by organic matter, mineral skeleton, aqueous solution and gas. The experiments show that the specific heat of soil is the weighted mean of specific heats of all these material components. However, gas content and specific heat of the gas are so small to be negligible.

The specific heat of soil skeleton mainly depended on mineral composition and organic matter content and is not related to temperature. The specific heat of organic matter is greater than that of mineral. When organic matter content is high, therefore, the specific heat of soil is obviously increased.

The specific heat of water decreases with increasing temperature, but the specific heat of ice increases with temperature rising, and their change rates are very small. In general, frozen soil temperature changes within a small range in almost cases, and then the impact of temperature change on specific heat cannot be considered.

3. Frozen soil temperature

The temperature profiles in frozen soil can be detected by the borehole observation. From temperature measurements at different times at a certain depth interval in a borehole, three forms of temperature curve can be obtained: ① the temperature varies with depth at different times; ② the temperature changes with the time at a certain depth; ③ the temperature contour map. The temperature gradient caused by the heat flow from the earth's interior to the ground surface is called as the geothermal gradient. The inverse of the geothermal gradient is called as the geothermal rate, namely, the vertical distance of the temperature change by 1 °C. The temperature below the ground surface changes seasonally and its variation amplitude decreases with increasing depth. At a certain depth, the temperature change within a year is no more than ±0.1 °C, and this depth is called as annual variation depth of ground temperature, and the ground temperature at this depth is called as the mean annual ground temperature.

The phase change of water in a frozen soil layer has an important influence on the temperature distribution in the soil because even a small phase change can produce a great latent heat. In addition, the phase change also causes the change of thermal parameters. It can be said that the complexity of the frozen soil temperature field is caused by the phase change of water within the soil.

The seasonal and multiyear freezing (thawing) depth of a soil layer can be calculated by the mode of active layer thickness of frozen soil (see Chap. 10). Because soil parameters change with the temperature, as well as the moving of phase change front, the calculation of the temperature field of frozen soil is a nonlinear problem, and its analytical solution is difficult to be obtained, and so the numerical solution is generally obtained by using numerical methods.

4.3.3 *Thermal Properties of Snow Cover*

1. Thermal parameters

Snow density has a large range from lower than 100 kg/m^3 in fresh snow up to higher than 600 kg/m^3 in firn on a glacier, but mostly is between 200 and 500 kg/m^3 for old snow. Therefore, it is important to investigate the relation of thermal properties with density of snow. If taking the specific heat and the melt latent of snow as values of unit mass, they are independent of density, and thus it is emphasis to investigate the relation of thermal conductivity with density of snow.

Since existing air between snow grains, heat convection and radiation and water vapor diffusion caused by sublimation and condensation have some effects on the measured thermal conductivity. Therefore, the value derived from experimental measurements is called the effective thermal conductivity. Numerous investigations show that the effective thermal conductivity of snow increases with increasing density although the scattering of data exists (Fig. 4.9), which is believed due to the variation of sample conditions such as aging and grain size distribution and the different techniques employed. Summarizing many research results (Yen et al. 1991, 1992) suggested the relation as follows:

$$\lambda_{se} = 2.224\rho_s^{1.885} \tag{4.11}$$

where λ_{se} is the effective thermal conductivity of snow and ρ_s is snow density in mg/m^3. If temperature changes less, its influence can be ignored, but when temperature has a large range, temperature effect should be considered. Yen et al. (1991, 1992) developed an expression to correlate the effect of temperature as well as density on λ_{se} by:

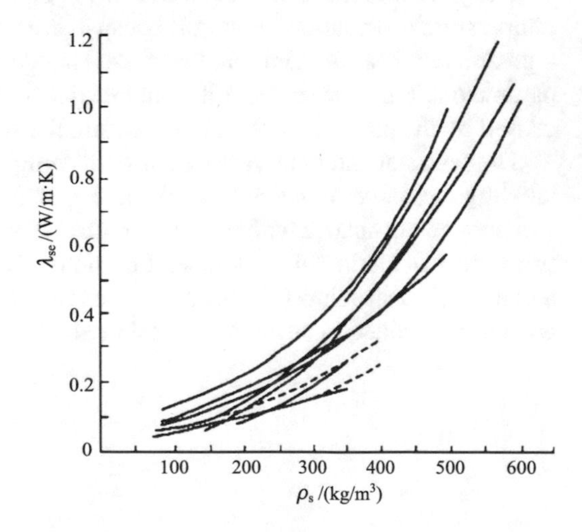

Fig. 4.9 Effective thermal conductivity of snow (λ_{se}) versus density (ρ_s) revealed by various experiments (from Yen et al. 1991, 1992)

$$\lambda_{se} = 0.0688 \exp(0.0088T + 4.6682\rho_s) \quad (4.12)$$

Here T is temperature in Kelvin. According to these formulae, the effective thermal conductivity of snow is about 0.05 W/(m K) at density of 100 kg/m^3, similar to that of glass filament insulator, 0.13 W/(m K) at density of 300 kg/m^3 and 0.44 W/(m K) at density of 500 kg/m^3, similar to that of a brick.

2. Temperature in snow cover

Since climatic and topographic conditions vary on both regional and local scales, snow cover duration and thickness change place to place. Temperature in snow cover are not only various in different sites but changes with time at a same site due to changes in thickness, surface temperature, snow structure and so on.

Temperature in snow cover is mainly influenced by temperatures at the snow cover surface and on the ground surface beneath. If surface temperature is below 0 °C and changes in the harmonic which can be described by a sine or cosine function, the propagation velocity of surface temperature waves is slightly less than 0.5 m per day when the heat conductivity Eq. (4.6) is used and 0.13 W/(m K) of heat conductivity of snow is taken at density of 300 kg/m^3 from the previous statement about thermal properties of snow. In general, the amplitude of temperature wave attenuates with depth rapidly in snow cover. At 0.1 m depth, it is about one third of that at surface, and it is about 11% and 1% at depths of 0.2 m and 0.4 m, respectively. Plus other heat transfers, such as radiation and convection, the depth at which temperature amplitude is about 1% of that at surface is about 0.5 m. When snow cover is thicker than 0.5 m, therefore, temperature at depths less than 0.5 m changes with surface temperature, and temperature gradient in the deeper snow layer is dependent on difference between daily mean temperatures at surfaces of snow cover and ground.

When snow cover surface is melting, the meltwater percolation and refreezing latent heat can raise temperature in snow cover significantly. Especially in the case of that snow cover is wet, temperature is basically at 0 °C, which is often seen in the humid climate regions and in the period of snow cover melting.

4.3.4 Thermal Properties of Sea Ice, River Ice and Lake Ice

1. Thermal parameters

The thermal parameters of river ice and freshwater lake ice are close to those of pure ice. Salt water lake ice is similar to sea ice, and its thermal parameters can be referred to sea ice. Sea ice contains salt and bubbles. Part of the salt is stored in solid form and the other is dissolved in water. Therefore, sea ice is a heterogeneous mixture of solid ice and salt, brine and gas. Its thermal parameters are quite different from those of pure ice.

There is no fixed freezing point for sea water, and the sea water with salinity of 3.25% begins to freeze from –1.5 °C, but it is not completely frozen even it is cold to –53.8 °C. The salinity of sea ice is lower than that of sea water, and the salinity of most sea ice is 0.3–0.5%. The age of ice is more than one year, and the salinity is usually only 0.1%. However, this small change can also cause obvious changes on the physical properties of sea ice.

The melting latent heat of sea ice is smaller than that of pure ice, and varies with temperature and salinity. If the temperature and salinity are expressed by T (°C) and S (‰), the latent heat of sea ice (L_{si}) in the range of 0 to –8 °C is expressed as (Yen et al. 1991, 1992):

$$L_{si} = 4.187\left(79.68 - 0.505T - 0.0273S + 4.3115\frac{S}{T} + 0.0008ST - 0.009T^2\right) \tag{4.13}$$

The specific heat capacity of sea ice is considered as the sum of latent heat of ice, brine, condensed salt and the latent heat of phase change caused by the change of temperature. Therefore, the relationship among the specific heat of sea ice, temperature, and salinity is very complicated. Different researchers have different formulas according to the experiment results at different temperature intervals.

There are often air bubbles in sea ice, and the thermal conductivity of air is obviously lower than that of ice. Therefore, the thermal conductivity of sea ice is not only influenced by temperature and salinity, but also related to the air content (which can be reflected by porosity or density), as it shown in Fig. 4.10. According to experiment results, the relationships between thermal conductivity and temperature and between thermal conductivity and salinity can be fitted respectively. For the influence of temperature, some experiments gave (Yen et al. 1991, 1992):

$$\lambda_{si} = 1.16\left(1.94–9.07 \times 10^{-2}T + 3.37 \times 10^{-5}T^2\right) \tag{4.14}$$

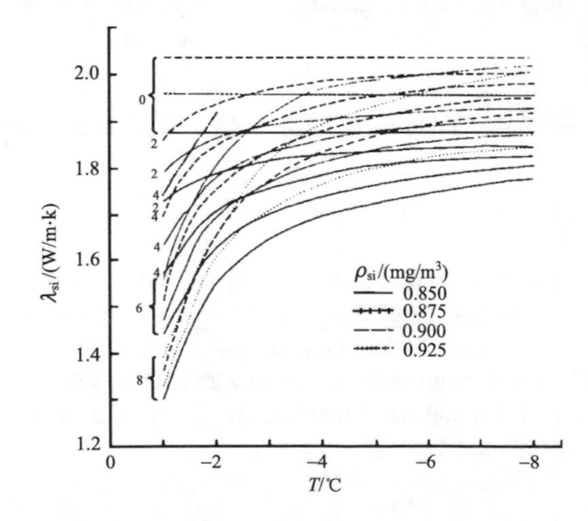

Fig. 4.10 Effective thermal conductivity of sea ice as a function of temperature for various salinities and densities (from Yen et al. 1991, 1992)

where λ_{si} is the thermal conductivity and T is temperature in °C.

There are also simple fitting formulas for the common effects of temperature and salinity, such as (Untersteiner 1961):

$$\lambda_{si} = \lambda_i + 0.13S/T \tag{4.15}$$

where, λ_i is the thermal conductivity of the pure ice.

2. Temperature

A feature common to sea ice, river ice and lake ice is that their surfaces are mainly controlled by atmospheric condition after they formed. Although all sea ice, river ice and lake ice experience interaction with water beneath them, situation is simpler at base of lake ice. Basically, lake ice contacts intimately with water and so its surface temperature changes with air temperature and base temperature is fixed to the freezing point of lake water. The propagation and the attenuation of surface temperature waves with depth are determined by ice thermal properties and the fluctuation period (or frequency) and amplitude of surface temperature. If taking thermal parameters of pure ice, the propagation velocity of daily temperature wave is about 1.08 m per day and the amplitude decreases to about 10% at 0.4 m depth and to about 0.34% at 1.0 m depth. This means that, if daily temperature change amplitude is 20 °C, the 1 m depth temperature changes within 0.7 °C correspondingly.

River ice is more complex due to water flow beneath and thus thermal physics of it must be involved with river hydraulics. Although the thermal conductivity and thermal diffusivity of all kinds of ice are larger than those of water, the ice on the surface of water prevents the propagation of the cold wave from air into the water, and also prevents the evaporation of the water surface and reduces the energy exchange between the water and the atmosphere.

4.4 Other Properties of Main Cryospheric Components

4.4.1 Albedo

1. Albedo of snow cover

The cryosphere occupies a large area in earth surface, thus the estimation of its albedo is extremely important to surface energy balance. Among all elements of the cryosphere, snow has the largest extent and largest spatial–temporal variation in albedo. Factors influencing the snow albedo can be concluded into internal characters and external conditions.

If simply divides snow into fresh snow, clean and compact dry snow, coarse aged snow, wet snow and polluted snow, their corresponding albedo can be approximately 70–90% or even higher, 80–90%, 50–70%, 30–50%, 20–30% or lower, respectively.

Impurities, especially impurities on snow surface, have a fundamental influence on albedo, because most of these impurities has a high optical absorption comparing with snow, e.g., dust, black carbon, organic matter, etc.

External factors influencing snow albedo are generally sun's zenith angle, topography (shade), weather conditions (e.g., clouds, aerosols), etc.

At present, it is common to monitor albedo using remote sensing technique. However, satellite sensors typically provide narrow-band data, calculations are needed to convert it into broad-band data. Also, data from satellite sensors is the combination of surface radiation and atmospheric radiation from a certain direction. Therefore, the correction for atmospheric effects and the effects of anisotropic reflectance are also needed before band conversion. Certainly, the resultant broad-band albedo must be validated against enough ground observations before its applications.

2. Albedo of sea ice

Similar to snow, sea ice is widely distributed and also has a very large variability. The albedo of sea ice varies greatly due to the difference in surface conditions, as shown in Fig. 4.11. In general, albedo of thick ice covered by snow (multiyear ice) is the highest, while + albedo of the melting ice is the lowest. Table 4.1 lists the total differences of albedo between different types of sea ice on clear and full cloudy days.

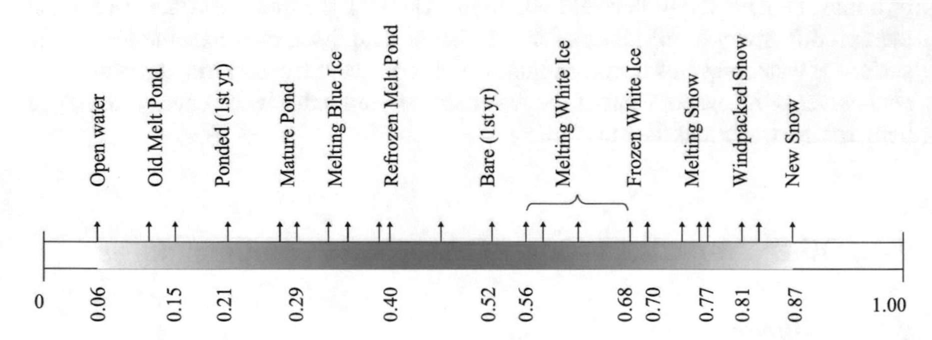

Fig. 4.11 Variations in total albedo observed for various sea ice surface features (Grenfell and Maykut 1977)

Table 4.1 Total albedo values in sunshine and cloudy weathers for various sea ice types

Ice type	Clear	Overcast
Melting old snow	0.63	0.77
Melting white ice	0.56	0.70
Melting blue ice (first year)	0.25	0.32
Mature melt pond (mid-summer)	0.22	0.29

Grenfell and Maykut (1977)

3. Albedo of glaciers

Comparing with snow cover and sea ice, the area coverage of glaciers in a drainage basin or region is relative small except Antarctic and Arctic regions. On the regional scale, however, glacier meltwater has important effects on water resource. For a single glacier, albedo variation in temporal and spatial is one of key factors influencing glacier melting.

If the glacier is covered completely by snow from terminus to top, the spatial variation of albedo is relative less. In fact, however, the glacier surface characteristics are different in the various elevations or different positions at the same elevation at most time. The differences in surface characteristics include snow coverage, snow type and water content, impurities and topographic factor, etc.

Under the background of climate warming, the albedo variation plays a vital impact on glacier melting. After the increasing of air temperature, glacier melting enhances, the ablation area expands, and average surface albedo decreases. The impurity enrichment also decreases the glacier albedo. The decreasing albedo in turn enhances glacier melting and results in the intensified mass loss. Therefore, the feedback between glacier melting and albedo is one of the main causes leading to glacier accelerated retreat.

4. Albedo of frozen soils

Although the change of frozen soils albedo is not as large as the glacier in a local space, the surface conditions have a large difference due to the huge area of the frozen soil regions. The surface condition of frozen soils regions depends on vegetation cover and vegetation types. Under the conditions of bare soil or very sparse vegetation, the factors affecting the surface albedo are mainly soil types and surface soil moisture. There are many factors influencing vegetation surface albedo and its mechanism is very complex. Vegetation form (available to characterize roughness) and physiological functions (such as leaf area size) is very important except soil moisture.

Snow cover is one of the most important factors caused the variation of surface albedo in frozen soils regions. Due to the extreme high albedo of snow cover and a large temporal and spatial change, the albedo change in the frozen soils region is very complicated. Therefore, the frozen soils albedo is closely associated with the change of snow cover and its albedo.

4.4.2 Electrical Properties

1. Electrical properties of glaciers and snow cover

Based on the difference in dielectric property between ice and other materials, and between ice with different structures, the ice penetrating radar equipment developed according to the principle of radio echo-sounding has been widely applied in glaciers and ice sheet. It can be used to measure ice thickness and bedrock topography and to identify the bottom debris layer, englacier and subglacial streams, temperate ice layer, ice structure saltation layer and englacier impurity enrichment belt, etc.

According to the conductivity property difference between ice and other materials and the influence of ice structure on conductivity, electrical conductivity measurement (ECM) of an ice core in the field can not only distinguish vertical variation in material composition, but also realize the ice physical characteristics. The liquid conductivity measurement in laboratory can reveal the environmental factors causing the variation of conductivity.

The electromagnetic property of snow is influenced by density, particle diameter, moisture, impurity content, temperature, etc. Therefore, these parameters can be estimated by measuring the electromagnetic wave absorption, reflection and penetration of snow and its emissivity.

2. Electrical properties of permafrost

Resistivity of frozen soils is the main index in electronic properties of frozen soils, which is much larger than that of thawing soil. The resistivity of thaw soil depends on the mineral composition, specific surface area and shape, porosity, water content and mineralization degree of pore water, and so on. But, the resistivity of frozen soils also depends on ice content and structure of frozen soils except the mentioned factors. During the process of water freezing in soils, soil expansion caused by the water crystallizes in soils change the spatial structure of the pore and the spatial distribution of soil particles. The ice crystals in soils reduce the amount of water in the pores, increase the mineralization degree and ice content, and thus change the resistivity of the frozen soils. The main factors influencing the resistivity of frozen soils include temperature, structure and water content of frozen soils and the mineralization degree of pore water, etc.

Generally, soil temperature has a large influence, and the resistivity of frozen soils increases with temperature decreasing.

Due to the low conductivity of ice, the distribution of ice in the frozen soils has an important influence on resistivity.

The mineralization of pore water results in the increase of mineralization lead the increasing of the number of conductive cations, and decreasing of the resistivity of frozen soils.

According to the difference of the electrochemical properties between frozen soil and thaw soil, as well as between ice and other substances, techniques of geophysical survey can be used to acquire the important information such as the permafrost thickness, ground ice and the structure of frozen soils.

3. The electromagnetics properties of sea ice, river ice and lake ice

Sea ice is a mixture of pure ice and other impurities. The ice and these impurities have different dielectric properties. To accurately determine the dielectric constant of certain type of sea ice, we must study the dielectric properties of salt containing substances and other impurities and bubbles respectively. Relatively speaking, the effects of brine and bubbles in sea ice are most important. For a year or younger sea ice, the effect of brine is more prominent; for multiyear sea ice, the effect of bubbles is more important because of the decrease of salinity. Therefore, we can use the pure ice—air model and the pure ice—brine model to describe the influence of the two parameters on the dielectric properties of the sea ice. The sea ice dielectric mixture model is the superposition of the two models.

Usually, sea ice structure (c axis orientation, grain size, distribution of bubbles and salts, surface characteristics, etc.) and thickness may vary greatly with location. Therefore, the heterogeneity of sea ice structure and thickness must be taken into account in the application of remote sensing radar technology to the detection of large range of sea ice characteristics.

The electromagnetics of river and lake ice are relatively simple compared to sea ice, which can be directly referred to related research results on pure ice or relatively homogeneous salt water ice.

Extended Readings

Representatives

Nye J. F. and Glen J. W.

They both are British physicists and glaciologists with great achievements in research on mechanical properties of ice and glacier dynamics. During a long time before 1940s, the mechanism of ice flow and glacier movement had been in various hypotheses, although X-ray crystallography was applied to determine ice crystalline structure in early 1900s. In late 1940s, J. F. Nye firstly explained ice flow mechanism by assuming that ice was a perfectly plastic solid and described the glacier flow mathematically. Soon after, deformation experiments of J. W. Glen revealed that ice was neither a linear viscous nor a perfectly plastic material and his work triggered a campaign of experimental studies on ice flow law. Late on, Nye summarized these experimental results and applied the ice flow law to establish the glacier flow theory. Up to now, Glen's Law and Nye's theory are still regarded basic to describe mechanism of ice flow and glacier movement.

Classic Works

1. *The Physics of Glaciers (4th Edition)*

Authors: Cuffey K. M., Paterson W. S. B.

Publisher: Elsevier, Amsterdam, etc., 2010

The physics of Glaciers authored by W. S. B. Paterson (1924–2013), a British and Canadian glaciologist, explains systematically the physical principles understanding the behavior of glaciers and ice sheets and also introduces other relevant contents such as glacial mass balance, hydrology, climatology and ice core study. Although covering a wide range of topics, this book puts emphasis on introduction of basic concepts and principles so that most of its contents are intelligible to senior undergraduate students. Since it appeared at the first edition in 1969, revised versions was given in 1981, 1994 and 2010 successively. Every reversion has much updates and expansion relative to the previous edition. The fourth edition published in 2010 is co-authored by K. M. Cuffey and W. S. B. Paterson and adds some new chapters such as ***Ice Sheets and the Earth System*** and ***Ice, Sea Level and Contemporary Climate Change***. It also concerns the major advances in most aspects.

2. ***Ice Physics***

Author: Hobbs V. P.

Publisher: Clarendon Press, Oxford, 1974

The book titled *Ice physics* is authored by Peter V. Hobbs (1936–2005), a British-born American Scientist of atmospheric physics. This classic monograph provided the first comprehensive account of the physics and chemistry of ice. This book places emphasis on the basic physical properties of ice (electrical, optical, mechanical, and thermal), the modes of nucleation and growth of ice, and the interpretation of these phenomena in terms of molecular structure. Applied aspects of ice physics are also discussed. It was first published in 1974 and reprinted in 2010.

Chapter 5
Chemical Characteristics of the Cryosphere

Lead Authors: Shichang Kang, Junying Sun

Contributing Authors: Tingjun Zhang, Qingbai Wu, Cunde Xiao, Zhijun Li

Cryospheric chemistry is one of the most important research aspects of Cryospheric Science and has important impacts on the climatic and environmental changes. Studies of the chemical characteristics of the cryosphere have mainly focused on the spatial–temporal distribution, sources and processes of chemical components in the continental cryosphere (glaciers, ice sheets, snow cover, lake ice, and river ice), marine cryosphere (sea ice, submarine permafrost, and ice shelves), and aerial cryosphere along with their climate and environmental effects. Atmospheric chemical species emitted by anthropogenic activities and natural processes can be transported globally by atmospheric circulation and deposited into the cryosphere through wet and dry deposition. Furthermore, after physical, chemical and biological transportation and transformation, chemical components can be archived in the cryosphere (e.g., glaciers and ice sheets) and act as indices of paleo-climatic and environmental changes, while some components are released into the atmosphere and hydrosphere. The transportation and transformation processes of chemical components in the cryosphere have great impacts on the climate and environment.

This chapter introduces glacio-chemistry, permafrost chemistry, and chemistry of river (lake) ice and sea ice. Our understanding of the chemistry and processes related to different cryospheric components is variable. Glacio-chemistry is one of the most extensively studied and well understood science. It is important in the research of paleo-global changes. However, due to seasonality and mobility in environments of permafrost, sea, river and lake, others are less understood than glaciochemistry.

5.1 Sources of Chemical Components in the Cryosphere

Water, which continuously circulate globally, is the most active species in the cryosphere. Precipitation, evaporation, freezing, melting, thawing and condensation are all components of the hydrological cycle. Surface water evaporates into the atmosphere, condenses into clouds, rain, snow, or hail and precipitates back to land and the ocean. Water that flows over the ground is referred to as surface runoff, while some proportion of runoff seeps into the ground to form a water layer or stream

D. Qin et al. (eds.), *Introduction to Cryospheric Science*, Springer Geography,
https://doi.org/10.1007/978-981-16-6425-0_5

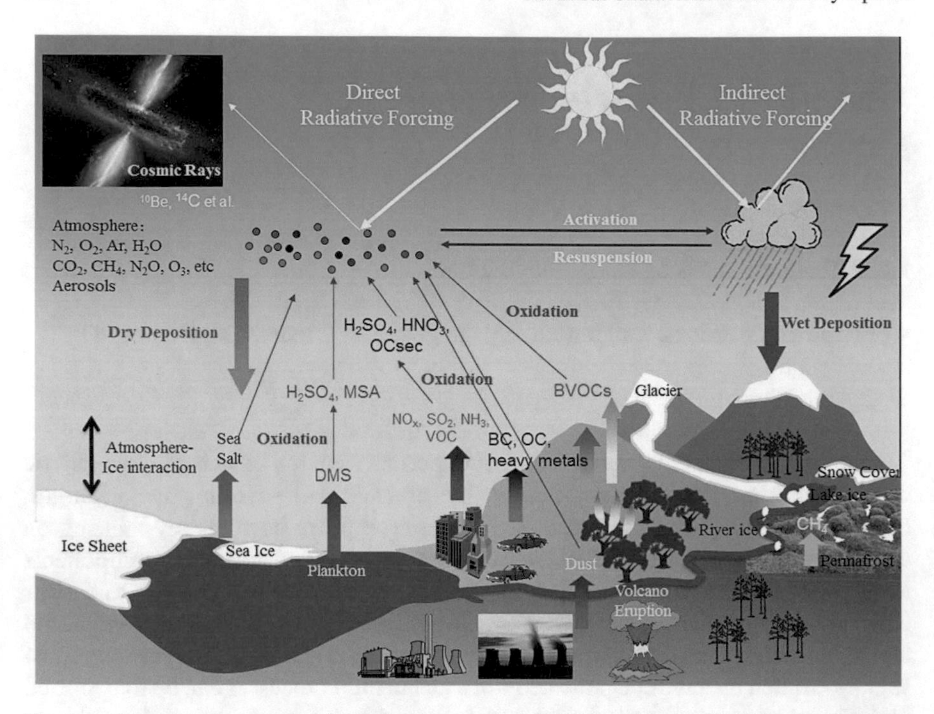

Fig. 5.1 Sources and processes of chemical components in the cryosphere. DMS (dimethyl sulfide); MSA (methanesulfonic acid); SOC (secondary organic carbon); OC (organic carbon); BC (black carbon); BVOCs (biogenic volatile organic compounds); VOCs (volatile organic compounds). By courtesy of R. J. Delmas, Helence Cachier and modified

(also known as subsurface water flow). Surface runoff and subsurface water flow sometimes transform into each other, and overtime, the water returns to the sea or inland lakes. Along with the water cycle, biogeochemical cycles of various chemical compositions occur (Fig. 5.1).

Glaciers and snow cover are the products of precipitation. The chemical composition of precipitation mainly originates from the dissolution and washout of atmospheric aerosols. The chemical composition of precipitation in different regions exhibits substantial regional differences and seasonal variations and primarily contains chemical ions (e.g., HCO_3^-, SO_4^{2-}, Cl^-, NO_3^-, Na^+, Ca^{2+}, Mg^{2+}, K^+, and NH_4^+), inorganic elements, organic components, etc. These chemical components mainly originate from natural emissions from physical, chemical and biological processes, such as volcanic activity, dust storms, ocean waves, lightning, land and sea animals and plants, cosmic dust, etc. and anthropogenic emissions from industrial and agricultural activities.

The main source of water in rivers is derived from precipitation, which has a low salt content. The chemical composition of river water is related to the geological and climatic conditions of the basin and thus has substantial diversity and variability. River water is not only the primary water source for human societies and industries but also the main pathway for human sewage; therefore, river chemistry is most affected by anthropogenic activities.

The chemical composition of lake water is mainly affected by the quantity and quality of runoff, solar radiation, evaporation intensity, etc. The composition is also related to the lake parameters, such as the area and depth of the lake. When the water flow amount in and out of the lake is large and the water evaporation amount is relatively small, the salt content in the lake is relatively low, creating a freshwater lake. However, when the lake is closed and evaporation is strong, dissolved salt accumulates to form a salty lake.

The ocean is the largest ecological system on the earth and contains a complex system of physical, chemical, biological, geological and other processes, which make the chemical composition of marine water significantly different from that of terrestrial water. The high salt content is a major feature of seawater. The average ocean salinity is approximately 35‰, with slight differences in different regions. The main soluble chemical components of seawater (Cl^-, Na^+, SO_4^{2-}, Mg^{2+}, Ca^{2+}, K^+, HCO_3^-, Br^-, Sr^{2+}, F^- etc.) account for 99.8–99.9% of the dissolved salts, with Cl^- and Na^+ alone accounting for more than 80% of the dissolved salts. The concentration of these soluble chemical components is usually stable, except for that of HCO_3^- and Ca^{2+}.

The chemical composition of frozen ground (permafrost) is mainly affected by soil properties and physicochemical processes associated with freeze–thaw and biological processes, and the chemical composition can be divided into organic and inorganic compounds. The organic compounds include soluble amino acids, humic acids, carbohydrates and complexes of organic-metal ions. The inorganic compounds include Ca^{2+}, Mg^{2+}, Na^+, K^+, Cl^-, SO_4^{2-}, CO_3^{2-}, NO_3^-, NH_4^+, $H_2PO_4^-$ and small amounts of iron, manganese, copper, zinc and other salt compounds, as well as a variety of gases in soil pores.

5.1.1 Main Processes of Atmospheric Chemical Constituents into the Cryosphere

There are two main processes, namely, dry and wet deposition, through which atmospheric chemical constituents deposits into the cryosphere. Dry deposition refers to the transport of atmospheric chemical constituents to the surface of the cryosphere in the absence of precipitation, whereas wet deposition refers to the process by which the chemical components are mixed and deposited with precipitation.

Dry deposition of gases and particles is generally represented by three steps: ① transport down through the atmospheric surface layer to a very thin layer of stagnant air adjacent to the surface induced by turbulent diffusion; ② molecular (for gases) or Brownian (for particles) transport across this thin stagnant layer of air, which is called the quasi-laminar sub-layer, to the surface itself, where the transport mainly occurs by diffusion and sedimentation; and ③ uptake at the surface, which mainly occurs by absorption, solubility, adherence, etc. The transport rate of different chemical constituents varies in each of these steps. For instance, dry deposition in the Greenland ice sheet is essentially controlled by turbulent diffusion (step 1). The dry deposition rate of SO_2 is larger than that of surface, while it is several times lower than that of HNO_3.

Wet deposition refers to the natural processes by which atmospheric chemical constituents are scavenged by atmospheric hydrometeors (cloud and fog drops, rain, and snow) and consequently deposited on the surface of the earth, and these processes mainly include nucleation, in-cloud scavenging and below-cloud scavenging.

Solid precipitation is predominant in the polar regions and high mountain areas at middle and low latitudes. Substantial differences in the scavenging of chemical constituents do not occur during the formation of raindrops and snowflakes but do occur during precipitation. As a raindrop falls through air, it collides with and collects aerosol particles and undergoes water evaporation and the absorption and reemission of trace gases. Scavenging by snowflakes is very weak due to low temperatures. Therefore, below-cloud scavenging is not important in the polar regions. Knowledge on the properties of gases and particles within and below clouds, such as the gas and particle concentrations, cloud condensation nuclei characteristics, ice crystal size distribution, and frost properties, is needed to fully understand and quantify wet deposition.

Wet and dry depositions are the ultimate paths by which trace gases and particles are removed from the atmosphere. The characteristics of wet and dry deposition for different chemical compositions and regions is quite diverse. The relative importance of dry deposition compared to wet deposition depends on the particular species, the species solubility in water, the amount of precipitation in the region, the surface cover properties, etc. The distribution of precipitation over time is also an influencing factor for deposition. For the same annual precipitation, if the precipitation is concentrated within a short period, the relative importance of wet deposition may decrease.

5.1.2 Effects of Cryospheric Chemistry on the Climate and Environment

Cryospheric chemistry plays an important role in climate systems at different time scales (day, season, interannual, decadal, and centurial), which are mainly achieved by affecting water cycle on the Earth's surface, through the processes such as the radiation balance (snow-albedo and ice-albedo feedback mechanism) and the exchange of chemical constituents in the cryosphere with the other spheres. Glaciers (ice sheets), snow cover, river ice, lake ice and sea ice have high albedo, and their spatiotemporal variations significantly affect the global energy balance and water cycle process and change the climate dynamic process at a regional or global scale, thus affecting climate changes.

The albedo of each component in the cryosphere is affected by its chemical composition. When the concentration of light-absorbing impurities (such as black carbon, and dust) on the surface increases, the albedo of snow and ice significantly decreases, which speeds up snow and ice melting and thus changes the energy budget and water cycle. In addition, the ion elution of glaciers and snow, especially during the early stages of melting, results in ion pulsations in the melting water, which could affect the river water chemistry. During the freezing of sea ice, some salt exists in the form of brine, which significantly increases the albedo of sea ice.

If the salinity of sea ice is low, increasing salinity in the underlying seawater will drive thermohaline circulation in the ocean. Variations in permafrost not only affect the climate system by changing water-exchange and heat-exchange processes of the Earth-atmosphere but also influence the global carbon cycle and climate change by the transformation of natural gas hydrates and as carbon sources or sinks. Because of seasonal freezing and thawing of permafrost, soluble salts in frozen ground accumulate at the surface, resulting in permafrost salinization. In short, the chemistry of the cryosphere has an important influence on the climate and environment. Understanding the chemical characteristics of the cryosphere is an important component of global change research.

5.2 Glacio-chemical Characteristics

Glacio-chemistry is a branch of cryospheric science that investigates the chemical compositions of glacier and ice caps and their environmental significance via physical, chemical and biological transportation and transformation of these constituents. Snow and ice, which are unique environmental media, are regarded as ideal natural archives of atmospheric compositions because these media are direct receptors and excellent reservoir for atmospheric dry and wet deposition. Studies on glacio-chemistry of snow/ice provides a unique opportunity to monitor current global environmental processes and reconstruct paleo-climatic and environmental changes. Snow and ice chemistry records offer direct or indirect information for many research fields of global change, such as climate change, biogeochemical cycling, human activities, and geological and cosmic events. Glacio-chemistry is interdisciplinary and an important component of cryospheric science. Snow and ice samples retrieved from Antarctica, Greenland and glaciers at low and middle latitudes are very informative and highly reliable and consequently can accurately reflect climate and environmental processes. Therefore, it is important to improve our understanding of modern glacio-chemistry processes (e.g., seasonal variations and geographical distribution patterns of chemical compositions in snow and ice) to elucidate the role of the cryosphere in the global biogeochemical cycle and potential environmental effects on future cryospheric changes.

A wide variety of chemical components exist in glaciers, and the environmental indicators in snow and ice vary for different components. Since 1960s, research on glacio-chemistry of snow and ice in polar and low-latitude and mid-latitude alpine regions has been advancing. First, the relationship between hydrogen and oxygen stable isotopic ratios in snow and ice and the ambient temperature was established to evaluate moisture sources and reconstruct paleo-climatic changes. As the field of glacio-chemistry advanced, further research enhanced our understanding of cryospheric science. For example, the particulate matter concentrations in snow and ice could be used to investigate environmental changes due to historical volcanic eruptions and modern atmospheric dust. In addition, radioactive elements in snow and ice could be used to monitor anthropogenic pollution (e.g., nuclear tests). Currently, great progress has been made in some aspects of glacio-chemistry, such as major ions, low-molecular-weight organic acids and trace heavy metals in snow and ice. Recently, many studies on organic carbon, black carbon and persistent organic pollutants have been conducted. With rapid improvements in analytical and laboratory testing technology, our current understanding of glacio-chemistry will continue to advance in the future.

5.2.1 *Inorganic Components*

1. Conductivity and pH

Conductivity is a comprehensive index for the total ion content in snow and ice and is a sensitive indicator of atmospheric environmental changes in the continental and marine cryosphere, thus indicating general atmospheric processes. Variations in conductivity mainly reflect changes in the chemical characteristics and concentration levels of chemical components in snow and ice. Investigating the relationship between major ions and conductivity is very important for understanding the factors that affect conductivity variations in snow and ice. For instance, the ocean is the main source of chemical components in the Antarctic ice sheet. Therefore, the conductivity is positively correlated with SO_4^{2-}, NO_3^-, and Cl^- but negatively correlated with crustal ions, which indicates that sea-salt ions are the primary factors that affect the chemical characteristics of snow and ice in the Antarctic ice sheet. Many studies on the conductivity of snow and ice in Antarctic and Greenland ice sheets have revealed that conductivity has a positive linear relationship with pH values, which could be used to reconstruct historical volcanic eruption events. The linear relationship between conductivity and pH values suggests that acidic ions (e.g., Cl^- and SO_4^{2-}) are the primary factors that affect the glacio-chemistry in the polar regions.

Terrestrial alkaline aerosols largely contribute to the chemical components in snow and ice within glaciers on the Tibetan Plateau. Different from the patterns observed in the polar regions, alkaline ions are the primary factors that influence conductivity variations in snow and ice. This phenomenon is supported by the linear relationship between conductivity and most alkaline cations (e.g., Ca^{2+} and Mg^{2+}), suggesting that alkaline mineral salts (e.g., Ca^{2+} and Mg^{2+}) dominate the snow and ice chemistry over the Tibetan Plateau.

2. Major ions

Major anions (Cl^-, SO_4^{2-}, and NO_3^-) and cations (Ca^{2+}, Mg^{2+}, K^+, Na^+, and NH_4^+) are the main components of soluble chemicals in snow and ice. In terms of the spatial distribution patterns of these components in the polar regions, the snow and ice in the central part of the Antarctic ice sheet have the lowest concentrations of major ions because this region is located at the endpoint of air masses along the western Antarctic transport channel, and also the farthest reach for the long-range transport of air pollutants and matter from terrestrial ecosystems. Thus, chemical components in snow and ice in central Antarctica can preserve background information on the global atmospheric environmental conditions in the tropopause and lower stratosphere. However, the source regions and transport processes of atmospheric aerosols to the Arctic vary greatly because of complex land and sea distributions in this area. Therefore, the snow and ice in the Arctic has greater glacio-chemical heterogeneity than the Antarctic ice sheet. For example, the snow and ice in the northern parts of Greenland and Canada are slightly contaminated, while the central part of the Arctic has been suffering from Arctic haze because of the long-range transport of polluted air masses, resulting in much higher major ion concentrations in snow and ice in the central Arctic than in the surrounding areas. Consequently, the chemical components of major ions in the snow and ice of the central Arctic represent the atmospheric environmental conditions in the lower Arctic troposphere, while the major ions in the snow and ice of the Greenland ice sheet could represent background information of the atmospheric environment in the central Arctic troposphere.

The general spatial distribution patterns of the major ion concentrations in the snow and ice of the northern Tibetan Plateau are far higher than those of the southern Tibetan Plateau (e.g., the Himalayas). This phenomenon could be attributed to the very frequent dust events during winter and spring, which can largely contribute mineral dust to the glaciers in the central and northern Tibetan Plateau and northwestern China. Meanwhile, the major ion concentrations in the northern Tibetan Plateau glacier are generally the highest among all the remote glaciers in the world. Moreover, the major ions (e.g., Ca^{2+}, Mg^{2+} and SO_4^{2-}) in the snow and ice of the central and northern Tibetan Plateau mainly originate from terrestrial sources, indicating that transported Asian dust has very important effects on the atmospheric environment of the Tibetan Plateau. However, the atmospheric environment of the southern Tibetan Plateau is influenced by both natural (i.e., terrestrial and marine) and anthropogenic sources, and thus, the major ion concentrations in the snow and ice of the southern Tibetan Plateau are comparable to those of the Arctic, suggesting a difference in the spatial distribution patterns of the atmospheric environments over the Tibetan Plateau.

Most of the major ions in the polar snow and ice show seasonal fluctuations, especially for ions such as Na^+, Cl^- and Ca^{2+}. For example, Na^+ and Cl^-, which are excellent tracers for sea-salt aerosols, show clear seasonal fluctuations in the South Pole and Greenland Summit, and the concentration of Na^+ in snow and ice in winter is 5–10 times higher than that in summer because of the frequent intrusion of winter marine air masses in the polar regions.

Unlike Na^+ and Cl^-, no clear significant seasonal variations were observed for Ca^{2+} in the South Pole; however, Ca^{2+} has the highest concentration in the snow and ice of Greenland during the spring. The differences in the variations in Ca^{2+} in the snow and ice of the Arctic and Antarctic ice sheets could be attributed to their unique and different source regions. Furthermore, Ca^{2+} in the snow and ice of the Greenland ice sheets mainly originates from terrestrial (i.e., crustal) sources, and the peaks in the spring are caused by frequent dust events over the Northern Hemisphere. In contrast, the Antarctic ice sheet is located far from the terrestrial source regions of the Northern Hemisphere with highly concentrated Ca^{2+}, and these crustal ions rapidly diminish after long-range transportation to the South Pole, resulting in low concentrations and insignificant variations in Ca^{2+} in the snow and ice. Except for the ions from marine (Na^+ and Cl^-) and crustal sources (Ca^{2+}), ions such as NO_3^- and SO_4^{2-} in the snow and ice of the Antarctic and Greenland ice sheets also show peaks, but the peak values are not very prominent during spring or summer. Moreover, sea-salt aerosols in the Antarctic and Greenland ice sheets predominantly originated from the surrounding oceans, so the Cl^-/Na^+ ratio in snow and ice is very close to that in standard seawater (1.17), representing inputs from oceanic sources. Therefore, the Cl^-/Na^+ ratio in snow and ice can be used to quantitatively determine the relative contributions from different sources in the polar regions. Generally, if all the Na^+ ions in snow and ice are assumed to originate from the ocean, the relative contributions from sea-salt and non-sea-salt ion sources could be quantitatively estimated according to the ratio of Na^+ in snow and ice to that in standard seawater:

$$\text{nss}A = A - \text{Na}(\text{ss}A/\text{ssNa}) \tag{5.1}$$

where A and ss$A$ are the concentrations of designated ions in snow and ice and in standard seawater, respectively, and Na and ssNa are the Na^+ concentrations in snow and ice and in seawater, respectively.

In the polar regions, non-sea-salt Ca^{2+} (i.e., nss Ca^{2+}) is generally considered an index to reflect atmospheric dust inputs, and non-sea-salt sulfate (nss SO_4^{2-}) is often used as an indicator for volcanic eruptions. For example, nss SO_4^{2-} peaks recorded in ice cores could be used to reconstruct major volcanic eruptions on a global scale over previous centuries.

On the Tibetan Plateau, most of the major ion concentration peaks occurred during the non-monsoon season (i.e., winter and spring), and the lowest concentrations always appeared during the monsoon season because of heavy and intensive precipitation events, which wash out the aerosol particles. A very clear and significant seasonal variation in the ions (Ca^{2+} and SO_4^{2-}) was observed, and the Ca^{2+} concentration in snow from Mt Qomolangma in the Himalayas was an order of magnitude higher during the non-monsoon period than during the monsoon period. This significant seasonal difference in the Ca^{2+} concentrations could be explained by the frequent and high inputs of sandstorms during winter and spring and the scavenging effect of precipitation on the crustal aerosols in the summer monsoon season. The major ion concentration peaks in snow and ice from the Antarctic ice sheet are mainly affected by the inputs of sea-salt aerosols, and those from the Arctic ice sheet are greatly influenced by crustal dust and long-range pollutants (e.g., Arctic haze) during winter and spring. On the Tibetan Plateau, these substantial dust layers in the snow and ice predominantly originated from atmospheric dust deposition during the nonmonsoon season. The general pattern for the glaciochemistry over the Antarctic, Arctic and Tibetan Plateau indicates that the seasonality of chemical components in snow and ice largely varies among the regions. Moreover, because these patterns are largely influenced by the land-sea distribution, atmospheric circulation at a global scale, and human activities, the atmospheric environments of the Antarctic, Arctic and Tibetan Plateau are heterogeneous and characterized by special geographical features. Therefore, the glaciochemistry over the Antarctic, Arctic and Tibetan Plateau is regarded as a good indicator of environmental and climatic change, and seasonal variations in major ions can serve as a powerful tool to date ice cores.

Major ions in snow and ice are not immutable, and complex physical and chemical processes (e.g., migration and transformation) affect the glaciochemical components at the snow and air interface. For instance, atmospheric NO_3^- deposited onto the Antarctic ice sheet can be released back into the atmosphere after a short period, resulting in a rapid reduction in the NO_3^- concentration in snow and ice. This phenomenon is mainly associated with a variety of sources of nitrogen compounds in the atmosphere. This complex process after atmospheric deposition can cause NO_3^- in fresh snow to be released back into the atmosphere because of photochemical decomposition and/or re-evaporation effects. Therefore, numerous postdepositional processes occur after chemical components are deposited onto glaciers from the atmosphere. For example, the percolation and refreezing processes of meltwater can alter the original chemical compositions of snow and ice, which is known as the major ion leaching effect (wash out or elution of ions). For mountain glaciers that have been significantly influenced by the leaching effect, a large amount of major ions could be released into runoff with the loss of snow and ice meltwater. Thus, the original chemical compositions in the snow and ice may be greatly redistributed because of postdepositional processes, such as the leaching effect. Accurately understanding how the leaching effect influences the original records of chemical components is crucial to interpreting paleoenvironmental and paleoclimatic records in ice cores. Thus, under modern environmental conditions, understanding the influences of physical, chemical and biological processes on chemical components from the atmosphere to snow and ice provides a very solid foundation for ice cores dating. In summary, glaciochemistry research greatly improves the accuracy and credibility of ice-core studies and thus enhances the reliability of the conversion between ice-core signals and environmental changes. Therefore, the above factors are very helpful for the reconstruction of past climatic and environmental changes from ice core records.

Ion pulses are defined as major ions in snow and ice and are rapidly released into runoff during glacier ablation. Generally, a small amount of meltwater (less than 10% of the entire snow meltwater equivalent) will release more than 80% of the soluble chemicals into runoff after several hours to a few days, resulting in instantaneous peaks of chemical components in the runoff. Therefore, "ion pulses" in runoff directly indicate the rapid release of major ions during snow melting. For example, a significant "ion pulse" occurred in the snowmelt runoff at the cirque basin of the Urumqi headwater regions. The major ion concentrations were the highest in the snowmelt runoff at the beginning of glacier ablation, higher than the concentrations in the snowmelt during spring, in the rainfall runoff during summer and in the near-surface runoff during early winter.

3. Heavy metal elements

Generally, heavy metals are widespread in nature at very low background levels. However, increasing anthropogenic emissions have resulted in heavy metal pollution at a global scale. Therefore, changes in the concentration levels of heavy metals in polar and mountain glaciers are good indicators to evaluate the impacts of human activities on pristine atmospheric environments.

The concentrations of heavy metals (e.g., Pb, Cd, Zn and Cu) in the snow and ice of the Greenland ice sheet showed distinct seasonal variations, with low values in fall and winter and high values in late winter and early spring. Similarly, these heavy metals exhibited significant seasonality in the snow and ice on Dollema Island in the Antarctic, and the Pb concentrations were the highest in fall and winter and the lowest in summer. The spatial distribution pattern of heavy metals in the Greenland ice sheet showed the highest Pb concentrations in the northern region, while the Cd, Zn and Cu concentrations were higher in the central region than in the southern region. In the Antarctic ice sheet, the concentration levels of heavy metals in snow and ice showed an upward trend along a west–east transect across the Antarctic ice sheet (i.e., from Seal Nunataks station to Mirny station). The concentration levels of Pb in the western part of the transect indicate the background values of Pb in the ambient air, while the high Pb concentrations in the eastern part may be influenced by regional human activities (anthropogenic influences). Moreover, in Queen Maud Land (i.e., Asuka-S16 and S16-Dome Fujii) in Antarctica, the deposition flux of heavy metals (e.g., Cu) in snow and ice significantly decreased with increasing distance from the coast to inland.

The heavy metals in the snow and ice of the Tibetan Plateau are mainly influenced by the inputs of crustal materials and anthropogenic emission sources, and the spatial distributions of heavy metals in snow and ice exhibit significant differences. For instance, the contribution of Pb from anthropogenic sources decreased from 59.3 to 10% with increasing elevation and distance from industrial and agricultural activities to regions without these activities (Fig. 5.2). The contribution from anthropogenic sources to Pb was less than 30% for all most of the sampled glaciers (Yu et al. 2013).

In general, the concentration levels of heavy metals in snow and ice on the Tibetan Plateau are higher than those in the Arctic and Antarctic, and higher heavy metal concentrations in snow and ice were observed during the non-monsoon season than during the monsoon season (Fig. 5.3). Moreover, the spatial distribution pattern of heavy metals in snow and ice on the Tibetan Plateau was greatly influenced by the proximity to the source regions of dust and human activities. For example, the concentrations of total Hg (Hg_T) in glaciers over western China were below 15 pg g^{-1} and were significantly higher than the reported concentrations for Antarctica (Zhang et al. 2012). Therefore, the Hg_T concentrations in glaciers on the Tibetan Plateau

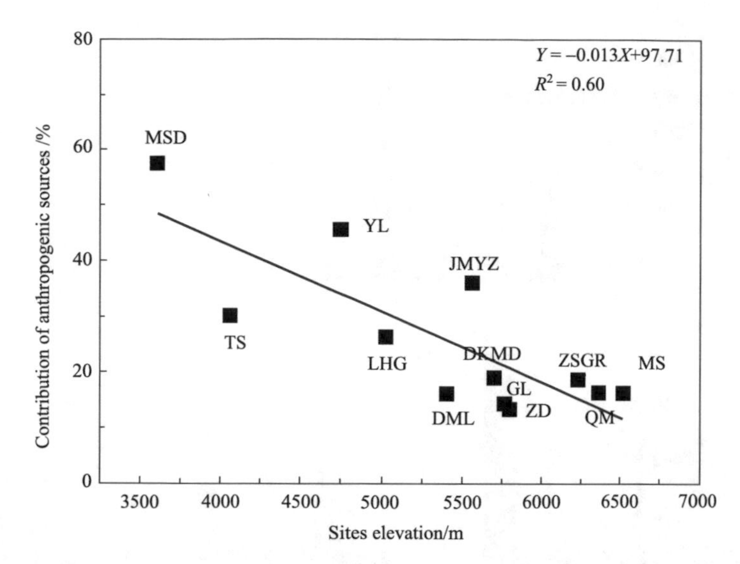

Fig. 5.2 Relationship between anthropogenic Pb contribution fraction and the sampling elevation of snow and ice on the Tibetan Plateau (Yu et al. 2013). MSD: Musidao Glacier; TS: Urumqi No. 1 Glacier; YL: Jade Dragon Snow Mountain; LHG: Laohugou No. 12 Glacier; DKMD: Dongkemadi Glacier; DML: Demula Glacier; JMYZ: Jiemayangzong Glacier; GL: Guoqu Glacier; ZD: Zhadang Glacier; MS: Muztag Glacier; ZSGR: Zangsegangri Glacier; QM: East Rongbuk Glacier. The trend line was derived by linearly fitting the data from all the sampling sites

are representative of the levels of Hg in mountain glaciers at a global scale. The Hg_T concentrations in the glaciers over western China showed significant seasonal variations and were high in the non-monsoon season and low in the monsoon season. For the spatial distribution pattern, the Hg_T concentrations in snow and ice were high in the south and low in the north over western China. Moreover, significant positive relationships were found between the Hg_T concentrations and insoluble particulates, indicating that the transport and deposition of atmospheric Hg over western China was most likely in the form of particulate-bound Hg (Hg_P). Generally, the spatial distribution patterns of heavy metals are governed by the proximity to the source regions of dust and human activities. Consequently, the spatiotemporal variability of heavy metals provides an opportunity to evaluate the impacts of human activities on atmospheric heavy metals over the different regions of western China.

The relative contribution of heavy metals in snow and ice from natural and anthropogenic sources can be estimated by using the crustal enrichment factor (EF_X). Thus, the influence of human activities on heavy metal pollution (EF_X) in snow and ice can be quantitatively determined according to the following formula:

$$\mathrm{EF_X} = \frac{(C_\mathrm{X}/C_\mathrm{R})snow/ice}{(C_\mathrm{X}/C_\mathrm{R})crust} \tag{5.2}$$

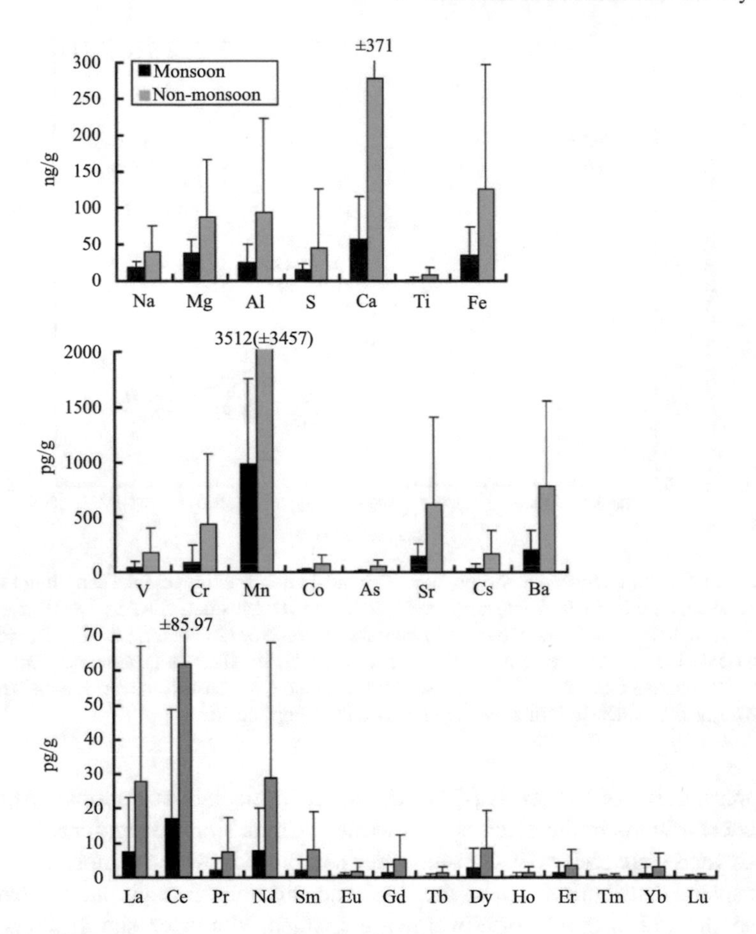

Fig. 5.3 Comparisons of the element concentrations between monsoon and non-monsoon seasons in the East Rongbuk firn core, Mt. Everest (Kang et al. 2007)

where C_X is the concentration of the element of interest and C_R is the concentration of the reference element. "Snow/ice" refers to the concentration of the element in snow and ice, and "crust" refers to the concentration of the element in the crust. The average elemental compositions in the upper continental crust (UCC) are usually regarded as the elemental compositions in the crust (i.e., reference crustal material). Elements with an EF value close to unity (<10) indicate a strong influence from natural components, whereas high values of EF (>10) may suggest potential anthropogenic origins. Numerous previous studies on the EF_X have indicated that heavy metal elements (e.g., Pb, Zn, and Cu) in the snow and ice of the Arctic, Antarctic and mountain glaciers over western China have been influenced by anthropogenic activities.

Many processes, such as the exchange, evasion and accumulation of heavy metals, occur at the snow/ice-atmosphere interface. For example, atmospheric mercury depletion events (AMDEs) were observed in the 1990s, and a large amount of atmospheric Hg was deposited through dry and wet deposition onto the Arctic and Antarctic ice sheets, indicating that the polar regions could act as an important sink for atmospheric Hg at a global scale. Therefore, investigating the mechanisms that affect the exchange, accumulation and chemical reactions of Hg at the snow/ice-atmosphere interface is very important. Although Hg_T concentrations in Arctic snow and ice have significantly increased after AMDEs, the deposited Hg could be re-emitted back into the atmosphere within a short period because of the photoreduction of Hg. In contrast, the atmospheric deposition of Hg over the Tibetan Plateau is mainly associated with particulates and is dominantly in the form of Hg_P, which is more stable than reactive gaseous mercury (RGM) in redox (i.e., photoreduction) reactions after deposition onto snow. Therefore, the deposited Hg could largely be retained in the glaciers on the Tibetan Plateau, indicating that the cryospheric region in western China could be a more important sink for Hg than the Arctic and Antarctic glaciers.

5.2.2 Organic Components

The records of organic compounds in glaciers not only provide important information on climate change and biological activities but also act as indicators of environmental changes in mountain regions. Studies of organic compounds in glaciers include two major aspects. The first involves organic compounds in snow and ice that originate from natural sources (e.g., fatty acids, dicarboxylic acids and fatty hydrocarbons) and can be used to investigate the origins and evolution of organic matter by analyzing organic compositions, carbon numbers, and odd–even predominance. The second is the study of worldwide organic compounds from anthropogenic emission sources, such as persistent organic pollutants (POPs), which have received special attention.

POPs were first detected in Arctic glaciers in the 1970s. The polycyclic aromatic hydrocarbon (PAHs) records in the snow and ice of Greenland showed significant seasonal variations, and the contents of most PAHs peaked in spring and winter. The physicochemical properties of PAHs are relatively stable, so these materials are considered excellent proxies for tracing anthropogenic environmental changes. For example, the PAHs concentrations in the snow and ice of Site-J in the Greenland ice sheet began to rise in the early 20th century, and the PAH levels in the late 20th century were 50 times higher than those in the 18th century. POPs (dichlorodiphenyl-trichloroethane, DDT) in the snow and ice of an Antarctic glacier were first reported in the 1960s. However, published data regarding POPs in Antarctic snow and ice are still very scarce, and thus, the types of organic pollutants have been much less reported in the Antarctic than in the Arctic. DDT was used worldwide as an important organic pesticide in the mid-20th century and could be transported over long distances and deposited onto Antarctic snow and ice through global atmospheric circulation patterns.

For example, the POPs levels (e.g., DDT, polychlorinated, biphenyls and hexachlorocyclohexane) in the Antarctic firn during the mid-20th century were higher than those in modern Antarctic snow and ice, indicating that these organic pollutants may have been transported to the Antarctic since the 1960s. Moreover, phenanthrene, anthracene and other low molecular weight PAHs with trace contents have been detected in the East Antarctic ice sheet in recent decades. PAHs are relatively volatile and could be easily transported over long distances because of their existence in gaseous forms in the atmosphere. This phenomenon indicates that many forms of PAHs could be transported over long distances to the Antarctic through atmospheric circulation.

Mountain glaciers are closer to industrial and agricultural regions than polar glaciers, so environmental signals from human activities could be directly preserved in the snow and ice in high mountain glaciers, and thus, organic pollutants in snow and ice can better indicate environmental changes. The concentration levels of POPs in middle-latitude glaciers are generally higher than those in polar glaciers. Moreover, POPs that originate from South Asia have been detected in glaciers on the southern Tibetan Plateau because of the atmospheric circulation of the Indian monsoon. For example, the amount of n-alkanes in organic compounds (e.g., pristane, phytane, long chains of tricyclic terpenes of C_{19}–C_{29}, C_{24} tetracyclic triterpenes, C_{27}–C_{35} alpha beta hopanes, and C_{27}–C_{29} steranes from petroleum residues) from the Dasuopu Glacier on Mt. Xixabangma, Tibetan Plateau (6400–7000 m a.s.l.), suggests that the Himalayas have been affected by anthropogenic emissions of organic pollutants and the Gulf War. The spatial distribution pattern of *n*-alkanes in glaciers gradually decreased from the northeast to the south over the Tibetan Plateau (Qiyi Glacier, Yuzhufeng Glacier, Xiao Dongkemadi Glacier, and Gurenhekou Glacier). The levels of *n*-alkanes in snow and ice were comparable to those in the Belukha and Sofiyskiy Glaciers in the Altai in central Asia but significantly higher than those in the Antarctic and Arctic ice sheets. Moreover, the contributions of *n*-alkanes from anthropogenic sources are significantly higher than those from natural sources, suggesting that the chemical compositions of POPs in glaciers on the Tibetan Plateau have been greatly affected by rapid industrialization and urbanization.

5.2.3 Insoluble Particles

1. Dust

Dust originating from arid regions can be transported over long distances via atmospheric circulation and deposited onto glaciers, which can change the mass and energy balance of glaciers and consequently accelerate glacier melting because of albedo reduction. Studies on the characteristics of dust in glaciers mainly include the spatiotemporal distribution of dust concentrations and fluxes, physicochemical properties (e.g., grain size, morphology, and chemical composition) and sources of dust, which can indicate environmental and climatic changes, etc. The average concentrations of dust in the glaciers in western China are generally higher than those in the Arctic (e.g., Penny and Devon ice cap), Greenland (e.g., Summit) and Antarctic (e.g.,

Dome A and Dome C). Moreover, the spatial distribution of the grain size and size distribution of dust is significantly different in glaciers worldwide. In general, dust in glaciers in western China features high modal values and single distribution patterns for grain size and thus differs markedly from the dust in the snow and ice over the Antarctic and Arctic regions. For example, the grain size of dust in the snow and ice of the Tienshan glaciers ranges from 3 to 25 μm, with a single-peak distribution pattern, while the grain size of dust ranges from 1 to 2 μm in the snow and ice of Greenland and the Arctic, with a bimodal distribution pattern.

The dust concentrations in the snow and ice of Greenland and Antarctica exhibited clear and significant seasonal variations and were generally higher in winter than in summer. The peak dust concentrations in glaciers in western China occurred from April to June because of the frequent Asian dust storms at that time. Studies of the physicochemical characteristics, seasonal variations and source tracing of dust could elucidate the spatiotemporal distribution pattern of dust in glaciers at a global scale. For example, previous studies from Urumqi Glacier No. 1 suggested that both the grain-size distribution and major ion concentrations (e.g., Ca^{2+}, which is representative of crustal sources) showed maximum values during frequent dust storm events. In summary, the spatiotemporal distribution of dust in snow and ice exhibits significantly different patterns among the Antarctic, Arctic and western China regions, and this phenomenon is mainly governed by the proximity of each region to dust source regions.

2. Black carbon

Black carbon (BC) is an important component of atmospheric aerosols and can significantly decrease the albedo of glacier surfaces, thus accelerating glacier melting (8 kiles et al. 2018). Studies of BC in the snow and ice of the polar regions first appeared in Europe and America in the early 1980s. BC concentrations from 0.1 to 30.34 ng g^{-1}, with an average of 0.2 ng g^{-1}, were observed in Antarctic snow and ice. However, the BC levels in Arctic snow and ice were approximately 10 times higher than those in Antarctica, ranging from 2 to 3 ng g^{-1}. Moreover, the average concentration of BC in fresh snow was approximately 4 ng g^{-1} on the sea ice in the Arctic Ocean. BC tends to accumulate in granular ice crystals in dust layers as snow and ice melt, with an average concentration of 8 ng g^{-1} in surficial crystals and 18 ng g^{-1} in interior crystals. Furthermore, the concentrations of BC in the snow and ice of the European Arctic were significantly higher than those in the Canadian Arctic and Arctic Oceans, mainly because the European Arctic is located near anthropogenic emission sources.

In recent decades, research on BC in glaciers in the middle-latitude to low-latitude regions (e.g., the Tibetan Plateau) has made substantial progress (Zhang and Kang 2017). Surprisingly, the levels of BC in glaciers range from tens of ng g^{-1} (fresh snow and snow pits) to thousands of ng g^{-1} (aged snow and ice and superimposed ice) in the cryosphere of western China. The high concentrations of BC in snow and ice are mainly caused by the enrichment of BC during glacier ablation. Moreover, BC in glaciers significantly decreased from east to west and from north to south over western

China. In terms of glaciers around the world, the BC levels are very low in remote regions worldwide, such as the Antarctic, and are representative of the background values of snow and ice BC on a global scale. However, the BC levels in the snow and ice of the Tibetan Plateau are markedly higher than those of the Arctic and Antarctic because of the influence from anthropogenic activities. Similar to the spatiotemporal distribution pattern of dust, BC in snow and ice is also related to factors such as the state of glacier melting, geographical conditions, source regions of anthropogenic emissions, and atmospheric circulation. Moreover, significant seasonal variations in BC occur in glaciers worldwide. Specifically, BC in snow and ice generally peaks in winter in the Antarctic and Arctic regions. On the Tibetan Plateau, the highest BC concentrations usually occur during the nonmonsoon season, and the lowest BC concentrations occur during the monsoon season. Furthermore, the BC in the glaciers between the northern and southern Tibetan Plateau exhibits opposite seasonal patterns.

5.2.4 *Stable Isotopes*

Water molecules consist of hydrogen (H) and oxygen atoms (^{16}O) and can easily evaporate from the ocean surface into the atmosphere. However, when water vapor condenses into droplets, water molecules with heavy isotopic compositions (D and ^{18}O) preferentially fall to the surface of the earth as rain from the atmosphere. Thus, the isotopic fraction of water molecules can lead to a significant distribution pattern of isotopic compositions in natural water, including snow and ice. Accurately determining the isotopic compositions of water molecules from different scenarios and processes is very important for investigating the behavior of isotopic fractionation. Generally, measuring the relative abundance of water molecules is more accurate and practical compared to measuring the absolute abundance. Therefore, the ratio of the heavy isotope abundance to the light isotope abundance ($R = {}^{18}O/{}^{16}O$ or D/H) in various water molecules can be expressed as the deviation (δ, ‰, difference or ratio) of a measurement of "standard mean ocean water" (R_0) from its true value as follows:

$$\delta = \frac{R - R_0}{R_0} \times 1000 \tag{5.3}$$

where R is the ratio of the heavy isotope abundance to the light isotope abundance, R_0 is the isotope ratio of standard mean ocean water, and δ is the deviation of a measurement of R_0 from its true R value. The stable isotope ratios of δD and $\delta^{18}O$ in snow and ice are widely applied as proxies for the reconstruction of past climate and environmental changes, and this technique has greatly contributed to our understanding of global change. Spatiotemporal variations in δD and $\delta^{18}O$ in the surface or shallow snow and ice of glaciers are the foundation for the interpretation of paleoclimatic and paleoenvironmental changes, as reconstructed by ice cores. According to the

Rayleigh fractionation model, Dansgaard (1964) summarized the main factors that could influence the stable isotopic ratios in snow and ice worldwide, such as temperature, sources of water vapor, latitude, and continental effects. At middle to high latitudes, especially in the polar regions and high Asian glaciers, the stable isotope ratios of water molecules in snow and ice are mainly controlled by the temperature and amount of precipitation. In particular, seasonal changes in δD and $\delta^{18}O$ values in the snow and ice of polar regions are dominated by temperature effects, with high isotopic values during the summer and low values during the winter.

On the Tibetan Plateau, the factors that affect seasonal changes in δD and $\delta^{18}O$ value in snow and ice can be divided into two groups. Specifically, the variations in δD and $\delta^{18}O$ values in snow and ice from the northern Tibetan Plateau are positively correlated with temperature. However, the variations in $\delta^{18}O$ values in snow and ice from the southern Tibetan Plateau are negatively correlated with precipitation because of the significant influence of the Indian monsoon to summer rainfall. For the geographic distribution pattern of hydrogen and oxygen isotopes, δD and $\delta^{18}O$ values are high in the western Antarctic and low in the eastern Antarctic. For example, the ratio of δD to the west of Vostok Station is 40‰ higher than that to the east of Vostok Station at the same temperature. Generally, the spatial distribution pattern of $\delta^{18}O$ values in the snow and ice of the Antarctic ice sheet is highly correlated with the condensation temperature of rain droplets, and the drop in temperature is caused by geographical factors, such as latitude and altitude effects and the distance from the coastline to inland area of the Antarctic. In summary, diverse sources of water vapor and different underlying surfaces exist in the Arctic and Antarctic, resulting in a heterogeneous distribution of stable isotope ratios. Thus, the variations in δD and $\delta^{18}O$ values in the snow and ice of the Antarctic and Arctic are relatively affected by factors such as the sources and transport processes of water vapor, snowfall formation, and seasonal changes and postdepositional processes of snowfall. The above factors could affect the reliability and accuracy of paleotemperature reconstruction from ice cores.

Generally, the fractionation of stable isotopes of heavy metals is less influenced by physical and/or biological processes. Therefore, the isotopic ratios of heavy metals offer an effective and powerful tool for tracking the sources and transformation processes of trace metals in snow and ice. The isotope ratios of some important heavy metals (e.g., Pb, Sr, Nd, Cu and Zn) in snow and ice have been widely used to trace source regions and determine changes in the atmospheric environment. For example, the fractionation of stable Pb isotopes is much less affected by transformation and geochemical processes, and the isotopic abundance of Pb is high and stable in different sources. Thus, the four types of stable Pb isotope ratios in snow and ice could be regarded as fingerprints of potential sources and the relative contributions from natural and anthropogenic emissions. For example, the contents of radiogenic Pb isotopes in the glaciers of the southern Tibetan Plateau are higher than those of the northern Tibetan Plateau, and anthropogenic Pb predominates in low-altitude glaciers and in glaciers that are surrounded by intensive human activities. Moreover, the distribution of Sr–Nd isotopic compositions in the crust has a horizontal zonation, which cannot be easily altered during transformation and deposition. Hence, the isotopic compositions of Sr–Nd can be used as a very good proxy for tracing dust sources in snow and ice. For example, the Sr–Nd isotopic compositions in the dust layers of the East Rongbuk glacier snow pit (Mt. Everest) are consistent with the distribution pattern of the local/regional crustal materials, while the compositions of Sr–Nd isotopes in the other layers of the snow pit are comparable to the distribution pattern of isotopes in the arid regions of northwestern India. Likewise, the stable isotopes of Sr–Nd in particles from the western Qilian Mountains (i.e., Laohugou Glacier No. 12) are similar to the pattern of Sr–Nd isotopes in mineral dust from the Badan Jaran Desert, which is the most likely source region of the dust in Laohugou Glacier No. 12.

5.3 Chemical Characteristics of Frozen Ground

5.3.1 Chemical Processes in Frozen and Freezing Ground

1. Chemical processes in freezing and thawing ground

Essentially, the chemical reactions that take place in soil during freezing and thawing and in the frozen state are the same as those in unfrozen materials. These reactions are solution, hydration, substitution, oxidation–reduction, ion exchange, etc., but these processes have a number of specific features in cold regions. For example, some salts dissolve at a very slow rate at low temperature. Apparently, because of the low temperature, the permafrost regions contain considerable amounts of products from chemical interactions between dissolved substances and water molecules, that is, hydrates and crystalline hydrates. Cation exchange reactions are likely important for frozen ground because unfrozen water is a rather concentrated solution, and the ions in unfrozen water actively interact with those in mineral surfaces.

At the beginning of freezing, water turns into ice, creating a new mineral. Gravitational, capillary and loosely bound nonsaline water crystallizes at the temperatures below 0 °C. Film water normally freezes within a wide range of negative temperatures, as determined by the curve of unfrozen water content. Salt water, with a mineralization of more than 30 g L^{-1}, crystallizes at temperatures from approximately –2 to -1.5 °C, whereas brines may remain liquid at –20 °C and even lower temperatures. The freezing of water usually causes the distinct fractionation of salts between the solid and liquid phases. A portion of the salts dissolved in water is enclosed in the ice, some of the less soluble salts precipitate, and some easily soluble salts are squeezed into lower water layers, thus increasing their mineralization. During freezing, in accordance with the degree of solubility at negative temperatures, the most insoluble salts of $CaCO_3$ precipitate first (at temperatures from –3.5 to –1.5 °C), followed by Na_2SO_4, $CaSO_4$, etc. (at temperatures from –15 to –7 °C); these salts form so-called crystal hydrates. Consequently, cryogenic layers are enriched with gypsum ($CaSO_4 \cdot 2H_2O$), $Na_2SO_4 \cdot 10H_2O$, and calcite ($CaCO_3$). Below the freezing boundary, water is highly mineralized because of the easily soluble salts expelled from the frozen layer (chlorides of calcium, magnesium, and sodium and hydrocarbonates of sodium). Highly mineralized subground water (200 g L^{-1} and higher) forms because of this cryogenic concentration.

Ground water normally has a high carbon dioxide content because the solubility of gases (including CO_2) and the organic matter content rapidly increase with decreasing temperature. The concentration of hydrogen ions in ground water from permafrost regions can increase by a factor of several hundred, which might cause an acidic reaction in the medium. The nature of many chemical reactions and the behavior of soil components largely depend on the pH of the medium. An acidic medium is more aggressive and chemically active, decomposing silicates and strengthening hydrolysis reactions, than neutral and alkaline media. Thus, frozen materials in permafrost regions are predominantly reducing media with high contents of divalent ferrous Fe^{2+}. Ferrous oxide colors soil in bluish-gray shades, creating gley soil. This soil is predominantly fine grained, reducing and acidic in nature.

The breakdown of organic matter also differs. Because of slow biological and biochemical reactions in the frozen ground, the transformation of vegetal and zoogenic remains into organic matter is reduced, and decomposition of the remains (formation of humus) terminates at a less mature stage. Thus, light-colored fulvic acids form rather than humic acids (a product of the final stage of decomposition). In tundra soils, the content of fulvic acids may reach 70%, while humic soils contain only 10–15% humus matter. Fulvic acid destroys minerals because of its high acidity, homogeneously saturating the soil and forming a massive compact layer. Meanwhile, more viscous and less mobile humic acids in soil produce lumpy, nutty structures, such as chernozem.

2. Chemical processes in frozen ground

Since the liquid water in the frozen ground is almost invisible, it was considered that the soil is in a state of inactive chemical reactions for a long time. This misconception underestimates the role of unfrozen water and would automatically lead to the application of Van't Hoff's law, which states that with a drop in temperature of 10 °C, the rate of chemical reaction is reduced by half. Although the absence of free water in frozen ground would seem to preclude the escape of chemical components from the unfrozen water, mass-transport processes are sufficiently intense because of the diffusion of ions in unfrozen water, which adjusts the concentration of dissolved substances. In frozen ground, flow is also active in unfrozen water films, inducing the convective transport of ions and soluble matter with migrating water. During this process as described above, phase transitions between pore ice and films of unfrozen water will take place in accordance with increasing or decreasing ion concentrations.

3. Chemical processes during the repeated freezing and thawing of ground

Chemical reactions in seasonally thawing ground are much more intense and periodic than those in permafrost. The interaction between soil minerals and water (both free and bound) is a fluctuating process, and the phase transitions of water into ice and its reverse process should result in marked intensification of chemical weathering in seasonally frozen ground. Moreover, an intense chemical transformation in seasonally frozen ground commences at the very first stage of weathering under the effects of hydrolysis, leaching, oxidation, hydration, and colloid migration, thus forming neogenic clay and other minerals. A study in Antarctica has shown that oxidation takes place and MnO and Fe_2O_3 accumulate in the 10–15 cm surface layer of soils if the oxygen supply is sufficient, coloring the ferric and manganese extractions on rock fragments in ochre-rusty or orange-red hues. Lower in the layer, where manifestations of wash-out products and carbonization exist, more mobile products of weathering accumulate. Studies of surface weathering crusts under the microscope have established the decomposition stages of the original minerals. At the initial stage, chlorite disappears, followed by hornblende and biotite, i.e., banded and stratified silicates are the first to disintegrate. Feldspars become covered with yellowish-brown, fine silty aggregates, i.e., secondary clay minerals.

In cold tundra and taiga regions, non-gley cryogenic soils are dominant followed by non-gley soils with poor drainage conditions, The chemical elements in non-gley soils are arranged according to their migration capacity: Si > Fe > Ti > Al. Silicate forms, which appear because of hydrolysis, are fairly mobile in an acidic medium and are evacuated from the soil profile. Iron, titanium and aluminum in an acidic medium have slow solubility and normally remain in the soil as oxides and hydroxides. During humification in permafrost regions, fulvic acid appears as one of the most aggressive and mobile forms of humus. This acid moves downwards with the soil solution and destroys hydroxides and silicate minerals by forming different types of organic-mineral compounds (oxalates, chelates, fulvates, and adsorbed organic-mineral compounds).

Fulvates and oxalates, which are more mobile compounds, once are removed from the soil profile, whereas chelates and adsorbed organic-mineral compounds will soon lose their mobility and remain in the illuvial horizons. During this process, brown-colored Al–Fe-humus or coarse silty-clay horizons are appeared. At the same time, real humus horizons and horizons of Al–Fe-humus and titanium compounds are formed. Compounds of titanium, aluminum, iron (Ti–Al–Fe) and humus accumulate in these alluvial horizons, and this process is typical of cryogenic soil formation. The leached horizon is depleted in Fe and Al hydroxides and oxides; therefore, a relatively high SiO_2 content and a bright shade of coloring occur because of the decomposition and removal of dark-colored compounds and minerals.

Chemical and physicochemical processes are slightly different in typical gley soils of the northern European part of the former USSR and Siberian coastal lowlands. The profiles of gley soils do not normally show distinct illuvial horizons, but the contents of Fe_2O_3 and Al_2O_3 decrease concurrently with relative silica enrichment in gley and gley-podzol soils, such as heavy sandy-silty clay. The higher mobility of iron is a result of its transition under reducing conditions into the monoxide form $Fe(OH)_2$, which does not precipitate from solution until the pH reaches approximately 5–6. The blue-gray monoxides of iron color the profile of gley soil with typical gray and blue-gray hues. This phenomenon is also promoted by the presence of fulvic acids, which are the immature forms of humus and are not brown but light gray, thus making the color of gley soils less intense than that of non-gley soils.

The chemical differentiation of weathering products, which is closely associated with the mobility of chemical elements, is particularly important in geochemical processes in permafrost regions, especially in soils with cyclic freezing and thawing. Mobile chemical elements are intensively removed by underground and surface runoff; other elements, by contrast, are practically immobile and accumulate in watershed areas and on slopes, increasing their relative concentrations. For example, potassium, calcium, magnesium, sulfate and chloride ions in permafrost regions are very mobile and migrate in all waters in a purely dissolved state. The silicate form of silicon migrates mostly as mono-silicic and poly-silicic acids, which are removed from solutions by ground water. A certain amount of silicic acids (up to 40%) can be transported as gels and colloids in combination with organic matter. Non-silicate SiO_2 in permafrost regions is practically immobile, which is determined by the formation process of calcareous soil. The low mobility of silica is attributed to the extremely low solubility of SiO_2 in a highly acidic medium, which is typical of tundra and taiga soils. Up to 70–90% of aluminum migrates through permafrost regions as colloids and complex compounds with humic acids. Iron (Fe^{2+} and Fe^{3+}) has low mobility outside permafrost regions; for instance, approximately 90–98% of the total iron content migrates as highly mobile colloids under cold and humid conditions. Under the boreal conditions, some other micro-components (Ti, Zn, Cu, Ni, etc.) also become more mobile and are transported not as simple ions but as colloids or complex ions, which form with participation from high molecular mass organic matter.

5.3.2 *Natural Gas Hydrates*

Natural gas hydrates are ice-like crystal compounds that consist of water and gas molecules under high-pressure and low-temperature conditions. Most of the gas components include methane, ethane, propane, and their hydrocarbon homologues, carbon dioxide, nitrogen, and hydrogen sulfide, and so on. Gas hydrates appear similar to ice (Fig. 5.4A), generally white or light yellow, and can be directly ignited. Water molecules form a hydrogen-bonded polyhedral ice-like clathrate crystal lattice, which encases small gas molecules, then form gas hydrates with single or complex components. For the most common gas hydrates in nature, the main component of the gas is methane. Because the methane content exceeds 99% of the gas stored in gas hydrates, which are generally called methane hydrates. Gas hydrates are naturally widely distributed around the world in sediments under permafrost regions, ocean regions near continental margins and the bottoms of deep lakes.

Gas hydrates are extremely unstable. Thus, as the global temperature rises, permafrost degradation destroys the temperature and pressure conditions under which gas hydrates can stably exist. Gas hydrates will likely dissociate and release massive amounts of methane. Therefore, methane hydrate has been considered a potential

Fig. 5.4 Clathrate hydrates of natural gases synthesized in the laboratory

source of greenhouse gases under climate change. Some samples have been found under the permafrost regions of the North Slope of Alaska, USA, and Mackenzie Delta, Canada. As revealed by gas and oil investigations under the permafrost regions of Russia, abundant gas hydrate resources have been found in permafrost regions. However, the resource reserves of gas hydrates on land are much smaller than those in the ocean. Gas hydrates on land mainly occur in layered and bulk configurations, with most being methane hydrate. Thus, gas hydrates on land generally have higher gas saturation and economic exploitation values than those in the ocean. The development of permafrost has a close relationship with the existence of gas hydrates. The temperature and pressure conditions for the formation of gas hydrates predominantly occur in permafrost. Furthermore, permafrost is a type of geological body with extremely low permeability. Therefore, permafrost can effectively impede gas migration from deep layers, which is helpful for the aggregation of natural gas and forms the required trapping conditions for the formation of gas hydrates.

Permafrost is widely distributed on the Qinghai-Tibetan Plateau, and most of the permafrost regions possess low-temperature and high-pressure conditions, which are favorable for gas-hydrate formation. The first investigation of gas hydrates in permafrost regions was conducted in 2007. Subsequently, exploration drilling was performed in the permafrost region of the Muli mine at Qilian Mountain in 2008, and some gas hydrate samples were successfully collected at a depth of 130 m. A second exploratory drilling operation was conducted at the same location in 2009, and additional gas hydrate samples were successfully collected at depths from 130 to 260 m. Approximately 50% of the gas stored in the hydrates is methane, in addition to other types of heavy hydrocarbon gases. Successful drilling of gas hydrates on land was performed in China. Drilling exploration and well-logging studies were conducted in the saddle-back base of Kunlun Mountain on the Tibetan Plateau in 2013. Geophysical well logging, the abnormal release of gas from borehole cores, and the geochemical analyses of the gas confirmed that gas hydrates exist under the permafrost regions of Kunlun Mountain. In addition, these findings indicate that gas hydrate formation and accumulation may occur under the permafrost regions in the interior of the Qinghai-Tibetan Plateau. After America, Canada, and Russia, China has become the fourth country where the gas hydrate samples have been found in permafrost regions. These gas hydrates will play an important role in the energy resources, environment and climate of China.

5.4 Chemical Characteristics of River and Lake Ice

River and lake ice, as major elements of the cryosphere, are mainly distributed in regions of high latitudes and altitudes. Factors that affect the chemical characteristics of river and lake ice include the geology, geomorphology, water-discharge sources, atmospheric deposition, and physical, chemical and biological processes. However, due to limited practical applications, few studies have been conducted on river and lake ice chemistry. In this section, several processes will be listed to show the mechanisms which affect the chemical characteristics in ice and water.

5.4.1 Changes in Hydrogen–Oxygen Stable Isotope Ratios Between Ice and Water

The fractionation of hydrogen and oxygen stable isotopes in ice and water is defined as the change in isotope (heavy and light) ratios in ice with the freezing of water. Heavy isotopes ($^{1}H_2{}^{18}O$ and $^{1}HD^{16}O$) preferentially enter ice during freezing. However, no fractionation between ice and water occurs during melting because of the slow migration of hydrogen and oxygen isotopes in ice. According to the thermal balance equation, hydrogen and oxygen fractionation is controlled by the reaction rate constant, that is, the equilibrium isotope fractionation factor:

$$\alpha^* = \frac{R_I}{R_W} \tag{5.4}$$

When $\alpha^* = 1$, no isotope fractionation occurs; when $\alpha^* > 1$, isotope fractionation occurs during natural freezing. R is the ratio of $^{18}O/^{16}O$ or $D/^{1}H$. The subscripts I and W represent ice and water, respectively. The transition between ice and water can be regarded as equilibrium fractionation, so the degree of fractionation can be indicated by the separation factor (ε *):

$$\varepsilon^* = 1000(\alpha^* - 1) = \delta_I - \delta_W \tag{5.5}$$

where $\delta_I = \left(\frac{R_I}{R_S} - 1\right)$ and $\delta_W = \left(\frac{R_W}{R_S} - 1\right)$.

In freshwater systems, $\delta^{18}O$ and δD range from 2.8 to 3.1‰ and from 17.0 to 20.6‰, respectively, at an icing speed of < 2 mm h^{-1} when the temperature approaches 0 °C.

The formation of river and lake ice depends on changes in the external environment. Once ice forms, an independent system is created. Therefore, the external environment significantly influences the amount of hydrogen and oxygen stable isotopes between the ice and water systems. Generally, the semi-closed system, which forms by slow freezing, is the most common process that takes place in rivers and lakes. $\delta^{18}O$ values exhibit four stages in ice and water depending on the freezing process: ① a closed system with slow freezing: $\delta^{18}O$ values in the ice and water system present a nonlinear, rapid downward trend with increasing freezing depth but they have a tendency to approach; ② a semi-closed system with slow freezing: $\delta^{18}O$ values in the ice present a nonlinear decrease at a stable rate with increasing ice depth; ③ an open system with slow freezing: $\delta^{18}O$ values in the ice-water phase remain constant as the ice thickens; and ④ an open system with rapid freezing: $\delta^{18}O$ values in the ice non-linearly increase and then gradually approach a constant value as the freezing depth increases, while the $\delta^{18}O$ value in the water is always constant.

5.4.2 *Trace Gas Distribution in River and Lake Ice*

The distribution of trace gases in ice and water is controlled by the type of ice, chemical reactions, biological respiration and photosynthesis taking place in water, while gas exchange would be impeded by the presence of ice. Therefore, the distribution of CH_4, CO_2, O_2 and N_2 in lake ice varies with increasing lake-ice depth. For instance, the CO_2 mixing ratio tends to increase with increasing ice depth; however, the O_2 mixing ratio changes in an opposite manner, while changes in the CH_4 mixing ratio are complicated, exhibiting significant differences in lake ice. This phenomenon occurs because when lake ice forms, photosynthesis is blocked, and plants and animals consume O_2 and produce CO_2 during respiration. The mixing ratio of N_2 in lake ice is approximately 78%, similar to that in surface water. In general, the total trace gas content in lake ice is very low, and the solubility of gas in water is approximately 100 times that in ice.

5.4.3 *Exclusion and Optical Properties of Colored Dissolved Organic Matter in River and Lake Ice*

Colored dissolved organic matter (CDOM), which is mainly produced by rotting substances, is optically measurable and usually found in natural water environments. Dissolved organic carbon (DOC) is the carbon content that characterizes the concentration of dissolved organic matter. In general, the CDOM and DOC concentrations in river and lake ice are lower than those in submerged water. The exclusion factor is usually used to quantify the effect of freezing on organic and inorganic contents. The CDOM exclusion factor is the extinction coefficient at ultraviolet wavelengths, while the ion exclusion factor for inorganic species is based on the ratio of water conductivity. The CDOM and ion exclusion factors substantially differ with the ice profile (Fig. 5.5). With the existence of white ice, the CDOM and ion exclusion factors are larger in black ice. In general, the CDOM exclusion factor is greater than the ion exclusion factor, and the CDOM exclusion factor consistently changes in lakes despite the existence of certain variability. When ice forms by slow freezing without snow, the CDOM exclusion factor is larger at the surface, while the ion exclusion factor is larger at the bottom of the ice, which is consistent with high ion repulsion efficiency. In the Romulus Salt Lake (Fig. 5.5D), the CDOM and ion exclusion factors peak at the top of the ice in the absence of snow, and CDOM has a larger

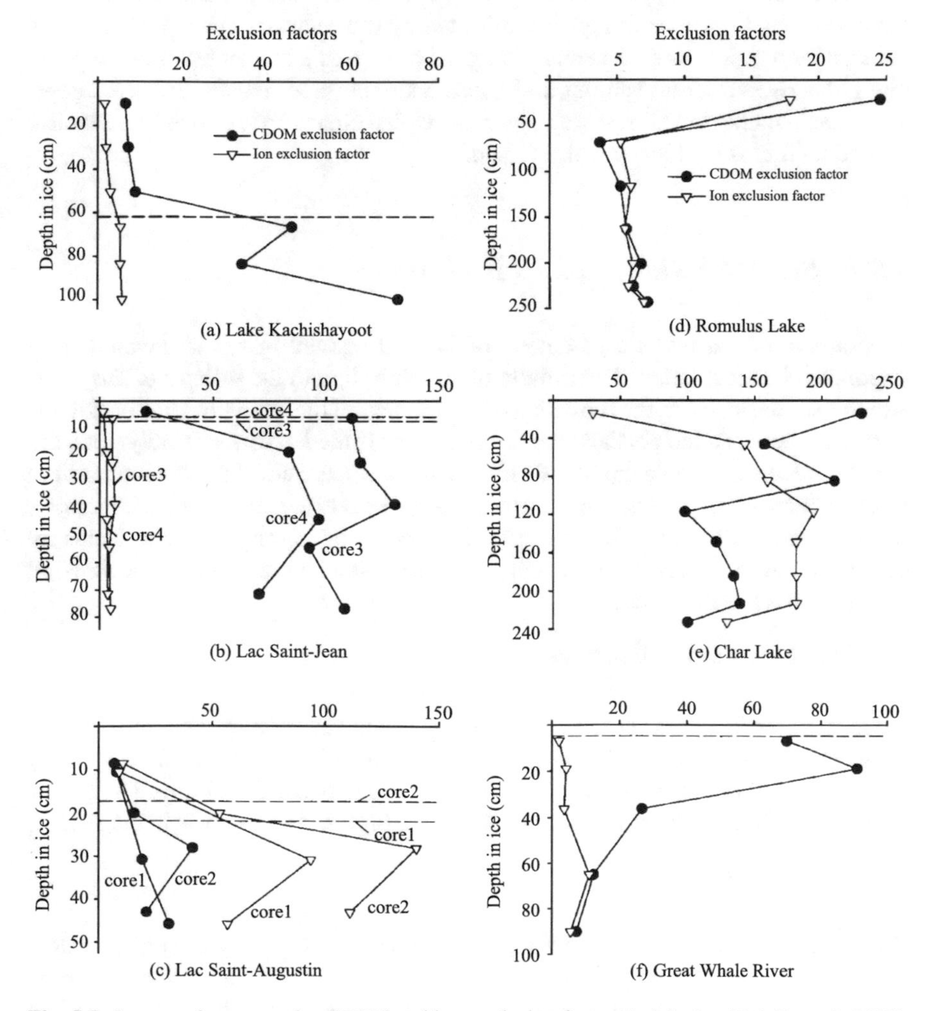

Fig. 5.5 Impact of snow on the CDOM and ion exclusion factors in lake ice (Belzile et al. 2002). The dotted lines indicate the dividing line between white ice and clean ice

exclusion factor than the ions. However, CDOM has a smaller exclusion factor than the ions in other parts of the ice. Therefore, the presence of surface snow affects both the CDOM concentration and inorganic ions in lake ice. In addition, mountain areas exhibit a substantial exclusion effect in lake ice because of inorganic salts, which regulate the salinity of lakes on a seasonal scale.

5.5 Chemical Characteristics of Sea Ice

Sea ice accounts for approximately 7% of the earth's surface, and the chemical characteristics of sea ice are greatly reflected by the seawater chemistry, which is affected by physical, chemical and biological processes between water and ice and input from rivers. Sea-ice salinity, major ions, nutrients, trace metals, dissolved gases and organic matter are all research topics in sea-ice chemistry. Among these topics, sea-ice salinity is the most widely studied.

5.5.1 Sea-Ice Salinity and Its Evolution

The salinity of sea ice is a parameter of the salt content in sea ice, which is an important chemical index. The salinity of sea ice refers to the salinity of the water after the sea ice melts. In the formation of sea ice, part of the interstitial fluid separates from gaps between ice crystals into the seawater. If the ice forms rapidly, then the gaps between ice crystals are rapidly filled with new ice. Part of the interstitial fluid is not drained in time and is trapped in brine bubbles between ice crystals. Therefore, sea ice is a mixture of solid ice and brine. Moreover, the salinity of sea ice is mainly determined by three factors: the salinity of the seawater before freezing, the freezing speed and the age of the ice.

1. Salinity of ocean before freezing

No matter how fast the seawater freezes, some salt always precipitates from the ice, so the salinity of the sea ice is always lower than that of the seawater from which it formed. In general, the salinity of sea ice is mostly between 3 and 8‰. The higher the salinity of the seawater is before freezing, the higher the salinity is in the sea ice.

2. Freezing rate

When the temperature of seawater decreases to the freezing point or slightly below the freezing point, water molecules first freeze and precipitate salts. However, some salts remain in ice crystals, which gradually form salty brine bubbles. During sea-ice formation, the lower the air temperature corresponds to the higher the freezing rate, and eventually the ice thickness increases. When salts cannot precipitate fast enough, the amount of salt bubbles and the sea-ice salinity increase. The surfaces of sea ice directly contact seawater and cold air, so the freezing speed is high, and thus, salts do not easily precipitate. However, the growth at the bottom of ice is slow, and the ice needles are vertically aligned, which facilitates the outflow of interstitial fluid. Therefore, the distribution of salinity in the ice decreases with depth.

3. Age of sea ice

The salinity of sea ice is also significantly related to the age of the ice. The salinity of freshly formed ice is very high. Over time, the salinity of the ice constantly decreases. When the water temperature rises, the melting of the sea ice begins with needle crystals or salty bubbles, and to a certain extent, the interstitial fluid between adjacent salt bubbles slowly drains away.

The volume of brine in sea ice is determined by the salinity and temperature of the ice. As the salinity and temperature of sea ice increase, the volume of brine also increases to maintain the phase equilibrium between sea ice and brine (Fig. 5.6). Usually, the following empirical formula provides a simple method, which can be applied to calculate the brine volume fraction (υ_b). This parameter is important in physical, biological and chemical research of sea ice.

$$\upsilon_b = S_i(45.917/T + 0.930) \quad -8.2 < T \leq 2.0\,°\mathrm{C} \tag{5.6}$$

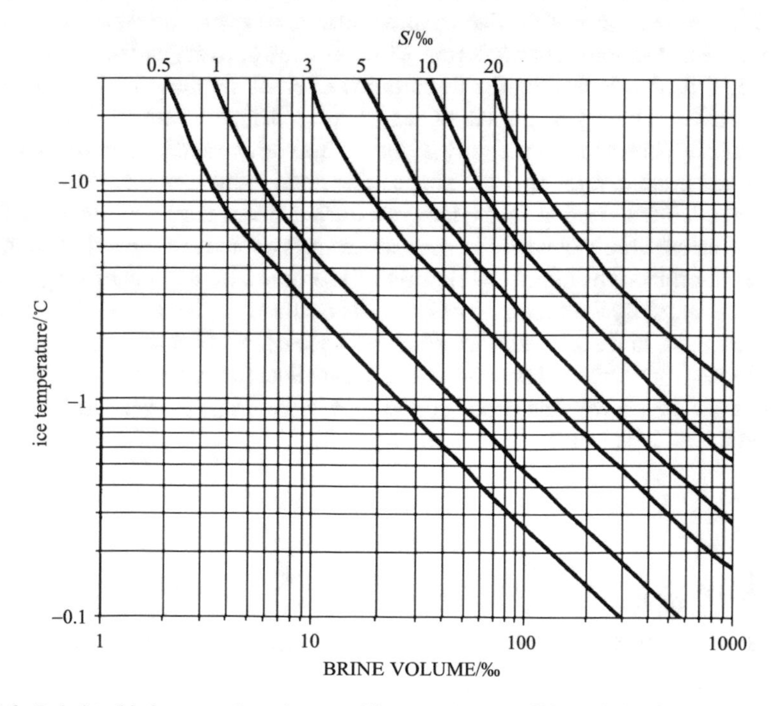

Fig. 5.6 Relationship between the volume and ice temperature of bittern in ice for a certain salinity range (Kämäräinen 1993)

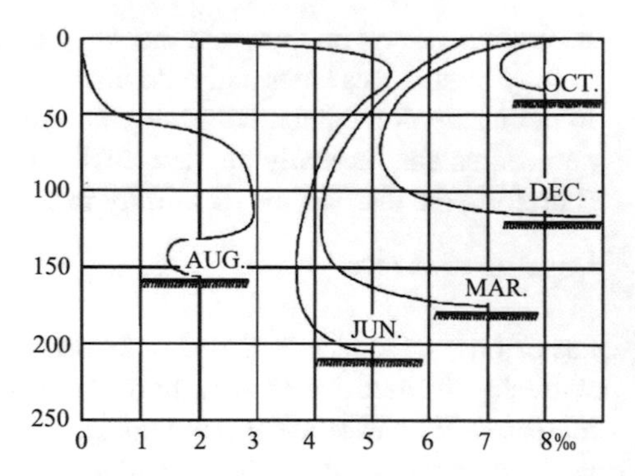

Fig. 5.7 Evolution profile of salinity in sea ice during winter and melting seasons in the Arctic (Thomas and Dieckmann 2003)

$$\upsilon_b = S_i(43.795/T + 1.189) \quad -22.9 < T \leq -8.2\,°\text{C} \tag{5.7}$$

where T is the sea-ice temperature and S_i is the sea-ice salinity.

4. Evolution of sea-ice salinity

Annual sea-ice salinity changes with depth form a "C" shape, but the salinity of the sea-ice surface substantially decreases during the melting season. At present, most large-scale sea-ice models assume that the salinity of sea ice is constant, which does not reflect the response of sea ice to atmospheric or ocean boundary conditions. Temperature and salinity have important effects on the porosity and pore microstructures of ice, further indicating the importance of studying the evolution of sea-ice salinity profiles. During the growth of sea ice (Fig. 5.7), the vertical distribution of sea-ice salinity shows that the salinity is high at the surface and bottom of the ice and lowest in the middle, forming a "C" shape. The main influences include salt separation and sea-ice desalination during ice growth. Generally, the most important factor that controls the salinity of newly formed sea ice in winter is salt separation along the ice-water interface. The initial distribution of salt in the ice and underlying water is further influenced by the expulsion of brine. The slower the ice grows, the less brine that accumulates at the ice-water interface because of diffusion and convection.

The following empirical formula is used to calculate the salt separation coefficient k_{eff} based on the growth rate of ice v_{i} and then to estimate the sea-ice salinity $S_{i,0}$ based on k_{eff} and the salinity of seawater S_{w}:

$$S_{i,0} = k_{\text{eff}} S_{\text{w}} \tag{5.8}$$

where

$$k_{\text{eff}} = \frac{0.26}{0.26 + 0.74\exp(-7243 v_i)} \quad v_i > 3.6 \times 10^{-5}\,\text{cm/s}$$

$$k_{\text{eff}} = 0.8925 + 0.0568 \ln v_i \quad 3.6 \times 10^{-5}\,\text{cm/s} > v_i > 2.0 \times 10^{-6}\,\text{cm/s}^{\cdot}$$

$$k_{\text{eff}} = 0.12 \quad v_i < 2.0 \times 10^{-6}\,\text{cm/s}$$

The actual observed salinity profile is different from the predicted results based on the growth rate of ice and the separation of salt mainly because of the loss of salt during ice consolidation and aging. Essentially, two different types of desalination mechanisms exist: ① During the growth stage of winter ice, the surface of the ice is cooled, driven by the temperature gradient. This process includes gravitational brine expulsion, brine migration and other processes. ② The desalination of warm ice in the presence of low-salinity water on the surface or at the bottom of the ice.

The sea-ice temperature gradient controls the migration of brine bubbles; that is, brine bubbles migrate from areas of low temperature to areas of high temperature. At the micro-scale level, the movement of a single halogen bubble under the temperature gradient is significant, but this phenomenon barely affects the overall salinity profile.

One effective desalination mechanism under the conditions of ice formation is gravitational excretion. When growing sea ice is cooled from above, the lower the sea ice temperature the greater the salinity and density of the brine. The growth of ice layers occurs along a positive temperature gradient and unstable brine density profile, resulting in the reverse convection of ice brine, that is, high-density brine inside the ice is exchanged with the underlying salt water. The amount of gravitational discharge depends not only on the temperature gradient of the ice but also on the permeability of the ice. The desalination rate of the gravitational discharge is a function of the local temperature gradient and the volume fraction of the brine. The pressure gradient associated with the gravitational discharge is very small, so this process is mainly related to the pore size. If the volume fraction of the brine is lower than a certain critical value, such as 50–70‰, then gravitational discharge stops.

Another important desalination process of cold ice is the deportation of brine, which occurs during the formation or growth of ice because of decreasing sea-ice temperatures. When sea ice cools, the water in the brine freezes, increasing the concentration of the brine. The volume of frozen water is approximately 10% larger than that of liquid water, so a portion of the brine is expelled. Brine expulsion is mainly affected by volume changes during ice formation, which is related to the difference between the brine and ice densities. This process may be an important desalting mechanism in the initial stage of ice formation and growth, although not as effective as gravitational drainage.

Ice-water salt separation, gravitational discharge, and brine expulsion can be used to explain the "C" salinity profile of new ice. When the growth rate of sea ice is low, the separation coefficient is reduced as a whole, so the salinity of the newly formed ice is reduced.

The desalting process is most effective in the summer melting season. During this time, the porosity and permeability of ice are generally high, and the low salinity at the surface and bottom of the ice can replace the high-salinity brine inside the ice. During melting, desalination is driven by the net water potential produced by melting snow or ice on the surface of the ice. The surface of the ice has a low salinity, and the melt water permeates into the ice body to replace the high-salinity brine. The use of different tracer substances has shown that the longitudinal and lateral transmission of melting water changes with seasonal variations and is related to the permeability of the ice. As much as 25% of the annual melt water from Arctic sea ice is preserved in the pores of the ice. At the same time, the diffusion and convection of fresh water at the bottom of the sea ice causes the salinity of thin sea ice to be close to zero.

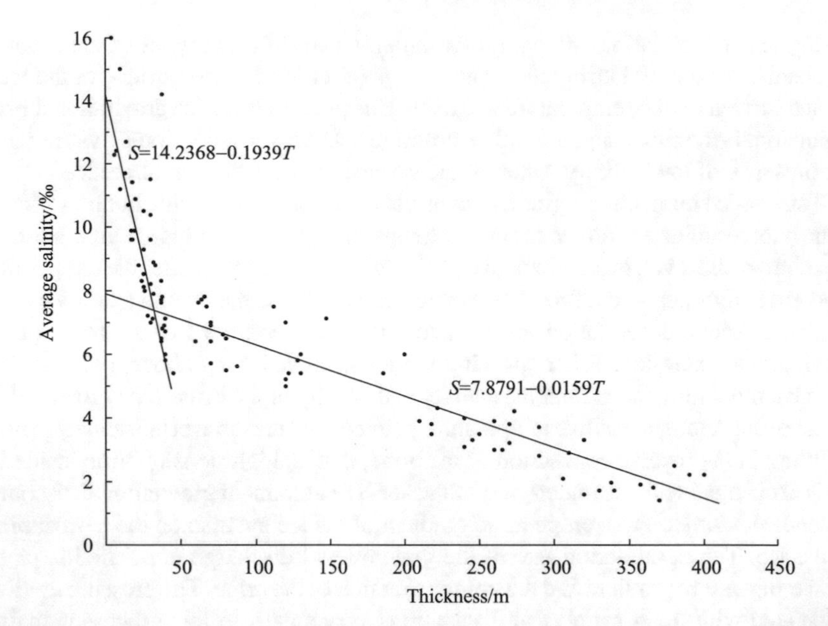

Fig. 5.8 Relationship between the thickness and salinity of sea ice during growing periods (Kämäräinen 1993)

During ice growth, the average salinity of the ice layer has a certain relationship with the ice thickness (Fig. 5.8). When the ice thickness is less than 40 cm, the average salinity linearly decreases. When the ice thickness is greater than 40 cm, the average salinity is still linearly correlated with the thickness of the ice but decreases with thickness.

5.5.2 Sea-Ice Phase Diagram

Sea ice is a mixture of solid ice crystals and brine bubbles. To understand the changes in ionic components as seawater freezes, we need to understand the physical and chemical phase change relations. To simplify the problem, Fig. 5.9 shows the main features of phase transitions in standard seawater ($Na^+ + Cl^-$: 85%, SO_4^{2-}: 8%, and $Mg^{2+} + Ca^{2+} + K^+$: 6%). In a closed system, when seawater with a salinity of 34‰ is cooled below the freezing point (–1.86 °C), the amount of ice continuously increases with decreasing temperature. The main dissolved salt in the seawater does not enter the lattice, so the salinity of the seawater increases and the freezing point decreases. Below –5 °C, the mass fraction of the ice can reach 65%, and the salinity of the seawater increases to 87‰. Below –8.2 °C, the sodium sulfate in seawater becomes saturated and begins to precipitate. If the temperature continues to decrease, sodium sulfate continues to precipitate. The precipitation of other salts as seawater freezes,

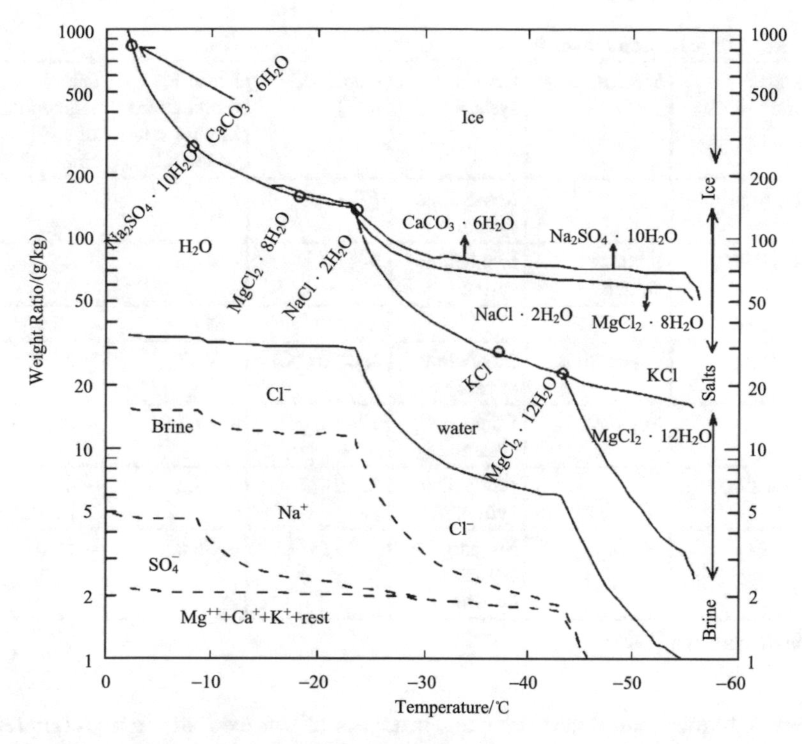

Fig. 5.9 Phase diagram of standard sea ice (Kämäräinen 1993)

such as $CaCO_3 \cdot 6H_2O$ and $NaCl \cdot 2H_2O$, and their distribution and mineralogy in sea ice are not well known. $NaCl \cdot 2H_2O$ is estimated to start precipitating below –22.9 °C. When the mass fraction of water is reduced to 8%, a small amount of liquid still exists below –30 °C or even –40 °C. The existence of unfrozen water at low temperatures greatly affects the survival of microbes in winter sea ice.

Table 5.1 shows the properties of various solid salts in sea ice, including the initial crystallization temperature during precipitation, the eutectic temperature in pure salt solution, and the density and crystal system of each salt.

5.5.3 *Gas in Sea Ice*

Air in sea ice can be found in the form of closed bubbles during freezing. Some potential formation pathways include that when there is dynamic action on the surface of the water or there are non-dissolved gases or gases produced by animals and plants, these gases will be enchsed in the growing ice. The typical gas volume concentration ranges from 0.5 to 5%. The composition of the gas is generally close to the pure air component. The proportions of general N_2, O_2, and CO_2 are 82%, 17% and 0.4%, respectively. The oxygen content is lower than that of pure air, and the carbon dioxide

Table 5.1 Salt parameters in sea ice

Molecular formula	Mineral name	Crystal system	Density/(kg m^{-3})	Eutectic temperature in pure salt water/°C	Initial temperature of salt in brine/°C
$CaCO_3 \cdot 6H_2O$		Monoclinic system	1771	/	– 2.2
$Na_2SO_4 \cdot 10H_2O$	Sodium sulfate	Monoclinic system	1464	– 3.6	– 8.2
$MgCl_2 \cdot 8H_2O$				– 33.6	– 18.0
$NaCl \cdot 2H_2O$	Cryohydrate	Monoclinic system	1630 (0 °C)	– 21.1	– 22.9
KCl	Sylvine	Cubic system	1984	– 11.1	– 36.8
$MgCl_2 \cdot 12H_2O$		Monoclinic system	1240	– 33.6	– 43.2 (unstable)
$CaCl_2 \cdot 6H_2O$	Marble	Six-party crystal system	1718 (4 °C)	– 55.0	– 55.0

Source Kämäräinen (1993)

content is higher than that of pure air. In layers of sea ice that are not completely frozen, typical gas pores only comprise 0.5% of the volume. Meanwhile, the air in pores in the freeboard section near the surface generally comprises 1–5% of the volume.

Because sea ice encloses brine and air, its density is different from the density of pure ice without bubbles (0.917 g cm^{-3}). However, in practice, the density effect of brine and bubbles is often ignored. The typical density of sea ice is 0.915–0.920 g cm^{-3}, which is very close to that of pure ice. The density of the freeboard section may be lower than that of the sea ice, measuring 0.89–0.92 g cm^{-3}. Table 5.2 shows the relationship of sea-ice density with salinity and temperature.

Table 5.2 Density of sea ice at different salinity and temperature values (unit: g cm^{-3})

Salinity/‰	Temperature/°C							
	– 2	– 4	– 6	– 8	– 10	– 5	– 20	– 23
2	0.924	0.922	0.920	0.921	0.921	0.922	0.923	0.923
4	0.927	0.925	0.924	0.923	0.923	0.923	0.925	0.925
6	0.932	0.928	0.926	0.926	0.926	0.925	0.926	0.926
8	0.936	0.932	0.929	0.928	0.928	0.928	0.929	0.929
10	0.939	0.935	0.931	0.929	0.929	0.929	0.930	0.930
15	0.953	0.944	0.939	0.937	0.935	0.934	0.935	0.935

5.5.4 Effects of Biological Processes on Sea-Ice Chemistry

As sea ice melts, the salinity of seawater decreases at the ice-water interface, and the vertical stability of the seawater is enhanced. The algae that inhabit the ice breed in large quantities. A large number of algae and bacteria exist in sea ice, which greatly influences the sea-ice chemistry through photosynthesis and oxygen respiration.

With abundant ice algae and relatively high primary productivity, the dissolved inorganic carbon and CO_2 gas in sea-ice brine significantly decreases, the pH increases (up to 10), and O_2 becomes over-saturated. This significant decrease in dissolved inorganic carbon and CO_2 gas and O_2 over-saturation indicate that photosynthesis by algae exceeds the net respiration. In fact, this relationship occurs only when algae reproduce in large quantities. If a large amount of algae die and bacteria reproduce, this trend will be the opposite.

CO_2 gas and carbonates in seawater form a buffer system, which can significantly change the sea-ice brine, the interstitial water trapped in ice or the pH of honeycomb ice. The difficulty in collecting undisturbed sea-ice brine samples and the complexity of pH measurements under high salinity limit the work in this area. In general, the pH decreases with increasing ionic strength, which increases the solubility of calcium carbonate. However, this trend is concealed by the effects of photosynthesis in the ice. In a closed sea-ice system or for a large amount of organic matter in the sea ice, the increase in pH is mainly caused by photosynthesis, which decreases the amount of dissolved inorganic carbon.

The absorption of carbon during photosynthesis leads to the biological isotopic effect of stable isotopes, which makes the organisms rich in ^{12}C. The isotope ratio effect of photosynthetic algae is approximately –27‰; that is, the organic carbon from photosynthesis is enriched in ^{12}C compared with the available CO_2. However, the total biological isotope fractionation is influenced by many factors, such as the dissolved CO_2 concentration, carboxylesterase type, growth rate, cell size, and cell structure. During biological respiration, the isotope effect is very small, and the CO_2 produced by respiration has the same stable isotope ratio as organic carbon.

The limitation of diffusion in sea ice leads to the accumulation of some gases produced by sea ice, such as dimethyl sulfide (DMS), which has been increasingly investigated. DMS is mainly derived from the decomposition of dimethylsuophoniopropionate (DMSP). The concentration of DMSP in sea-ice organisms is an order of magnitude higher than that in deep water and open seawater. The production of DMSP is affected by light, temperature, nutrient supply, ultraviolet radiation, etc. However, the salt content in sea ice is an important factor that affects the production of DMSP in algae. The synthesis and accumulation of DMSP in cells occur under high-salinity conditions. When the salinity of the environment is reduced, DMSP decomposes and releases DMS. Therefore, sea-ice brine bubbles with pH values up to 10 can also promote this reaction. The distribution of sea-ice DMSP greatly varies, which reflects changes in the species of the organism community. The release of large amounts of DMS in sea-ice areas is usually associated with the melting of sea ice. Decreasing seawater salinity is beneficial to the decomposition of DMSP.

A large increase in marginal herbivores on the edges of ice usually increases the concentration of DMS in the seawater, increasing the release of DMS from the sea to the atmosphere. DMS is not the only volatile gas released by sea-ice algae; the tropospheric enrichment of reactive bromine is also closely related to sea ice. The high concentration of short-term BrO in the troposphere is caused by the autocatalysis of Br_2 released by sea ice and sea salt. Polar sea-ice algae can also produce bromine in halogenated compounds, such as large amounts of bromoform, dibromomethane, bromochloromethane, and methyl bromide. These substances can be converted to active bromine by photochemistry, which has important significance for polar chemical cycle.

The most important concern in sea-ice chemistry is inorganic salts, such as nitrate, nitrite, ammonium, phosphate, and silicate. In an abiotic system, when ice is formed, the concentration of these inorganic salts changes conservatively, which is proportional to the changes in salinity. The major ions (Na^+, K^+, Mg^{2+}, Ca^{2+}, Cl^-, and SO_4^{2-}) in sea ice of different types and ages basically follow the theoretical dilution line for changes in ice salinity, whereas nitrate, nitrite, ammonium, phosphate, and silicate significantly deviate from this predicted line. As with the dissolved gases discussed earlier, these deviations are closely related to biological activity in ice, which increase the spatial variability of these components.

During the formation of sea ice, the incorporation of water-soluble organic matter into the ice body is conservative, following changes in inorganic salts and dissolved gases. Only low molecular weight molecules can be retained in the ice. In the sea-ice microbial networks, algae are the sink of soluble inorganic nutrients. Isoxic protozoan and multi-cellular animals excrete soluble organics, and isoxic bacteria use organic and inorganic nutrients to maintain growth. Sea ice is also an important reservoir of soluble organic matter. The soluble organic matter in sea ice is considered to be an important source of organic material to enhance microbial activity under ice through melting ice or ice-water exchange. Because of the large dilution of sea ice, this effect only occurs in the water near the ice.

Questions

1. How can black carbon affect cryospheric changes?
2. What are the main sources of chemical components in glaciers?
3. What are the impacts of the carbon cycle in permafrost on the climate?

Extended Readings

Classic Work

***Atmospheric Chemistry and Physics: From Air Pollution To Climate Change*(Second Edition).**

Authors: Seinfeld J. H., Pandis S. N.

Publisher: New York: JOHN WILEY & SONS, INC, 2006

Atmospheric Chemistry and Physics (second edition) is a classic monograph on atmospheric chemistry and physics and is a widely recognized textbook for atmospheric sciences. This book provides rigorous and comprehensive information regarding atmospheric chemistry and physics, including the characteristics of aerosols and atmospheric pollutants, the interaction between aerosols and atmospheric pollutants, the effects of gas and atmospheric particles, the chemical components of the atmosphere, and mathematical calculations of atmospheric transport models. This book includes discussions of the atmosphere and trace components in the atmosphere, the chemical kinetics, atmospheric radiation and photochemistry of the atmosphere, the atmospheric chemistry of the stratosphere and troposphere, the aqueous-phase chemistry of the atmosphere, the physicochemical properties of atmospheric aerosols, the kinetics and thermodynamics of atmospheric aerosols, aerosol-radiation interactions, mesoscale meteorology, cloud physics, atmospheric diffusion, atmospheric circulation, the global sulfur and carbon cycles, chemical transport models of the atmosphere, and statistical models. Compared to the first edition, the second edition of *Atmospheric Chemistry and Physics* introduces the atmospheric chemistry of the stratosphere and troposphere, the formation, growth and dynamics of aerosols, air pollution meteorology, cloud formation and cloud chemistry, atmospheric chemistry and its interaction with climate, aerosol radiation and its effect on climate and chemical transport models, and other factors.

Chapter 6
Climatic and Environmental Record in Cryosphere

Lead authors: Tandong Yao, Ninglian Wang

Contributing authors: Baiqing Xu, Liping Zhu, Xiaohong Liu, Huijun Jin, Qingbai Wu, Xiaoxin Yang

Reconstruction of climatic and environmental changes in the past is fundamental to understanding cryospheric evolution, as well as interpreting present climatic and environmental changes and projecting future scenarios. This chapter will introduce the climatic and environmental significance of various paleo-proxies in the cryosphere (Fig. 6.1), with a special emphasis on high-resolution climatic and environmental reconstructions from the ice core, frozen ground, tree ring, and lake sediment records. The temporal resolutions and time spans of these records are showed in Fig. 6.2.

6.1 Ice Core Record

Ice cores are snow and ice samples of the vertical cylindrical column retrieved from glaciers (including ice sheet, ice cap, and other types of glaciers). Ice core study was first conducted on Greenland and Antarctic Ice Sheets, and later in the lower-latitude alpine regions. Polar ice cores have helped unravel high-resolution climatic and environmental history of Earth in the past 800,000 years, and will continue refresh our understanding of climate evolution and related processes in the earth system.

Ice core study dated back to 1930 when Ernst Sorge stayed at a station in Greenland over the winter, he quantified the density, ice layers and deep-frost layers of a 15-m bore hole in the snow and proposed that precipitation seasonality could be recorded in the snow. Ice core study in the modern sense was first proposed in 1954 by Henri Bader, as the then chief scientist and director of US Army Snow, Ice and Frozen ground Research Establishment (now renamed as US Army Cold Regions Research and Engineering Laboratory), who suggested that "snowflakes fall to Earth and leave a message". Henri then led ice core drilling projects on Site-2 in Greenland in 1956 and 1957 summers, and retrieved two ice cores, 305 m and 411 m in depth respectively. He also led the retrieval of a 309 m-long ice core from Byrd Station, Antarctica, in 1957/1958, and a 264 m-long ice core from Little America V at the Ross Ice Shelf, Antarctica, in 1958/1959. Willi Dansgaard, a pioneer of ice

D. Qin et al. (eds.), *Introduction to Cryospheric Science*, Springer Geography,
https://doi.org/10.1007/978-981-16-6425-0_6

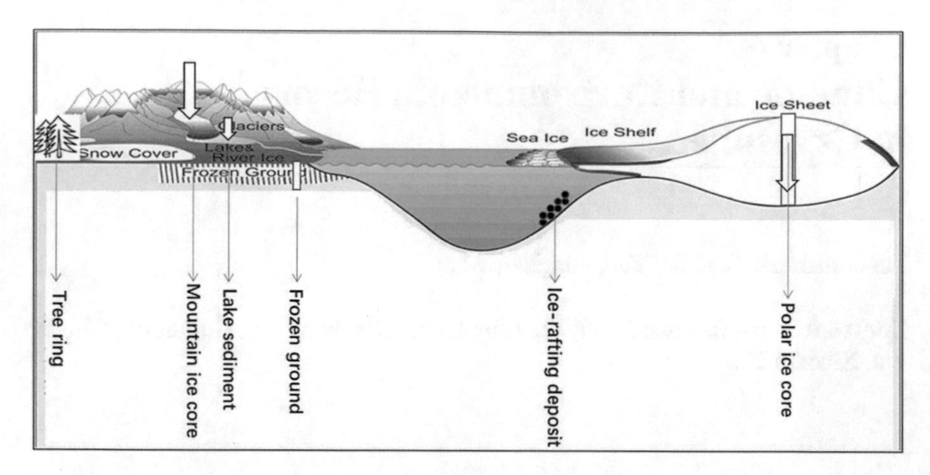

Fig. 6.1 Distributions of global cryosphere and its paleoenvironmental reconstruction media

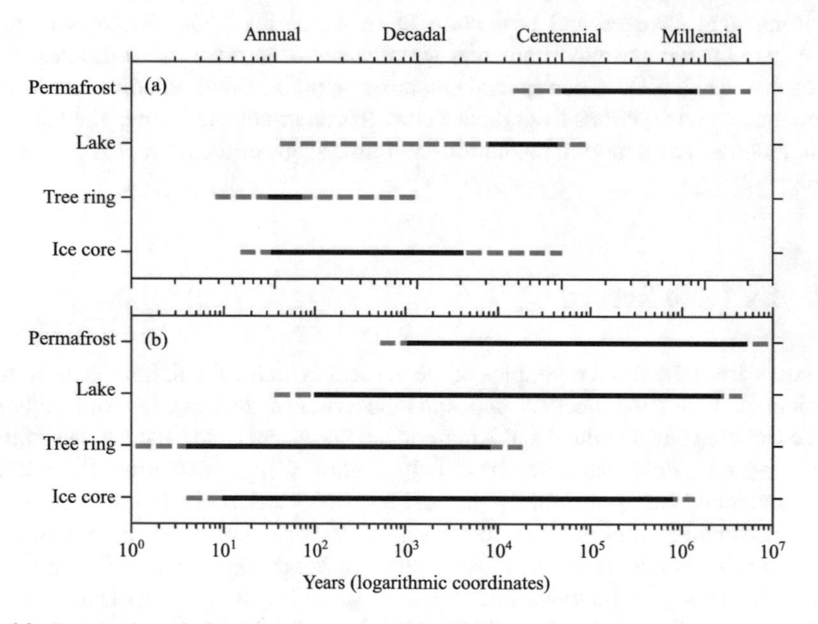

Fig. 6.2 Temporal resolution (**a**) and time span (**b**) in different climatic and environmental records

core paleoclimate study, quantified the relationship between stable water isotopes in precipitation and temperature after studying stable isotopes in precipitation and Polar Ice Sheet's surface snow samples. This finding has laid a physical basis for paleotemperature reconstruction from stable isotopes in ice cores.

The first ice core to bedrock (1,388 m in length) was retrieved from the Camp Century, Greenland, in 1966. Deep ice cores retrieved at the earlier stage also involved those from the Byrd Station (1968), Dome C (1979), and Vostok (the early 1980s) in

Antarctica and Dye 3 in Greenland (the early 1980s). In the early 1990s, eight nations within the European Union accomplished the Greenland Ice core Project (GRIP) at the summit of the Greenland Ice Sheet, and simultaneously an US research team also successfully launched the Greenland Ice Sheet Program 2 (GISP-2) project to retrieve ice cores 30 km away west from the summit, thus ushering a new era in ice core study. In the past three decades, more ice cores have been retrieved from the Antarctica (e.g., EPICA Dome C, Dome Fuji, EDML, Law Dome, Talos Dome, WAIS) and Greenland (NGRIP and NEEM).

The first long tropical ice cores were retrieved in 1983 from the Quelcaya ice cap, a largest tropical ice cap, which is 5670 m above sea level in the southeastern Andes of Peru. On the Third Pole (TP), the first deep ice core was drilled on Dunde Ice Cap in Qilian Mountains, Qinghai, China in 1987. In 1992, another deep ice core was drilled to bedrock on the Guliya ice cap in West Kunlun Mountains with a length of 309 m, which is the deepest and oldest ice so far acquired from the mid-latitude and low-latitudes. The Dasuopu ice core drilled on Mt. Xixabangma in central Himalayas in 1997 is the highest elevated ice core so far retrieved on Earth. Other ice cores retrieved from the TP included the East Ronbuk ice core from Mt. Qomolangma, the Malan and Puruogangri ice cores from central Tibetan Plateau, and several marine-type warm glaciers from the southeastern TP. Detailed information of major ice cores retrieved from the TP, Greenland and Antarctica can be found in Table 6.1 and Fig. 6.3.

6.1.1 Paleoclimatic Proxies in Ice Core

Ice cores acquired from accumulation zones on glaciers could preserve the past snowfall continuously, and thus recorded the past climate and environment changes reflected by the various proxies (Table 6.2).

$\delta^{18}O$ or δD in ice cores is widely regarded as a proxy for air temperature. Because isotopic fractionation of water is closely related to air temperature during evaporation and condensation, variations in $\delta^{18}O$ or δD in precipitation thus record could implicate temperature changes.

Ice core net accumulation is used as a proxy of precipitation. Precipitation is usually in solid form in glacier accumulation zones. Thus annual net accumulation in ice core (after calibrating from ice flow model) could be used to represent the variations in annual precipitation, regardless of mass loss caused by sublimation and wind scouring. Annual net accumulation reconstructed from ice cores drilled from dry snow zones could well imply annual precipitation variations at the ice coring site.

Air bubbles in ice cores are considered as the fossil of paleo atmosphere. As glaciers are formed through the long-term accumulation and snow metamorphism, air in the firn layer will gradually be trapped in bubbles when the compacted snow turns to ice. So far, among the all media using for reconstructions of palaeoclimate

Table 6.1 The characteristics of deep ice coring sites throughout the globe

	Location	Latitude	Longitude	Altitude/m	Accumulation rate/(mm/a)	Borehole basal Temperature/°C	Depth/m
Antarctica	Komsomolskaya	74° 5′ S	97° 29′ E				885
	Vostok	78° 28′ S	106° 48′ E	3490	23	– 55.5	3766
	Taylor Dome	77° 48′ S	158° 43′ E	2365	50–70	– 43.0	554
	Byrd	80° 1′ S	119° 31′ W	1530	100–120	– 28.0	2164
	Law Dome	66° 46′ S	112° 48′ E	1370	700	– 22.0	1195.6
	Dome F	77° 19′ S	39° 40′ E	3810	23	– 57.0	3035.2
	Talos Dome	72° 47′ S	159° 04′ E	2315	80	– 40.1	1620
	EPICA Dome C	75° 6′ S	123° 21′ E	3233	25	– 54.5	3259.7
	Siple Dome	81° 40′ S	148° 49′ W	621	124	– 24.5	1003
	EDML	75° 00′ S	00° 04′ E	2822	64	– 44.6	2774
	WAIS	79° 28′ S	112° 05′ W	1766	22	– 31	3405
	Dome B	77° 5′ S	94° 55′ E				780
	Berkner Island	78° 18′ S	46° 17′ W	886			181
	D47	67° 23′ S	154° 3′ E	1550			145
	Dronning Maud Land	75° 00′ S	00° 04′ E	2892	80		120
	DT263	76° 23′ S	77° 01′ E	2800	106		82.5
	Plateau Remote	84° S	43° E	3330	49		200
Greenland	Milcent	70° 18′ N	45° 35′ W	2410	530	– 22.3	398
	Dye 2	66° 29′ N	46° 33′ W	2100	374	– 17.2	100.2
	Dye 3	65° 11′ N	43° 49′ W	2038			2490
	Camp Century	77° 10′ N	61° 8′ W	1885	380		1387

(continued)

Table 6.1 (continued)

	Location	Latitude	Longitude	Altitude/m	Accumulation rate/(mm/a)	Borehole basal Temperature/°C	Depth/m
Greenland	Crete	71° 7′ N	37° 19′ W	3172	298	– 30.4	404
	GISP	65° 11′ N	43° 49′ W				2037
	GRIP	72° 35′ N	37° 38′ W	3238	230	– 31.7	3029
	GISP 2	72° 35′ N	38° 29′ W	3214	248	– 31.4	3053
	NGRIP	75° 6′ N	42° 19′ W	2917	190	– 31.5	3085
	NEEM	77° 27′ N	51° 4′ W	2450	220	– 29	2540
	Humboldt-M	78° 32′ N	56° 50′ W	1995	197		146.5
	Renland	71° 18′ N	26° 43′ W	2340			324.35
The Third Pole	Noijin Kangsang	29° 2′ N	90° 11′ E	5950	470	–	55.1
	Malan	35° 50′ N	90° 40′ E	5680	230	– 6.5	102
	Muztagata	38° 17′ N	75° 06′ E	7010	564	– 23	41.6
	Puruogangri	33° 55′ N	89° 05′ E	6070	342	– 7	170.42
	Guliya	35° 17′ N	81° 29′ E	6710	200	– 2.1	308.6
	Dunde	38° 06′ N	96° 24′ E	5325	400	– 4.7	139.8
	Dasuopu	28° 23′ N	85° 43′ E	7200	1000	– 13.8	167.7
	Nyainqêntanglha	30° 24.59′ N	90° 34.29′ E	5850			124
	East Rongbuk	27° 59′ N	86° 55′ E	6500	400	– 9.6	80.36
	Inylchek	42° 15.6′ N	80° 25.2′ E	5100	1440	– 11.2	165
	Tanggula	33° 06′ N	92° 04′ E	5645	189	– 7.7	190.5
South America	Huascaran	9° 07′ S	77° 37′ W	6048	130	– 5.2	166.1
	Sajama	18° 06′ S	68° 53′ W	6542	440	– 10.3	132.8

(continued)

Table 6.1 (continued)

	Location	Latitude	Longitude	Altitude/m	Accumulation rate/(mm/a)	Borehole basal Temperature/°C	Depth/m
	Quelcaya	13° 56′ S	70° 50′ W	5670	1200	– 3	168.68
Others	Kilimanjaro	3° 04.6′ S	37° 21.2′ E	5893	920	– 0.4	50.9
	Belukha	49° 48′ 26″ N	86° 34′ 43″ E	4062	560	– 14.2	139
	Elbrus	43° 20′ 53.9″ N	42° 25′ 36.0″ E	5115	1455	– 17	182
	Colle Gnifetti	45° 55.74′ N	7° 52.58′ E	4455	1400	–	80.2

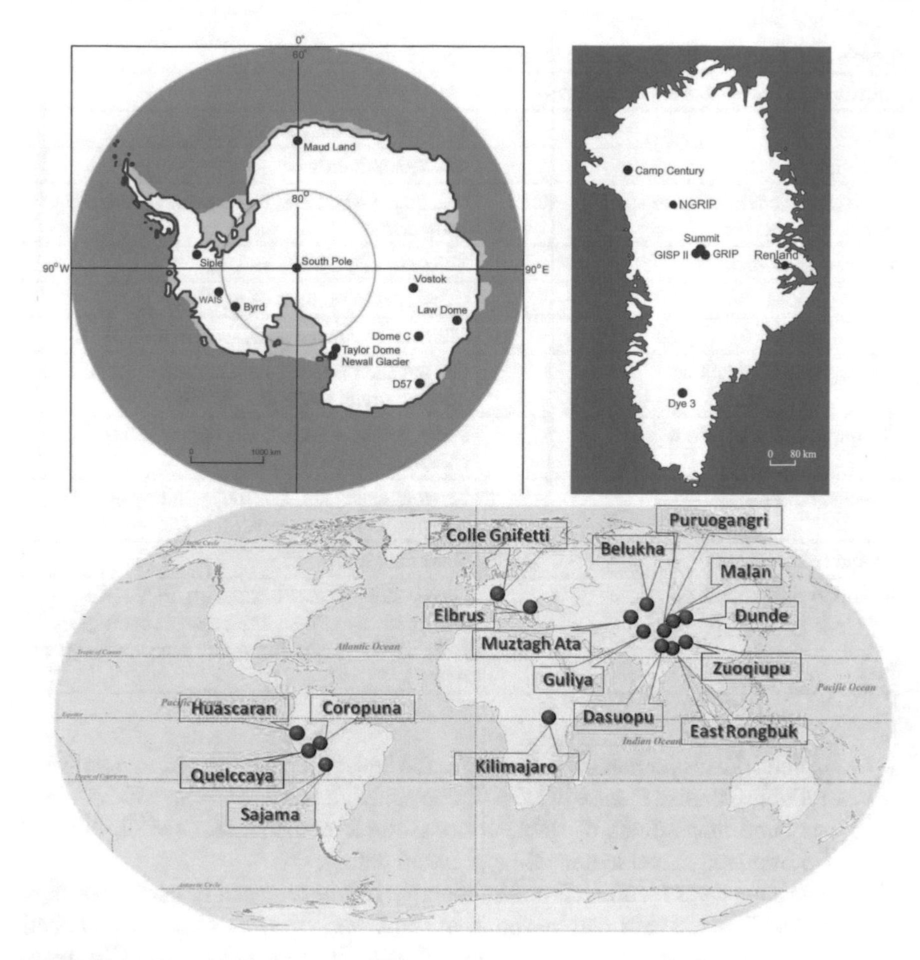

Fig. 6.3 Map of global deep ice core drilling sites

and/or palaeoenvironment, ice core is the only one that can be able to reveal past atmospheric greenhouse compositions at high resolution.

6.1.2 Ice Core Dating

Accurate dating is key to paleoclimatic reconstruction. Given different accumulation rates and ice core depth, ice core dating requires different approaches in order to achieve a chronology as accurate as possible. Major dating approaches include the following:

(1) Annual layer counting. This relies on the seasonality of some physical or chemical parameters in ice cores. For the mid-latitude and low-latitudes' alpine

Table 6.2 Proxies of climatic and environmental parameters in ice core

Climatic and environmental parameters	Major proxies
Temperature	$\delta^{18}O$, δD, melt layers
Precipitation	Net accumulation
Atmospheric chemical components (natural and anthropogenic)	Gas (e.g., CO_2, CH_4, N_2O) content, chemical components
Volcanic eruptions	Volcanic ash, ECM, nss SO_4^{2-}
Solar activities	Cosmogenic isotopes (e.g., ^{10}Be)
Sea ice extent	Methane sulphonic acid, sea salt ionic density
Atmospheric circulations	Glacial chemical component (major ions), micro-particle size and concentration
Variation in area of arid region	Particle concentration, contents of terrestrial chemical compositions
Biomass burning	Laevoglucose, ash, low molecular weight organic acids (e.g. oxalate), black carbon, K^+
Altitude of ice sheet	Gas content
Anthropogenic activities	Heavy metal (e.g., Pb, Cu, Hg), POPs (e.g., DDT), industrialized inorganic products (e.g., NH_4^+, SO_4^{2-}), anthropogenic greenhouse gases release

glaciers, dust layer generally formed at the end of the melting season is usually a reliable physical feature in ice core dating. High-resolution chemical components and fluctuations of stable water isotopic compositions ($\delta^{18}O$, δD) are also frequently used in annual-layer based dating.

(2) Reference layer identification. Radioactive matter released in nuclear tests in the 1950s and 1960s can be found in polar ice cores, as well as in tropical alpine ones, through the measurement of deuterium concentration or β horizon, thus the layers can be identified as a reference layer. Large quantity of sulfur dioxide (SO_2) released into the atmosphere during large-scale volcanic eruptions also forms numerous reference layers in polar ice cores. This record can be compared with documentary records of major volcanic eruption events in the past hundreds of years, thus offering important reference layers for ice core dating. In practice, appearances of odd peaks of non-sea-salt SO_4^{2-} (nss SO_4^{2-}) in ice cores, usually with nss SO_4^{2-} exceeding the average value over 2 times the standard deviation, correspond to volcanic eruptions.

(3) Radioactive isotope method. Taking advantage of the half-life and the comparison between the original concentration and the remnant during measurement, radioactive isotopes can be used for ice core dating. Radioactive isotopes in the atmosphere are reserved in ice cores through dry or wet deposition with aerosols on the ice surface, or being trapped in air bubbles when compacted snow turns to ice. Radioactive isotopes in ice cores are mainly derived from three sources, cosmogenic (e.g., ^{12}S, ^{37}Al, ^{14}C, ^{36}Cl, ^{10}Be, ^{81}Kr and etc.),

nuclear tests (e.g., ^{3}H,^{137}Cs, ^{90}Sr and etc.) and other nuclear industry (e.g., ^{210}Pb and etc.). Among those, ^{210}Pb, ^{10}Be and ^{36}Cl are most frequently used radioactive isotopes in ice core dating.

(4) Ice flow model. This approach requires the establishment of a depth-age quantitative relationship from ice flow models in order to identify the year at a particular depth.

(5) Orbital tuning. This approach is based on the theory that paleoclimate variation and its driving factors (e.g., solar insolation) have cycles during the Quaternary, and the cycles can be applied to the identification of time scales with a resolution of thousands of years or even longer. A target curve of climatic drivers is first identified based on earth inclination and precession, and then interpolated on a curve of equal time interval from the original control time in the climatic proxy curve, to filter for a curve corresponding to the variation cycle of climatic drivers. The comparison between the target curve and filtered curve finally yields a chronology with acceptable uncertainty.

(6) Cross-dating with similar time series. With the pretext that significant climate phases and/or events are global with a huge regional impact, this approach relies on the comparison of ice core variation series with a known climatic curve to construct depth-age relationship in ice cores.

6.1.3 Ice Core Records in Greenland Ice Sheet

A total of six deep ice cores drilled to bedrock in Greenland during the past 50 years contain ice from the entire Holocene, the last glaciation, and the Eemian (Fig. 6.4), thus confirming the fact that most of Greenland was covered by ice during the Eemian. Comparison of present day $\delta^{18}O$ values with those observed in the Eemian ice facilitates the quantification of changes in size of the Greenland Ice Sheet during the Eemian. There is a rather uniform (3‰) increase in the six ice cores in Eemian $\delta^{18}O$, including the small Renland ice cap which is only 330 m thick and whose Eemian ice remained 3 m thick till today. The contradiction of the ice thickness close to bedrock in Renland to the ice flow model imposed narrow constraints on thickness changes at Camp Century, GISP2, GRIP, and NGRIP as well. The 2.5–3.5‰ Eemian $\delta^{18}O$ increased in most Greenland ice core $\delta^{18}O$ should therefore be attributed to temperature change. The NGRIP and Camp Century Eemian $\delta^{18}O$ increases a bit less than the Renland $\delta^{18}O$, which implies that the southern part of the Greenland Ice Sheet decreased more significantly than the northern part during the Eemian. Latest effort featuring North Greenland Eemian Ice Drilling NEEM succeeded in retrieving a 2540-m-long ice core, thus allowing for a complete record of the Eemian interglacial (130,000–115,000 years ago). The NEEM ice core $\delta^{18}O$ increased significantly at 126 kyr, which implies warming by 8 ± 4 °C than the average of the recent millennium and representative of the maximum temperature in Greenland after the onset of Eemian. This peaking in temperature is followed by cooling during the interglacial, which is attributed to the strong local summer insolation decreasing trend. Despite

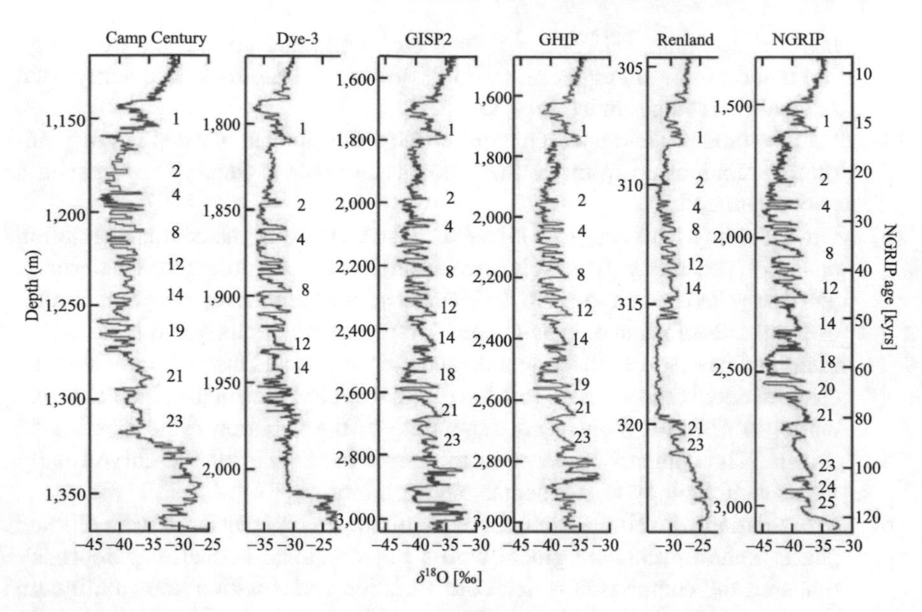

Fig. 6.4 Early Holocene, Glacial, and Eemian δ^{18}O records from six Greenland ice cores (John Sen et al. 2001)

that, reconciliation of the NEEM data with Greenland Ice Sheet simulations points to a modest contribution (≤2 m) to the observed 4–8 m Eemian sea level high stand.

All Greenland ice core δ^{18}O showed abrupt shifts in the glacial parts, which are identified as Dansgaard-Oeschger (D-O) events. Different ice coring sites, however, observe different intensity in δ^{18}O increase, with the Camp Century core showing the largest δ^{18}O changes whereas the changes at Renland features the smallest. Such a heterogeneity is guesstimated to be due to a split jet-stream by the giant Laurentide ice sheet and geographically dependent moisture trajectories during the glacial period.

The Holocene in Greenland ice core is generally represented by a period of remarkably stable δ^{18}O values. Synchronization of δ^{18}O in Dye-3, GRIP, and NGRIP records using common volcanic reference horizons showed significant common decreases at about 8,200, 9,300 and 11,400 years ago, known respectively as the 8.2 and 9.3 k cold evens and the Preboreal oscillation. Those events are also observed in the records of ocean sediment cores, tree rings, and stalagmites from all over the Northern Hemisphere, and they are believed to be associated with very large freshwater outbursts from the decaying Laurentide ice sheet.

6.1.4 Ice Core Records in Antarctic Ice Sheet

The time scale of ice cores from the Antarctic Ice Sheet can be extended back to the past 800 ka (Fig. 6.5), consisting of eight glacier-interglacial cycles. The Antarctic

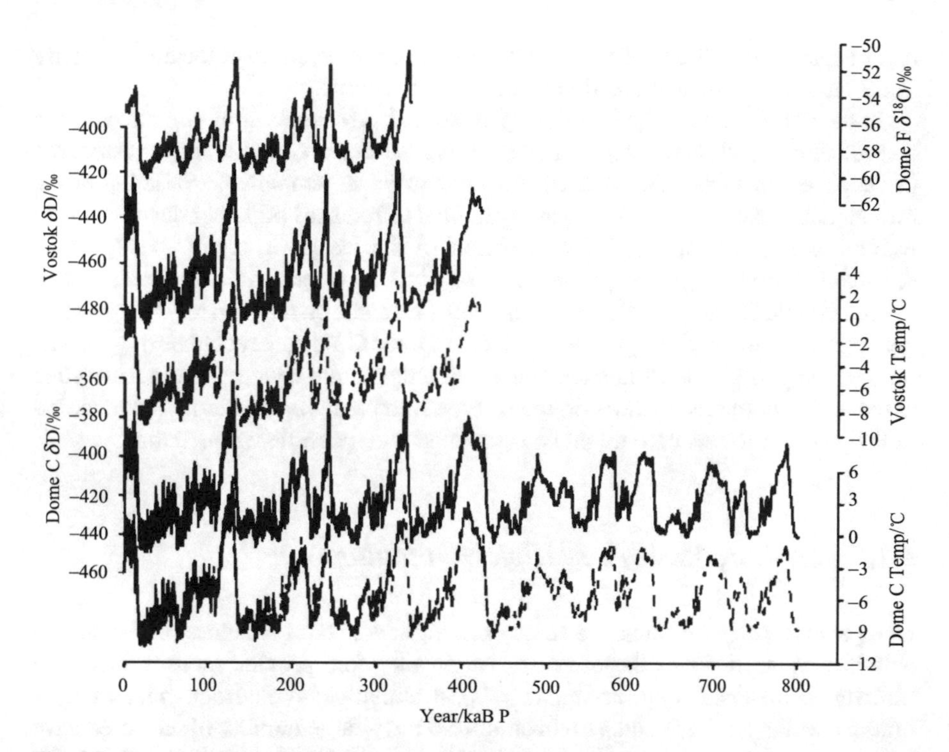

Fig. 6.5 The time series of stable isotopes and reconstructed temperature from Dome F, EPICA Dome C and Vostok ice cores in Antarctica (EPICA community members, 2004)

ice core records show that, under the influence of earth's orbital forcing, the glacial-interglacial cycle is characterized by the composition of three cycles of 100 ka, 40 ka and 19–23 ka, of which the dominant cycle period is 100 ka. Temperature fluctuation in 800–430 ka B. P. was less intense than that after 430 ka B. P. The glacial period usually take up over 80% of the total glacial-interglacial cycle, leaving the interglacial period to take up less than 20% by spanning 10–30 ka. Comparison of stable isotopes in ice cores drilled from the Antarctica shows good consistence of climate changes over the past 400 ka (Fig. 6.5). Integration with the Milankovitch theory reveals that the major driver for glacial-interglacial climate variation in Antarctic ice core records is the shift of the summer solar insolation to high latitudes regions in the northern hemisphere.

The most noticeable feature of the glacial-interglacial cycle in the Quaternary is that there is a significant transition of the dominant cycles occurring at ca 900 ka B.P., from 40 ka cycle to 100 ka cycle (also known as the mid-Pleistocene climate transition). Long-term ice core records are crucial for understanding this climate transition, however, only a weak signal has been observed in the Dome C ice core record. For this, Dome A area of Antarctica features an annual minimum temperature as low as –58 °C, in addition to low accumulation rate (<25 mm w.e./a), negligible ice flow velocity and thick ice, meeting all necessary requirements for ice core record spanning millions of years. It might therefore be an ideal location to study mid-Pleistocene climate transition from ice core records.

In the circumstance of glacial-interglacial cycle, the forecast of climate changes on millennium scale and abrupt climate events is particularly important. δ^{18}O records in deep ice cores from the Antarctica show consistent temperature variation at the millennium scale, though with distinct regional differences probably due to varying water vapour sources, altitudinal evolution of the ice sheet, and/or precipitation seasonality/periodicity. Composite curve of δ^{18}O time series in Law Dome, Talos Dome, Simple Dome, EDML and Byrd ice core records on the Antarctic edge shows coherent variation with δ^{18}O series in EPICA Dome C ice core record during the last interglacial period, confirming the consistent temperature changes in the Antarctica at the millennium scale. However, the reversed temperature variation trend recorded in the Talos Dome ice core might be associated with possible wrong dating.

6.1.5 Ice Core Records in High Mountains

Compared to polar ice cores, the cores from high mountains are closer to the human activities area. Study of these records might therefore provide more information directly on understanding the impact of past climatic and environmental changes on the development of human civilization. So far, a large number of ice cores have been retrieved from high mountain regions in the mid- and low-latitudes (Fig. 6.3), particularly on the Tibetan Plateau where expansive glacial coverage constitutes an important area for ice core study. Comparison of those ice cores retrieved from different mountains (Fig. 6.6) can shed light on glacial growth and variations under different circumstances. The following section will focus on the study of an ice core retrieved from Mt. Xixabangma in central Himalayas, as it helps uncover past climatic and environmental changes in the mid-latitudes and low-latitudes.

Located on Mt. Xixabangma in central Himalayas, the Dasuopu glacier (28° 23′ N, 85° 43′ E) is a valley glacier with a length of 10.5 km and covering 21.67 km^2 in area. Its snow line altitude is 6200 m a.s.l. The ice coring site is located in an ice and snow platform (7000 m a.s.l.) within the accumulation zone, where annual net accumulation is estimated to surpass 700 mm. The 10 m bore hole temperature is close to –14 °C and the ice temperature at the bottom of the core is –13 °C. In 1997, a China-U.S. joint expedition team succeeded in retrieving three deep ice cores from the site, two of which reached the bedrock and measured, respectively, 160 and 150 m, in length. Lab results of δ^{18}O and ion (anion) concentrations in the Dasuopu ice core showed clear seasonality, thus allowing for ice core dating for the upper part by counting annual layers. The chronology was further verified by the β activity measurement that clearly identified the ice layer affected by the 1963 nuclear test. The bottom of the ice core was dated using the ice flow model.

Net accumulation reconstructed from the Dasuopu ice core is a good proxy of Indian summer monsoon precipitation (Fig. 6.7), and it reveals a centennial variation rhythm of the monsoon precipitation in the past 400 years. Specifically, the ice core

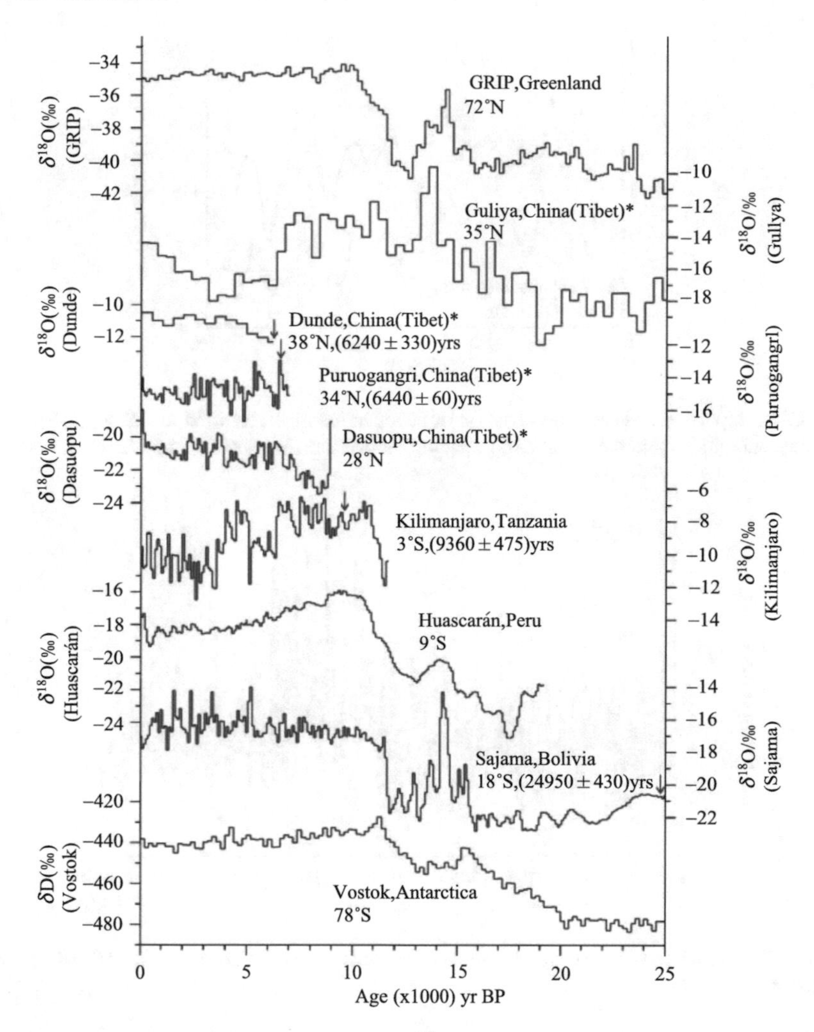

Fig. 6.6 Comparison of ice core records of $\delta^{18}O$ over the areas from Antarctica to Arctica (Thompson et al. 2005)

record (Fig. 6.8) shows monsoon precipitation began to increase in the early 17th century, peaking during 1650–1670, but turned to decrease after the Little Ice Age (LIA). Monsoon precipitation was shown to be low throughout the 18th century. The period between 1820 and 1920 appeared to be wet, but was followed by a decrease in precipitation.

The SO_4^{2-} concentration of the Dasuopu ice core reflected atmospheric SO_4^{2-} concentration in South Asia in the past 1000 years. It featured low values with small standard deviations prior to 1870, followed by clear increase afterwards with an accelerated increase since 1930. It is also found to increase in phase with the industrial emission of SO_4^{2-} to the atmosphere by South Asian nations.

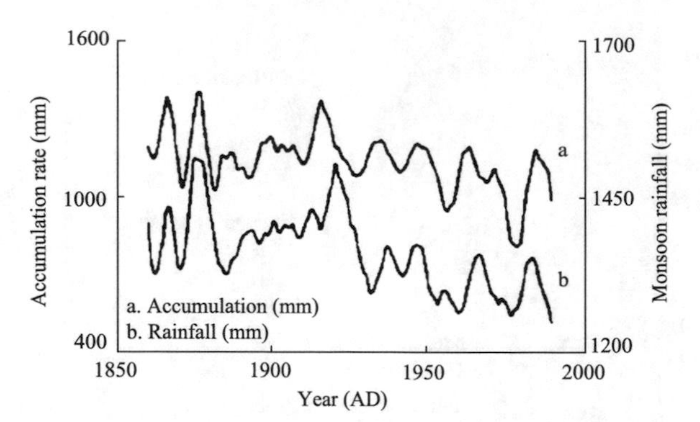

Fig. 6.7 Comparison between the variation of net accumulation recorded in Dasuopu ice core in Himalaya mountains and the amount of precipitation over northeast of Indian (Yao et al. 2000)

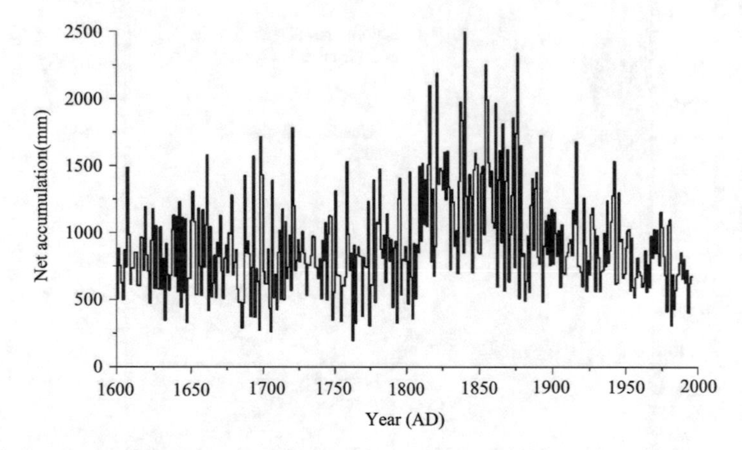

Fig. 6.8 The variation of net accumulation recorded in Dasuopu ice core from the Himalayas over the past 400 years (Yao et al. 2000)

The CH_4 concentration recorded in the Dasuopu ice core in the past 2000 years averaged around 825 ppbv, the value higher than that in the Antarctic and Greenland ice cores by, respectively, 160 ppbv and 120 ppbv during the same period, thus it verified tropical wetlands as an important source of atmospheric CH_4. It showed accelerated increase from 1850 onward and reached a value 1.4 times higher in the past 150 years, which is believed to reflect the strong influence of human activities on atmospheric CH_4 concentrations. The CH_4 concentration in the Dasuopu ice core hit record low during the LIA, and showed a negative increase in anthropogenic CH_4 emission in the interim of the two World Wars in the 20th century.

6.2 Frozen Ground Record

Frozen ground is the result of cold climates, and records past climatic and environmental changes through not only its temperature and active-layer thickness, but also its concurrent periglacial environment, geomorphology and fauna and flora. The paleoclimatic significance of periglacial geomorphology was proposed as early as the early 20th century. Later, German scientists studied ice-wedge casts, sand-wedges and involution resulted from the freeze–thaw cycles, and thereupon identified the southern limit of the paleo-permafrost in the European plain during the Würm Glaciation in the Late Pleistocene. Chinese scientists also discovered paleo-permafrost remnants in various areas in China, including frost-cracking wedge structures, paleo-cryoturbations, thermokarst landforms and others resulted from the past freeze–thaw cycles. Four frost-cracking wedge structures can help identify the occurrence of past permafrost, namely, soil-wedges, sand-wedges, and ice-wedges and ice-wedge pseudomorphs. As a common landform in the frozen ground area, frost heave is an upward swelling of soil upon freezing as a result of the increasing presence of ice as it grows towards the surface, upwards from the depth in the soil of where freezing temperatures have penetrated into the soil (the freezing front or freezing boundary). Frost heaving landforms are also important indicators for the occurrence of past permafrost.

Permafrost evolution since the Quaternary, the Mid-Pleistocene and Late Pleistocene in particular, can be directly or indirectly deduced and reconstructed from paleo-permafrost landforms and their distribution. Generally speaking, paleo-permafrost evidence can indicate two trends: permafrost aggradation and degradation. Direct evidence for aggradation can be found from paleo-permafrost table and pingo scars; direct evidence for permafrost degradation, on the other hand, can be found with thermokarst lowlands, cryoturbations, inactive rockglaciers, among many others. The most crucial and typical among these indicators are ice wedges, ice-wedge pseudomorphs, rockglaciers and palsas.

6.2.1 Ice-Wedges

Born from cold climates, ice wedges are shaped by terrestrial variables such as hydrothermal condition, ambient soil types, microreliefs and vegetation. As reliable indicators for the occurrence of past permafrost, ice wedges and the consequent ice-wedge pseudomorphs are widely used in the reconstruction of paleo-climate and paleo-environment.

An ice wedge is a wedge-shaped ice chunk with foliations, which gets thinner downward and can measure up to dozens of meters into the ground (Fig. 6.9).

In cold season, the active layer above the permafrost table freezes and the upper soil layer contracts to form frost cracks. When ground warms up, meltwater from the active layer fills the cracks and re-freezes upon contacting with the permafrost

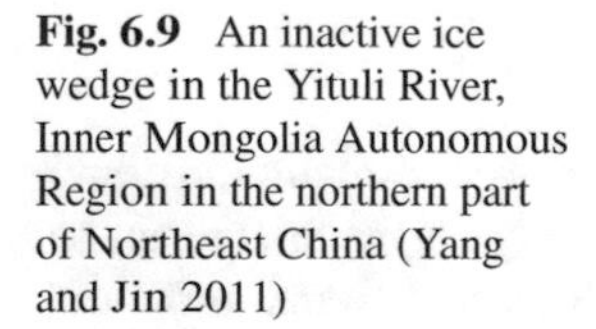

Fig. 6.9 An inactive ice wedge in the Yituli River, Inner Mongolia Autonomous Region in the northern part of Northeast China (Yang and Jin 2011)

to form ice veins. The refreezing processes in cold seasons result in a volumetric expansion, which in turn increases the width and depth of the fracture, or frost cracking. With more meltwater seeping into the cracks and refreezing afterwards, the ice veins grow. As this cycle repeats year after year, the ice wedges grow to form ice wedge polygons. Ice wedges can be classified as epigenetic and syngenetic depending on whether deposition is occurring before or simultaneously with the ice wedge formation. They can also be categorized as active and inactive in terms of their growing status. The relative position of polygon centroids to land surface can also be used to group ice wedges, as there are high-center and low-center ice wedge polygons.

In areas of past permafrost, ice wedges have melted and are filled with, instead of ice, sediments and dirt from the surrounding walls. Such ice wedge casts are known as ice wedge pseudomorphs, which can be used to estimate past climate. A number of ice wedge pseudomorphs have been discovered across Europe and America since the 20th century, providing vital evidence for the reconstruction of permafrost distribution and paleo-climate and environment since the Last Glaciation (or even as early as Early to Middle Pleistocene). Considerable ice wedge pseudomorphs have also been found in modern permafrost and seasonal-frost regions across China, typically in the Northeast, North and Northwest China, as well as on the Qinghai-Tibet Plateau and in the periphery high mountains. These are precious evidence for evolution of permafrost in China and East Eurasia. Quaternary strata in permafrost regions have preserved various wedge structures of ground freezing-related processes of different ages, including sand wedges, soil wedges, ice wedges and ice wedge pseudomorphs. Identification and chronological sequencing of these structures are the prerequisites for reliable paleo-environment reconstruction.

Ice wedges are most likely to develop in clusters on relatively flat fields. Wedge profiles become narrower from top to bottom (Fig. 6.10). The shape of wedge profiles vary with frost-cracking depth and freeze–thaw cycles. The profiles with narrow bottoms and wider tops indicate frost cracking deep enough to go through the active layer into the upper soil horizons of permafrost. The more spread-out profiles, on the other hand, are signs of frost cracking within the active layer, where frequent and intense freeze–thaw cycles give rise to a tongue-shaped wedge. In this sense, double layers and tip at the lower end are distinctive of ice wedge pseudomorphs. Sediments inside the ice wedge pseudomorphs are vertically layered, as most wedge walls have been deformed.

In China, ice wedges are not as widely distributed as they are across the Arctic coasts. Currently, all ice wedges found in China are inactive. Arid climate on the Qinghai-Tibet Plateau may have repressed the growth of ice wedges. As a result, it is rare to have in-depth studies on climate and environment being based on ice wedge records in China.

A group of inactive ice wedges were found on the first terrace of the Yituli River in the middle Great Khingan Range, Northeast China. Covered by peat layer and surrounded by peaty silt, one of the wedges is 1.4–1.6 m in height, 1.1 m in width with an observable height of 1.45 m. The top of this wedge is found to be consistent with existing permafrost table (Fig. 6.11). The ice wedge is transparent with clear

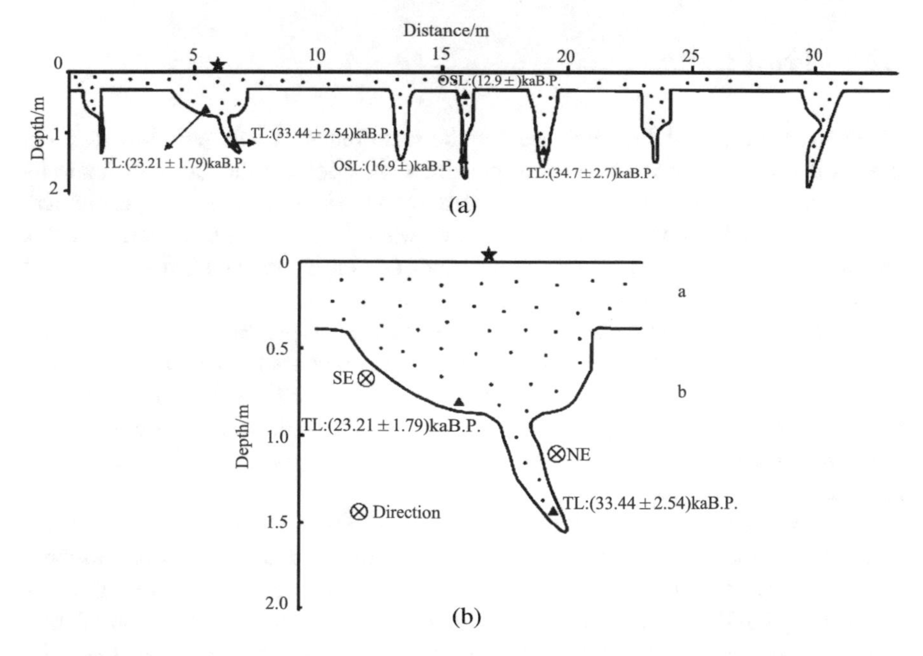

Fig. 6.10 The pseudotype of ice wedges 14 km south of Uxin Banner, Ordos Plateau, North China (Cui et al. 2002). **a** Vertical profile; **b** second profile to the left is a sand wedge with its upper and lower parts from different ages, as indicated by different wedge wall inclination: a represent the yellow layer of fine sand, and; b stands for the purple layer of gravel dirt

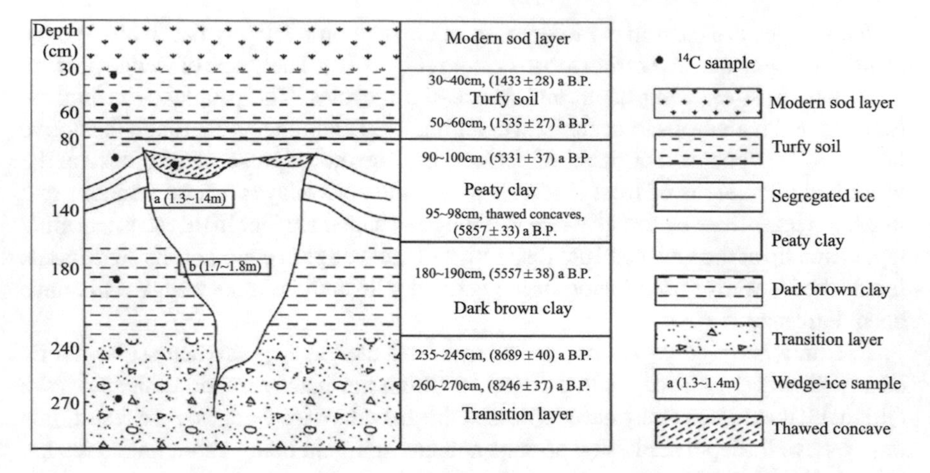

Fig. 6.11 The profile of ice wedge in Yituli River, Eastern Inner Mongolia, NE China (Yang and Jin 2011)

vertical bedding. Gray silt is scattered between its foliations. A ^{14}C dating has put the formation of this wedge at between 3,300 and 1,600 a B. P.

6.2.2 *Frost Mound*

Pingo scars are irrefutable evidence for the occurrence of past permafrost. Frost mound is a product of frost heaving, presented as a conical mound with ice at its core. The landform demands cold climate/environment as well as certain hydrological and hydrogeological conditions to take shape, whereas the presence of permafrost is not essential. The pingo scar preserves climate and environmental information that requires careful examination.

Large-scale pingos in closed systems are observed in permafrost regions across Northern Russia, Canada and Alaska. Large-scale pingos are more common in areas of discontinuous permafrost. Pingos found in the Great and Lesser Khingan Mountains in Northeast China are generally small in scale, with closed systems and of seasonal duration. Those found in permafrost regions on the Qinghai-Tibet Plateau are more diverse in type.

Frost mound study aims at clarifying the connections between the water source and its surrounding rocks. By determining the age of the mound formation, research in this line can depict how the mounds develop, evolve and decay. Frost mounds on the Qinghai-Tibet Plateau can be divided into two groups on the basis of their water sources: one fed by confined (artesian) water under the permafrost (the sub-permafrost water); and the other by water above the permafrost (the supra-permafrost water). The former group often comes with open systems while the latter, closed one.

Modern frost mounds are typical micro-landmark for permafrost regions. If frost/peat mounds are found in swamps, especially those on the edge of permafrost regions, it is safe to confirm the presence of past permafrost there. Pingo degradation happens along with expanded fractures with collapsed tops, while degradation of palsas take the form of substrate creeping.

Many collapsed pingos are discovered along the Qinghai-Tibet Highway and Qinghai-Kang Highway, as well as in the source area of the Yellow River. There, lowlands from pingo scars are formed either around beaded streams or in a U shape. Each lowland has an outlet. The largest lowland has a diameter of 200 m and a depth of 5–6 m. The ^{14}C dating of humus from the top of one pingo scars has put it at about 3,925 a B. P. The humus from the center of the lowland is dated at 720 a B. P. It thus can be inferred that the cluster of paleo-pingos were formed in the cold period of the late Holocene and then collapsed into lowlands in the warm period of the late Holocene. As such, sediment from pingos and lowland centers can be used as an evidence for reconstructing permafrost evolution and environment changes.

6.3 Tree Ring Record

Variations in tree ring width which reflect the dynamics of tree growth are determined largely by climates and ambient environments, thus tree-ring width can be used to infer the climatic and environmental history. Tree ring density has particular advantage in revealing intra-annual variations in various climatic factors, such as the seasonal extremely climatic events and/or persistent climatic event. Stable isotopic ratios in tree rings (e.g., C, H, O and N) are sensitive indicators, and record the biophysical responses of isotopic fractionation to climatic and environmental changes during tree growth. Climatic factors and atmospheric compositions affect the stomatal conductance of leaves, thus determining the photosynthesis assimilation rate and the degree of carbon isotope fractionation and their ultimate isotopic ratios in plants. Stable carbon isotopic composition ($\delta^{13}C$) in tree rings mainly records the balance between stomatal conductance and photosynthesis, while stable oxygen and hydrogen isotopes ($\delta^{18}O$ and δD) mainly record changes in water source and the leaf evaporative enrichment fractionation. As a result, tree-ring $\delta^{18}O$ and δD mirror temperature signal during precipitation processes and atmospheric humidity signal during plant evapotranspiration.

Alpine tree rings serve not only as proxies for climate parameters such as temperature and precipitation, but also as effective recorders of cryospheric changes such as glacier advancing/retreating, permafrost environmental variation, and snow cover changes etc.

6.3.1 Major Climate Events Recorded by Alpine Tree Rings

Alpine trees can survive under low temperature environment with the slow growth rate and the longevity. In this case, temperature is usually the major limiting factor of tree growth. Tree-ring width tends to be more sensitive to temperature variability, and thus tree-ring width is a reliable proxy for long-term temperature reconstruction. In arid regions, however, water availability can replace temperature to become the limiting factor, in which cases tree rings are used to track the precipitation or river runoff changes.

The Medieval Climate Anomaly (MCA), the Little Ice Age (LIA), and the 20th Century Warming are the most significant and influential climate events over the last 1000 years. The climate change over cold regions is shaped by above major events over the last 1000 years.

Climate anomaly during the MCA (900–1300 AD) has been revealed by tree rings over different climatic regions. Tree-ring width chronology of Qilian juniper (*Sabina przewalskii Kom.*) from the alpine Qilian mountains reveals that the period 1050–1150 AD was a warmer period, while temperature-sensitive tree ring chronology from upper tree line in the Qaidam Basin suggests that the period 1050–1150 AD to be relative cold. Temperature series of Tibetan Plateau over the last 1000 years indicate a cold phase during the 11th Century. Thus, tree-ring records from different regions revealed on the various starting time of LIA. Tree-ring records from the Qilian Mountain reveal the LIA occurred from 1440 to 1890; records from northeast Tibetan plateau placed the LIA occurrence from 1599 to 1702. Overall, tree-ring record in Tibetan Plateau also indicate the LIA extremum in the 17th Century when the plateau was commonly cold. Temperature series reconstructed from alpine tree-ring records indicate that the 20th century is the warmest century over the last 1000 years, which has been confirmed by records of tree-ring stable isotopes. However, fertilization effect induced by rising CO_2 concentration may have amplified the temperature increase inferring from tree ring width in certain regions.

6.3.2 Glacier Fluctuations Recorded by Alpine Tree Rings

With yearly resolution and absolute dating, tree-ring records are used to reconstruct glacier advance and retreat in many regions. The basic principles are as follows: trees standing in the way of a glacier's advance/retreat will get injured or even killed; on the other hand, during a glacier retreat, trees will grow in places previously occupied by the glacier. All such changes may record by tree rings, which can be used to reconstruct the glacier movement.

To calculate the minimum age when a glacier starts to retreat, previous studies would plus the age of the oldest living tree over the glacier moraines to the settlement period of the specific tree species. In this case, the pith of the living tree serves as an archive of its life processes, which, together with a calibrated settlement period, can infer the latest possible time when the glacial deposits takes shape; the age of the trees is determined by tree-ring counting. Relic wood in moraines is also crucial for the reconstruction of glacier advance, because the tree-ring timeline can be extended by cross-dating the relic wood with living trees. Unusual rings found on relic wood can reflect the occurrence of extreme environmental events, and provide crucial value for cross-dating.

In China, tree-ring-based glacier researches are focused mainly in the area of the Tibetan Plateau. Pith analysis of tree ring samples of populus pseudoglauca and picea balfouriana from the end and lateral moraines of Midui Glacier reveal the area's LIA started around 1767 AD (Fig. 6.12). The Midui Glacier moves quietly with spatial consistence compared to other glaciers in China and across the broader Northern Hemisphere. At interdecadal scales, there is an 8-year lag between temperature change and glacier movement. In addition, tree-ring samples of larix griffithii and picea balfouriana have also been obtained from glacial fore-field of Gawalong Glacier and Xincuo Glacier in southeast Tibet. Reconstruction of the glacier movement during the LIA based on these samples indicate different glacial deposits formed at different times. After the LIA, glacial movement in the 20th Century also varies among the glaciers from different geographical regions. The Gawalong Glacier has advanced in the late 20th century and it reached the same size as during the LIA.

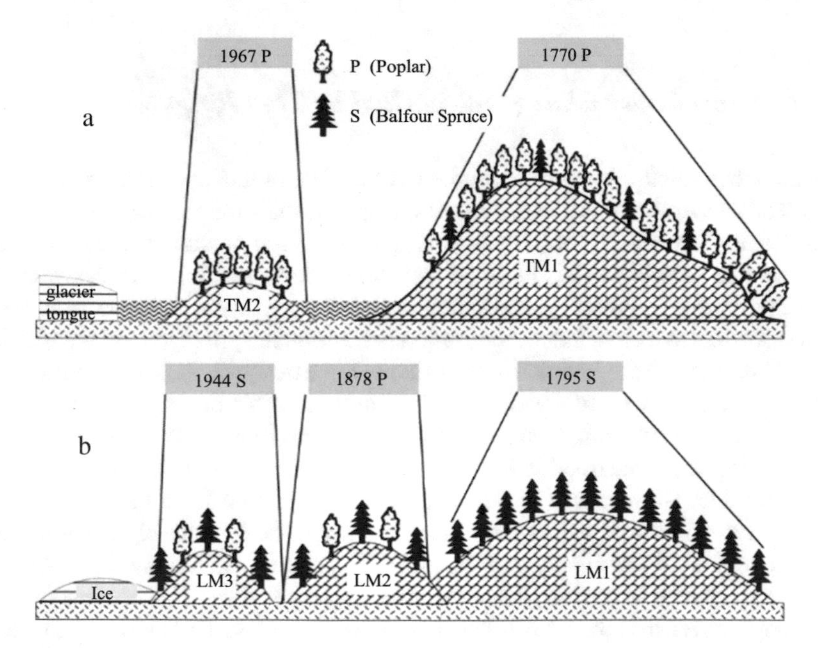

Fig. 6.12 Pith age distribution of the oldest trees over moraines of the Midui Glacier (Xu 2012) (a. trees on the terminal noraines; b. trees on the lateral moraines)

6.3.3 *Environmental Changes in Permafrost Region Recorded by Tree Rings from High Altitude Regions*

As global warming intensifies, the temperature of permafrost rises and the active layer of permafrost grows ever thicker and consequently transfers more water downward. These alterations can remarkably influence the growth environment for forests growing in northern high latitudes. As the soil surface lifts with freezing and drops with thawing, trees affected by the freeze–thaw cycles become tilted, and this impact can be recorded as a phenomenon known as compressed wood. In permafrost regions, the increment of tree growth is closely related to active layer temperature and water during the growing season. Warmer soil in winter could speed up the soil thawing in spring, which further influences the time that the trees start to grow. Studies based on tree-ring width index indicate that warming-induced active layer thickening will enhance the productivity of larch forests in permafrost regions, where the limiting factor of tree growth will shift from temperature to water available. according to records on larch (*Larix*) tree-ring width and stable isotopes, correlation between $\delta^{13}C$ and $\delta^{18}O$ have turned from negative to positive since 1960 in the north of central Siberia. Shift like this indicate in permafrost region, soil moisture became gradually decreased over the last half of the 20th century.

6.3.4 *Snow Cover Changes Recorded by Tree Rings*

Tree growth in subalpine regions is limited by spring snow cover and summer temperature. The deeper the snow cover becomes in spring, the stronger negative impacts it does to tree growth in the current growth year. Research on subarctic forest-tundra based on tree growth models revealed that more winter precipitation (snow cover) can delay snow-cover melting, and thus affecting forest growth. Because snow-cover takes more time to melt, cambium activation is consequently delayed and the growing season shortened. This alteration not only slow down tree growth, but also make trees less sensitive to temperature variability. Researches in this line indicate that, under certain conditions, tree rings can be used as effective proxies for snow-cover changes.

According to reconstruction based on tree rings from Gunnison River Basin in West Colorado, US, changes in snow water equivalent (SWE) in the 20th Century are within the threshold values over long-time scale. Years with high or low SWE peaks did not distribute evenly over the last 400 years. By examining snow cover on the northern Rocky Mountains, however, studies show that the 20th century suffered more snow-cover loss than any other periods over the last 1000 years. The snow-cover loss are relevant to winter moisture, especially the storms, from the Pacific to the Rocky Mountain.

El Niño–Southern Oscillation (ENSO) is indicated by some researchers as influential in winter climate across the western US. A decline of snow cover after El Niño is detected on the northern Rocky Mountains, whereas a snow-cover increase is observed at the same time across the snowy Southwest US. For the source of the Colorado River, SWE records over the last 1000 years reveal low snow cover only between 1300–1330 A.D. and 1511–1530 A.D., where the averaged SWE are considerable with that of the early and the late 20th Century. In contrast, SWE values are quite high from the 1650s to the 1890s on the northern Cordillera Mountains, which is consistent with the glacier advance during the Holocene. SWE also has been reconstructed for Sierra Nevada from 1500 to 1980 based on tree rings. In the study, tree-ring width from 1505 precipitation-sensitive oaks across 33 sites in the central California was used to examine the winter precipitation anomolies. At the same time, tree rings from temperature-sensitive trees are also studied to determine California temperature from January to March. The combined reconstruction of winter precipitation and temperature can capture the low SWE values accurately, which suggests that 2015 was the year with the lowest SWE (Fig. 6.13). It is also inferred that SWE values are closely related to decrease in precipitation and increase in temperature.

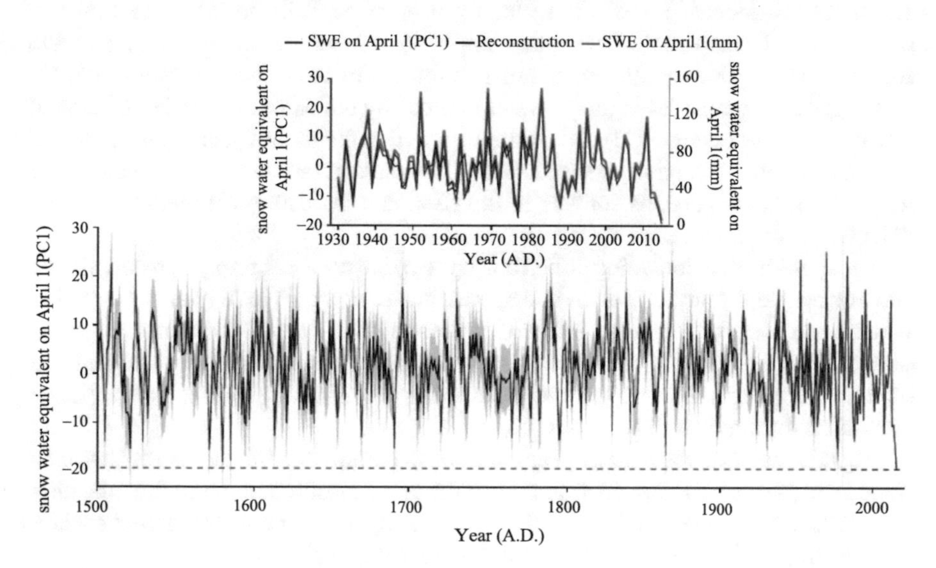

Fig. 6.13 Sierra Nevada April 1 snow water equivalent (SWE) reconstruction over the period of 1500–1980 (Belmecheri et al. 2016). The upper panel: the 1 April SWE average value (cm) of the 108 stations (blue line) against the instrumental (black) and reconstructed (red) first Principal Component (PC1) of 1 April SWE. The PC1 is obtained from 108 observation stations across the Nevada state, with an explained variance of 63% for the period 1930–1980. The lower panel: instrumental (red line; 1930–2015 A.D.); and reconstructed (black line; 1500–1980 A.D.) PC1 of 1 April SWE; the grey shadows indicate the error estimation. The dashed line represents the 1 April SWE value in 2015

6.4 Lake Sediment Record

There are three major types of proxies in lake sediment record: ① physical proxy, e.g., grain sizes, magnetic properties and etc., ② chemical proxy, including total organic carbon (TOC), total nitrogen content (TN), atomic N/C ratio, organic carbon isotopes, minerals, concentrations and ratios of chemical elements, and biomarkers, and ③ biological proxy, including spore-pollens, ostracods, diatom and midges etc.

The grain-size distribution of lake sediments has long been used as a proxy for a lake's hydrological conditions associated with regional climatic and environmental changes (Hakanson and Jansson 1983). Hydrological conditions of lakes on the Tibetan Plateau and the Arctic, however, are affected by not only precipitation, but also glacial melt due to temperature changes.

The TOC concentration refers to the percentage of organic matter that has not been further mineralized in the sediment. It is determined by the preliminary productivity and latter subsidence. The TN is generally used as an indicator of organic matter content of a lake. Atomic C/N ratios imply the source of organic matter in the sediment, suggesting aquatic organic matter sources by the values between 4 and 10, and suggesting terrestrial organic matter sources by the values above 20. Chemical elements such as titanium (Ti) and aluminium (Al) in the sediment are mainly caused by heterogeneous sediment accumulation. Their concentrations can help reconstruct inputs from heterogeneous sources in the past, and indicate precipitation to some extent.

The spore-pollen (including pollen and spores) is a classical proxy in paleoclimate and environment study, as pollen and spores in nature are huge in quantity, small in volume and easy to maneuver with a long preservation time. The features of pollen of seed plants, spores of cryptogam, as well as the microplantlet (i.e., algae) can be detected using a microscope, and referred to reconstruct vegetation and climate conditions during their depositions.

Aquatic fauna and/or flora such as ostracods, diatoms and midges are each accustomed to a particular living environment. Through deposition in the sediments, their cells or heads got preserved in fossils, whose analysis can help reconstruct species combination in the geological time and thus revealing paleoenvironment.

Lakes in cold regions are mainly distributed in high-latitude and high-altitude areas. Most lakes in these areas are located remotely with little direct impact from human activities. Lake sediments can effectively record the climatic and environmental changes in the cryosphere. On the basis of accurate dating, environmental proxy records such as physical, chemical and biological information in lake sediments are extracted to explain the climatic and environmental changes on different time scales.

Sediments of Lake El'gygytgyn in the Arctic well reveal the climatic and environmental changes at orbital scales. The $\delta^{18}O_{diatom}$ in core LZ1024 of Lake El'gygytgyn reflects the precipitation variation in the Arctic. Its result is in accordance with the $\delta^{18}O$ record of deep-sea sediments core LR04 and the δD record of EPICA Dome-C.

Both of them show the Milankovitch theory, reflecting the influence of precipitation variation in orbital eccentricity in the Arctic.

In the middle and lower latitudes, lake records on the Third Pole reflect the influence of the monsoon climate on sedimentary process. Due to the special geographical location, the Tibetan Plateau is affected by two major atmospheric circulations: the Asian monsoons and the Westerlies. The multi-elemental analysis of sediment cores in Qinghai Lake reveals that an anti-phase relationship is shown between the Westerlies and the Asian summer monsoons during the glacial-interglacial and glacial periods on the millennial scale. The Westerlies prevailed during the glacial periods, while the Asian summer monsoons prevailed during the Holocene. The alternation of Westerlies and Asian monsoons may be the main climate driver in the region since Quaternary.

Although sediments research of lakes in cold regions have gained a lot of achievements in rebuilding the climatic and environmental changes of the Holocene, there are differences in climatic and environmental records of different lakes in a larger area, which may reflect the regional differences of environmental changes in the historical period. Therefore, if lakes are widely distributed in the cold region (such as the Tibetan Plateau), a large range of core samples from lakes should be drilled in consideration of climatic zones and geographical differentiation, so as to construct climatic and environmental changes that reflect the geographical differentiation.

6.5 Other Climatic and Environmental Proxies in Cryosphere

In addition to the above proxies in the cryosphere, other proxies that are also used to reconstruct climatic, environmental and ecological changes in the cryosphere, such as glacial varve, glacial deposition landforms, lichens and mosses, animal remains (such as penguin dropping-amended soil and hair of seal), borehole temperature, among many others.

1. Glacial Varve

The term glacial varve is used to describe seasonal changes of sediments input into some periglacial lakes due to the seasonal glacial melting. The thickness, color and composition of sediments deposited in different seasons of a year are different. In summer, the plenty of meltwater runoff can bring large amount of sediments into lakes, and sand particles sink to the bottom of lakes quickly, while the most smaller clay particles are still floating in lake water. In winter, those clay particles eventually deposits slowly above the bottom sand layer. Year after year, the sediment layers are formed with clear textures alternated with one coarse and one fine. The sand layer is light in color, consisting of quartz and feldspar, while clay layer is darker.

2. Glacial deposition landforms

As the direct evidence left by past glacier activities, the depositional landform is an important basis to study the former glacier extent and reconstruct paleo-geographical environment. The maximum height of lateral moraine can be used to estimate the snow-line altitude (ELA) at that time. It is an important method to reconstruct paleoclimate from glacial remains by using geomorphological method to restore the glacier equilibrium-line altitude (ELA). The other geomorphological methods of estimating the height of the equilibrium-line altitude include: Accumulation Area Ratio (AAR) method, Area-Altitude Balance Ratio (AABR) method, Maximum Elevation of Lateral Moraines (MELM) method, Terminus to Head Altitude Ratios (THAR) method, Cirque-Floor altitudes (CF) method), glaciation threshold, among others. The AAR method is most commonly used to estimate the glacier equilibrium-line altitude, providing that glacier is in a stable state. The ratio of the accumulative area to the total glacier area is a fixed value. The ratio is determined by the glacial climate type, glacial landform type, topographical features in glacier area, debris-covered glaciers and others. The AAR of typical mid-latitude and high-latitude glaciers is between 0.55–0.65. Therefore, choosing the appropriate ratio is key to estimating equilibrium-line altitude. The THAR method is based on the assumption that the ELA is located at a certain height between the cirque headwall and glacier terminus. In most cases, THAR = 0.4–0.6. The advantage of AAR is that it is easy to calculate the ELA if knowing the height of glacier terminus and glacier source region; the difficulty and the problem lie in the proper ratio. The MELM method is suitable for geomorphology, especially for the glacier with better reserved trimline and lateral moraine. The result will be better if MELM method combines with AAR method.

3. Lichens and mosses

Lichen is a composite organism that arises from algae or cyanobacteria living among filaments of multiple fungi in a symbiotic relationship. Lichenometric dating is an effective method of dating, which is mainly applied to the dating of late Holocene. Lichens grow very slowly with a very long life span in areas above the alpine treeline in frigid zone, where is lack of other organic materials available for age dating. Lichenometric dating is widely applied in till dating of glacier front in polar and alpine areas. The combination of lichenometric dating and tree-ring dating allows the study on glacier retreats since LIA. Lichenometric dating itself also has some limitations. The indirect method needs to know the age of the substrate and therefore cannot be used in areas where the age of the substrate is indefinite. For the direct method, the growth curve obtained by directly monitoring the annual growth rate of lichens may not be representative on a long-term scale because the current global climate is in a warming trend.

Due to relatively simple structure, mosses can grow and thrive in alpine, high temperature, arid and low light environments where other terrestrial plants cannot survive. Lacking of real root and vascular tissues, mosses with large surface area of leaves are 10 times more sensitive to environmental factors than seed-bearing plants.

Mosses are one of the major plant species distributed in the cryosphere. Due to global warming and snow-line rising, significant changes of bryophytes distribution and genetic variation may take place in Qinghai-Tibet Plateau. Therefore, by investigating the changes of moss distribution law in periglacial regions of the Tibetan Plateau and northwest mountains in China, and studying on changes in the genetic structure of bryophyte species and its correlation with environmental factors through molecular biology method, global climatic changes can be indicated.

4. Dropping-amended soil and hair of seal

Bird droppings contain bio-phenotypic elements as well as carbon and nitrogen isotopes that can be used to reveal climate and ecosystem changes. The study on the bio-phenotypic elements of penguin ornithogenic sediments in Antarctic Ardley Island indicates that the depth-duration-frequency curves of acid-soluble $^{87}Sr/^{86}Sr$, $\delta^{13}C$ and $\delta^{15}N$ and chitinase gene in the penguin ornithogenic sediments were similar to those of bio-phenotypic elements, which speculated the historical changes of penguins. The research on bird dropping-amended soil has been extended to the Arctic and the Xisha Islands of South China Sea, etc.

The primary sources of food for Antarctic seals is krill. Only when supplement of krill is insufficient, Antarctic seals will consume other fishes. Therefore, using hair of Antarctic seals as a research medium, the relative changes of population density of Antarctic krill can be identified. Based on nitrogen isotope analysis of dropping-amended soil containing hair sequence of seals in the Fildes Peninsula, King George Islands, Antarctica, the relative change in the population density of krill can be deduced from the proportion changes of Antarctic krill in the seal diet composition. By analyzing the stable isotope in bones and feathers of Ardley penguins in the Vestfold Hills of East Antarctica, it is considered that the changes in krill abundance during the Holocene are related to regional climatic changes, i.e., the krill abundance is higher at lower temperatures. In addition, the $\delta^{15}N$ ($^{15}N/^{14}N$ value) value of the modern Ardley penguins is low, due to the massive hunting of seals and whales fed in krill by humans, resulting in the recent increase in krill quantity. The study provides a unique perspective on changes in food chain of regional ocean and contributes to the conservation and management of Antarctic marine living resources.

5. Borehole Temperature Inversion

The borehole temperature in glaciers and frozen ground is influenced by both surface temperature variations and underground steady-state heat fluxes. Hence, by analyzing borehole temperature records can reconstruct past earth's surface temperature change. As the surface temperature fluctuates downward, the amplitude of temperature fluctuations exhibits an exponential decay with increasing depth. The amplitude of short-term surface temperature oscillations (such as diurnal variation and seasonal variation) decays faster than the amplitude of long-term oscillations with increasing depth. This makes deep temperature profile available to reverse long-term trend of surface temperature. Borehole temperature records can reconstruct the history of temperature changes over a variety of time scales, usually from decades to centuries. At present, borehole temperature records are used to study on temperature changes in local, regional, hemispheric and even global scales.

Questions

1. How does ice core preserve the information of climatic and environmental changes?
2. Try to outline the advantages and disadvantages of the paleoclimate reconstructions by the different media in cryosphere?

Extended Readings

Representative

Willi Dansgaard (1922–2011) Danish paleontologists, firstly recognized that the Greenland Ice Sheet is the Historical Archives of Global climatic changes. He studied meteorology in Copenhagen University. After graduation in 1947, he started working on geomagnetic observations in Greenland. In 1951 Dansgaard begun teaching and research work in Copenhagen University. His first job is to install mass spectrometer installation and engaged in stable isotope analysis. One day in June, 1952, he collected precipitation samples from his own lawn using an empty beer bottle and funnel, in order to analyze the stable isotope ratio change in precipitation. As a result, it was found that the oxygen stable isotope ratio in precipitation changed significantly with the Warm and Cold Fronts. Oxygen stable isotope ratio in precipitation is closely related to the height of precipitation (Temperature decreases while altitude increases). This discovery opened up a new chapter for isotope meteorology research. Later, he conducted a systematic study on stable isotope ratio in global precipitation. After discovering the correlation between the seasonal variation and temperature of $\delta^{18}O$ in mid and high-latitude precipitation, he realized that $\delta^{18}O$ in accumulated ice and snow of Greenland region contains information of past climatic changes and proposed study on ice core in 1954. Through research on Camp Century and Dye-3 ice cores in Greenland, he systematically revealed the climate change records since the last glaciation and found that there was a millennium-scale fast changing in climate during the last glaciation, which is now called the Dansgaard-Oeschger (D-O) event. In 1976, he was awarded Seligman Crystal by the International Glaciological Society for his significant contribution to the ice core research. He received the Crafoord Award from the Swedish Royal Academy of Sciences in 1995 and the Tyler Award in 1996.

Classic Works

1. *The environmental record in glaciers and Ice sheets*

   Authors: Oeschger H., Langway C. C.

   Publisher: Wiley-Interscience Publication, John Wiley & sons, 1989

   In March 1988, a workshop on the Environmental Record in Glaciers and Ice Sheets was held in Germany. The main purpose of the workshop was not to focus on the "known" knowledge of scientists, but to encourage them to share their "unknown" knowledge. Rather than trying to solve a problem or arbitrating a certain point of view, the workshop aimed at identifying and discussing the most cutting-edge and most important scientific issues of the time in order to point out the future research direction. The book was published on the basis of this workshop, including four parts: ① how glaciers record and preserve environmental processes; ② anthropogenic impacts recorded in glaciers; ③ establishing ice core chronology; ④ ice-core record of long-term global changes in the environment. The book was the integration of collective Intelligence in ice-core research and was of great significance for understanding and studying ice core.

2. *Tree rings and climate*

Author: Fritts H. C.

Publisher: Academic Press, London, 1976

Tree Rings and Climate deals with the principles of dendrochronology, with emphasis on tree-ring studies involving climate-related problems. This book looks at the spatial and temporal variations in tree-ring growth and how they can be used to reconstruct past climate. Factors and conditions that appear most relevant to tree-ring research are highlighted. Comprised of nine chapters, this book opens with an overview of the basic biological facts and principles of tree growth, as well as the most important terms, principles, and concepts of dendrochronology. The discussion then shifts to the basic biology governing the response of ring width to variation in climate; systematic variations in the width and cell structure of annual tree rings; and the significance of tree growth and structure to dendroclimatology. The movement of materials and internal water relations of trees are also considered, along with photosynthesis, respiration, and the climatic and environmental system. Models of the growth-climate relationships as well as the basic statistics and methods of analysis of these relationships are described. The final chapter includes a general discussion of dendroclimatographic data and presents examples of statistical models that are useful for reconstructing spatial variations in climate.

Chapter 7
Cryospheric Evolutions at Different Time Scales

Lead authors: Shangzhe Zhou, Yuanqing He, Shiyin Liu

Contributing authors: Zhao Lin, Jiahong Wen, Linjuan Ma, Tingfeng Dou

The paleo-cryosphere is the cryosphere of the earth in the past geological time. The history of the Earth covers about 4.6 billion years (Ga). However, there are only 542 million years (Ma) since the Cambrian. The long geological period before the Cambrian covers about 4 Ga and is called the Precambrian. The Precambrian can be subdivided into the Hadean (4.6–4 Ga), Archean (4–2.5 Ga), and Proterozoic Eons (2.5–0.542 Ga). After the Hadean and Archean Eons, the early Earth ended the long molten stage of its geological history, and the lithosphere, atmosphere, hydrosphere, and biosphere formed. At the end of the Archean Eon (2.8–2.5 Ga), the surface temperature of the earth was about the same as that at present. Since then, the temperature has been fluctuating around this baseline and the earliest cryosphere appeared, taking the Early Proterozoic Glaciation as indicator. Cryosphere has changed at different time scales in the earth's history, such as hundred million years, ten thousand years, millennia, and century time scales. They have different causes and respectively display different cyclicity. The current cryosphere is resulted from those various time scale cycles. In this chapter, we introduce the cryospheric evolution on the four different time scales and its possible causes.

7.1 Cryospheric Variations on Tectonism Scale

It is well known that global glaciations occurred on the earth during Early Proterozoic Eon, Late Proterozoic, Ordovician–Silurian, Carboniferous-Permian and Quaternary (or Late Cenozoic). Among them, the maximum include the Late Proterozoic, the Carboniferous-Permian, and the Quaternary glaciations. We will just introduce these three great glaciations here.

D. Qin et al. (eds.), *Introduction to Cryospheric Science*, Springer Geography,
https://doi.org/10.1007/978-981-16-6425-0_7

7.1.1 The Late Proterozoic Glaciation

1. Precambrian moraine rock and its characteristic

Since the second half of the 19th century, extensive Late Proterozoic tillite and bedrock with glacial facets has been found on almost every continent. These ancient geological evidences demonstrate that an old glaciation happened on our planet. This glaciation is named the Varangian glaciation. On the basis of U–Pb ages, the Varangian glaciation has been dated to the Late Proterozoic, and can be divided into 4 stages, which are known as the Kaigas (770–735 Ma B.P.), Sturtian (715–680 Ma B.P.), Marinoan (660–635 Ma B.P.) and Gaskiers (585–582 Ma B.P.) glaciations. These local glaciations are named after typical localities: the Kaigas glaciation was named after Namibia, the Sturtian and Marinoan glaciations after South Australia and the Gaskiers glaciation after Newfoundland. Nantuo and Luoquan tillites in China belong to the Late Proterozoic, and the dating results indicate that the Nantuo tillite was deposited during the Marinoan glaciation. Because of the extensive distribution of Late Proterozoic tillite, the Cryogenian Period has been named by International Commission on Stratigraphy (ICS), for the period spanning 850–635 Ma B.P. The Late Proterozoic tillites have the following features:

Geographical location: Scientists calibrated the location of the tillites using paleomagnetic methods and found an unexpected result; the tillites were deposited around the equatorial area during the Varangian glaciation.

Laterite Formation: It has been found that the most Late Proterozoic tillites are red mixed diamictite, which overlyings red rock formation (Fig. 7.1). The hematite content is very high in these formations. For example, there are about 50 billion tons of iron ore in the Jacadigo formation in South America, and the iron content is more than 50%. A similar formation is also found in southern Australia, where the iron content is locally more than 40%, and about 300 million tons of iron ore could be mined. The above geological evidence demonstrates that the thick red formation was deposited before the Varangian glaciation. During that geological period, the temperature of our planet was very high, and weathering was very strong, therefore,

Fig. 7.1 Red tillite of Precambrian in Namibia (left: from Hoffman et al. 1998) and glacial striated surface in Oman (right: from Allen and Etienne 2008)

a thick red formation developed. The ancient glaciers advanced across these red formations, which is the reason why the tillites are red. Of course, some red tillites formed by this glaciation have been reworked by later chemical weathering.

Cap carbonate: This is another outstanding feature of the Late Proterozoic tillites. Commonly, the overlying formation is composed of dolomite, limestone or other carbonates. However, at some sites, such as Irkutsk, the tillite was formed together with halite or evaporite. All these rock types, and especially their abnormally low value of $\delta^{13}C$, imply that high temperatures and an arid climate. Geologists associate this with a perturbation of the carbon cycle and environmental variation after glaciation.

Marine and terrestrial environments: the majority of the Late Proterozoic tillites were found in oceans or shallow seas. Therefore, the ice-rafting events were very common at that time. According to the matrix of the tillites, some scientists speculated that the quick rise of sea level accompany with gale and surge behaviour during the ice sheet melting and receding. This represents another major environmental change during the transition from a glacial to interglacial stages in late Varangian glaciation.

2. Snowball Earth hypothesis

The distribution and sedimentological features of these Late Proterozoic tillites indicate that they were deposited around equator. To explain this, the Snowball Earth hypothesis was proposed by Kirschvink in 1992. Later, the cap carbonate covering Late Proterozoic tillites in Namibia was found by Hoffman and the extremely low content of $\delta^{13}C$ in these rocks was used to further support the snowball Earth hypothesis.

The snowball Earth hypothesis presumed that a completely frozen earth occurred during the Late Proterozoic, and that a several thousand meter thick ice sheet covered all continents and oceans, even in the areas around the equator. On the basis of the distribution of the red tillites, it is reasonable to speculate that the continents were assembled around the equator (The Rodinia supercontinent) and that a large obliquity occurred during that period. However, why the ancient glaciers developed around the equatorial areas? In 1975, Williams proposed a reasonable explanation, i.e., the obliquity was as high as 54°–126° during that period. If that was the case, then the equatorial areas would have been a preferential zone to develop glaciers, and the seasonality will be less differentiated around the globe. Weak zonality is also an advantage for forming warm water deposition and lateritic weathering, however, this hypothesis has not been confirmed. An exceptional explanation proposed by Schrage et al. in 2002 is that a supercontinent around the equator was the base for the snowball to form. They suggested that strong chemical weathering happened on the Rodinia supercontinent, and that atmospheric CO_2 dropped-down by this process. This would reduced the greenhouse effect caused by CO_2, decreasing the temperature of our planet and causing the ancient glaciers to develop. Once the ice and snow accumulated, a positive feedback would be initiated until the earth was frozen completely and, finally, a snowball was formed. Hoffman demonstrated that snowball conditions were maintained for a long geological period. The snowball ended because extensive volcanic eruptions increased the content of CO_2 in the atmosphere up to 350 times the present value, leading to a strong greenhouse effect. During this process, the methane produced by the disintegration of organics that were sealed under the ice sheet was also released, further enhancing the greenhouse effect. After the ancient glacier melted out, the precipitation brought the CO_2 back onto land and decomposed the rock, until finally carbonate precipitated and covered the tillites. The evidence for this is the extremely low content of $\delta^{13}C$, about 10–14‰ in the cap carbonate. These $\delta^{13}C$ values are some of the lowest that have been found from 1.2 billion years before the snowball Earth event up to the present. Hoffman assigned this geological phenomenon to photosynthesis, for more ^{12}C was absorbed by creatures. However, the extremely low content of $\delta^{13}C$ in carbonates demonstrated that the snowball blocked the transparency of the oceans and decreased primary productivity. Photosynthesis almost ceased, causing the ^{12}C content and low $\delta^{13}C$. All this geological evidence has been used to support the snowball Earth hypothesis. One consequence of this dramatic variation is that the natural selection of the original microorganisms (such as algae and fungus) and, as a result, the Cambrian explosion, which happened as the Earth entered the Paleozoic Era.

There is intense debate about the snowball earth hypothesis. On the basis of celestial mechanics, Pais and his colleagues denied the more than 54° obliquity presumption. Hoffman also acknowledged that the increased obliquity could not be used to explain the cap carbonates and their extremely low $\delta^{13}C$ values. Conversely, Christie and his cooperators thought that the abnormally low content of $\delta^{13}C$ in cap carbonate was questionable. Allen and his colleagues suggested that the frozen ocean was inconceivable because the exchange between seawater and the atmosphere has never been cut off, and the hydrological cycle was always active on Earth. Allen therefore preferred the "Slushball Earth" that proposed by Christie to the snowball hypothesis. Maybe the snowball existed on our planet once, but it is very difficult to explain how the snowball ended and no interpretation has been accepted by a majority of scientists yet. Although it is well known that the Late Proterozoic glaciation happened around equator, more research work need to be done to test the "snowball" hypothesis.

7.1.2 Carboniferous-Permian Glaciation

After the Late Proterozoic glaciation, the Earth entered another extensive glaciation in the late Paleozoic, known as the Carboniferous-Permian glaciation. The tillites of this glaciation have been found on almost all continents. The following are the main features of this glaciation: ① The main distribution areas are South Africa, Antarctica, Australia, and South America. However, the tillites of this glaciation were only found in low latitudes, such as the Arabian Peninsula and southern Asia in northern hemisphere. Typical and famous tillites among them are Dwyka formation on the Kalahari highlands in Africa, Wiajid formation in the Arabian Peninsula, Metschel formation in Victoria district in Antarctica, Talchir formation in India, and the Talaterang formation in the Sydney basin in Australia. The paleo-magnetism data indicate that this glaciation developed in high latitudes in the southern hemisphere at that time, including the tillites distributed in modern low latitudes in the northern hemisphere. In other words, this extensive glaciation developed in Gondwanaland during Carboniferous-Permian period. ② Almost all the tillites were deposited and distributed on land, and only few of them were found in oceanic environments. ③ The duration of this glaciation was from the late Carboniferous to early Permian. This means that the Carboniferous-Permian glaciation developed at high latitudes in southern hemisphere, contrasting with the late Precambrian glaciation that initiated at equatorial latitudes. Coal-bearing formations that contain abundant glossopteris flora fossils, which are called the Gondwana coal strata, covered the Gondwana tillites, after which the Gondwana Formation was named. There is also evidence to demonstrate that Africa, Australia, Antarctica, South American and southern Asia were once a unified supercontinent, that was named Gondwanaland (Gondwana is a local site in central India). However, the rest of the continents, such as Europe, North America and northeastern Asia, formed another supercontinent, called Laurasia. No tillites have been found in Laurasia yet, except in a few localities in southern Europe. Therefore, it is reasonable to infer that the Laurasia was situated in the low latitudes of the northern hemisphere during this period.

7.1.3 Quaternary Glaciation

Extensive glaciation occurred again in the Quaternary, at least eight glacial cycles have been revealed by the Antarctica ice core records during the past 800 ka. The Last Glacial Maximum (LGM) is the most recent glaciation, and its geological evidence and climate proxies are well preserved. These indicate that the global average temperature decreased about 10 °C, and that several ice sheets developed, and mountain glaciers expanded. The Laurentide Ice Sheet and the Cordilleran Ice Sheet, with an area of about 16×10^6 km^2, developed in North America; The Eurasian Ice Sheet was composed of the Scandinavian Ice Sheet, the British Ice Sheet and the Barents Ice Sheet with a total area of about 7×10^6 km^2 occurred on Eurasian continent; The Greenland Ice Sheet expanded, with an area about 3×10^6 km^2 at that time; A small ice sheet has also been developed in the Alps, with an area of around 2×10^5 km^2. Therefore, the total area covered by ice was about 26×10^6 km^2 in northern hemisphere during the LGM. In southern hemisphere, the area of the Antarctic Ice Sheet was bigger at the LGM than at the present (14×10^6 km^2). At the same time, mountain glaciers in Qinghai-Xizang plateau and its surrounding mountains, in the Andes and other ranges were also expanding. About 30% of the land was occupied by glaciers (ice sheets) during the LGM, compared with about 10% at present. Because large amounts of seawater are transported to the mainland and stored as ice sheets and glaciers, the sea level was 130–150 m lower than present, and the continental shelves were widely exposed. In addition, the areas of snow cover increased and tundra expanded. The distribution of species that prefer the warm and wet environments shrank to low latitudes, whereas, the species that are adapted to cold climate developed and expanded rapidly, such as the mammoth and woolly rhinoceros, whose population increased and became distributed across wide areas. You could imagine that our planet during the LGM was a very different place. This glaciation terminated at 11.7 ka, then, the Earth came into the Holocene which is characterized by a warm climate. During this optimal period, human beings began to make fine stone tools and the agricultural civilization began to develop. Before the last glaciation, there are marine oxygen isotope stage (MIS) 6, MIS 12 and MIS 16 glaciations, and the extent of glaciers during these glaciations was greater than that of the LGM.

1. The initiation of Late Cenozoic ice age

After the carboniferous-Permian glaciation, the Earth entered a high temperature period from the Mesozoic to the early Cenozoic. The temperature reached a maximum during the late Jurassic and Cretaceous. In the Eocene in the early Cenozoic, the temperature of the Earth began to decrease, and at the end of the Eocene, an instable ice sheet developed in Antarctica. Beginning in the late Miocene, extensive glaciers began to develop in the northern hemisphere. Stable ice sheets began to form in northern hemisphere at the beginning of Pleistocene in the Quaternary, which signaled that our planet entered the recent glaciation. Based on the records of marine sediments, the geological evidence of the early glaciations in the Cenozoic were revealed in the seas around Antarctica in southern hemisphere, as well as in the north Atlantic area, such as the Barents sea, Norwegian sea, northern and southeastern seas around Greenland, seas around Iceland and North America in northern hemisphere. The ice-rafted deposits indicated that the outlet glaciers of ice sheets must have entered the adjacent seas at that time. The ice-rafted deposits and their ages demonstrated that an ice sheet began to develop in the eastern Antarctic at about 35 Ma in the late Eocene, however, it only reached its stable status at about 14 Ma. In other areas, such as Alaska, Greenland, Iceland and Patagonia, the glaciers began to develop at about 8 Ma in the Miocene; ancient glaciers occurred in Bolivian Andes and Tasmania in Pliocene; and in the early Pleistocene in the Alps and New Zealand. The MIS curve indicates that 2.7 Ma is an important transition point; since then, the global ice volume reached its maximum in each glaciation. Therefore, the beginning of Quaternary was specified at 2.6 Ma by international research communities, and it seems that the Quaternary Period is synonymous with glaciation. The Quaternary glaciations in the middle and low latitudes are different from those in the above areas, especially on the Qinghai-Tibetan Plateau and in its surrounding mountains. The coupling between the tectonic uplift and glacial climate is a prerequisite for glaciers to develop in this region and the earliest glacial deposits found as of yet on the Qinghai-Tibetan Plateau and its surrounding mountains have only middle-Pleistocene ages.

Table 7.1 Glaciations during Pleistocene

Alps	North Europe	England	USA	Tibetan Plateau
Würm	Weichsel	Devensian	Wisconsin	Dali
Riss	Warthe	Gipping	Illinoian	Guxiang
Mindel	Saale	Lowestoft	Kansan	Zhonglianggan
Haslach	Elster	Beeston	Nebraska	Kunlun
Günz	Menapian	Baventian		Sishapangma
Donua				
Biber				

2. The cryospheric evolution in Quaternary

Glacial-interglacial cycles are the main characteristics of the Quaternary glaciation. The cold stages with glacier expansion and low sea level were suggested to be glacial periods, whereas, the warm stages between them were assigned as interglacial periods. On the basis of the glaciofluvial deposits along four tributaries of the Danube in southern Germany, Penck and Brückner proposed Günz, Mindel, Riss and Würm glaciations in their masterpiece—"Ice Age of the Alps" in 1909. Since then, more glacial evidence has been found in Europe and North America, and the glaciations named by Peck and Brückner in Europe were referred to in the following studies. During the early stage of Quaternary glaciation research, many local glaciations were named and glacial sequences established. These sequences have been compared with these four typical glaciations in the Alps (Table 7.1). In addition, later studies revealed another three glaciations in the Alps, they are Biber, Donua and Haslach glaciations. In China, more than 100 local glaciations have been named, especially on the Qinghai-Tibetan Plateau and its surrounding mountains. In 2008, Mr. Shi Yafeng summarized them and suggested that these local glaciations could be classified into five major glaciations (Table 7.1).

The characteristics of glacial remains demonstrated that it is difficult to obtain an integrated glacial history from them because the glacial deposits from one glaciation will be destroyed or reworked by more extensive subsequent glaciations. However, the glacial landforms indicate the clear and exact extent of ancient glaciers and are therefore the best evidence for reconstructing paleo-glacier extents. In the 1970s, the records of deep-sea cores became available that implied that the ratio of oxygen isotope of planktonic foraminifera could be used to reconstruct the global ice volume (sea water volume). The famous oxygen isotope curve of V28-238 is known as the Rosetta stone in climate change research. In addition, ice cores derived from Antarctic and Greenland, loess of Chinese Loess Plateau, lacustrine deposits around world have also verified a similar record as the deep-sea cores. In 2005, Lisiecki and his colleagues compiled and calibrated 57 oxygen isotope records and synthesized them into an integrated curve that covers the last 5.3 Ma. These records give us a clear and new insight into the cryospheric evolution in Quaternary and even in the Pliocene (Fig. 7.2).

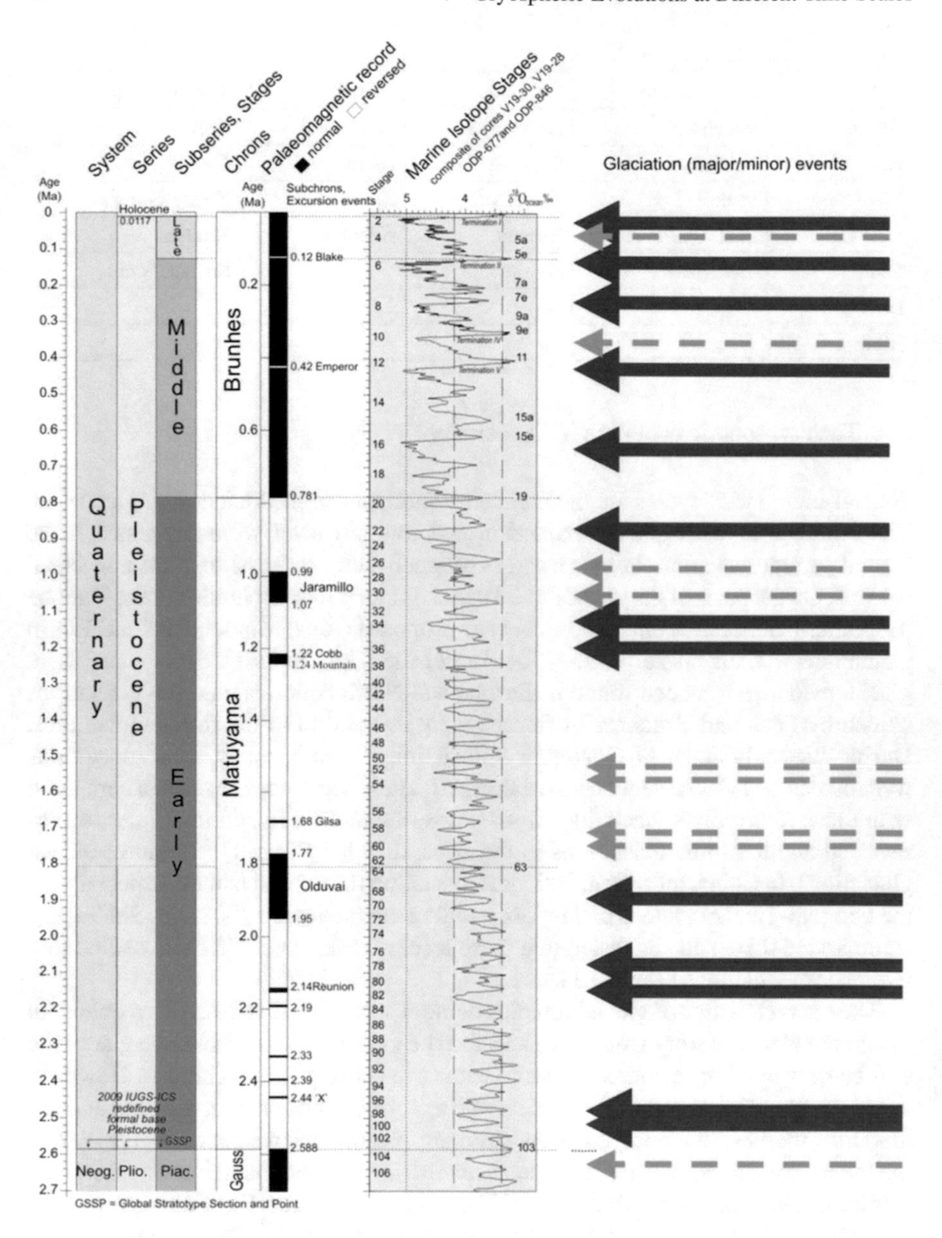

Fig. 7.2 The sequence of Quaternary Glaciations (Ehlers et al. 2007)

7.1.4 The Causes of Three Ice Ages

Three extensive glaciations (Late Proterozoic, Carboniferous-Permian and Quaternary glaciations) occurred on our planet. Scientists have attempted to find the causes and some plausible explanations have been proposed, such as the revolution hypothesis of the solar system, however, the convergence and divergence of supercontinents and their locations has a stronger relationship with these glaciations. The Late Proterozoic glaciation developed on a supercontinent that was located at low latitudes, which then separated. Two supercontinents, Gondwana and Laurasia were formed during Carboniferous-Permian on our planet, after that, they broke apart into several smaller continents. The present distribution and pattern of seas and continents gradually formed in the Cretaceous and Cenozoic. Scientists have proposed several reasonable interpretations for the initiation of the late Precambrian glaciation that developed on a supercontinent around the equator, however, no general consensus has been obtained as of yet. The Carboniferous-Permian glaciation happened in the southern hemisphere on the Gondwana Supercontinent, which was located in the polar region during that period. It is also believed that the Cenozoic glaciation initiated in the Antarctic. Therefore, the periods of the three ice ages (Figs. 7.3 and 7.4) have a strong relationship with tectonic movement of the continents on Earth.

In addition, it is worth noting that older tillites have been found before the Late Proterozoic glaciation in South Africa, in the Great Lakes area in North America, Fenoskandia, and Australia with ages of 2.2–2.45 billion years. These tillites were deposited in the early Proterozoic. Of course, very scattered tillites with Ordovician age were also discovered in some sites. The sites and distribution of deposits of these two glaciations demonstrate that their extents are limited (Fig. 7.4). Moreover, there is a strong relationship between glaciation and tectonic uplift in the middle and low latitudes, such as the Qinghai-Tibetan Plateau and its surrounding mountains, the coupling between this huge and young plateau and glacial climate caused it to enter into the cryosphere in Quaternary. From then on, glacial-interglacial cycles occurred with advance and retreat of glaciers.

7.2 The Cryospheric Evolution in Orbital Scale—Pleistocene Climatic Changes and Milankovitch Theory

The onset of the great Late Cenozoic ice age was associated with the formation of Antarctic continent resulted from the plate movement. However, the plate hypothesis is difficult to explain the glacial–interglacial cycles during Quaternary. For these cycles, the updated interpretation is the astronomical theory, namely, the Milankovitch theory, for the seminal work of M. Milankovitch, a Serbian scientist. The theory explained the climatic change in Quaternary successfully using the

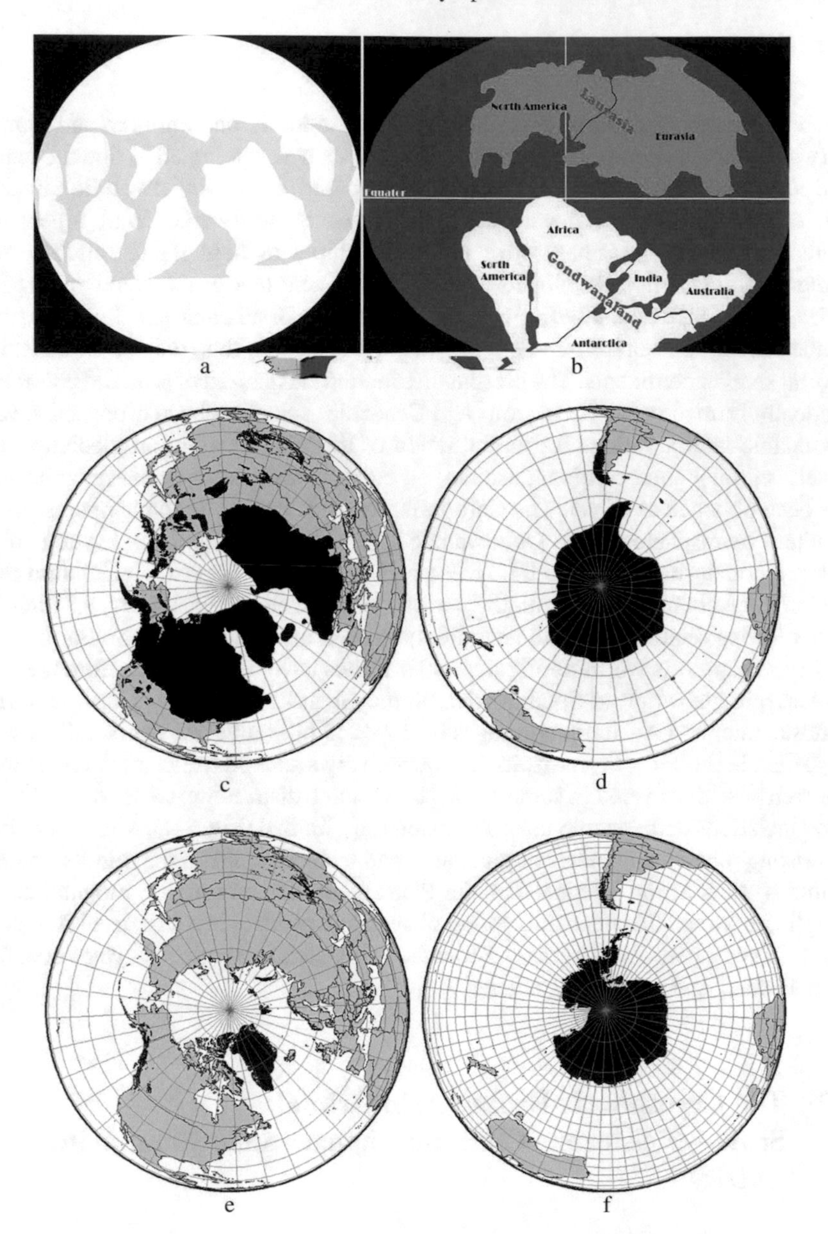

Fig. 7.3 Sketch maps for distribution of continents and ice sheets during the great three ice ages. **a** The Rodina supercontinent and the snowball earth during Late Proterozoic; **b** the Gondwana supercontinents and ice sheets during Carboniferous-Permian; **c**, **d** the ice sheets and glaciers on both two hemispheres during Quaternary ice age; **e**, **f** the ice sheets and glaciers on both two hemispheres at present

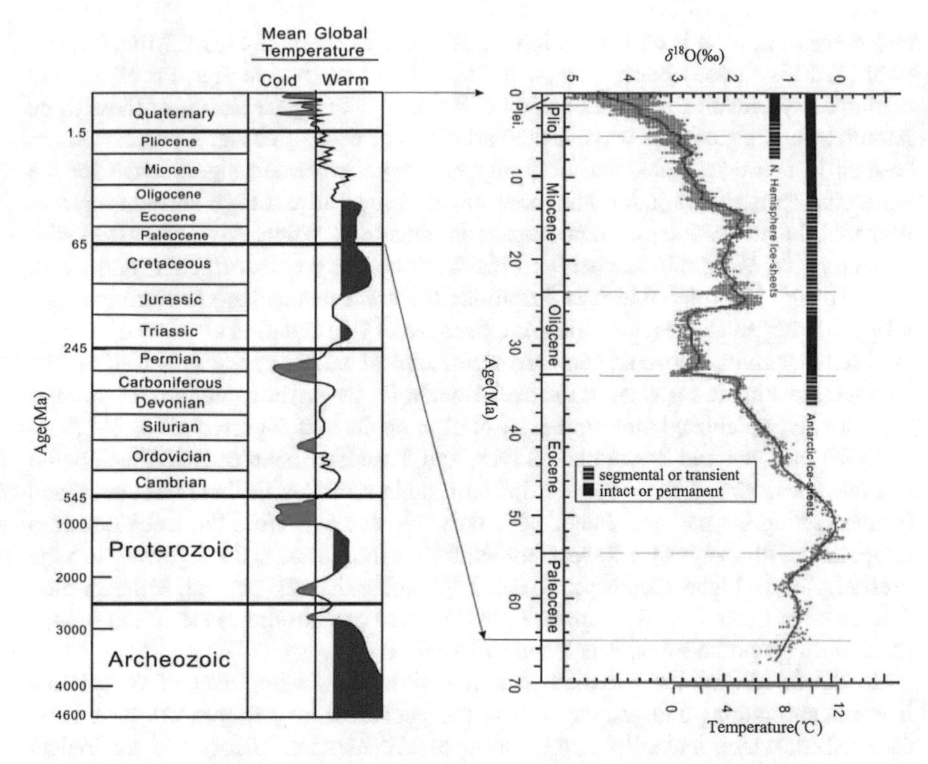

Fig. 7.4 Temperature changes of the earth's history (the left, from Frakes 1979) and since tertiary (the right, from Zachos et al. 2001)

secular variation of earth's orbital elements, so that the glacial–interglacial cycles were called the orbital scale evolution.

7.2.1 The Establishment of the Astronomical Theory of Ice Ages

A French scholar, Joseph Alphonse Adhemar published "the revolution of the sea" in 1842. In this book he tried to attribute the ice ages to the earth's orbital variations. His theory was based on the second movement law of Kepler, and the precession of the earth's axis that was found by Hipparchus, an ancient Greek astronomer. In1843, U. Le Verrier, a French astronomer found the variations of earth orbital eccentricity is 0–6% and earth's axis obliquity is 22°–25°. J. Croll, a Scotch scholar found that the eccentricity period is about 100 ka and with different amplitudes, a longer period is about 400 ka. On the basis of these findings, Croll developed the astronomical theory of ice ages. In 1901, S. Newcomb, an American astronomer found the obliquity variation period is 41 ka. So, the periods and amplitudes of the eccentricity, obliquity, and precession have been known in details since 1901. In 1941, M. Milankovitch published his famous book "Canon of Insolation and the Ice Age Problem". He explained systematically the formation of ice ages with the variations of these three parameters. His study shows that the variations of eccentricity and precession are enough to cause ice ages, and obliquity has more important significance for ice ages. The obliquity variation has more impact on polar and high latitude regions, however, the precession has more impact on equatorial region. After discussed with meteorologist Wladimir Koppen and his work contrary with Adhemar and Croll, Milankovitch identified that it is propitious for north hemisphere to form ice ages when summer solstice is near aphelion, because of this situation consists of a long-cool summer and a short-cold winter in north hemisphere. He stressed the feedback of ice sheets to climate and established a relationship between the snowline and summer radiation. He calculated the summer insolation on the earth at intervals of 10° from 5° to 75° latitudes and drawn their curves, and these have been called Milankovitch Curves. For example, the curve for 65° latitude in north hemisphere has succeeded in interpreting the past ice sheets. Milankovitch also converted the insolation into temperature, illustrating the lowest temperature in the ice age is 6.7 °C lower than the present, and the highest temperature is 0.7 °C higher than the present. Milankovitch Curves were quoted by W. Koppen in his book to explain the cause of the 4 Alps glaciations proposed by A. Penck and E. Brückner.

In the middle of the twentieth century, with the development of radioactive isotopes chronology and geomagnetism, the geochronology framework have been established, which make the combination of Milankovitch theory and geological records to be possible. In 1947, H. C. Urey demonstrated the possibility theoretically that oxygen isotope amount in the calcium carbonate of skeletons deposited in the sea could be used to calculate the sea water temperature. In 1955, C. Emiliani analyzed the samples of eight cores from the Caribbean Sea and published a seminal paper "Pleistocene temperature", indicating that seven glacial-interglacial cycles could be obtained from these records since 300 ka, and the temperature in ice age was 6 °C lower than that of today. In 1968, W. S. Broecker et al. dated three high reef terraces in Barbados, Hawaii and New Guinea to be 125, 105, and 82 ka using thorium isotope technique, the ages coincide completely with the 45° N Milankovitch curve. In 1969, J. Imbrie and N. Shackleton pointed out simultaneously that it is continental ice volume rather than sea water temperature for making a directly $\delta^{18}O$ change. After that, Broecker's study on V12-122 core of Caribbean Sea in 1970 and G. Kukla's study on Czech loess in 1975 revealed the cycle of 100 ka. Therefore, in 70th of twentieth century, in order to test the Milankovitch theory, J.D. Hays and J. Imbrie initiated a proposal that was called Climap, and organized many scientists and laboratories in the world to study geological records of terrestrial and oceanic sediments. Finally, they chose the V28-238 core in the west Pacific Ocean and the RC11-120 core in the south Indian Ocean that were dated by geomagnetism to reconstruct the isotope variation curves since B/M boundary in the past 700 ka. By a spectral analysis, they found astonishingly that these two curves display three cycles of 100, 41 and 23 ka (together with cycles of 19 ka), which are consistent with the periods of eccentricity, obliquity and precession in the Milankovitch theory (the precession contains double cycles of 23 and 19 ka because of the distance variation between the earth and the sun). From then on, the Milankovitch theory was testified and the V28-238 core was called The Rosetta Stone of late Pleistocene. By decades after that, these cycles have also been revealed from other long records of oceanic sediments, loess, Antarctic ice core, lacustrine deposits, speleothem etc., which made Milankovitch astronomical theory of ice ages to be established and became an up-to-date solution in deciphering Pleistocene climatic changes.

7.2.2 The Fundamental of the Astronomical Theory of Ice Ages

1. Eccentricity and its climatic significance

Kepler First Law indicates that the orbits of all planets are elliptical and the sun is situated at one focus of the orbits. So the earth-sun distance varies between Perihelion and Aphelion in a year. The received radiation on the earth from the sun is inversely proportion to the square of the earth-sun distance, and the following formula could be used to calculate the received radiation.

$$I = I_0/\rho^2 \sin h = I_0/\rho^2 \cdot (\sin\varphi \sin\delta + \cos\varphi \cos\delta \cos\omega) \tag{7.1}$$

where I is the solar radiation on the top of the atmosphere, I_0 is the solar constant, ρ is the earth-sun distance, h is the solar altitude, φ, δ and ω are geographical latitude, the declination of the sun and the hour angle.

The eccentricity, the ratio of the focal distance to major axis is 0.0167 at present, the higher value of e, the more elliptical of the orbit is. Kepler Second Law shows that the radius vector, drawn from the sun to the planet, cover equal areas in equal times. So the revolution velocity for a planet is faster near perihelion than near Aphelion (Fig. 7.5).

2. Obliquity and its climate significance

Obliquity shows the tilt between earth's axis and ecliptic axis, i.e., the tilt between equatorial plane and ecliptic plane. The obliquity is 23° 27' now. The higher value of obliquity, the more solar radiation on the poles and high latitude is vice versa. So the obliquity is responsible for the distribution of the solar energy on a planet.

3. The precession and its effect

The cycle of earth move around the sun (360°) is called sidereal year (365 d 6 h 9 min and 9.5 s). While the earth revolution takes the vernal equinox as the reference point (359° 09′ 35″), this cycle is called tropical year (365d5h48min 46 s). Therefore, a tropical year is 20 min and 23.5 s shorter than a sidereal year. This is because of the

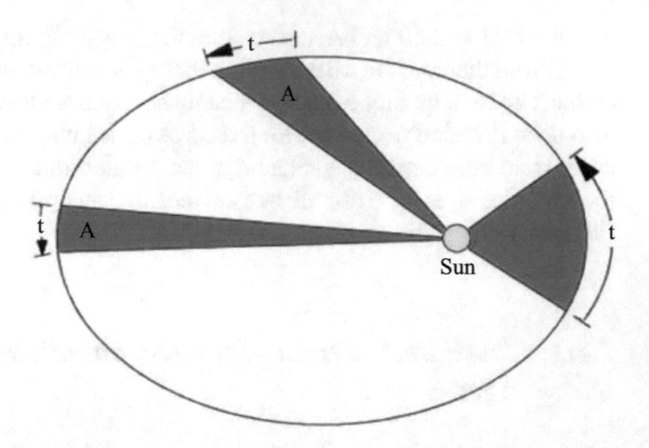

Fig. 7.5 Sketch map of Kepler law 2

precession of the earth's axis (Fig. 7.6). The earth's axis precession causes vernal equinox to shift 50.25″ westward each year. This is called precession cycle and the period is 25,800 year. However, the Perihelion (i.e., the major axis of the orbit) shifts eastward very slowly to meet the vernal equinox (Fig. 7.7), making the effective precession cycle is about 22,000 a. So the Perihelion position to the vernal equinox, called Perihelion ecliptic longitude, is the key element for deciding the length of summer and winter seasons on the two semi-hemispheres.

The Milankovitch theory divided a year into two halves, from Vernal equinox to Autumnal equinox is summer half year and from Autumnal equinox to Vernal equinox is winter half year. The difference length between them is given by the following formula:

$$T_s - T_w = 4T/\pi \times e\sin\lambda \approx 1.273Te\sin\lambda \tag{7.2}$$

where T_s is the length of summer half year and T_w is the length of winter half year, e eccentricity, λ Perihelion ecliptic longitude.

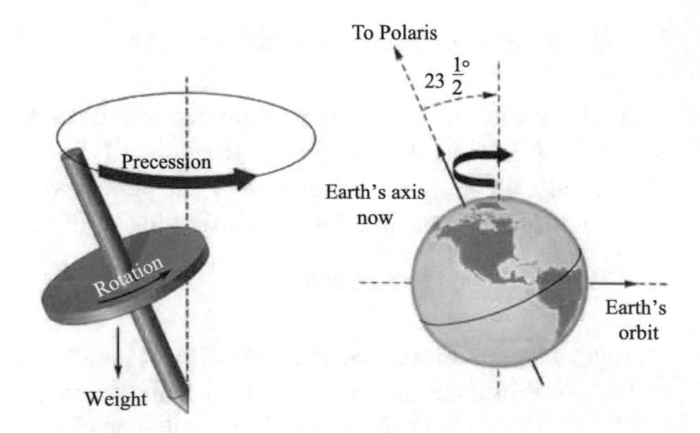

Fig. 7.6 The map showing precession of the earth's axis

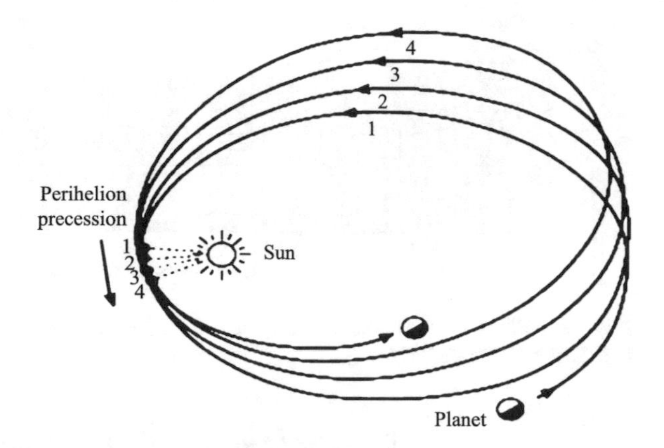

Fig. 7.7 The perihelion advance of the earth's orbit

Calculating by this formula, the current summer half year on north hemisphere is seven days longer than that of winter half year (on the contrary, the winter half year on north hemisphere is seven days longer than that of summer half year). When the eccentricity varies up to 0.07 and the summer solstice near the Aphelion, the summer half year will be 32.6 days longer than that of winter half year on northern hemisphere.

Synthesizing the variation of these three orbital parameters and their effects on the solar radiation, it is clear that a long cool summer half year and a short warm winter half year will be occurred on the northern hemisphere with high eccentricity and the summer solstice goes near the aphelion, however, under the condition of high eccentricity and the summer solstice goes near the Perihelion, a long cold winter half year and a short hot summer half year will be happened on the northern hemisphere. The condition is contrary on the southern hemisphere to the northern hemisphere, namely, when the summer solstice goes near the Aphelion, a hot summer half year and a long cold winter half year will be presented on the southern hemisphere. When the summer solstice goes near the Perihelion, a long cool summer half year and a short warm winter half year will be appeared. When the Vernal equinox and Autumnal equinox move to the Perihelion or Aphelion, the lengths of summer half year and winter half year are equal on both hemispheres. Milankovitch's hypothesis is opposite to being in opposition to Croll's, considering that a long cool summer half year and a short warm winter half year is a precondition for ice age, because short warm winter half year is favorable to snowfall on high latitude region and a long cool summer half year is an advantage to keep the snow cover from melting. As mentioned above, the eccentricity changes in the amplitude of 0–0.007 with a short periods of 100 ka and a longer one is 400 ka. The obliquity variation amplitude is 22°–25° with a period of 41 ka, and the precession cycle is 22 ka. So the insolation for any latitude and any season on earth can be calculated by the following formulas:

$$Q_s = TI_0 \Big/ 2\pi\sqrt{1 - e^2} \cdot (b_0 + \sin\varphi \sin\varepsilon)$$

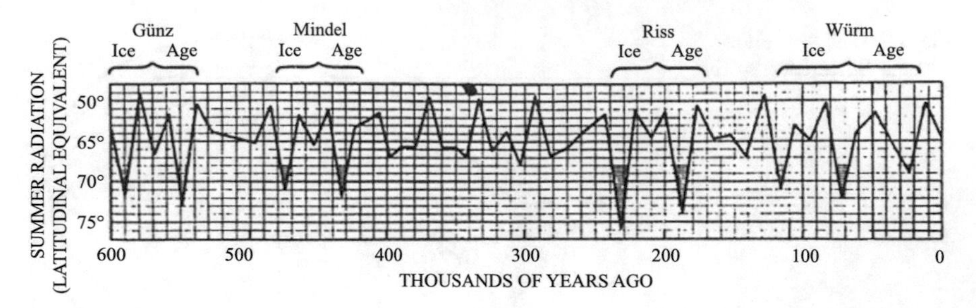

Fig. 7.8 Milankovitch summer radiation curve on 65° (for example, the radiation on 65° 226 Ka ago equate to that on 75° today)

$$Q_w = TI_0 \Big/ 2\pi\sqrt{1-e^2} \cdot (b_0 - \sin\varphi \sin\varepsilon)$$

$$Q_y = TI_0 \Big/ 2\pi\sqrt{1-e^2} \cdot b_0 \tag{7.3}$$

where T is the period of earth revolution, I_0 is solar constant, e is eccentricity, φ is latitude, ε is obliquity, and b_0 is a constant related to latitude.

Milankovitch has calculated the summer insolation variations of different latitudes since 600 ka. The insolation curve on 65° was used to interpret the glacial-interglacial cycles in Quaternary (Fig. 7.8).

7.2.3 Amendment of the Astronomical Theory of Ice Ages

The cycles of eccentricity, obliquity and precession have been affirmed by various geological records. However, it was found that glaciations had happened in low value of the eccentricity rather than the high one (Fig. 7.9), In order to resolve this problem,

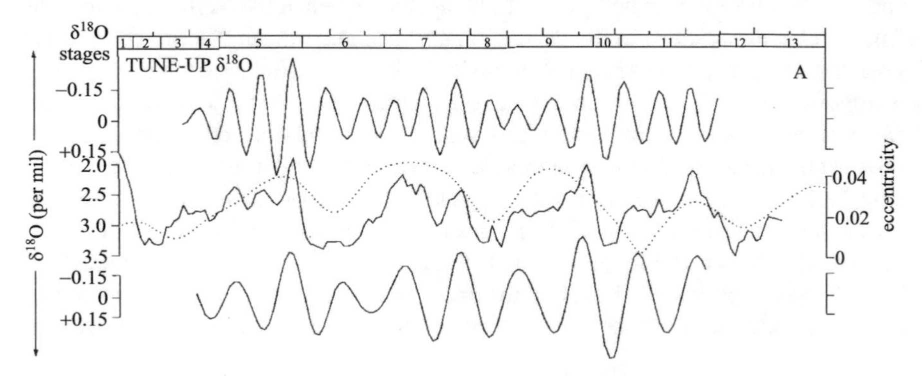

Fig. 7.9 Marine $\delta^{18}O$ record showing ice ages occurred during low eccentricity values (Hays et al. 1976). The middle solid line is the $\delta^{18}O$ curve and the dotted line is the eccentricity values

some scientists restudied the Milankovitch theory and found a major omission that the orbit axis doesn't change while the eccentricity varies. Consequently, the mean annual earth-sun distance becomes larger during low eccentricity value than that of high value, finally, this will cause global solar radiation decrease and resulting in ice age. Hence, the 100 ka cycle was attributed to the solar radiation rather than the solar radiation on definite latitude of one hemisphere that has been proposed before. This amendment, in a way, resolved other two puzzles. One is favorable condition for ice age, the summer solstice near Perihelion or Aphelion. The other is the synchroneity or asynchroneity of ice ages on the northern and southern hemispheres. This is because the precession has a weak effect with a low value of the eccentricity during a cycle of 100 ka. On the contrary, the precession effect is striking with a high value of the eccentricity in a 100 ka cycle. This situation has been confirmed by various isotope curves obtained from different geological archives. For example, 5a, 5b, 5c, 5d and 5e of marine isotope stage 5 with a high eccentricity value display sharp rise and fall, while in low eccentricity value it becomes stable. Therefore, if we want to test the asynchronism of glaciations and climatic changes on both hemispheres, we should compare the timings of glaciations during high eccentricity values, i.e., mountain glacial advances at 5b and 5d and warm proxy records at 5a, 5c and 5e.

7.2.4 The Challenges of the Astronomic Theory of Ice Age

The astronomic theory of ice age is so successful in explicating Quaternary climatic and environmental changes, therefore, the Quaternary science was regarded to enter a new research stage. However, there are still some puzzles between the geological records and the astronomic theory need to further explore and study.

A. Berger's calculation shows that the variation of the three orbital parameters adhere to the same law at least since 6 Ma. However, a 5.3 Ma stack of benthic $\delta^{18}O$ records from 57 global sites presented by Lisiecki and Raymo aligned by automated graphic correlation algorithm, revealing the records exhibits piecewise response to obliquity and precession since 5.3 Ma. For example, obliquity response appears directly forced with exponentially increasing sensitivity from 5.3 to 1.4 Ma, whereas, the 41 ka glacial cycles become approximately constant after 1.4 Ma. Precession response is directly forced over the entire 5.3 Ma, but its sensitivity with exponential increase does not begin until 2.5 Ma. As Lebreiro pointed out in 2013 that MIS11 is a stage with much low eccentricity value, while the precession response was strongest, this is the reason why a warmest interglacial occurred with a low eccentricity value (in fact, this situation is similar to current Holocene interglacial). In addition, the shape of the $\delta^{18}O$ curve displays distinct asymmetry from that of eccentricity cycles. It rises rapidly from trough (glacial) to peak (interglacial), while descends relatively slow with about two precession cycles from interglacial to glacial. In particular, why the Northern hemisphere glaciations didn't happened until 2.7 Ma? Why 100 ka eccentricity cycles had not responded until B/M boundary, and before that the dominant cycle is 41 ka? This is so-called Middle Pleistocene Transition.

These puzzles have different interpretations, which promote the progress of Quaternary climatic change research and great achievements have been obtained. However, these puzzles are from the complex response processes of the earth system rather than from astronomic theory of ice ages itself.

The astronomic theory of ice age also has very important significance in forecasting glaciations in the future. On the basis of this theory, A. Berger points out that "extrapolation has been made for the next 100,000 years assuming no human interference, next ice age is expected to occur before 60,000 years AP. The maximum amount of ice to be expected in the northern hemisphere is 27×10^6 km^3 representing a 70 m sea-level drop". Some 2000 years from now, a distinct cooling trend will begin. After approximately another 1000 years, the longest Pleistocene interglacial will be terminated. Global climate will then start a long-term deterioration until 23,000 years from now, the earth will once more find itself in another cycles of a new ice age.

7.3 The Cryospheric Evolution in the Sub-orbital Scale

Since the late Pleistocene, geological records show that there are not very distinct cycles of 10 ka and even the millennium cycles, which can not be explained by the Milankovitch theory. These cycles are known as the sub-orbital scale variations and its causes may be more complex.

7.3.1 Some Important Events of Climate Change

On the basis of many archives, including the records of marine sediments, lacustrine deposits, ice cores, loess and paleosol sequence, pollen and tree-rings, some important climate change events have been revealed by scientists throughout the world. These findings are fundamental requirement in studying the cryospheric evolution.

1. Dansgaard-Oeschger Events

In 1993, on the basis of Greenland ice core records, Dansgaard et al. found that the climate in this region had undergone a series of millennial, rapid, and dramatic cold-warm changes, which are called the Dansgaard-Oeschger events, and sometime are referred to as the D-O Oscillation during the last glacial cycle. In these D-O cycles, each warm period was rapidly followed by a cold period (Fig. 7.10), and temperature could fluctuate 5–7 °C within several decades, commonly, these cycles last 1000–3000 years. In the subsequent related studies, the D-O cycles were also revealed in the oceanic sediments in North Atlantic Ocean, loess and stalagmites.

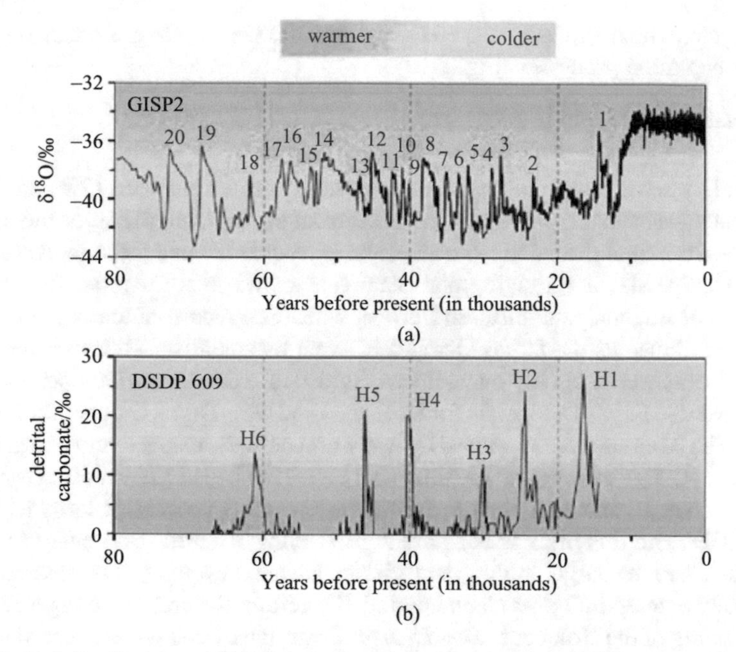

Fig. 7.10 D-O Oscillation and Heinrich Events. Grootes et al. (1993) (**a**); Bond and Lotti (1995) (**b**)

2. Heinrich Events

In 1988, Heinrich first reported that there were six layers of ice drift debris in the core that drilled from the North Atlantic and the time covers the last glacial cycle. The main features of these layers are high content of coarse particles and low proportional foraminifera. The upper 1–5 layers corresponding to MIS (marine oxygen isotope stage) 2–4, and the 6th layer corresponding to the transition between MIS 4 and 5. Heinrich considered that during the last glacial period, 6 large ice shelf-icing fractures caused a large amount of ice to enter the North Atlantic, causing a sharp temperatures drop of surface seawater and the dumping a large amount of debris into the ocean floor deposition. This 6 ice-floe events are so called as H event. Bond and others used accelerator 14C dating and deposition rate extrapolation to determine the age of these events were occurred in 14.3 ka, 21 ka, 28 ka, 41 ka, 52 ka and 69 ka respectively. It has been speculated that the H event reflects the dynamic phenomenon of the Ice shelf portion of the northern hemisphere when it has grown to large enough to break into an ice floes, which is not necessarily caused by cold weather, but by cutting off the warm-salt circulation (THC) by the temperature of the Atlantic water, which in turn affects climate. So the H event is often cited to explain the records of ice cores, stalagmites, and loess. H events are diametrically opposed to D-O events, where H events occur during the coldest periods of the D-O cycles, marking the end of a climatic cycle, and the subsequent rapid warming represents the beginning of a new cycle, which shows that the H event and the D-O cycle are not two isolated climatic evolution processes (Fig. 7.10).

3. Younger Dryas and B-A events

Younger Dryas was a short climatic cooling event occurred between 12.9 and 11.5 Ka B.P., which was a last rapid cooling event during the rapid warming of the climatic transition from final glacial to Holocene, also known as the late Ice Age. Andromeda (Dryas Octopetala) is an eight-petal plant that grows in the Arctic, found in the sediments of England and northern Europe, which is a record of temperature degradation and named as the Dryas Octopetala event by scientists. Because the similar cooling events were not only once, it was divided into Old Dryas Octopetala (oldest Dryas), Middle Dryas Octopetala (older Dryas) and New Dryas Octopetala (Younger Dryas). The 3 fairies were separated by two warm periods, respectively called Bøling events and Allerød events, and the time was 14.7–12.9 ka B.P. The oldest Dryas took place between 18 and 14.7 ka B.P during the shrinking process of large ice sheets of thc LGM, and this process ended with the Bøling-Allerød (BA) event as a sign. After that, another large cooling and glacier advance events (YD) occurred, then completely entered into post glacial period. Therefore, the end of Younger Dryas as the beginning of the Holocene. The Younger Dryas have been widely recorded in ice cores, tree rings, stalagmites and other climatic carriers, and there also are mountain glacier deposition as evidence. After Younger Dryas, the Holocene started. The temperature of the Holocene from 11.7 Ka has been maintained at a relatively high level, but it still fluctuates under a generally stable background.

4. The cooling event at 8.4 ka B. P.

The event experienced from 8.4 to 8 ka B.P. with the intensity of half YD, and ended rapidly at a climatic event warmer and weter than current climate. The evidence of the event can be found in North Pacific, Europe and North America.

5. The Megathermal in Holocene

The most warm period in Holocene (Megathermal) is also known as the High-temperature period (Hypsithermal) or the most suitable climate (Climate Optimum). In 1976, Hafsten first presented the concept of the Holocene Megathermal which occurred between 8.2 and 3.5 ka B.P. with the temperature about 2 °C higher than the present and corresponding precipitation increase. The warm and humid climate has brought a great development of the human agricultural society.

6. Neoglaciation

The newer Ice Age (Neoglaciation) usually refers to the colder climatic stage after the Holocene Megathermal, and mountain glaciers generally advanced in the regions of middle and low latitude. The time began roughly in 3.5 ka B.P.

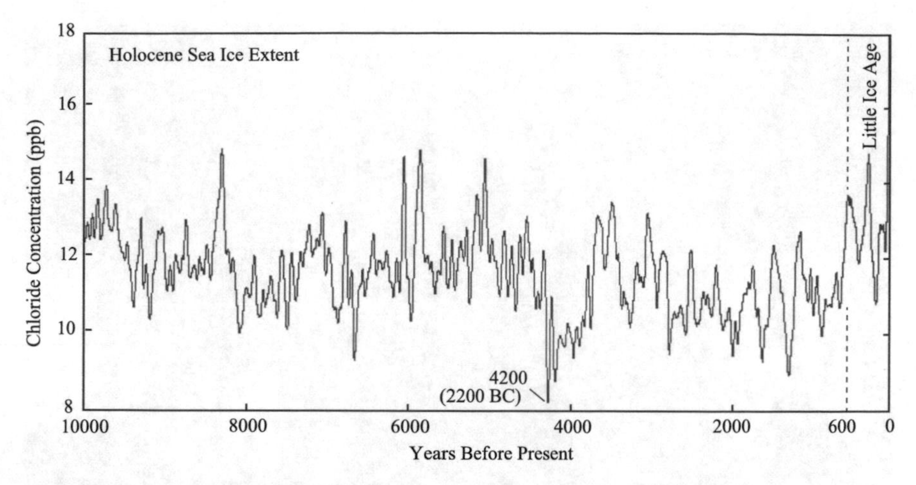

Fig. 7.11 Holocene sea ice extent changes from chloride concentration (Bond et al. 1997)

7. Little Ice Age (LIA)

The Little Ice Age (LIA) is a concept proposed by Matthes in 1939 to describe glacial advances since 4 ka, including the aforementioned new Ice Age. In 1972, Lamb limited the small ice Age from 1550 to 1850 during that period of relatively cold climate. Through the analysis of the $\delta^{18}O$ sequence of the GIPS2 ice core drilled from the Greenland Ice Sheet, some scholars pointed out that the LIA should occur between 1350 and 1800 AD, during that stage almost all mountain glaciers were advanced and left intact fresh moraines beyond them. In fact, in addition to the new Ice Age and LIA, other minor fluctuations were very frequent (Fig. 7.11), the causes of these millennium or even shorter-term climate variations are not very clear.

7.3.2 Cryospheric Evolution Since the Last Glaciation

1. Glacial evolution since the Last Ice Age

Ice Sheet Retreat: During the LGM, the area of ice sheets on northern hemisphere was more than 26×10^6 km^2, the ice sheet in North American covered Canada and most land on north of Missouri River and Ohio River, as well as northern Pennsylvania, entire New York State and New England. The Baltic Sea is the center of ancient ice sheet in Europe and it covers the whole Scandinavian Peninsula, it also covers the entire Barents Sea and Kara Sea eastward, linking the British Ice Sheet westward through Norwegian Sea. Covering the middle Germany southward and the distance to the Alpine Ice Sheet was only about 200 km. The Antarctic Ice Sheet has also expanded dramatically, covering the Ross Sea and the Weddell Sea of the West Antarctic, the East Antarctic Ice Sheet also expanded. Totally, about 30% of

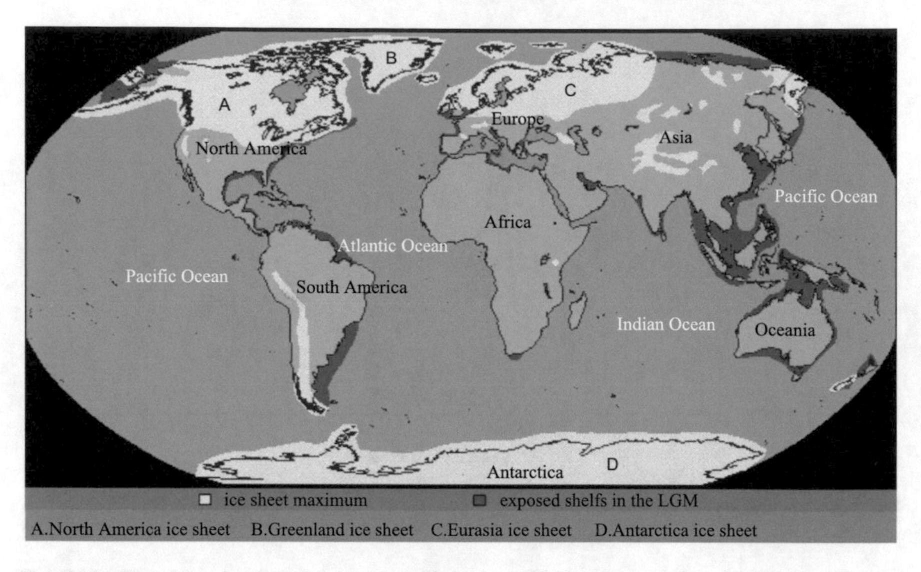

Fig. 7.12 Global ice sheets and exposed shelfs in the LGM

the land was covered by ice sheets during that period (Fig. 7.12). After reaching its peak in 20 ka B.P. the ice sheets were shrank, now, only Antarctic Ice Sheet with an area of about 14×10^6 km^2 and Greenland Ice Sheet with an area of about 17×10^5 km^2 are remained. Glacial deposits show that ice sheets retreated in an intermittent discontinuous way and researchers have reconstructed the size of ice sheets in different sub-stages.

Change of mountain glaciers: almost all mountain glaciers advanced during the LGM, resulting in abundant large-scale glacial landforms formation due to its longer duration (Fig. 7.13 left). The end of the last glacial cycle is called deglaciation.

Fig. 7.13 Moraines formed in glaciation period. Left: The Last glacial moraines in the southeast of Tibet; Right: the small Ice Age glacial ridge before the No. 1 Glacier in the Head Area of Urumqi River, Tianshan Mountains

Since then, glaciers were shrank periodically with occasional advances, however, the extent of these advances are smaller compare with the previous ones. Therefore, in most U-shaped valleys, several terminal-lateral moraines were formed between the terminal of modern glacier and the terminal-lateral moraine of LGM, commonly, these moraines were formed during late-glacial, Neoglaciation and Little Ice Age. For example, late-glacial moraine of Rongbuk Temple Stage is located 8 km beyond the modern glacier at Everest area. The evidences of glacial advances during late-glacial in more than twenty sites have been studied on the Qinghai-Tibetan Plateau. Moraine complexes of Neoglaciation were distributed hundreds to thousands meters beyond the modern glaciers. These moraines are integrated, incipient weathering, a thin soil has been developed and covered with grass or even shrubs and trees on them. The glacial landforms of Neoglaciation have been observed in almost all sites in the Qinghai-Tibetan Plateau and its surrounding mountainous areas. The moraine complexes of the Little Ice Age were distributed between the terminuses of modern glaciers and glacial landforms of Neoglaciation, commonly, they are consist of one to three end moraines and are hundreds to thousands meters from modern glaciers. The main characteristics of them are fresh, no weathering, no soil developed and grass covered, however, some pioneering plants, such as moss and lichen have been colonized on these end moraines, and shrubs even be observed in some sites (Fig. 7.13 right).

Sometimes, mountain glaciers response quickly to the abrupt climate change than the ice sheets in the polar areas. For example, there are very spectacular hummocky moraines formed in the Boduizangbu valley in the southeastern Qinghai-Tibetan Plateau, and the Kangxiwa valley between the Muztagta and Gonger Mountains in the eastern Pamir. These landforms were formed due to a sudden rise of equilibrium-line altitudes (ELAs) and a melting of stagnant/dead ice, which contain abundant debris. Several abrupt climate change events have been recorded in the Greenland ice core too, dramatically showing a significant temperature variation over a short-term. For example, 5 °C rising (14.6 ka B.P.) only took 3 years and less than 1 ka Younger Dryas has experienced 5–10 °C temperature changes during the deglaciation. In China, the Guliya ice core record demonstrates that a 6 °C rising between 12 and 16 ka and a cold-dry climate in Younger Dryas stadial. Stalagmite records also show that there was a one thousand years relative warm and humid climate between 15 and 13 ka, while 3 short-terms of cold and dry climate mutation occurred during this period too. These three climatic variations are called BA events that occurred during the oldest Dryas, older Dryas and younger Dryas.

Holocene is modern interglacial in the Quaternary. The climate of Holocene is also unstable, with at least 6 distinct glacial advances. In 1997, Bond and others established the North Atlantic cold events chronology, which is an important progress of abrupt climatic events research in Holocene. Mayewski and others collected 50 high-resolution proxy data around the world and confirms the universality and uniformity of abrupt climate changes in Holocene. According to the Chinese stalagmite record, the average interval of each cold event last about 1.5 ka in the high latitude regions (Fig. 7.14). A distinguish warm climate occurred in 5–6 ka B.P. the temperature in most regions was 2–3 °C higher than present, and even 4–5 °C higher in some areas

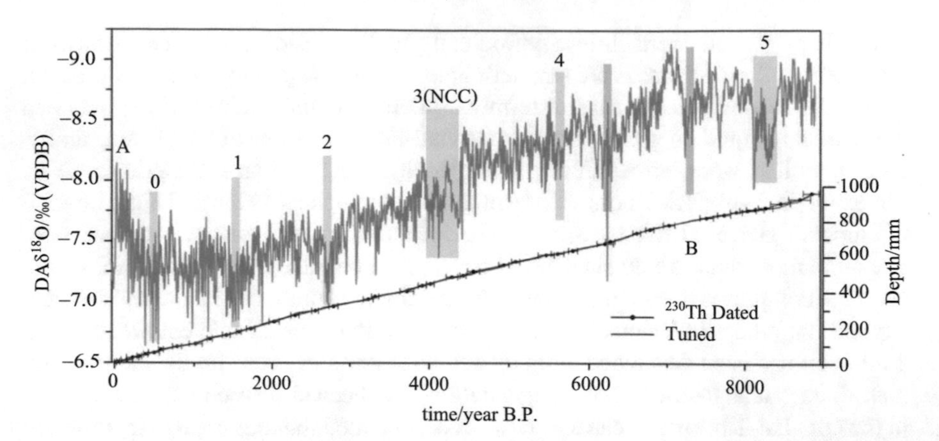

Fig. 7.14 Climatic changes recorded by stalagmite of Donggedong (from Wang et al. 2005) (Yellow indicates the drought period corresponding to the NCC (Neolithic Culture of China) event, grey indicates a dry period not corresponding to NCC, and the Red Line is the revised year record)

of the Qinghai-Tibetan Plateau. Guliya ice core record and proxy data also display clearly of this warm climate. The main features are stronger Asian summer monsoon, higher water level in inland lakes and luxuriant vegetation. Therefore, this sub-stage is also called Holocene Megathermal. However, the cryosphere of that time was smaller than that of present.

2. Permafrost changes since the Last Glaciation

The modern permafrost is about 22.55×10^6 km^2 in the Northern Hemisphere, accounting for about 23% of the land area. The maximum thickness of permafrost is 1500 m that located in the Villyuai River basin in the middle of the Lena River, Siberia. The permafrost was also expanded dramatically during the glacial period. The relief shaped by permafrost is called periglacial landform, which developed during and after the glacial period (and even exposed to be the continental shelf at the glacial period). These paleo periglacial landforms and their features could be used to reconstruct the permafrost distribution and climatic characteristics in glacial periods. For example, the gelifluction terrace, the gelifluction tongue, the gelifluction slope, the gelifluction fan, the rock-glacier, the rock stream produced by the Freeze–thaw creep-gravity effect; the stone ring, stone belt, gravel spot produced by freeze–thaw sorting effect; ice piton, frost mound, self-spray frost mound, peat mound, frozen heaving stone, frozen heaving grass ring, freeze–thaw creased slope, earth wedge, sand wedge, ice wedge produced by frost heaving and cracking effect; collapse, hot melt depression, hot melt lake, hot melt gully produced by hot melt effect. In addition, the periglacial environment also has its specific flora and fauna, such as tundra and rhino-mammoth fauna relict etc.

The main distribution areas of modern high latitudes permafrost are Alaska, northern Canada and most parts of Siberia. During the LGM, there are no great ice sheets developed in Siberia and Alaska. A large number of buried mammoths remains

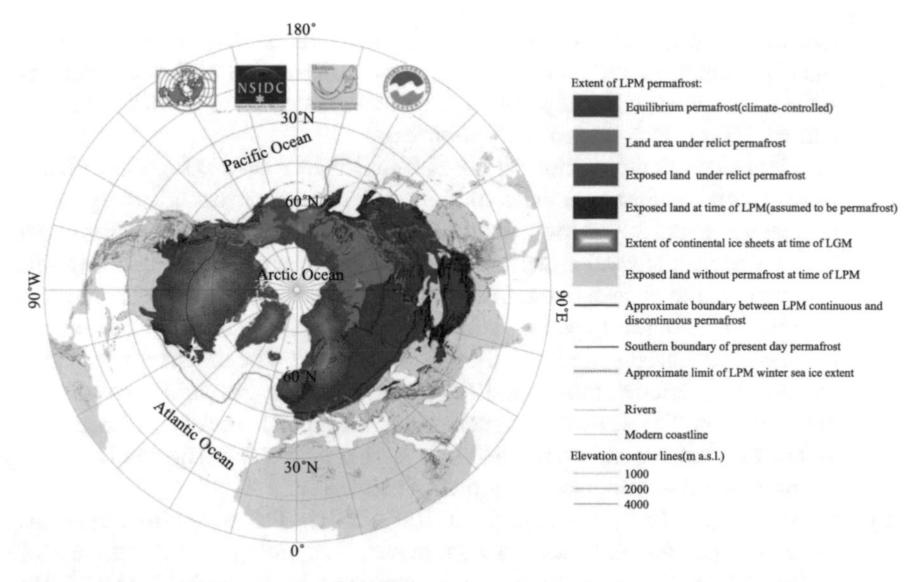

Fig. 7.15 Permafrost extent of north hemisphere in the LGM reconstructed using past periglacial relics (after Jef Vandeberghe et al. 2014)

were found in the permafrost indicates that the prevailing climate was severe-cold and permafrost extended dramatically. The ancient periglacial landforms in northern China also demonstrated that the permafrost expanded southward to the line of about 40° N and connected to the permafrost on the Qinghai-Tibetan Plateau through the Helan Mountains, the Altun Mountains and the Pamirs Plateau in the high latitudes at that time. In North America, especially in the periphery of the great ice sheets, the permafrost was also expanded during the Last Glaciation. The south boundary of the ice sheets is 40° N, while the boundary of the periglacial landforms is 33° N, extending for about 7° in latitude (Fig. 7.15).

The existing permafrost on the Qinghai-Tibetan Plateau was mainly formed in the Last Glaciation, and the extent and thickness of the permafrost should be much larger at that time. According to the distribution of the modern permafrost and the relics of the ancient on the Qinghai-Tibetan plateau, the permafrost evolution process and environmental changes since the Holocene can be divided into six distinct periods.

(1) Early Holocene climatic upheaval period (10.8 ka B.P. to 8.5–7 ka B.P). Permafrost formed during the Last Glaciation began to shrink. The lower boundary of permafrost in the fringe of the plateau generally increases by 300–400 m. Wetlands were formed in highland valleys and basins, and peat and thick humus layers begin to accumulate.
(2) Holocene Megathermal (8500–7000 a B.P. to 4000–3000 a B.P.): Most of the thick peat and humus layers of the plateau were formed in this period, indicating that the climate was warm and humid. Researchers have found fire ashes of human in many sites along Kunlun River between the Nachitai and Xidatan,

indicating that this area is suitable for human habitation. Due to the melting of shallow permafrost and underground ice, many thermokarst lakes and depressions are formed in the high plain. The main distribution characteristics of permafrost is island-shaped or deep-buried.

(3) Late Holocene Neoglaciation (4000–3000 a B.P. to 1000 a B.P.): The remains of periglacial landforms indicate that the ancient permafrost boundary in the plateau was about 300 m lower than the present and the temperature was about 2 °C lower than today. Permafrost reach the maximum area at the end of this cold period. The plateau permafrost area is 20–30% larger than that of present.

(4) Late Holocene warm-up period (1000–500 a B.P.): After the Neoglaciation, the climate experienced several small-scale fluctuations. The time is equivalent to the medieval period of the Sui and Tang Dynasties, and this period lasts several hundred years. The warm period makes the permafrost boundary increase 200–300 m and the temperature is 1.5–2.0 °C higher than today. The whole area of permafrost is 20–30% lesser than that of present.

(5) Late Holocene Little Ice Age (500–100 a B.P.): The permafrost area has expanded and thc thickness has increased. According to the remains of periglacial landforms, the boundary of permafrost in this period is 150–200 m lower than today and the temperature is 1.0–1.5 °C lower than present. The area of permafrost increases about 10%.

(6) Modern warming period (since 100 years ago): the conversation data show that the global average temperature has risen by 0.3–0.6 °C since 1880; in the past 30 years, the annual average temperature in the Qinghai-Tibetan Plateau has increased by 0.3–0.5 °C. The results show that the freezing depth decreases by 5–20 cm on average, while the depth of average seasonal thawing increases by 25–60 cm. The annual average ground temperature in permafrost generally increases by 0.1–0.4 °C. The lower boundary of permafrost increased 40–80 m, the total area decreased by 6–8%. It is estimated that the air temperature in the Qinghai-Tibetan Plateau may increase by 2.2–2.6 °C within the next 50 years. Permafrost degradation will be accelerated.

In the eastern China, the data show that permafrost covers the entire northeast in the last glacial age. The boundary extends from Liaodong Bay of 40° N to the south of the Yanshan Mountains, to the southern slope of Mt. Wutai, to Yongdeng in Gansu Province, then connected with the lower boundary of the Qilian Mountains and Qinghai-Tibetan Plateau. The permafrost relic in Great Khingan Mountains is also formed during the LGM. Since then, it has been degraded and re-developed several times.

3. Sea Level Changes since Last Glaciation

Sea level is negatively correlated with continental ice volume. From Last Glaciation to Holocene, global sea level rose significantly with the melting of glaciers, and the ocean area has expanded (Fig. 7.16). During the LGM, the sea level was 130–150 m lower than present and continental shelf was widely exposed. Such as the Bohai Sea, the Yellow Sea, the most part of China East Sea and part of the South China Sea

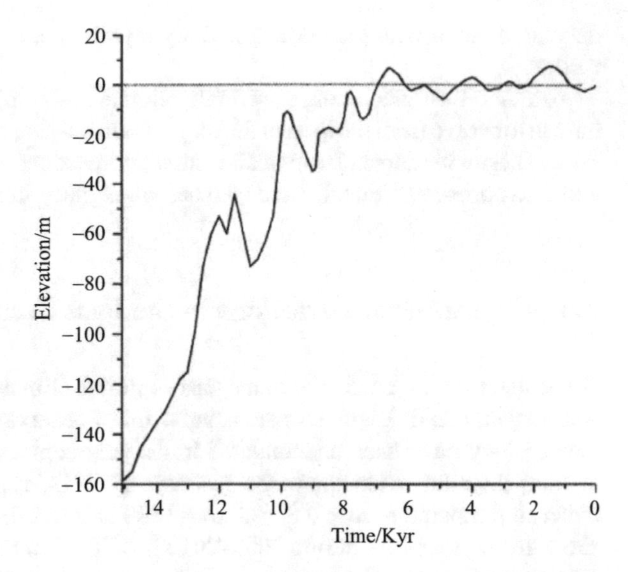

Fig. 7.16 Sea level change in eastern China Sea since 15 ka B.P. (Zhao Xitao 1996)

were all exposed, Hainan and Taiwan Islands connected together with the mainland of China, the coastline shifted significantly eastward, and the continentality increased in mainland. At 16 ka B.P., the sea level in eastern China has risen by about −100 m, the sea water has reached the vicinity of Jeju Island, about 2/3 of the East China Sea was submerged, the Kuroshio water from the surface to the thermocline depth at the same time enhance the impact on the Okinawa trough, leading to development of the Tsushima Current. Between 12 and 11 ka B.P., due to the YD strong cooling event, the transgression suddenly stopped when sea level rose to about −56 m and Kuroshio appeared a weaker process. To 11 ka B.P., sea level rose about −50 m the vast majority of China East Sea and the central trough of the Yellow Sea have submerged by seawater (Li Tiegang et al. 2007).

4. Changes of terrestrial ecosystems during the Last Glaciation

Glacial-interglacial cycles have changed the terrestrial ecosystems greatly in Quaternary, zonal direction on the earth has compressed toward to the equator during glacial stages, while it extended to high latitudes during interglacial ones. For example, during the LGM, tundra and mammoth-wooly fauna occupied from Siberia to the Yellow River drainage in China. From the beginning of Holocene to the Holocene Megathermal, forest vegetation expanded rapidly to high latitudes in the eastern China, forming a similar pattern of present. Temperate forests and grasslands have occupied again in the northeastern China, temperate grassland in northern China was replaced by warm temperate ones, and grass belt in the north migrated westward. The sandy land to the east of Mt. Helan was fixed, and active sand dune retreated to the west of Mt. Helan. Subtropical forests occupied a dominant position in the south of Yangtze River. Temperate and cold-temperate vegetations retreated to high altitude mountainous areas. After the Holocene Megathermal, the subtropical forests decreased again with the cooling and drying climate, and grassland range expands further.

After the Last glaciation, a warming climate and improving environment led the human to evolve from the time of the rough stone hunting to the Neolithic farming era. People began to crop cultivation and animal domestication, textile, pottery, smelting and then appeared. A qualitative leap has taken place in human society.

7.4 Cryospheric Variations in the Past Century

Great attention has been paid to the past centuries climate because of current global warming and its link with human being. A lot of observation data on glaciology and meteorology have been accumulated in the past century, so that we can use them to study the climate change in the past century. IPCC report pointed out that global mean air temperature rose 0.85 °C from 1880 to 2012. The longest observation data show that temperature during 2003–2012 is 0.78 °C higher than that of 1850–1900. The atmosphere and the ocean is warming, all aspects of the cryosphere have been changed dramatically. We will introduce those changes as following.

7.4.1 Centennial Variation of Antarctic Ice Sheet

The mass fluctuations in the Antarctic Ice Sheet arise from differences between the net snow accumulation and ice discharge. The thickness of the Antarctic Ice Sheet is up to 4897 m, and it could raise global sea level by 58 m when it melting out. Warm seawater enters the cavities beneath floating ice shelves drove faster basal melting and thinning them, and a reduction in the back-stress on the grounded ice upstream. The reduced buttressing will increase the flow of the ice streams feeding the ice shelves, and finally led to the retreating of grounding lines, including runaway retreat of glaciers grounded on bedrock that deepened inland. Observations of grounding line locations and ice stream velocities reflect that the 'marine ice sheet instability' is perhaps underway in West Antarctica. It was first proposed in the 1970s and supported by observations and modelling in recent two decades. The striking changes have occurred in Antarctic glaciers, ice shelves and the ice sheet. The amount and rate of sea level rise in the coming centuries depends on the response of the Antarctic Ice Sheet to the warming of atmosphere and ocean.

1. Ice Shelf imbalance

More than 80% of Antarctic ice drains into the Southern Ocean through ice shelves that fringe around the Antarctica continent. The ice shelves can provide mechanical support for the grounded ice sheet upstream through contact with confining side walls or sea mounts. The thinning or disintegration of the ice shelves triggered by

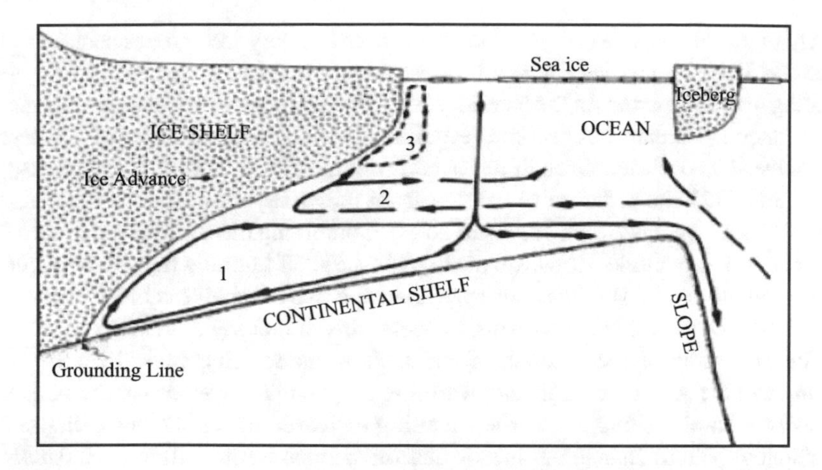

Fig. 7.17 The three models for explaining melting under ice shelf

the warming of the ocean or the atmosphere can weaken the buttressing capacity, resulting in the accelerated retreat of inland ice flows and the mass loss of the ice sheet.

The mass balance of the Antarctic ice shelf consists of four components: the ice shelf basal melting, icebergs disintegration, the ice draining into them, and the surface mass balance (the snow accumulation minus melting). The annual total disintegration of icebergs reach up to about (1321 ± 144) Gt and the basal melting is about (–1454 ± 174) Gt.

The melting of the Antarctic ice shelf is caused by three different patterns, and all of them are related to the relatively warmer seawater circulation (Fig. 7.17). The first pattern is related to the sea ice formation, a high-salt shelf water is formed during this process and sinks into the grounding line, causing the ice shelf bottom to melt strongly. While the oceans acquire abundant freshwater and mix with high-salt shelf water to form ice shelf water. Due to the buoyancy effect, the waters rise diagonally along the ice shelf bottom and move toward the front of the ice shelf. When it rises up to an ice shelf thickness of 300–500 m, the water body becomes super-cooled water due to the sharp rise of the pressure melting point, frazil ice is precipitated and adheres to the ice shelf bottom. The second pattern is the intrusion into ice shelf cavity due to the circumpolar deep layer of water. The third pattern is the mixture of coastal currents caused by tides and winds on the ice shelf front. These processes make a most noticeably ice shelf bottom melt near the grounding line and the ice shelf front.

The strongest thermal forcing and the maximum melting rate (over 40 m/a) occur at the grounding line of the Pine Island glacier in the Amundsen Sea, West Antarctica. Due to the warmer seawater intrusion, the ocean warms, the bottom melts and ice shelves appear to recede. And the reduction of the ice shelf buttressing effect causes Pine Island glacier flow increased by 34% during 1996–2006.

Although the major Ross, Filchner-Ronne, and Amery ice shelves remain stable since the 1990s, many ice shelves in West Antarctica have experienced long-term thinning over this period. In the locations where the ice shelves retreating or thinning, the grounded ice inland has become destabilized. Altogether, the volume of Antarctic ice shelves has declined through net overall thinning (166 ± 48 km^3 yr^{-1} between 1994 and 2012) and progressive calving-front retreating at the Antarctic Peninsula (210 ± 27 km^3 yr^{-1} between 1994 and 2008). Combining these losses amount to less than 1% of their volume. However, the highest ice shelf thinning rates have occurred in the Amundsen and Bellingshausen seas, where five ice shelves have lost between 10 and 18% of their thickness owing to ocean-driven melting at their bases.

The upwelling of the Amundsen and Bellingshausen Sea caused by the wind forcing and the Antarctic Peninsula warming are the main reasons for the Antarctic ice shelves basal melting, the surface melting increase, the ice shelves collapse and their spatial pattern changes. It means that the climate forcing affects the Antarctic Ice Sheet mass balance and thus the sea level at annual and decadal scales through the change of wind field.

2. Collapse of the Antarctic Peninsula ice shelves

Ice shelves at the Antarctic Peninsula (Fig. 7.18) are especially vulnerable, because they are situated at the most northerly latitude on the continent, where temperature

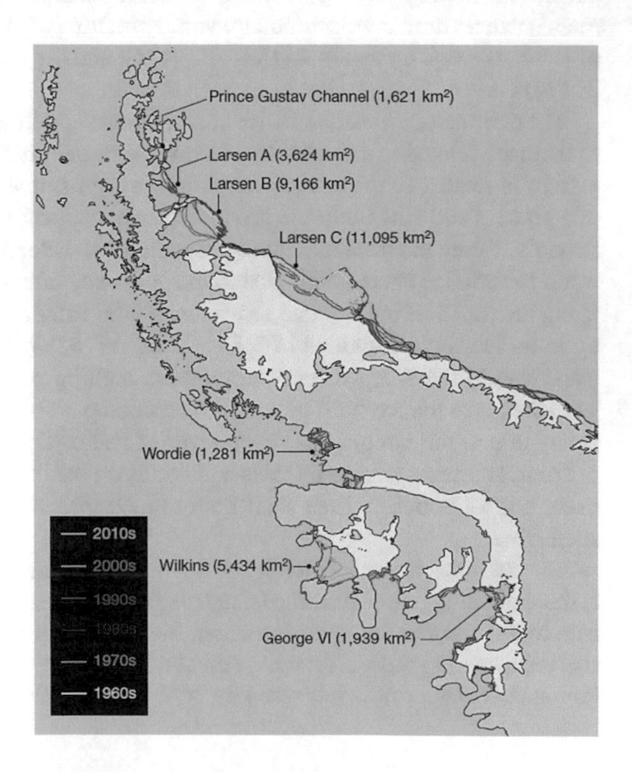

Fig. 7.18 Temporal changes in the location of ice shelf barriers at the Antarctic Peninsula. The outlines of the ice shelves (coloured lines) are as determined from satellite imagery since the 1950s, and the net reduction in area over the same period is shown in parentheses 14.73. Pale (beige) areas are ice shelves, darker (brown) areas are grounded ice (Shepherd et al. 2018)

is relatively high and summertime melting is common. In recent decades, part or entire ice shelves, including the Wordie, Wilkins, Prince Gustav, Larsen A, B and C have disintegrated (Fig. 7.18). Since the 1950s, 33,917 km^2, or 22% total area of the Antarctic ice shelves lost through retreat and collapse. Geological records have confirmed that the collapse events are unique during the Holocene period.

The extent of the Wordie Ice Shelf which on the western coast of the Antarctic Peninsula decreased from 2000 km^2 in 1966 to 700 km^2 in 1989. The Larsen Ice Shelf on the eastern coast of the Antarctic Peninsula collapsed about 9300 km^2 between 1975 and 1986–1989. The last remain section of Larsen A Ice Shelf was quickly disintegrated within a few weeks, and calved an iceberg of 2400 km^2 in 1995. From 31 January 2002, the Larsen B Ice Shelf collapsed 3250 km^2 in 35 days, and its collapse velocity was a terribly surprised. On July 12, 2017, a part of Larsen C Ice Shelf, about 5800 km^2 finally was apart away as iceberg. The ice shelf collapse continuously extends southward in the past several decades.

These events have been linked; warmer air temperatures lead to intensified surface melting, which is believed to cause hydraulic fracture of surface crevasses followed by ice shelf collapse. The Antarctic Peninsula is one of the areas with the rapidest mean annual air temperature rise in the world. The air temperature of the Antarctic Peninsula has increased 2.5 °C since the late 1940s, and this is several times higher than the global average, the rising rate is much higher than anywhere else in the Southern Hemisphere too. There may exists a climatic boundary for the ice shelves on the Antarctic Peninsula, and the mean annual isotherm of –5 °C could be taken as proxy for the limit of viability of the ice shelves. The isotherm associated with this limit having been driven south by the atmospheric warming over the past few decades. Many ice shelves have exceed the climatic limit due to recent atmospheric warming.

The collapse of ice shelves does not contribute directly to sea level rise, because they are afloat. However, there is an indirect effect: observations show that the grounded tributaries to the Larsen A and Larsen B ice shelves did speed up in response to the removal of the floating ice, which is presumed to have offered resistance. So far, ice shelf retreat and collapse has been restricted to those shelves situated at the Antarctic Peninsula, in relatively warm climates, and has not threatened those farther south, such as those on the fringes of the East and West Antarctic Ice Sheets. However, if the atmospheric warming continues with the same rate for the next 200 years, it is estimated that the impact of rising temperature on the largest ice shelf in Antarctic, the Filchner Ronne Ice Shelf and the Ross Ice Shelf will be have similar change to these on the Antarctic Peninsula. The Filchner Ronne and Ross Ice Shelves play a role on stabilizing the West Antarctic Ice Sheet, the water contained by them could raise the sea level by 5 m globally.

3. Grounded ice imbalance

The stability of Antarctica Ice Sheet can be assessed by tracking the movement of its principal glaciers and ice streams. Thanks to step increases in the quantity of high-resolution satellite image acquisitions with time, systematic surveys of ice flow across and around the continent reveal anomalous behavior in most part of Marie Byrd Land, and also at isolated sites at the Siple Coast and the Antarctic Peninsula, and in East Antarctica. In most part of these places, the velocity of ice flow has increased during the satellite image available era, and, when considered them as a whole, the rate of ice discharge from Antarctica exceeds inland snow accumulation.

Although most part of Antarctica Ice Sheet has remained stable over the past 25 years, an obvious imbalance in many coastal sectors—such as the thickening of the Kamb Ice Stream and the thinning of glaciers flowing into the Amundsen Sea and at the Antarctic Peninsula. These changes reflect imbalance between ice flow and snow accumulation within the surrounding catchments. The velocity of the Kamb Ice Stream is unusually low and has not changed in recent decades, but analysis of ice-penetrating radar data show that it stagnated over a century ago.

Glaciers that flowing into the Amundsen Sea of West Antarctica are particularly susceptible to climate forcing. The ice shelves in this sector have thinned by 3–6 m yr^{-1}, and its glacier grounding lines have retreated 10–35 km since 1992. According to analysis of the marine geological record, this rate has increased 20–30 times since LGM. In response to these perturbations, the grounded glaciers inland have sped up and thinned at faster rates too. For example, since the early 1990s, the ice flow rates of the Pine Island Glacier terminus have increased by about 1.5 km yr^{-1} and the ice thinning rates have risen to over 5 m yr^{-1}. Surface lowering has spread inland across the Pine Island drainage and Thwaites glaciers at speeds of between 5 and 15 km yr^{-1}, and the majority of their catchments are now in a state of dynamical imbalance, thus, the sector overall contributed 4.5 mm to global sea level rise between 1992 and 2013.

7.4.2 Glacier Changes

Glacier changes are dominated by climate change, and glaciers respond in scale through the coupled effect of mass balance. Due to the hysteresis of the dynamic adjustment of glaciers, the scale change are not completely synchronous with climate change. Specifically, the mass balance of glaciers can only manifest in glacier extent through movement and mass adjustment. The glacier evolution rate is determined by glacier scale. A large-scale glacier has a long response hysteresis. Secondly, the thermodynamic properties of glaciers are decisive factors. Temperate glaciers exhibit sliding behaviors at their base. The movement velocity of temperate glaciers driven by the changes of mass balance, and the velocity is higher and the range is bigger than those of cold glaciers with a same scale. These factors determine the different responses to climatic changes of glaciers with different scales in the same region and glaciers of the same scale in different regions.

There are few glaciers in the world have long-term mass balance monitoring data. The earliest available observation data could track back to the middle of the twentieth century. The previous mass balance was reconstructed mainly by observation-based models. Currently, accepted results of glacier change, such as mass-balance series of glaciers globally, except for ice sheets were reconstructed using the degree-day method and changes of glacier length. The results have demonstrated that glaciers globally, except for the Greenland and Antarctic ice sheets, are generally in a state of mass loss. The mass balance of global glaciers was 197 ± 24 Gt/a during 1901–1990, 226 ± 135 Gt/a during 1973–2009, 275 ± 135 Gt/a during 1993–2009, and 301 ± 135 Gt/a during 2005–2009. Mass loss has been intensified since the 1970s. Additionally, the mass balance of glaciers shows some differences in different regions. From 2003 to 2009, Alaskan glaciers suffered the greatest mass loss, followed by glaciers in the polar region of Canada and surrounding areas of the Greenland Ice Sheet. The glacier mass loss in Alaska, the north polar region of Canada, the southern Andes, and mountainous regions of Asia accounts for more than 80% of the global glacial loss.

1. Changes in glacier terminus

Glacier retreating has been dominant globally in the past 100 years and has intensified from the 1980s. However, there have been two 10-year stability stages and even an advance stage, especially glaciers at middle and low altitudes. For example, the first significant alpine glacier advance (Fig. 7.19) occurred in 1911. Additionally, the number of advancing glaciers increased sharply and the annual growth rate of the quantity of advancing glaciers was higher than 50% from 1916 to 1922. This process generally persisted to the middle and late periods of the 1930s. The second global glacier advance started after the mid 1960s, that was another glacier advance in the past 100 years. The annual growth rate of the quantity of advancing glaciers was higher than 50% from 1977 to 1985. According to follow-up monitoring, the growth

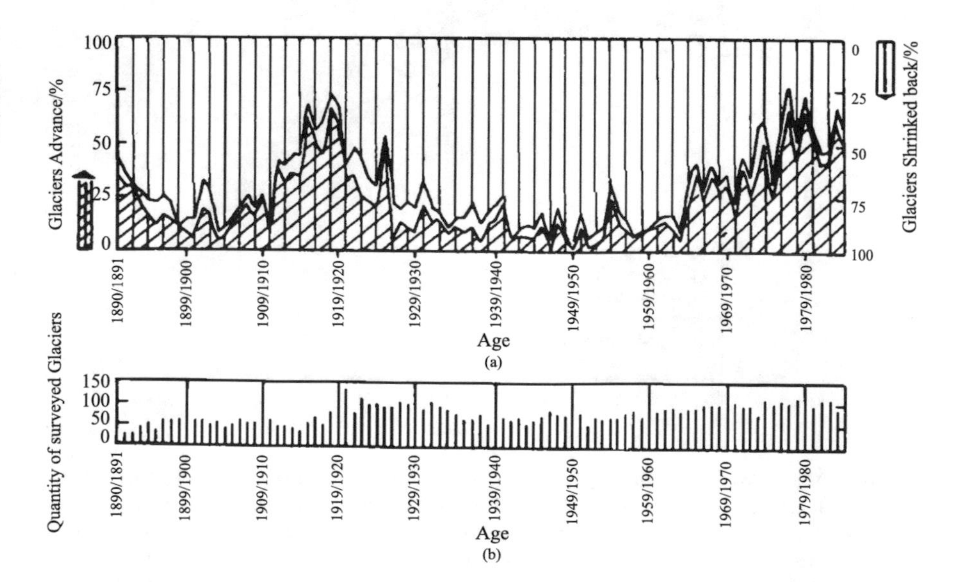

Fig. 7.19 Changes of alpine glaciers in Switzerland in the past 100 years (Aellen and Funk 1988)

rate of receding glaciers was higher than 50% after 1986. Comparing the glacier terminuses of these two advances, the position of the second period was far from that of the first period, indicating that glacier scale decreased gradually.

The glacier retreating trend can be recognized easily from a comparison of the lengths of 500 long-series glaciers around the world. Some large valley glaciers have receded thousands of meters in the past 120 years. The glacier recession rate in middle latitude regions varied from 2 to 20 m/a, and large glaciers have presented a continuous recession trend. Middle scale glaciers have shown generational stage changes, while the length changes of small-scale glaciers have shown high-frequency fluctuations under a general background of recession.

In the 1990s, glacier advances abnormally in Scandinavia and New Zealand, which may have been related to the unique climatic changes (increased precipitation in winter) in these two regions. In Iceland, the Karakorum Mountains, Svalbard and other regions in the world, the observation data show some glaciers often have poor dynamic stability (glacier surging). Glaciers' terminus disintegration showed that they have a rapid recession, however, some glaciers with thick supra-glacial till at the ablation zone remain relative stable.

The glacier length change reflect the impacts of long-term, low-frequency climatic changes. Fortunately, the monitoring of some glaciers can be traced back to the seventeenth century or earlier. Based 169 glaciers' length records and a regional classification homogenization of them, glaciers underwent the same evolution process from 1700 to 2000 in the world (Fig. 7.20). Generally, glaciers in different regions began to shrink at about 1800 and maintained a high recession rate from about 1850. The

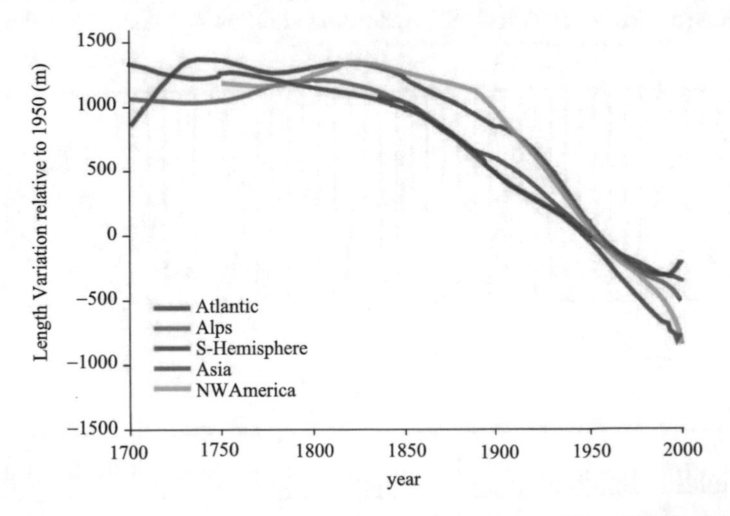

Fig. 7.20 Mean length changes of glaciers of the world (Oerlemans 2005). The Southern Hemisphere includes the Tropics, New Zealand, and Patagonia; the northwestern area includes North America (mainly the Rocky Mountains in Canada); the Atlantic area covers the southern parts of Greenland, Iceland, Jan Mayen Island and Svalbard in Norway, and the Scandinavian Peninsula; others include the Alps Mountains in Europe, Caucasus Mountains in Asia, and mountains in Central Asia

observation data also demonstrate that glacier recession decelerated to some extent from 1970 to 1990, but accelerated again after the 1990s.

Since 1900, The modern glacier changes can be divided into several stages in China. Most glaciers were relatively stable or advancing from the beginning of the twentieth century to 1930. From the 1940s to 1960s, most of them were receding, except for a few in relative stability or progression. From the 1970s to 1980s, there was a relative deceleration of glacial recession, some glaciers became stable or even advance again. Most glaciers entered a re-recession in the 1990s. Particularly, glacier terminus recessions in some mountainous areas are very faster, such as the Himalayas, the southeast Tibet, the Hengduan Mountains, the Pamirs, the Karakoram Mountains, the Tianshan Mountains and Anemaqen in the east segment of the Kunlun Mountains, and the Lenglongling in the eastern Qilian Mountains.

In short, the worldwide glacier changes shown a similar pattern in the past more than 100 years, especially in the recent decades. Glacial recession is the main trend, however, it was accompanied by two small stages of advances (1820s–1830s and 1960s–1970s or 1980s).

2. Changes in glacier extent

Only a small part of observation data related to glacier area change, and such monitoring began relatively late (Fig. 7.21). The available published data of 19 regions in the world show glacier area change have the following features: ① Glacier areas in all regions have been shrinking since the 1940s. ② The rates change in each region fall into generally similar intervals, but there are large differences among them. ③ Significant area shrinkage was observed in western Canada (Zone 2), middle Europe (Zone 11), and low-altitude regions (Zone 16). ④ The proportion of shrinking glaciers in all regions has increased continuously.

Glacier disappearance has been reported frequently. Reports include more than 600 glaciers in the polar region of Canada, the Rocky Mountains, north Cascade, Patagonia, the Alps in Europe and Asia, the Tianshan Mountains and some tropical mountainous regions etc. However, the actual number of glacier disappearances might be larger. These results prove that the equilibrium line altitude (ELA) of glaciers has been increased significantly.

3. Glacier changes in China

Relevant investigations have demonstrated that most glaciers in China have been shrinking since the 1950s, which is consistent with the glacier change in other regions of the world. According to observation data, Glacier No. 1 at the headwater of the Urumqi River, Tianshan Mountains, has generally been in negative mass balance state since 1959. The mass balance of the glacier fluctuated slightly from the late 1950s to the early 1980s. The mass balance level was relatively low and glacier recession was slow, subsequently, mass loss intensified. In particular, negative mass balance accompanied by obvious glacier shortening, has increased continuously since the mid 1990s (Fig. 7.22).

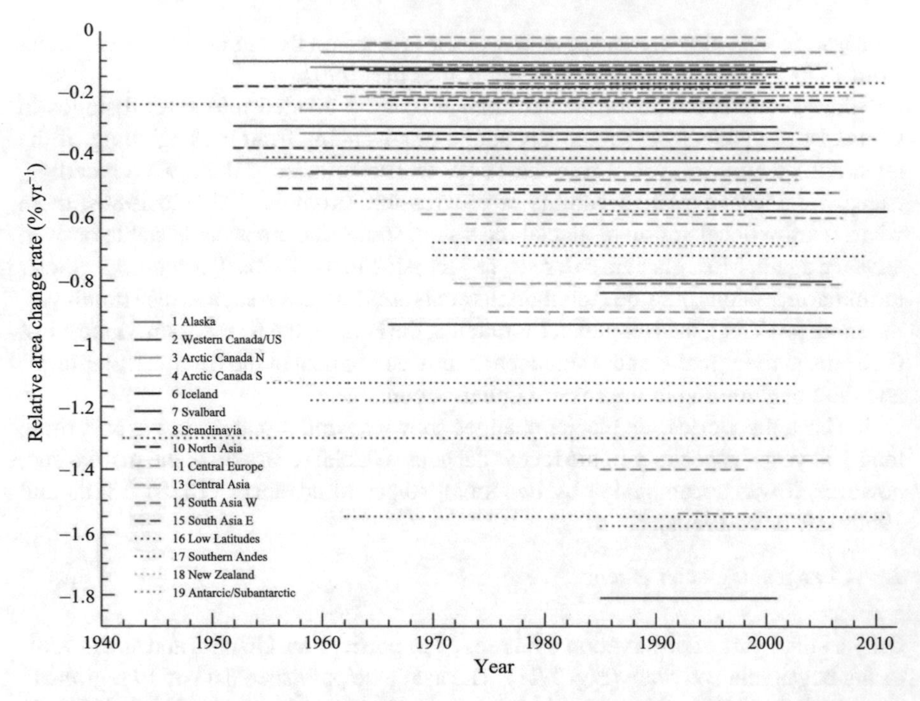

Fig. 7.21 Mean area variability of glaciers in 19 regions of the world. Each line represents the annual average area shrinkage of glaciers in the region. The length of each line reflects the average statistical period. Regions without numbers have no data of glacial change in the corresponding region

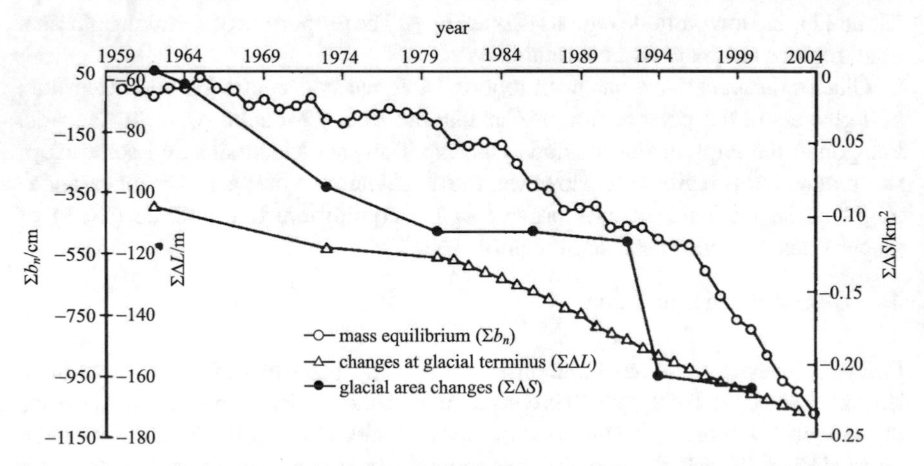

Fig. 7.22 Changes in length, area, and mass balance of Glacier No. 1 at the headwaters of the Urumqi River, Tianshan Mountains, during 1962–2004 (Xie and Liu 2010)

The study on glacier area change in different regions of western China have revealed that glaciers are generally shrank, but there were significant regional differences. The annual shrinkage rates in the source of the Yellow River and northern slope of the Tianshan Mountains were the highest, while the annual shrinkage rates of glaciers on plateaus and in the Kunlun Mountains were smaller than the other regions.

7.4.3 Global Changes of Permafrost

Permafrost is mainly distributed in the Arctic region, the Antarctic region and the middle-low latitudes with high altitude. In the background of global warming, the permafrost in all regions was degrading with an increasing temperature, and the active layer thickness is increasing globally. However, due to the influence of local factors, the permafrost changes in various regions show different trends.

1. Pan-Arctic regions

In the mid 1980s, early 1990s and early 2000s, the Alaskan region was relatively cold. According to drilling records, the permafrost temperature at the depth of 20 m was relatively stable and even showed cooling trend (Fig. 7.23). However, after 2007, two observation sites in the northern part of Alaska showed that the permafrost temperature at 20 m depth increased by 0.2 °C.

In discontinuous frozen zones of Mackenzie Corridor Region in Western Canada, the annual mean ground temperature has risen by 0.2 °C per 10 years over the past 25 years. Permafrost temperature in the southwest region remains stable. The active layer thickness of central region has increased by about 5 cm per year from 1998 to 2007. Permafrost temperature at the depth of 15 m in the eastern Ellesmere Island area has increased at the rate of about 0.1 °C per year, while increased by 0.1 °C per 10 years at the depth of 36 m for the past 30 years. In the La Loren mine area of Quebec, the permafrost temperature cooled in the first 50 years of the twentieth century, following by a warming trend, but also showed a cooling trend in the late 1950s to the late 1980s, and then warming, there has been a clear trend of thickening of the active layer since 1993. At Umiuzaq, the temperatures at 4 m and 20 m increased by an average of 1.9 and 1.2 °C since the 1990s.

In the northwestern part of Siberia in Russia, the surface temperature showed an increasing trend from 1974 to 2007 with an increase of 2 °C in cold frozen ground region and an increase of 1 °C in warm frozen ground region. Most of the warming occurred in 1974–1997. The permafrost temperature did not change in many areas from 1997 to 2005, and even some areas showed a cooling trend. The warming trend appeared in region with low temperature below –0.5 °C after 2005.

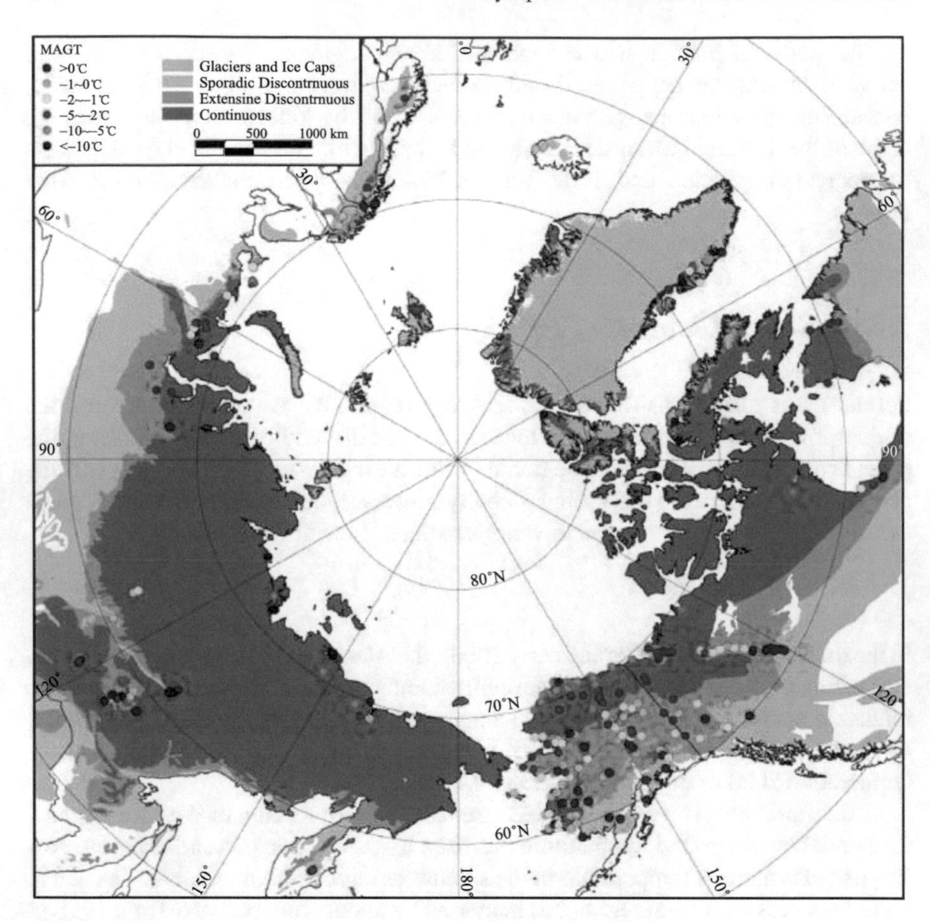

Fig. 7.23 The drill bores around arctic and mean annual ground temperature (Romanovsky et al. 2010)

2. Northern Asia regions

Permafrost research in northeastern China has focused on the Greater and Lesser Khingan Mountains in China. Studies have found that in 1970s, the southern limits of permafrost overlapped with the annual mean air temperature isotherm of –1 to 0 °C in the western part of the Greater Khingan Mountains, overlapped with the 0 °C isotherm in the Songnen Plain, and Overlapped with the 0–1 °C isotherm in the eastern part of the Lesser Khingan Mountains, which means that the overall southern boundary is between –1 and 1 °C isotherm. Currently, discontinuous and island permafrost areas have been reduced by 9×10^4 to 10×10^4 km^2, only 35–37% of the area in 1970s. In the northern part of the Greater Khingan Mountains, the thickness of the active layer was 50–70 cm in the 1960s and 1970s, but increased by 32 cm in 1978–1991 and the soil temperature at the depth of 20 cm increased by

0.8 °C. In the middle of the Greater Khingan Range in the early 1980s, the depth of seasonally thawing declined while the annual mean air temperature increased. In the 1990s, the maximum thawing depth increased from 1 to 1.2 m, and the annual mean air temperature increased from –5.5 to –3.0 °C.

3. North Europe regions

Researches in northern Europe were mainly concentrated in Iceland and Scandinavia. The inter-annual temperature difference in Iceland is smaller and the number of frozen days is relatively short for the temperate marine climate. Even in winter, some areas also appear hot melt phenomenon. The lower limit of Icelandic alpine permafrost gradually decreases from south to north. Large areas of permafrost are distributed above the altitude of 1000 m in the south and 800 m in the north. Snow cover is generally distributed in the area above the altitude of 1000 m, the adiabatic ability of snow cover is an important factor affecting the distribution of permafrost in Iceland, and the influence of seasonal snow cover distribution is more significant than summer temperature and precipitation on the degradation of permafrost. For the strong winds mainly come from the southeast region, which makes no snow cover in the southern plains and the snow cover is mainly on the northern slope. This made the soil temperature of southern slope to decrease rapidly in winter, while the cool spring and summer allow the permafrost in the northern slope to last longer, which offset the uneven distribution of temperature in the north and south. Continuous permafrost is distributed above the altitude of 1200 m. Observations over the past few years showed that the annual mean ground temperature was 0.5–1 °C higher than that in 1961–1990. The surface temperature of most observation sites was greater than 1 °C. The permafrost retreats at many observation sites, especially in drier areas. The shrinkage of permafrost is even more significant than other regions due to the high ground temperature in Iceland.

Long-term borehole observations in the Scandinavian region found that the ground temperature had a clear increasing trend in the late 1950s and early 2000s. The extreme ground temperature appeared in the summer of 2003, which was basically consistent with the data observed in Iceland.

4. Antarctic regions

The number of Antarctic permafrost observation sites increased greatly from 21 to 73 during the International Polar Year (Fig. 7.24). In general, the permafrost temperature decreases with the increasing altitude from the coast to the interior of the Antarctic continent.

The highest value occurs in the South Shetland Islands, which is slightly below 0 °C. In northern Victoria, the temperature of the permafrost varies from –18.6 to –13.1 °C, and varying from –22.5 to –17.4 °C in the Komodo Channel, whereas which reaches –23.6 °C in the high-altitude Rose Island region. From 1950 to 2000, the annual average air temperature in Antarctic was rising at a rate of 0.56 °C every 10 years, which also reflects the warming trend of the Antarctic.

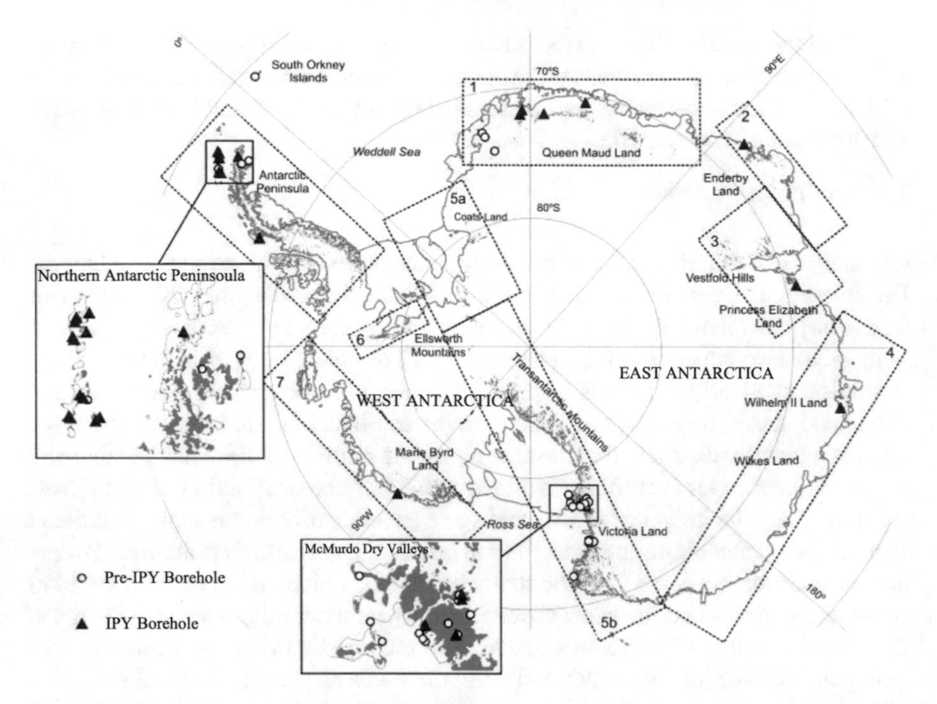

Fig. 7.24 Permafrost survey stations in Antarctica

5. Central Asia regions

The permafrost observation sites in central Asia are mainly concentrated in the Qinghai-Tibetan Plateau, Tianshan Mountains and Mongolia regions (Fig. 7.25). The observation data show that the permafrost temperature increased 0.02–0.19 °C per year from 1999 to 2006 in the Qinghai-Tibetan Plateau. The thickness of active layer in 2006 or 2007 is the largest, and the changes range from 35 to 61 cm. It can be seen that the permafrost temperature at 6 m increased from 0.07 to 1.02 °C per year from 1996 to 2002, but decreased in recent years, which may be related to the local climate change. In Tianshan and Kazakh region, the ground temperature at 14–25 m increased from 0.3 °C in 1974 to 0.6 °C in 2009. The active layer thickness was 3.2–3.4 m in 1970 and 5.2 m in 1992, the active layer thickness increased by 23% compared with that in the 1970s. At the observation site of China in the upper reaches of the Urumqi River, Tianshan Mountains, the annual mean ground temperature increased from –1.6 °C in 1992 to 1 °C in 2008, and the soil temperature increased by 0.4–0.9 °C at different depths in soil profiles. The depth of annual mean 0 °C soil temperature increased from 10 m in 1992 to 12 m in 2008, all of which indicate that in recent decades, the air temperature, ground temperature and active layer thickness of Tianshan Mountains have obviously increased.

In Mongolia, the data from the observation sites M1a and M3 showed that the rising rate of the active layer thickness was 20–40 cm per year, while which was only

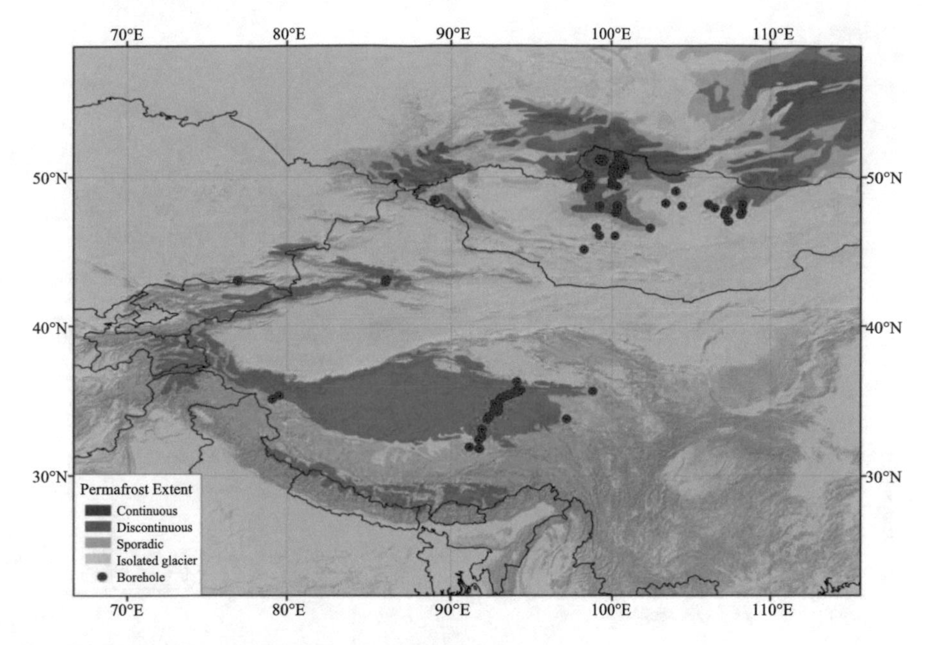

Fig. 7.25 Permafrost survey stations in the central Asia

0.5–2.0 cm per year in M6 a and M7a over the past few years. However, in 2009, the thickness of active layer observed in most areas showed a decreasing trend. Permafrost temperatures have been higher in the past 15–20 years than in the 1970s and 1980s (Fig. 7.26). This trend was very similar to that in central Asia and the European mountains, but compared with Eastern Siberia and Alaska, the warming trend of Mongolia was much smaller.

7.4.4 Seasonal Snow in the Northern Hemisphere

Seasonal snow cover, 98% of which lies in the Northern Hemisphere (NH), can reach a maximum coverage of 45.2×10^6 km^2. Snow cover is ongoing significant changes due to the continuous rising temperature and the modulation of atmospheric circulation caused by global warming.

A most direct and effective effect for snow cover comes from the spring temperature, which prevents snow from accumulation and accelerates snow melting. Averaged March–April NH snow cover extent (SCE) decreased significantly over the 1922–2012, and the decreasing rate speeds up since 1960s, especially after 1980s (Fig. 7.27). This related to not only the persistent overheating, but also the transition of atmospheric circulation around 1980. In March, SCE reduction over Eurasia contributes most to the SCE decreases in NH, while in April SCE in both Eurasia and North America decreases significantly. It's worth noting that SCE reduction in

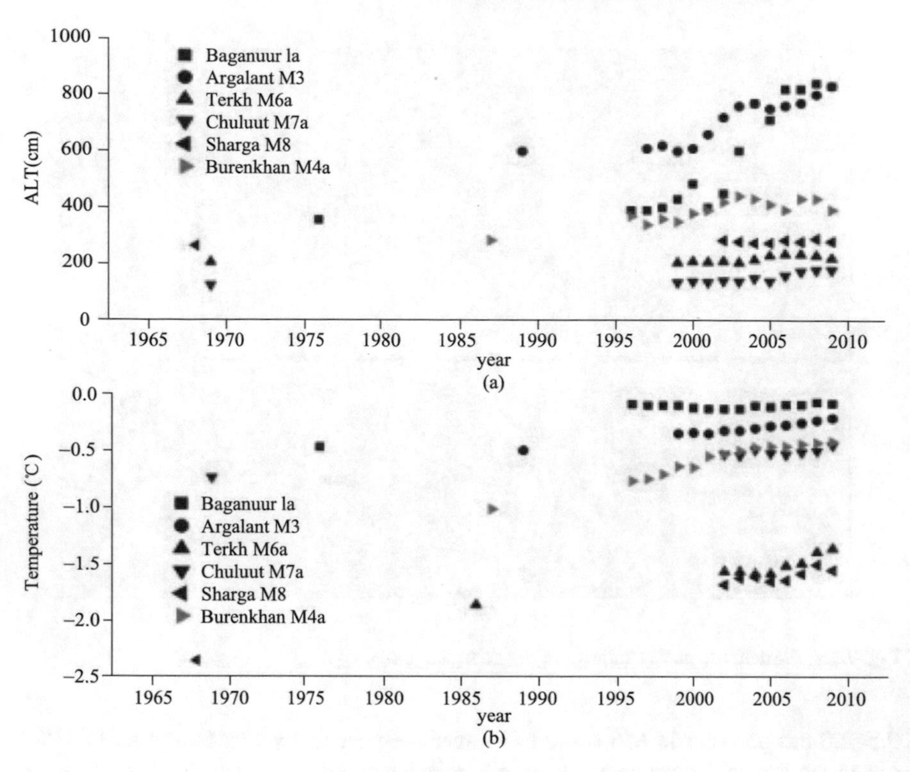

Fig. 7.26 The active layer thickness and mean annual ground temperature changes in Mongolia

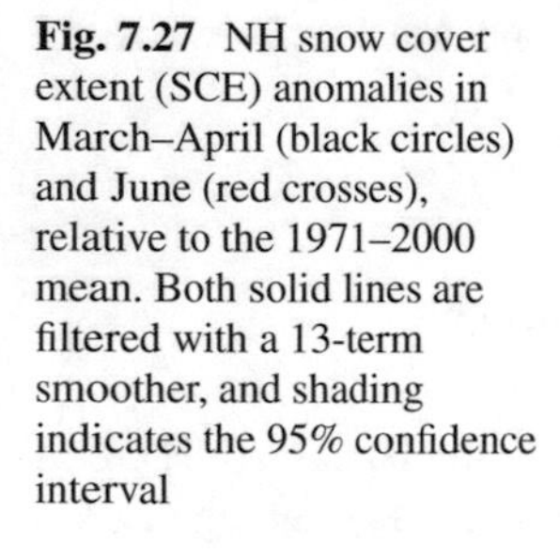
Fig. 7.27 NH snow cover extent (SCE) anomalies in March–April (black circles) and June (red crosses), relative to the 1971–2000 mean. Both solid lines are filtered with a 13-term smoother, and shading indicates the 95% confidence interval

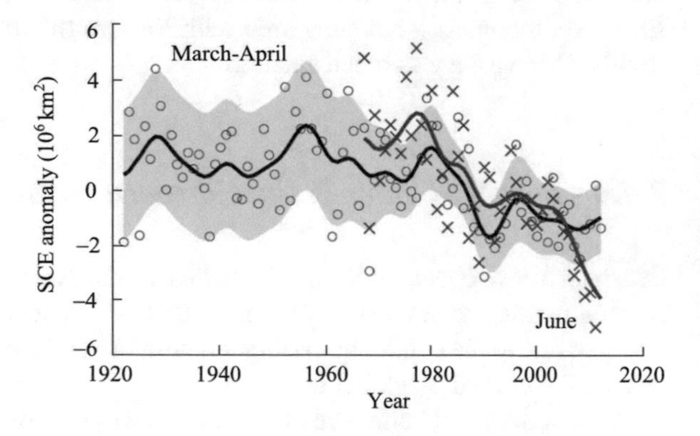

June, no matter absolute or relative are both beyond that shows in March–April. Additionally, the duration of the snow season averaged over NH grid points declined by 5.3 days per decade since winter 1972–1973, owing to earlier spring snowmelt. Over Eurasia, although winter snow accumulation increases significantly, significant trends toward a shortening of the snowmelt season have been identified over most of Eurasia and the pan-Arctic region since 1979, with a trend toward earlier melt of about 5 days per decade at the beginning of the melt season, and a trend of about 10 days per decade later at the end of the melting season.

In China, spring snow depth (SD) and snow water equivalent (SWE) both decrease significantly during 1951–2009, but with apparent regional differences. For winter SD, significant positive trends are found in northwestern China, while in northeastern China the most distinct characteristics are the increased amplitude of inter-annual variations. For the Qinghai-Tibetan Plateau, both winter SD and SWE decrease significantly, and keep less than normal since the beginning of the twenty-first century.

7.4.5 Sea Ice Changes in the Arctic and Southern Ocean

Since 1970s, Arctic sea ice extent has been decreasing significantly. There are a number of pre-satellite records, some based on regional observations taken from ships or aerial reconnaissance, while others were based on terrestrial proxies. These records suggest that the decline of sea ice over the last few decades has been unprecedented over the past 1450 years. The earlier study of the marginal seas near the Russian coastline using ice extent data from 1900 to 2000 found a low frequency multi-decadal oscillation near the Kara Sea that shifted to a dominant decadal oscillation in the Chukchi Sea. A more comprehensive basin-wide record showed very little inter-annual variability until the last three to four decades.

Figure 7.28 shows an updated record of the Walsh and Chapman data set with longer time coverage (1870–1978) that is more robust because it includes additional historical sea ice observations. A comparison of this updated data set with that originally reported by Walsh and Chapman (2001) shows similar inter-annual variability that is dominated by a nearly constant extent of the winter (January–February–March) and autumn (October–November–December) ice cover from 1870 to the 1950s. Sea ice data from 1900–2011 as compiled by Met Office Hadley Centre are also plotted for comparison. In this data set, the 1979–2011 values were derived from various sources, including the satellite data. Since the 1950s, more and more in situ data are available and have been homogenized with the satellite record. These data show a consistent decline in the sea ice cover that is relatively moderate during the winter but more dramatically during the summer months.

The rapid reduction in Arctic sea ice extent and thickness are synchronously. On the basis of the up-looking sonar data, it is found that the average sea ice thickness of the Arctic Ocean decreased from 3.1 m in 1958–1976 to 1.8 m in 1993–1997. In 2003–2007, the sea ice thickness in each sector of the Arctic Ocean reduced by more than half as compared with that of the 1958–1976.

Sea ice melting led to break-up of Arctic pack ice over a wide range, producing new sea ice region covered by large and small brash ice, which is similar with the nature of the traditional marginal ice zone. If these new formed regions are referred to as the marginal ice zone, the magnitude order of marginal ice zone width can reach 10^3 km, and the sea ice in these regions is more vulnerable and easy to melt. Around

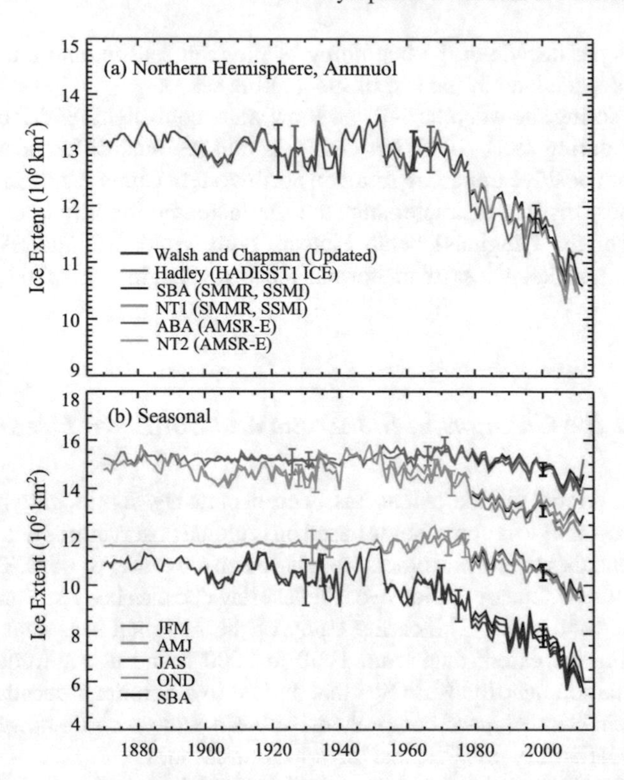

Fig. 7.28 Variations of Arctic sea ice extent in century time scale, from 1870 to 2011. **a** Annual ice extent and **b** seasonal ice extent using averages of mid-month values derived from in situ and other sources including observations from the Danish meteorological stations from 1870 to 1978 (updated from Walsh and Chapman 2001). The yearly and seasonal averages for the period from 1979 to 2011 are shown as derived from Scanning Multichannel Microwave Radiometer (SMMR) and Special Sensor Microwave/Imager (SSM/I) passive microwave data using the Bootstrap Algorithm (SBA) and National Aeronautics and Space Administration (NASA) Team Algorithm, Version 1 (NT1); and from Advanced Microwave Scanning Radiometer, Version 2 (AMSR2) using algorithms called AMSR Bootstrap Algorithm (ABA) and NASA Team Algorithm, Version 2 (NT2). In (**b**), data from the different seasons are shown in different colours to illustrate variation between seasons with SBA data shown in black

1980, the multi-year sea ice accounted for more than 75% of the total sea ice, while this value is only 45% in 2011. The Arctic multi-year sea ice is concentrated within a hundred kilometers near the northern slope of Canada. Every year, there are lots of ice losses over Canadian Arctic, in spring, some of which is transported to the western Arctic with the Beaufort Gyre and reaches farthest to the East Siberia Sea. The reduction of multi-year sea ice is a sure sign of a decreasing sea ice extent. If this trend remains, the Arctic ice free summer will come in the near future.

Unlike the Arctic, the sea ice around Antarctica is constrained to latitudes north of 78° S because of the presence of the continental land mass. The Antarctic sea ice is largely seasonal, with an average thickness of only 1 m at the time of maximum extent in September. Only a small fraction of the sea ice survives the summer minimum in February, and very little Antarctic sea ice is more than 2 years old.

According to the satellite observation data, the annual sea ice extent in the Antarctic increased at about 1.5% per decade. However, there are regional differences in trends, with decreases in the Bellingshausen and Amundsen seas, but a larger increase in sea ice extent in the Ross Sea that dominates the overall trend. Whether the smaller overall increase in Antarctic sea ice extent is meaningful as an indicator of climate is uncertain because the extent varies so much from year to year and from place to place around the continent. A recent study suggests that these contrasting trends in ice coverage may be due to trends of regional wind speed and patterns. Without better ice thickness and ice volume estimates, it is difficult to characterize how Antarctic sea ice cover is responding to changing climate, or which climate parameters are most influential.

Due to the high variability of Antarctic sea ice over each sector, and the significant regional differences in the trend of its extent and area, it's difficult to construct a uniform record to characterize the change in sea ice extent over the whole Antarctic region. Therefore, there is not an accepted conclusion of sea ice change in century time scale. Present studies mainly focused on a specific sector and investigated the long-term variations in sea ice extent or northern border over that area.

Questions

1. The North America and the North Europe were covered by great ice sheets during LGM, but there was no ice sheet to develop in the vast Siberia in same latitude, why?
2. What is the cause of MPT (Middle Pleistocene Transition)?
3. Can the records of MIS be taken as evidence for synchronous or asynchronous glaciations at both hemisphere?

Chapter 8
Interactions Between Cryosphere and the Other Spheres

Lead Authors: Yongjian Ding, Genxu Wang

Contributing Authors: Shiqiang Zhang, Yong Luo, Gengnian Liu, Jianfeng He, Baisheng Ye, Rensheng Chen, Bingyi Wu, Yongping Shen

The formation and evolution of cryosphere have profound impacts on the atmosphere, hydrosphere, biosphere, and lithosphere, which in turn have a large influence on the distribution and changes of the cryosphere. Understanding the interactions between cryosphere and the other spheres are important to expand the cryospheric science. At the same time, changes in cryosphere also affect the anthroposphere. In this chapter, the interactions between cryosphere and the other spheres were briefly introduced with an emphasis on the functions of the cryosphere in the other spheres. The impacts of cryosphere on the anthroposphere will be introduced in Chap. 9.

8.1 Cryosphere and Atmosphere

The exchange of heat, mass, and momentum occurs at the interface between the atmosphere and land surface covered by ice and snow. Interactions between ice and atmosphere have important influence and feedback on adjustment processes between the atmosphere and cryosphere in the climate systems. They affect the dynamic and thermodynamic characteristics of the boundary layers between the atmosphere and sea ice and the boundary layers between atmosphere and land surface covered by snow and ice, which include exchange of radiation (long-wave radiation, solar radiation and reflection), momentum, heat (latent and sensible heat), and mass (water vapors, gases, particulates, etc.). They also affect the movement of wind on the sea ice and snow cover. The interaction between cryosphere and atmosphere plays a critical role in the regional and global climate systems, as well as in their variability and changes. On one hand, cryosphere is a natural indicator of climate change because it is very sensitive to climate change (see details in Chap. 3). On the other hand, various components in cryosphere (such as ice sheets, mountain glaciers, snow cover, sea ice, lake ice, river ice, and the frozen ground) have great regulating effect on climate at different temporal and spatial scales through complex feedback processes due to the characteristics of high albedo, large phase change, low conductivity of ice and snow.

D. Qin et al. (eds.), *Introduction to Cryospheric Science*, Springer Geography,
https://doi.org/10.1007/978-981-16-6425-0_8

8.1.1 Snow and Ice Albedo Feedback

It is needed to introduce two confusing concepts about albedo and reflectivity before discussing the ice and snow albedo feedback. Reflectance is a ratio of the reflected solar radiation to the incident solar radiation at a certain wavelength, whereas albedo is the integral of reflectance at each wavelength. Thus, subtle changes in albedo can alter the energy balance relations in land–atmosphere systems, which can eventually cause climate change. Ice and snow have a large impact on energy absorption because of their high albedo. The albedo of the clean snow surface can reach up to more than 90%, whereas other land surface albedo is generally between 10 and 30%. The ocean absorption energy over the area covered by sea ice can be up to 9 times lower than that over open water area.

The albedo of snow and ice surface depends on the reflection properties of snow and ice, and air or sky conditions. Many physical properties affect the snow and ice albedo, such as snow grain size, density, water content, impurity, and so on. Sky condition, such as atmospheric water content and turbidity, cloud cover and form, affect snow and ice surface albedo by altering the incoming solar radiation and its spectral distribution.

The snow and ice albedo feedback is one part of interactions between cryosphere and atmosphere. The feedback is considered as a positive feedback between changes in albedo and surface temperature, which affected by the properties and spatial distribution of snow and ice. Comparatively, the ice and snow albedo feedback is a typical positive feedback in the climatic system. Increasing surface temperatures lead ice and snow to melt, which decreases the ice and snow extent and, thus, the surface albedo. Subsequently, solar radiation absorbed by the surface will increase. Conversely, decreasing surface temperatures lead to opposite changes, which increases the ice and snow extent, then increase albedo, which further enhances the decreasing surface temperature. This feedback can be applied to explain the small-scale variability of snow cover. When a small amount of snow melts, it causes the surface darkness and higher absorption rate of solar radiation, which leads to more snow melt. Net radiation is an important source of energy for glacier ablation, where the change of albedo in a small area of a glacier will cause a relatively large variation in the glacier ablation. This mechanism is also used to explain recent shrinkage of sea ice in the Arctic Ocean. It is one of the important reasons why changes in the sea ice cover in the Arctic Ocean are taken as amplifier and indicator of climate change.

8.1.2 Exchange of Sensible Heat and Latent Heat Between Ice and Air

In the upper boundary of ice and snow layer, the transfer of energy is affected by incident solar radiation, long-wave radiation, reflected shortwave radiation, scattering of longwave radiation by snow surface, turbulent sensible heat flux and latent heat flux in the vertical direction, and the thermodynamic processes of ice and snow. In the lower boundary of ice and snow, the basic factors controlling the heat transfer include the oceanic turbulent heat flux and latent heat flux of ice condensation or melting. The heat flux through the ice is controlled by the above factors. In the heat budget of multi-year ice during the winter, the strong radiation cooling effect and relatively small amount of heat transferred from the sea through sea ice make the surface temperature of sea ice lower than the ambient air temperature. This is the main reason for the stable stratification of the atmospheric boundary layer and a decrease of sensible heat flux, which is also the reason for cooling of the lower atmosphere. In winter, the contribution of turbulent latent heat flux to the heat budget is not obvious, because low air temperature leads to lower vapour content in the atmospheric boundary layer.

Cryosphere affects the climate system due to its huge cold storage and phase-change latent heat. When ice melts into water at 0 °C, the latent heat of the phase change is 33.4×10^4 J/kg. When solid ice transforms into the gaseous phase (sublimation), the latent heat of the phase change increases to 283×10^4 J/kg. During the melting/thawing process, the components of the cryosphere (such as glaciers, ice sheets, permafrost, snow cover, and sea ice) undergo a self-heating (to reach 0 °C) and then phase change processes. A significant sensible heat and latent heat exchange occur during melting process between the ice or snow surface and atmosphere, and also between the ocean and sea ice, which affects the weather and climate system at different spatiotemporal scales.

For quantitative analysis of the connection between cryosphere and climate, a cryosphere component model, such as a sea ice model, or parameterization of cryosphere component in the Land Surface Model, or sea ice model and parameterized cryosphere Land Surface Model coupled with the climate model, is typically required. Through the above models, refinement and parameterization of cryosphere components in the climate model can be improved, and the role of cryospheric components in climate change can be further analyzed.

Unfortunately, the parameterization of the cryosphere in the Land Surface Model is very complicated. For example, in terms of parameterization of the frozen ground, the parameterization of thermal conductivity, heat capacity, thermal diffusion coefficient, surface roughness, albedo, and Bowen ratio in the sensible heat and latent heat exchange process between frozen ground and air, are required. Some parameters are also needed to consider the different phase transition processes of freezing and thawing. As for thermal conductivity, the parameter in freezing and thawing processes are different.

8.1.3 Momentum Exchange Between Ice and Air

The momentum exchange between ice and air is mainly reflected in deformation, breakage, and accumulation of sea ice, which is caused by inhomogeneity in the sea ice motion field under the combinations of ocean current, wind, and waves.

It is generally believed that global warming contributes to decrease in Arctic sea ice. However, some studies suggested that the rapid decline of Arctic sea ice can be attributed to a shift of the surface wind pattern. The correlation between strong Arctic wind and the decline of Arctic sea ice extent has been determined by analyzing the relationship between the annual intensity and convert time between strong and weak Arctic winds, and changes of the sea ice extent since 1979. A study revealed that strong anticyclonic wind anomalies since the late 1990s have transported more sea ice into the Greenland sea, which influences the effect of momentum exchange over sea ice changes. After sea ice blown into the warmer waters of the North Atlantic Ocean, they change the salinity and temperature of the North Atlantic Ocean, which will affect the stratification of the ocean and marine ecosystems. The typical example of air, ice, and ocean momentum interaction is polynyas, which forms in the offshore region. Polynya is the open water in the offshore area of the Arctic which is covered by continuous sea ice, it can exist even in an environment with temperatures as low as of –50 °C. Polynyas are formed when warm water upwells from deep sea to the surface by ocean dynamical effect.

Moreover, the formation of wind-blown snow is also forced by atmospheric movement. The multi-phase flow of snow particles dispersed by the airflow from the ground or the snow conveyed by the wind is called wind-blown snow. Wind-blown snow can redistribute the snowfall, which has an important influence on the mass balance of snow and ice. At the same time, wind-blown snow has an important influence on the energy balance of snow and ice covering surfaces. Wind potentially causing the sublimation and abrasion of snow grains from the air and enhancing water vapour transport from snow to the atmosphere. Wind-blown snow can change the roughness of snow surface, affect the energy exchange between the air and snow surface. It also can alter the air visibility, and then change the air energy transmission.

The redistribution of snow in alpine regions plays an extremely important role in the formation of avalanches.

8.1.4 Cryosphere and East Asian Monsoon

The East Asian Monsoon (EAM) is driven by thermal gradients between the Asian Continent and Indian and Pacific Oceans. Interactions between the global ocean, atmosphere, and land have important impacts on EAM. There are many driving factors of EAM, and the interactions between driving factors and EAM are very complex. The driving factors include (Fig. 8.1): ① The thermal conditions of the equatorial Eastern Pacific and Western Pacific Warm Pool in the east direction, including ENSO events and the Pacific Decadal Oscillation (PDO), etc. ② The thermal conditions on Eurasia and the Qinghai-Tibetan Tibetan Plateau in the west direction, including snow cover, sensible and latent heat, etc. ③ The convection of tropic, thermal conditions of the South China Sea, the atmospheric circulation of the southern hemisphere, etc., in the south direction. ④ The sea ice in the Arctic Ocean, Arctic oscillation, and blocking high over East Asia in the north direction, which

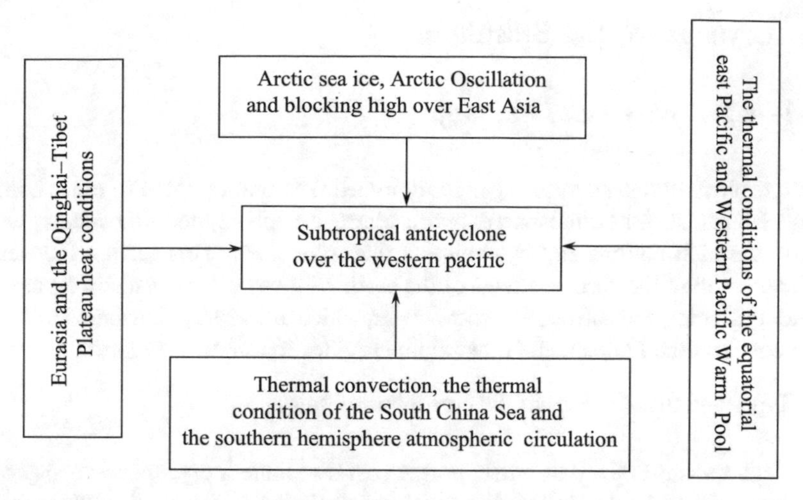

Fig. 8.1 The influencing factors of the East Asian Monsoon

reflect the influence of middle-latitude and high-latitude atmospheric circulation. ⑤ The subtropical anticyclone over the western Pacific in the center, which reflects the influence of subtropical atmospheric circulation. These five major factors determine the main thermodynamic and dynamic condition of the East Asian monsoon, namely the atmospheric circulation and underlying surface heat condition. In this concept model, the components of cryosphere including the Arctic sea ice and snow cover over Eurasia and Tibetan Plateau play important roles in these factors.

Besides the snow cover has a direct and important influence on the local atmosphere, continuous variations in snow cover can be transmitted by the planetary wave, resulting in a larger range of atmospheric circulation anomalies. The anomalies of Eurasian snow extent during autumn to early winter are significantly correlated with the Northern Hemisphere winter circulation, while the Siberian snow cover anomalies during autumn have a significant negative correlation with the Northern hemisphere Annular Mode (NAM). The positive anomalies in autumn-early winter snow cover in the Tibetan Plateau caused abnormal atmospheric circulation similar with the Pacific-North American teleconnection pattern (PNA) in the northern hemisphere. In addition, changes in the Eurasia snow cover is closely related to summer precipitation in China. When snow cover of the Eurasian continent in spring is positive anomalies, summer precipitation in China from South to North presents a distribution pattern from less to more and then to less. When the snow cover in the Eurasian is negative anomalies, the opposite pattern of summer precipitation in China is observed.

8.2 Cryosphere and Biosphere

8.2.1 Cryosphere and Ecology

All cryospheric ecosystems are generally interlinked and coupled to other components of earth, such as atmosphere, hydrosphere, biosphere, and lithosphere, which control the distributions and dynamics of the cryosphere. This section focuses on the interaction of these components of the earth with key components in cryosphere including glacier, permafrost, and snow cover, which are widely distributed in Arctic, Antarctic, Tibetan Plateau, and other alpine regions in middle-low latitude.

1. Types and distributions of the cryospheric ecosystem

If cryosphere was defined as where permafrost distributed, cryospheric ecosystems mainly consist of the Arctic, the Antarctic, and the Tibetan Plateau (i.e. "three poles" of the earth). The cryospheric ecosystem may also include a pan-arctic area to the north of latitude 50° N, where the terrestrial ecosystems are mainly comprised of boreal forest (Taiga) and tundra. Antarctic tundra ecosystem is the sole type of local terrestrial ecosystem in the Antarctic. In comparison with Antarctic and Arctic regions, ecosystems in Tibetan Plateau are more diverse; large variations in elevations created diverse terrestrial ecosystems including desert, meadow, steppe, and forest.

1) Types of the cryospheric ecosystem and its distribution in the pan-arctic region

Tundra and boreal forest are the key terrestrial ecosystems in the pan-arctic region. In the pan-arctic region, the vegetation types from south to north varied from Taiga, forest tundra, southern tundra, typical tundra, Arctic tundra, and continental desert. The similar pattern is also observed in North America as the following sequence: Taiga, forest tundra, shrub tundra, grass tundra, heather tundra, and polar desert.

Plant productivity and growth rate are generally low in the pan-arctic region. The main cause lies in the low decomposition rate of soil organic matter and hence high soil nitrogen deficit due to the presence of permafrost and cold weather. For the Taiga, forest tundra, and southern tundra (shrub tundra), the species diversity indices are 765, 446, and 180 with an average biomass, or productivity, of 115, 56, and less than 3.0 Mg ha^{-1}, respectively.

Tundra is dominated by shrub, grass, moss, and lichen, which widely distributed in the Northern Hemisphere and covers a large area in northern Eurasia and its adjacent islands. The largest tundra is located in north Siberia with an area of approximately 3×10^6 km^2. In the Southern Hemisphere, tundra is only distributed at the Malvinas Islands (Falkland Islands), South Georgia Islands, and the South Orkney Islands at the southern tip of South America. Moreover, there is sporadic Alpine tundra in the Alpine Belt around the world, such as the Changbai Mountains in China and the Rocky Mountain in North America. Tundra is underlain by continuous permafrost benefited from a constant daily soil temperature below 0 °C and a short and cold summer. Plant species richness in the tundra is relatively low with approximately 100–200 species in the vast northern tundra and 400–500 species at southern areas of the Arctic. Similarly, there are only 100–200 plant species thriving in alpine tundra. Vegetation in tundra was primarily dominated by plant families such as Ericaceae, Salicaceae, Cyperaceae, Gramineae, Ranunculaceae, Cruciferae, Compositae, as well as moss and lichen. The types of tundra in Eurasia are mainly comprised of shrub tundra, moss-lichen tundra, and arctic tundra from south to north, showing decreasing species richness, simpler structure, and low biomass towards the north. The productivity of tundra is low; rough estimates showed that mean ecosystem productivities in the southern shrub tundra, the middle moss-lichen tundra, and the northern Arctic tundra are 2.28, 1.42, and 0.12 Mg ha^{-1}, respectively.

Another key arctic ecosystem is the boreal forest, or Taiga, which is the largest forest ecosystem by area in the world. It covers approximate 11% of Earth's land surface which extends more than 1000 km from the forest belt in the southern Arctic tundra to the coniferous forest belt. Boreal forests in Eurasia are dominated by *Picea abies, Pinus sylvestris,* and *Betula pubescens*, and in Siberia by *Picea obovata, Picea sibirica*, and *Larix jezoensis.* Boreal forests of North America harbor more plant species than Eurasia and Siberia, which includes gymnosperms genera such as *Picea, Abies, Pinus*, and *Larix*, and angiosperms genera such as *Populus* and *Betula.* Similar to the tundra, most boreal forests grow on permafrost and experience cold and dry winters and short and cool summers. Plant species composition is simple in Taiga forest with a sole shrub layer, a grass layer, and some moss over the land surface.

2) Types of alpine ecosystems in Tibetan Plateau and their distribution

In natural regionalization, the Tibetan Plateau is traditionally divided into nine natural zonings as follows: Guoluo-Naqu Alpine shrub meadow region, Alpine meadow in southern Qinghai Province, Kunlun Alpine desert region, Qiangtang Plateau Alpine Steppe, Montane Conifer forest region in western Sichuan and eastern Tibet, Montane shrub-steppe in southern Tibet, Ali montane desert and semi-desert region, Chaidamu montane desert region, Qilian montane steppe region.

The first four regions are within the plateau sub-alpine zone, and the rest five regions are parts of the plateau temperate zone. In this chapter, we only focus on four typical natural regions that are closely related to the cryosphere, including alpine shrub, alpine meadow, alpine steppe and alpine desert.

The first region, alpine scrub-meadow, is a subclass of grassland colonized by phanerophytes, frost-resisting mesic evergreen, and summer-green shrub. These vegetation types are important components for vertical vegetation belts in Tibetan Plateau. Alpine scrub-meadow in the montane area is distributed above treeline and is mosaicked with alpine meadow. This type of vegetation in the montane area is usually colonized by some shrub with height 30–80 cm with a land cover of 20–40%, including *Salix oritrepha, Caragana jubata, Myricarica elegans, Rhododdendron thymifolium* etc.

Second, the alpine meadow is usually dominated by Kobresia genus, sedge plants, such as *Kobresia pygmaea*, *Kobresia humilis*, *Kobresia capillifolia*, and *Kobresia tibetica*. Majority of these plants are resistant to cold weather with a series of adaptive features, such as growing in clusters, low statues, small leaves with trichomes, short growing periods, vegetative reproduction, and viviparous reproduction. The alpine meadow is characterized by simple community structure and undistinguishable boundary of vegetation layers. The alpine meadows can be categorized as three types: alpine *Kobresia* meadow, alpine *Carex* meadow, and alpine forb meadow. The alpine *Kobresia* meadow, dominated by *Kobresia pygmaea* and *Kobresia humilis* with a high vegetation coverage of 50–80%, is widely distributed over a large region either elevation ally lowered alpine scree moraine of the Tibetan Plateau, or the well-drained floodplain, montane slopes area with elevation 3200–5200 m in the eastern Plateau.

The third region, alpine steppe ecosystem, features a sparse and simple community structure with a low vegetation coverage. The alpine steppe is a typical natural ecosystem of the alpine zone in middle Asia as well as cold regions across the world. The *Stipa purpurea* steppe and *Carex moorcroftii* steppe are two major types of alpine steppe in Tibetan Plateau with low vegetation coverage of 20–40%. These two steppe types are mainly distributed above an elevation of 4000 m in wide valleys between mountains, margins of plateau lake basins, palaeo terrace, paraglacial fans, river terrace, denudation high plains, and drought mountains.

The last region, alpine desert, is mainly located at plateau lake basins, wide valley, and stony slopes of valley bottoms in northwest of the Tibetan Plateau. The elevation of alpine desert varies from 4600 to 5500 m, which is underlain by continuous permafrost with a cold and drought climate. The alpine desert has simple community structure and a low vegetation coverage of 10%, which is usually associated with very few species including *Ceratoides compacta*.

Overall, topographic pattern of the Tibetan Plateau and characteristics of atmospheric circulation reveal that the regional combination of internal temperatures and water conditions have obvious zonality changes, showing a tendency to shift from a warm and humid climate in the southeast to cold and dry in the northwest. Therefore, the vegetation distribution shows zonality of forests, meadows, steppes, and deserts, which are very similar to zonal variations from the forest, grassland, to deserts from southeast to northwest China. Meanwhile, there is an obvious vertical zonal variation of alpine vegetation in the plateau. The vegetation distribution from southeast to northwest as the terrain elevation increases is: mountain forest belt (evergreen broad-leaved forest, cold temperature coniferous forest), alpine shrub and alpine meadow belt, alpine steppe belt (warm grassland in the depression at lower altitude), to alpine desert (warm mountain desert in the drought wide valley and valley slop at lower altitude). Thus, the distribution pattern of alpine vegetation in the Tibetan Plateau is the combination of horizontal and vertical zonality, where vertical zonality and horizontal zonal variation of vegetation is collectively known as alpine vegetation zonality.

3) Animals in the cold region

Despite the rigid environment, cryospheric ecosystems are not deprived of animals. Terrestrial herbivore mammals include *Lepus arcticus* (rabbit) and *Rangifer tarandus* (reindeer), and carnivores include *Ursus maritimus* (polar bear), *Canis lupus arctos* (Arctic wolf), and *Alopex lagopus* (Arctic fox). Aquatic mammals include species of seal, sea otters, walrus, whale, and beluga, and a variety of fish include grayling, northern dogfish, gray trout, herring, Arctic salmon, and so on. There are also around 120 species of birds in the Arctic, most of which are migrator with only about 12 species are not (e.g. *Anas acuta*, *Anas Penelope*, *Bubo scandiacus*, *Sterna paradisaea*). In fact, one in six birds in the Northern Hemisphere breed in the Arctic. Insect species are scarce in the Arctic (less than 10,000 species), much less than most area of the world. This is not a surprise as these cold-blooded creatures do not cope well with the cold climate. The majority insect species are flies, mosquitos, mite, grasshoppers, spiders, and centipedes, among which flies and mosquitos comprise 60–70% of the total numbers of insects.

Even fewer animal species live in the Antarctic than Arctic. There are only a few species that live in the ocean or on coastal lands, with elephant seals and sea lions being the most abundant mammals. More than 80 bird species have been recorded, among which only approximately ten species breed in Antarctica. The most abundant species are Emperor penguin (*Aptenodytes forsteri*), Chinstrap penguin (*Pygoscelis antarcticus*), and Imperial shag (*Phalacrocorax atriceps*). As to insects, only some extreme cold-tolerant mites and wingless insects were observed in the coastal refuges.

Tibetan Plateau supports a rich diversity in animal species, including many that can only be found in this region. There are more than ten mammal species, including *Equus hemionus*, *Ursus pruinosus*, and *Pantholops hodgsoni*, Rodent species prevail in the open alpine meadow and alpine steppes, such as *Ochotona curzoniae*, *O. tibetana*, *Marmota himalayana*, and *Myosoalax fantanieri*. Rich bird species exceed 100 species, such as *Buteo hemilasius*, *Gyps fabulous*, *Falco tinnunculus*, and *Columba leuconota*. Animals in alpine ecosystems usually have large habitats, moving around the alpine meadow, alpine steppe, and alpine shrubs.

2. The interaction between permafrost and vegetation

1) The effects of vegetation on permafrost

At the global and regional scales, the dynamics and distribution of permafrost are mainly determined by spatial variation of temperature and precipitation. As temperature and precipitation vary with altitude, longitude, and latitude, the distribution of permafrost may vary globally at these three dimensions. At the regional scale, the formation and distribution of permafrost are greatly determined by both the vegetation and topography. Numerous studies have shown the impacts of the vegetation on permafrost through modification of the water balance, snow cover, formation of soil organic matters, and soil textures by the vegetation. Moreover, vegetation-induced changes in soil organic matters and textures may change the energy balance at soil surface by changing soil thermal conductivity and energy balance.

Vegetation attenuates solar radiation, which can significantly reduce the net radiant flux reaching the soil surface and hence surface temperature. This process has a direct impact on the hydrothermal process of the frozen ground. For example, in the Larix forest in Great Khingan Range the net radiant flux reaching the lower vegetation canopy was only 60% of that upper part in the summer, so nearly 40% of the solar radiation was reflected and absorbed by the canopy. At the Alpine meadow of the Tibetan Plateau, however, the sensible heat and ground surface heat flux of the grassland with a coverage 30% were higher than those of grassland with a coverage of 90% by 19% and 41%, respectively, while the latent heat flux was lower by 47% in grassland with coverage 30%.

Vegetation effects on soil hydro- and thermo-state may further affect both permafrost formation and distribution; such processes are closely related to vegetation structures, land cover, and surface soil moisture. For example, soil temperature at a depth of 30 cm can be 7–9 °C higher in a well-drained forest than a poorly-drained forest based on a study in Alaska. In the Tibetan Plateau, decreases in vegetation cover will lead to a decrease in soil temperature and soil moisture in well-drained conditions at alpine meadows; in contrast, decreases in vegetation cover will lead to an increase in soil temperature and soil moisture in the poorly-drained alpine marsh meadow. Such different responses between alpine meadow and alpine swamp are attributed to two reasons. The first reason is that moss and lichen generally dominate the poorly-drained area, which would benefit the accumulation of organic matter and the development of the peat layer. The peat layer can reduce net solar radiation on the soil surface during summer while increase the loss of heat by increasing the thermal conductivity after freezing. When the heat loss in winter is greater than heat gain in summer, the energy balance (net loss of heat storage) is beneficial for the formation and preservation of permafrost. Secondly, as the heat capacity of water is 4–5 times of the mineral soil, ground cover in the poorly-drained area keeps a relatively stable temperature and a shallow thawing depth.

The impact of vegetation on the development of permafrost is also represented by the redistribution of precipitation and snow cover, which directly affects soil moisture and permafrost. Vegetation canopy intercepts a certain amount of precipitation which is then be evaporate or sublimate into the air. Therefore, the existence of vegetation reduces the amount of precipitation that landed on the ground and hence snow depth. When snow cover presents on the ground, as snow is not a good conductor of heat, it may alter the exchange of heat between the atmosphere and land surface. As heat insulation effect of snow gradually increases with increasing snow thickness, heat exchanges between ground and atmosphere are restricted at locations with snow thickness deeper than 80 cm, resulting in permafrost vanishing completely. In the pan-arctic region, the spatial distribution of snow varies greatly due to the effects of forests and shrubs on snow interception, blocking, and capture, which is explainable for the spatial heterogeneity of permafrost distribution.

In summary, vegetation and some environmental factors collectively facilitate the development of the permafrost. Therefore, a widely used classification of permafrost was established based on if the vegetation plays the key role or not on the permafrost formation (Fig. 8.2a). The formation of the continuous permafrost in the Arctic tundra

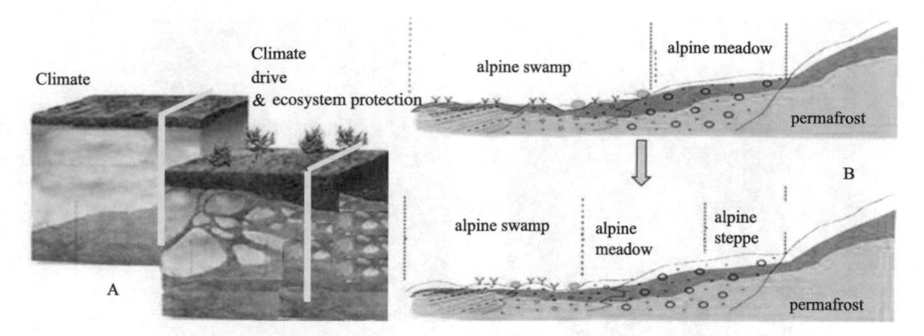

Fig. 8.2 Effects of vegetation on permafrost formation and distribution

and polar desert areas was mainly driven by climate factors; extremely cold climate limit the development of the vegetation. In the south of shrub tundra, the vegetation plays a key role in the formation of permafrost (Fig. 8.2b) by serving as insulators under climate warming. However, this is not the case in all areas. Increases in the vegetation cover and litter mass in some areas may promote heat transfer, which reduces the permafrost.

2) Effect of permafrost on vegetation distribution

Hydro-thermal characteristics of permafrost may have some effect on the species richness, composition and structure of vegetation community. A pronounced example is the interactions of vegetation with tundra polygons and polygon ponds in the Arctic area. Tundra polygons are polygon-shaped patterned ground that formed by ground material in stony surfaces; water often fills the polygon center and hence forms polygon ponds. The rims of the polygons are formed by ice wedges inside the soil. They were first frost cracks, where melted water entered and refreeze during spring and early summer. Repeated cracking and refreezing would enlarge the cracks and grow form ice wedges (Fig. 8.3). The rim above the ice wedges may protect vegetation from wind damage as ice crystals in the wind can cut through the cuticle layer of leaves, cause desiccation of leaves. Therefore, some tundra vegetation can survive inside tundra polygons where winds are less strong, including moss, lichen and. In contrast, in similar areas with strong wind but without ice wedges and dry weather, it is usually bare ground exist.

In the Taiga forests with permafrost, differences in the permafrost conditions led to varied forest types and productivity. In wetlands with ice-rich or shallow-buried permafrost of the Great Khingan Range Mountains in China, coniferous forests (Taiga) are widely distributed. However, mainly due to the shallow active layer, anoxia conditions in the rooting zone, and slow nutrient cycling, the growth and the productivity of these forests is extremely slow. The wet conditions also fostered forest swamps, shrub swamps, Carex swamps, and Sphagnum swamps in the Taiga area. Very different vegetation dominates the Tibetan Plateau. A large area of alpine

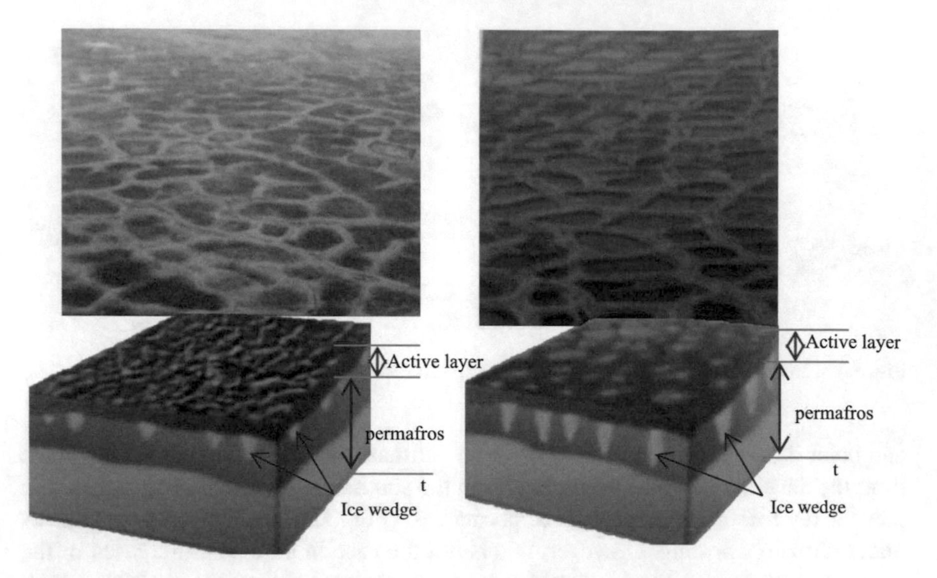

Fig. 8.3 Tundra polygons and polygon ponds that were formed by the by ice wedges inside the soil

meadow and alpine wetland is distributed in the arid and semi-arid cold zones along the Kunlun-Tanggula Mountains.

3. Permafrost and cryospheric ecosystems under a changing climate

In a warmer climate, cryospheric ecosystems with permafrost are expected to experience substantial changes as they are very temperature sensitive. We have already seen many dramatic changes in the sensitive permafrost area. In this section, we compile some examples of such changes.

The warming thaws a large area of permafrost in the Arctic area which accelerates the nutrient cycling and vegetation productivity. Increased air temperature improves the growth condition of plants that were originally limited by temperature; increased soil temperature improves soil microbial activity, accelerates the decomposition of organic matter, and improves the availability of soil nutrients (such as nitrogen). Moreover, melted ice not only provided sufficient water but also created a thicker active layer where plant roots could reach. All these factors increased the vegetation productivity. This can be represented by the Arctic tundra. Both the NDVI and vegetation cover increased with global warming, which was caused by the shrub expansion and changes in vegetation composition in the tundra. The longer growing season is another factor that benefits the increase of vegetation productivity in the permafrost area. The warmer climate changes plant phenology by advancing the leave sprout in the spring and delaying leaf senescence in the autumn.

One would expect the expanding of vegetation in the cryosphere due to the increased productivity, but this is not always the case. New factors may limit vegetation growth in a warmer climate, such as heat- and drought-induced vegetation mortality and unfavourable growth conditions due to the melt of permafrost. First, the elongated growing season and warmer temperature increase the evapotranspiration of vegetation. Given precipitation is unchanged, higher evapotranspiration would create drier soil and hence promote drought. Moreover, if absolute air humidity does not change, a higher temperature will increase the water vapour pressure deficit in the air, which would also increase the water stress. Heat-induced and drought-induced vegetation mortality events have been recorded throughout the world including arctic forests. Second, melted permafrost or ice wedges in the soil can dramatically change the growth condition of vegetation. One example is the "drunken forests" in Siberia, where trees tilted due to changed soil conditions after the melt of permafrost or ice wedges. Moreover, water melted from permafrost may not drain from the system, which may create anoxia conditions for roots as there may not be sufficient oxygen that diffuse to roots to keep their functionality.

Observations have shown that the permafrost region of the Tibetan Plateau has been degrading from 1967 to 2004. Despite increases in the vegetation cover since 2005 due to increased precipitation in some area, permafrost degradations were followed by decreases in vegetation cover and productivity in alpine meadow, and increases in desertification area in the alpine steppe. Model simulations indicated that warming did benefit the increase of biomass of dominant species in Tibet in the short-term, but alpine grassland tends to degrade in the long-term. The area loss of alpine grassland may be attributed to increased soil drought stress, shading from taller vegetation, higher soil moisture loss at the root, as well as snow cover reduction.

The change in permafrost may also change the nutrient and carbon cycling due to its large influences on the structure and functioning of the soil biological community. Increased temperature directly accelerates nitrogen mineralization rate and enzyme activity, and also modifies community composition of soil microorganisms. Meanwhile, carbon input, soil moisture, and nutrient availability may affect the soil microbial community indirectly. Such changes may subsequently influence the regional and global carbon cycle by altering the decomposition rate and emission of CO_2 and CH_4.

In summary, the effects of permafrost degradation on ecosystems varied in different regions. Therefore, further studies may focus on revealing the mechanisms of permafrost degradation and increasing the modeling capacity to simulate the degrading processes of permafrost in the future.

4. Permafrost Microorganisms

Microorganisms are an important component of the cryosphere. Unfrozen water in the brine trickle or crystals and ice wedges in the permafrost provide habitat for microorganisms. These permafrost microorganisms play a vital role in biogeochemical cycle of frozen ground, and to a certain extent, it could act as one sensitive indicator to climate change. Recently studies have shown that the permafrost microorganisms have a high diversity and a large spatial heterogeneity. Microbial community composition and species richness varied in different region or permafrost condition. The permafrost microorganisms in Tibetan Plateau are larger than that in the Antarctic, Arctic, and Siberia. In contrast, the total number of culturable bacteria in Tibetan Plateau is lower than that in the polar area but similar to that in Siberia. Moreover, common species of bacteria have been detected across permafrost of the earth. For example, two types of bacterias including *Arthrobacter* and *Planococcus* were found to commonly exist in permafrost with varied adaptive traits to cold environment. However, despite the large quantity of total bacteria in the permafrost, the quantity of culturable microorganisms is low.

Species diversity and population dynamics of permafrost microbe were usually determined by multiple environmental factors, such as soil moisture, temperature, organic matter content, and pH value. Species richness and diversity in arctic area sharply decreased along Taiga forest towards tundra. Moreover. Spatial heterogeneity of permafrost microbe may reflect variations of soil depth. Generally, distribution of permafrost microbes is determined how thick the active layer was, which usually reaches a depth of 60–100 cm in the upper soil layer. Moreover, microbes in the top 70 cm of soil account for approximately 70–86% of all microbes in the whole active layer in an alpine meadow of the permafrost region in the Tibet Platcau. Results in the tundra of Northern Canada showed that the number of microbes decreased with increasing soil, and the same trend was observed in microbial community composition and structure. Repeated freezing and thawing process of the active layer did not have significant impacts on permafrost microbial composition and structure in general. The dominant microbial species may also change with soil depth. For example, *Actinobacteria* and *Crenarchaeota* species dominated in the active layer, *Proteobacteria* and *Euryarchaeota* species dominated on permafrost table, and *Firmicutes* and *Crenarchaeota* species dominated inside permafrost.

Similar to the non-permafrost region, soil microbes in permafrost area serve a significant role in soil carbon and nitrogen cycles by acting a key role in the process of decomposition and nutrient cycling. However, permafrost microbes may be more sensitive to the climate than non-permafrost microbes. Increased temperature in permafrost regions may promote plant growth, increase soil organic matter and litter mass, reduce soil inorganic nitrogen content, and thus significantly increase the microbial species and activities. Thus, stronger effects of climate change on the metabolism of permafrost microbe could directly shift composition and structure of soil microbial community. Such changes may further change the C and N cycles between the atmosphere and soil.

As permafrost microbes are commonly anaerobic, permafrost metagenome data show that genes in the permafrost microbes may provide them metabolic potentials in the anaerobic carbon metabolism. These microbes may serve in the processes of fermentation and methane production (i.e. methanogenesis). Abundant methanogenesis strains (methanogens) create a good environment for methanogenesis. However, global warming has significantly changed the microbial communities of the cryosphere, increased the microbial biomass in the permafrost area, and accelerate decomposition within a short period. This will lead to increased emissions of the greenhouse gases, including CO_2 and CH_4. The increased emission of greenhouse gases will create a positive feedback on air temperature, which will further change the carbon balance and nutrient cycle in the cryospheric ecosystems.

5. Snow cover and cryospheric ecosystems

1) Interactions between snow cover and vegetation

Snow cover affects the vegetation through its impact on soil water and thermal conditions. Snow cover can increase the albedo of the land surface and reduce the absorption to radiant energy, thus leading to a lower snow surface temperature than air temperature. Meanwhile, snow, as a bad thermal conductor, can prevent heat loss in the ground in winter, thus inducing higher soil temperature than air. The thermal insulation of snow depends on its thickness and duration. Thinner snow usually provides cooling effect. Changes in snow cover have a greater influence on the soil temperature than the vegetation cover in the Northern Hemisphere, where more seasonal variations of snow thickness exist.

Snow cover could affect vegetation types and their distributions through hydro-effect and thermal-effect. As shown in Fig. 8.4, thickness and melting time of snow cover could affect vegetation type and its distribution. This is because that the hydro-effect and thermal-effect of snow could directly change soil nutrient availability. Moreover, thickness and melting time of snow cover may determine community composition and function traits of plants, such as canopy height, leaf area index, and biomass. Vegetation groups adaptable to local snow conditions including different snow thickness and duration of snow cover. For instance, *Kobresia myosuroides* is only distributed at the area with thinner or short duration of snow cover. Comparatively, *Carex pyrenaica* and *Trifolium parryi* are commonly distributed in areas with thick snow cover or long durations of snow cover. However, Species with wider adaption for snow thickness also exhibit significant differences in community abundance and structure. For example, *Acomastylis rossii* can grow at area with various thicknesses of snow cover, its abundance and cover varied in different area. Generally, snow cover can promote plant biomass accumulation and growth in the northern hemisphere, but only to a certain limit. Beyond the limit, thicker snow cover could decrease plant productivity possibly due to the longer melting time and excessive water that may cause shorter growing season and anoxia soil conditions (Fig. 8.4).

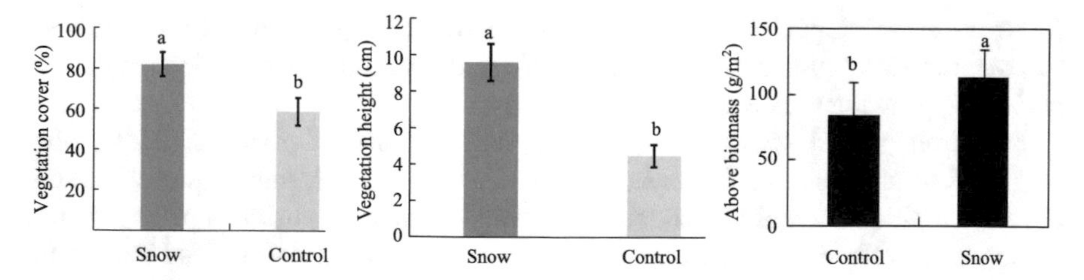

Fig. 8.4 Relationship between snow cover and vegetation

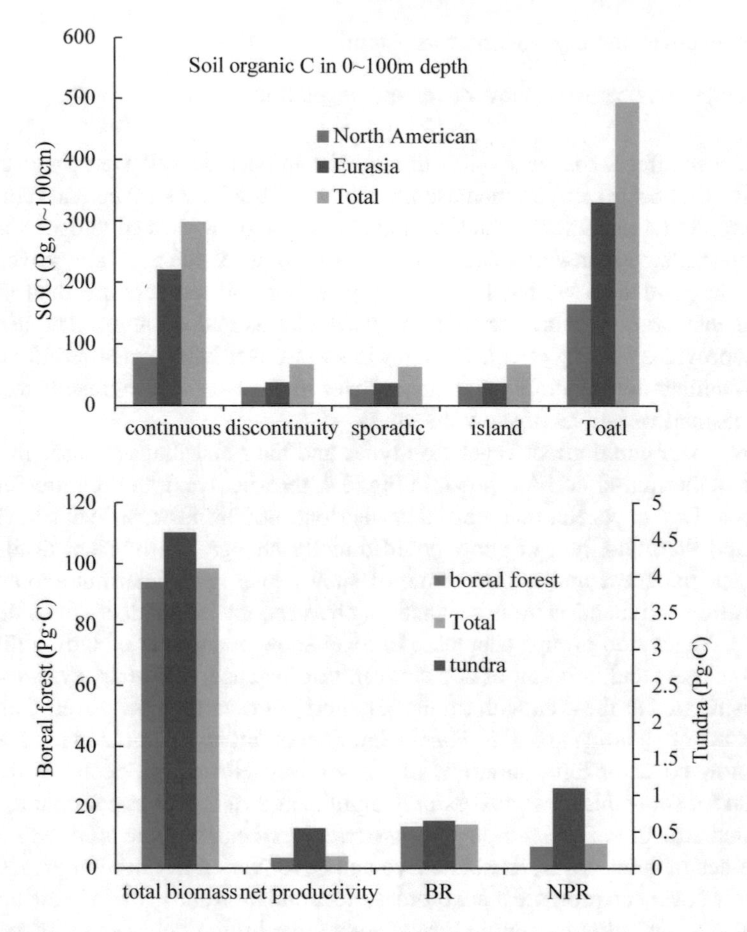

Fig. 8.5 Soil organic carbon in permafrost region and the net biomass of an arctic ecosystem

2) Effects of change in snow cover on cryospheric ecosystem

Snow cover accompanied the climate change and cast great impact on terrestrial ecosystems. Early snowmelt leaves the land longer snow-free time which may cause prolonged growing season and early flowering.

Moreover, varied species composition and diversity of plants could be also attributed to changes in soil moisture, elevated temperature, and changes in snow cover. In addition, increased productivity in cold region by extension of plant growing season firstly effectively increases absorption ability of carbon, thus lead increasing carbon sequestration, but gradual decrease in carbon sequestration was observed in recent observed results with gradually less snow and decreasing plant production.

For animals, the earlier snowmelt and higher temperature change life history for lots of invertebrates, such as shorter hibernation. Species richness in naphthalene invertebrates reduced because of advanced plant flowering and shorter flowering duration. Moreover, some invertebrates, such as spiders, exhibit obvious phenotypic variation. Change in snow cover also affects vertebrates also have significant responses to snow changes, for example, life history in some vertebrates has been changed due to the varied food chain, even some species numbers firstly increased and then decreased with change in snow cover.

8.2.2 Cryosphere and Its C/N Cycles

The biogeochemical cycle in cold region is an important component of mass circulations in the cryosphere. Different from other regions, the biogeochemical cycles in the cryosphere are closely and sensitively related to the freeze–thaw processes and the consequent heat and mass exchanges. This chapter is mainly focused on carbon and nitrogen cycles in the cryospheric regions. The phosphorus cycle is not included due to the scarcity of studies on this topic. Water cycles in the cryosphere are introduced elsewhere in this book.

1. Carbon storage and its pattern of distribution

The total Arctic area in (including subarctic and arctic) is 20×10^6 km^2, which accounts for 13.4% of the global land area (149×10^6 km^2). The total biomass in the Arctic is about 83.0–113.8 Pg C, which accounts for 12.1–16.5% of the global terrestrial biomass. The vegetation net primary production n in this region accounts for 6.1–10.1% of the global value, which is less than the global mean (Fig. 8.5). The biomass of the Taiga forests is 80–108 Pg C and the net productivity is 2.5–4.3 Pg C a^{-1}. The tundra has a low biomass with about 3.4–5.8 Pg C and a net productivity with 0.5–0.6 Pg C a^{-1}. Despite the low productivity, the Arctic area is a huge terrestrial carbon pool, which is mainly attributed to its relatively inactive soil carbon pool. However, different estimate methods for the carbon pool have been proposed due to difficulty in calculating carbon stocks for permafrost. The soil organic carbon content at a depth of 0–100 cm in the permafrost area of the Northern Hemisphere is shown in Fig. 8.5, which is based on the mostly latest estimation results. The density of soil organic carbon in most of the tundra and the boreal forest ranges from 10 to 50 kg/m^2, with higher values in the wet area of the permafrost region.

The soil organic carbon pool was estimated as 496 Pg C at the depth of 0–100 cm in the permafrost region of the Northern Hemisphere based on the result of 2009 (355 Pg C based on 1991 results). At the depth of 0–300 cm, the soil organic carbon pool increased to 1024 Pg C, among which 298.5 Pg C of the soil organic carbon pool was distributed in continuous permafrost regions, accounting for 60.2% of the total amount in permafrost regions. The distributions in the discontinuous, sporadic, and island permafrost regions were 67.3, 62.9, and 67.1 Pg C, respectively. Geographically, Eurasia permafrost was 331.1 Pg C, accounting for 66.8% of the total amount in the northern hemisphere. North America, including Greenland, was 164.7 Pg C, accounting for only 33.2%. The soil carbon at 0–100 cm depths in the Arctic and

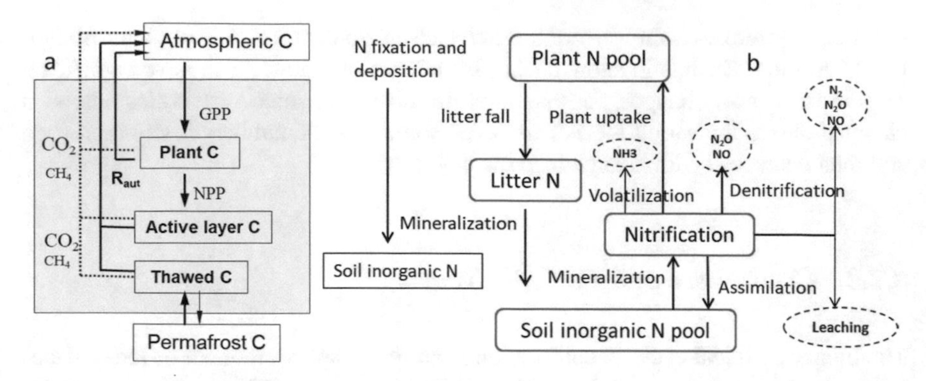

Fig. 8.6 Carbon and nitrogen cycle in permafrost regions

subarctic regions accounts from 33.1 to 45.1% of the global soil carbon, which was estimated at between 1100 and 1500 Pg C In addition, about 407 Pg C was deposited in deep layers of Siberian soils below 3 m, and about 241 Pg C was stored below 3 m in seven large Arctic alluvial fan deltas. It is estimated that the soil organic carbon pool in the permafrost region of the entire Northern Hemisphere was roughly 1672 Pg C, which is equivalent to 50% of the global terrestrial carbon pool. All the information above indicates the important role of the pan-Arctic area as a carbon sink. However, reliable estimation of soil organic carbon is always challenging due to the heterogeneity of soil carbon distribution, difficulties to estimate underground pool, and simply the large scale of the area. Therefore, it is necessary to constantly update the estimations when more advanced approaches become available.

2. C/N cycles in the permafrost region

1) C/N cycles in the permafrost region

Carbon cycles in terrestrial ecosystems are generally fulfilled through several major processes: photosynthesis and carbon sequestration, plant autotrophic respiration, litter fall and decomposition, soil organic matter accumulation, and heterotrophic respiration. However, the uniqueness in the carbon cycling of the permafrost area lies in the carbon storage in the permafrost and the release of carbon during permafrost thawing. Due to the low temperature, sequestered carbon slowly accumulated in the soil with little from decomposition involved during the process. However, as the global warming accelerates the melt of permafrost, carbon in frozen ground will be released to the ecosystem in both aerobic and anaerobic processes (Fig. 8.6a) depending on how much soil moisture exists in the active layer are. Soil carbon released in permafrost regions under climate change may be offset by the increase in ecosystem productivity at the regional scale. Here we take tundra and boreal forest in arctic region as an example. Litter fall in tundra vegetation is equivalent to its net productivity, which is approximately 0.5×10^9 t a^{-1}. Litter mass in boreal forest was estimated to be approximate 2.06 t hm^{-2} a^{-1} based on 29 studies worldwide, which is equivalent to a total amount of 2.47 × 10^9 t a^{-1} in the approximately 12 × 10^8 hm^2 of the pan-Arctic area.

The N cycle is a key factor to determine C cycles in ecosystems. It may control some key biological processes including plant growth, carbon use efficiency, water user efficiency, decomposition rate, and utilization of other nutrient. As shown in Fig. 8.6b, nitrogen in the atmosphere enters soil through deposition, N-fixation, and decomposition of plant litter. Nitrogen is then stored in the soil pools through the mineralization processes. Stored nitrogen may be released into the atmosphere as N_2O and NO through denitrification or leaches into the water as soluble NO_3^- through nitrification and denitrification processes. In permafrost regions, the majority of soil N is stored in organic matter under low temperature, and there is only a small account of nitrogen was fixed by N fixation and deposition. Lichens and bryophytes in the permafrost region of arctic and Tibetan Plateau are typical N-fixer with symbiotic cyanobacteria. The annual nitrogen fixation of these cyanobacteria is up to 0.8–1.31 kg N hm^{-2} a^{-1}, which accounts for 85–90% of the total nitrogen input in some Arctic watersheds.

In permafrost regions, the snow accumulation, N cycling, and vegetation growth may form a positive feedback loop. Snow accumulation provides insulation of underlying soils. Consequently, increases in soil temperature of the active layer significantly enhance the microbial activity and, in turn, the soil nitrogen mineralization rate in winter and spring, which further promotes the growth and invasion of shrub vegetation.

Higher plant and larger plant cover in shrub could capture more light for photosynthesis than those with lower height because most of their canopy could be exposed outside snow, thus promote shrub growing better. An increase in the shrub canopy is conducive to capture and block more snow, which further strengthens soil nitrogen mineralization in winter and spring, thereby accelerating the expansion of shrub vegetation. This positive feedback process is also related to the higher carbon to nitrogen ratio of woody shrubs. The biomass production of available nitrogen per unit area was significantly increased when grasses with low C/N ratio are replaced by woody shrubs with high C/N ratios.

2) The impact of permafrost degradation on C/N cycles

Carbon and nitrogen pools in permafrost regions are sensitive to changes in temperature and moisture. A higher temperature could melt permafrost and accelerate the decomposition of soil organic. The CO_2 released from the decomposition may provide one of the most significant positive feedbacks of the global terrestrial ecosystem to global warming. However, the released CO_2 may be offset by the increase in the productivity of plants in the permafrost regions, which is caused by elongated growing season, higher rate of nutrient cycling, and changes in species composition towards more productive species. Therefore, the balance between the effects of released CO_2 and increased C sequestration by plant will determine if the permafrost region will become a carbon source or not.

Due to the lack of observations and complexity in processes, the estimation of carbon release from permafrost largely relies on modeling studies, which create large uncertainties in the estimations. For instance, simulated results based on a group of numerical models show that carbon emissions in the Arctic permafrost will reach approximately 0.5–1.0 Pg C per year by the end of the twenty-first century under the constant warming scenario in the future. This is in line with the carbon emissions (estimated at 1.5 ± 0.5 Pg C per year) caused by the changes in global land use and land cover. Thawing frozen ground will cause the Arctic region to convert from a huge carbon sink to a huge carbon source under climate warming based on these simulations. However, if one considers the scenario that the forests replace tundra, the maximum carbon uptake could be 4.5 kg C m^2. In contrast, if woody plants were replaced by lower grasses, soil carbon emissions will be accelerated. In addition, the special snow-vegetation-soil water-heat coupling effect largely influences carbon and nitrogen emissions in the cold region. The change in winter emissions could attribute greater (about 14–30% of annual increment) to the increment of carbon emissions in the Arctic shrublands. Therefore, the complex couplings among vegetation, snow cover, and soil temperature, and soil moisture may drive the increased greenhouse gases emissions (CO_2, CH_4 etc.) in the Arctic region.

8.3 Cryosphere and Hydrosphere

There is a feedback relationship between cryosphere and hydrosphere during the climate fluctuates between colder and warmer. With a warmer climate, cryosphere shrinks, water cycle intensifies, and the global mean sea level rises. Meanwhile, cold fresh water from cryosphere flows into the ocean changing the ocean's salinity and temperature, thereby affecting the global thermohaline processes and the climate. Moreover, from a regional perspective, cryosphere changes have important impacts on middle and high latitude basins, where cryosphere meltwater is the main source of river runoff, and runoff changes will affect local water resources and ecosystems.

8.3.1 Characteristics and Roles of Cryospheric Hydrological Process

1. Characteristics of cryospheric hydrological process

Complexity of cryospheric hydrological process. The hydrological processes of the different components of cryosphere are complex and variable. The ice ablation, runoff generation, and runoff confluence processes of cryospheric components (such as glaciers, permafrost, snow cover, and sea ice) are very complicated. Take glacier as an example, ablation of the ice surface and confluence of subglacial channels are not only associated with glacier size, properties, and types, but also with superglacial feature, coverage of moraine, the development of ice crevasse, and so on, which makes observations and modelling of the glacier runoff very difficult.

The complexity of cryosphere hydrology is also reflected in the spatiotemporal scales in each cryospheric component. The response of glacial runoff to climate change varied with different sizes of different types of glacier, which is also closely related to the strength of climate change. At the same time, both large and small glaciers exist in one basin, which leads to a more complex response of glacial runoff process to the climate change. The response of permafrost to climate change is more complicated due to a long time. The hydrological changes of snow cover are mainly on the seasonal scale, where the distribution of snow cover, area, and mountainous terrain all impact the melting process of snow.

Uncertainty in observation of the cryospheric hydrological process. Observation of the cryospheric hydrological process is an important means of obtaining first-hand information and also a basis for cryospheric hydrological studies. Accurate hydrological observations of each component are inevitable choices to understand the dynamics, mechanisms, and the law of cryospheric hydrology. Because cryospheric components are mainly distributed in high altitude and cold regions, the difficulties concerning logistics and harsh environment during fieldwork, and the complexity of cryospheric hydrological phenomena, leading to more uncertainty in accurate observations of hydrologic processes.

Accurate observations of the glacier runoff: because glaciers are generally distributed in alpine areas or valleys, therefore, setting up of a section to observe all runoff generated from glacier melting is very difficult in practice. As shown in Fig. 8.7a, the selected observation basin of glacier runoff always includes an area without glacier, which makes glacial runoff hard to distinguish from rainfall runoff from bare rock. Runoff observed at the selected section R include total runoff from bare rock Rb and glacier surfaces Rg, where Rb is needed to observe when calculating

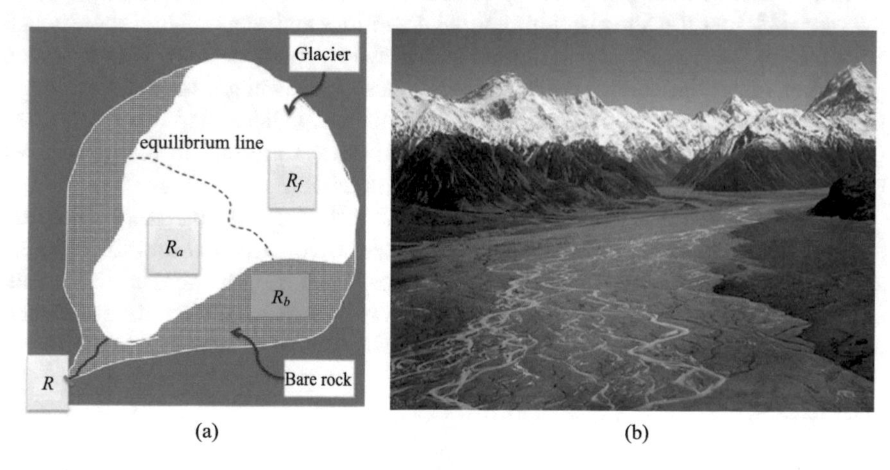

Fig. 8.7 The components of glacier runoff (**a**) and braided river channel at glaciated region (**b**)

Rg, *Rg* includes runoff *Ra* from the ablation area, *Rf* from the accumulation area, and deducted *Rb* from observed *R*. Generally, rivers located at a glacial terminal are always braided (Fig. 8.7b). Particularly in large glaciers, it is difficult to choose a suitable straight, controlled hydrological observation section. A section far from the glacier has to be chosen, which further increases uncertainties in the accuracy of glacial runoff observations, and extends the different calculations of *Rg*. This will be introduced in detail in later paragraphs.

Among the components of the hydrological cycle and water balance, precipitation and evaporation in the alpine region are difficult to observe and are estimated in most of the cases, which causes large uncertainty. Spatial difference of precipitation in mountains is very large and difficult to observe. Hydrological components including precipitation gradient (which is positive or negative based on altitude), the ratio between solid and liquid precipitation, evaporation from alpine glaciers, permafrost, snow cover, and evaporation (sublimation) from sea ice surface, are hard to directly observe and have great spatial variability in cold regions, which increases the uncertainty in research on cryospheric hydrology.

Similarities and differences in cryospheric hydrological components. The runoff formation of rivers in cryosphere is different from that in warmer regions. Due to the characteristics of solid water in the cryosphere, the ice-water phase change is the most significant characteristic. The solid–liquid conversion in the formation of runoff is a fundamental process in cryospheric hydrological processes, where runoff formations are closely related to heat conditions (temperature index). It is different from traditional hydrological runoff that depends mainly on precipitation.

Because glaciers have little changes in short periods, glacial runoff can be considered stable in one year. The glacier runoff mainly depends on heat conditions (high or low temperatures), while snowmelt runoff is mainly related to the snow cover extent and snow depth. Although snowmelt processes are controlled by temperature, the volume of the snowmelt runoff is also influenced by accumulation of snow cover. The snow cover extent is seasonally varied in contrast with the glacier. Therefore, the snowmelt runoff is the result of thermal conditions and accumulated snow cover. Runoff in the permafrost region is mainly affected by the impermeability of frozen ground, leading to higher surface runoff coefficients and lower groundwater recharge. In fact, due to ground ice content, thickness of the active layer, and other factors of permafrost, a certain amount of groundwater is always present. In the permafrost region, there is either little or no surface runoff in winter.

Diversity of river supply types. Considering different contributions of glaciers, snow, and rainfall to runoff, the supply of river can be classified into three types: snowmelt (Fig. 8.8a), snowmelt and ice meltwater (Fig. 8.8b), and rain, ice meltwater, and snowmelt (Fig. 8.8c). In mountainous basins in western China, the watershed always has a very distinct vertical zonation, where most rivers are supplied by a mixture of the glacier, rainfall, snowmelt runoff, and groundwater recharge, so the classification of river supply is more diverse.

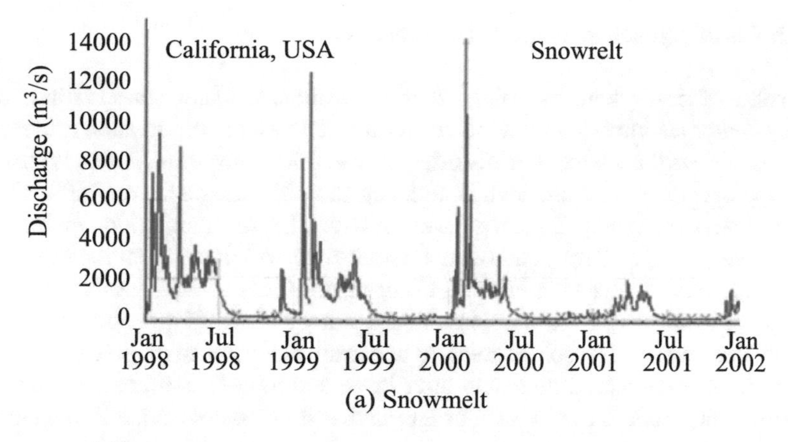

(a) Snowmelt

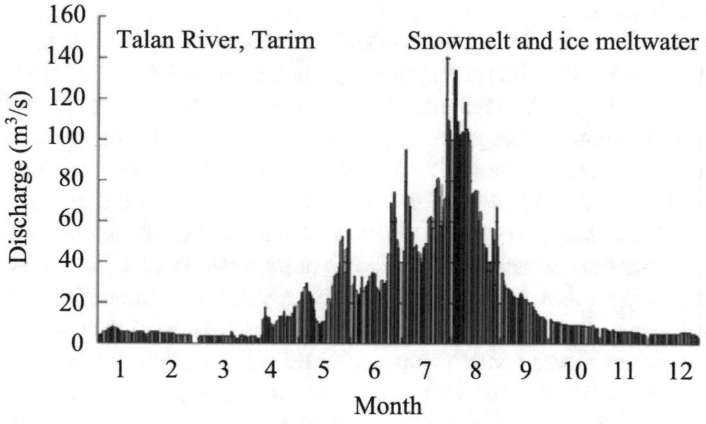

(b) Snowmelt and ice meltwater

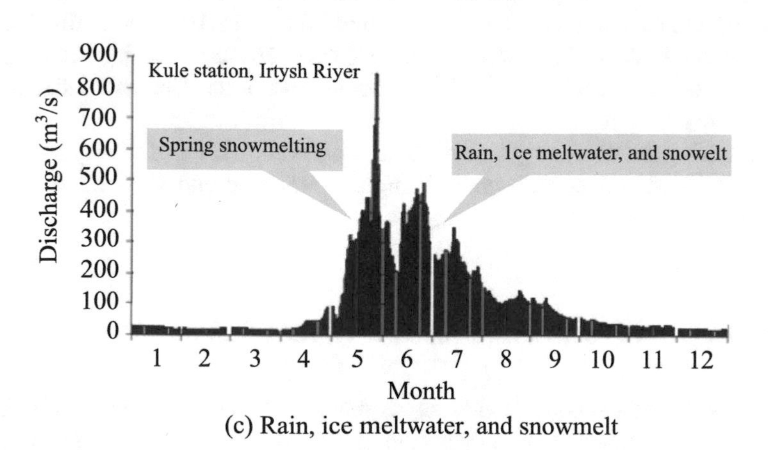

(c) Rain, ice meltwater, and snowmelt

Fig. 8.8 Different river supply types in the cryospheric basin

2. Roles of cryosphere hydrological process

The roles of cryosphere hydrological process include water conservation, water supply (water resources), and runoff regulation. The water conservation role reflects as cryosphere is the source of many large rivers around the world, due to cryosphere developed in alpine areas and high latitude regions. The Yangtze River, Yellow River, Tarim River, the Nujiang, Lancang River, Ili River, Irtysh River, Brahmaputra River, Indus River, Ganges River, and other famous rivers originate from the cryosphere over and around the Tibetan Plateau. Contrary to rainfall-runoff headwater region, cryosphere transforms solid water into liquid water and runoff probably includes the accumulated water. Cryosphere releases accumulated water to the river, even in the arid periods, and water drains only after large and long-term climate fluctuations. Therefore, cryosphere can be seen as an inexhaustible water source throughout the history of mankind.

The hydrologic role of cryosphere is widely recognized as water supply. As solid water, the cryosphere itself is an important water resource. The water resource property of cryosphere can be measured by two means: one is the total volume, and second is the annual recharge to river runoff. The recharge of cryosphere is an important part of surface runoff. The average annual glacier runoff in China is about 60.965 billion m^3 during the 1960s–2006, which is near to the annual runoff of Yellow River discharge to sea. In western China, about 58% of the total glacier runoff is mostly concentrated in the Tibet Autonomous Region, which is followed by Xinjiang Uygur Autonomous Region with about 33%. About 60% of the total glacier runoff contributed into outflow rivers, and about 40% contributed to the inland rivers. Permafrost stores a vast volume of solid water during the freezing process, which increases soil water storage, and suppresses soil evaporation and the formation of supra-permafrost water and supra-permafrost runoff. Thus, soil moisture variation in permafrost is unique compared with that in the soil in warmer regions. The average gravimetric water content at a depth of 10 m is 18.1% in permafrost regions of the Tibetan Plateau. The estimated annual water resource obtained from ground ice conversion into liquid water in permafrost regions of the Tibetan Plateau is about 50×10^8 to 110×10^8 m^3.

Considering the water supply and water conservation roles of the cryosphere, the runoff regulation role of cryosphere is highly important and will be introduced in detail in Sect. 8.3.4.

8.3.2 *Cryosphere and Macroscale Water Cycle*

From the hydrological point of view, cryosphere can be considered as a solid hydrosphere. Over the long course of evolution, the solid–liquid phase change processes between the ice and liquid water of the ocean have affected the global water cycle and have a profound influence on water resource, ecosystems, and climate change on global and regional scales. The expansion of cryosphere correlates with a decrease of liquid water in the ocean and weakening of the global water cycle. This change in water circulation through the solid–liquid phase change process is closely associated with changes in the atmosphere, ocean, land, and ecosystems and is a key factor of climate change.

1. Sources of freshwater and water balance in the Arctic and Antarctic

High latitudes tend to receive more freshwater, while sub-tropics receive less, due to impacts of the cryosphere. The freshwater in high latitude regions forcing to the ocean's surface in a non-uniform way across latitudes. However, as the rate of change of the freshwater forcing in the northern high latitudes is larger than the southern high latitudes than that in high latitudes, the question about a possible global redistribution of major water masses in the hydrological cycle and specifically in the ocean arises.

Both solid and liquid freshwater in the Arctic and Antarctic are very important since they change the hydrological cycle of the ocean. The average freshwater balance in the Arctic and Antarctic from 1960 to 1990 (Flavio et al. 2012) is shown in Fig. 8.9, where the number associated with arrows represents fluxes, and values in the box represent storage. The freshwater balance between the Arctic and Antarctic are through water transformation and cycle processes of atmosphere, ocean, land, and sea ice. It needs to point out that runoff entering the Arctic is mainly snowmelt runoff. Therefore, sea ice, snow cover and other components of the cryosphere in Arctic domain play an important role in the freshwater cycle. No terrestrial runoff

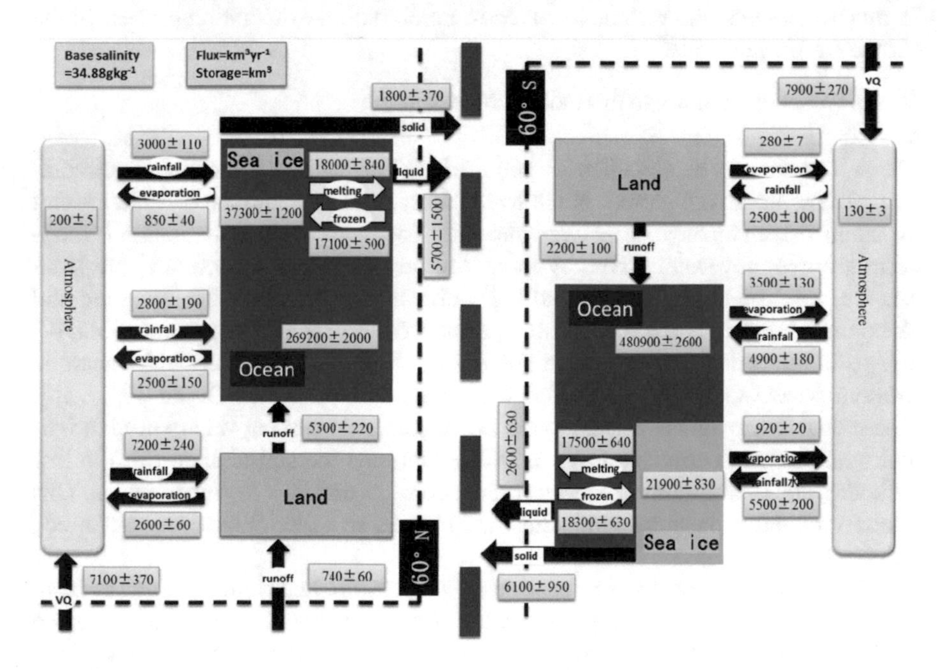

Fig. 8.9 The average freshwater budget between 60°–90° N and 60°–90° S during 1960 to 1990 (Flavio et al. 2012)

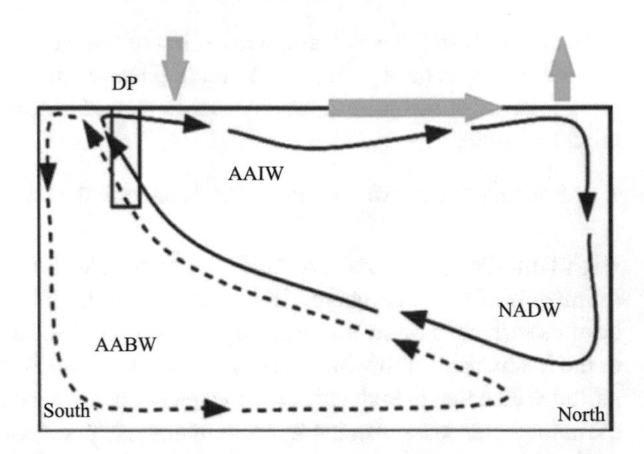

Fig. 8.10 The two main branches of meridional overturning circulation (modified after Ivanova 2009), AABM is Antarctic Bottom Water, NADW is North Atlantic Deep Water, AAIW is Antarctic Intermediate Water, DP is the Drake Passage

directly contributes to Antarctica, and only some of the exposed land surface runoff contributes to the Southern Ocean. It can be seen from Fig. 8.9 that freshwater storage in the Arctic Ocean and the Southern Ocean between 60° and 90° are 27×10^4 and 48×10^4 km^3, respectively. The second freshwater storage is in sea ice, which is 3.7×10^4 km^3 and 2.2×10^3 km^3 in the Arctic Ocean and the Southern Ocean, respectively. In the freshwater cycle, sea ice contributes the most freshwater, approximately 1.8×10^4 km^3, to the Arctic freshwater cycle via a freezing–thawing process. Snowmelt contributes about 0.5×10^4 km^3 a^{-1} to the freshwater cycle in the Arctic, which is much more than the volume involved in precipitation-evaporation process in the Arctic freshwater cycle.

2. Cryosphere and ocean thermohaline circulation

Ocean current can be classified as wind-driven current and thermohaline circulation. The wind-driven current is relatively short-term, and surface oceanic current is mainly driven by large-scale wind currents. The thermohaline circulation is long-term average movement driven by many factors, including temperature, pressure, sea ice, etc. The branches of global thermohaline current along the Antarctic and Arctic are shown in Fig. 8.10. The high density North Atlantic Deep Water (NADW) moves southward after sinking from the Arctic Ocean, then interacts with Antarctic Bottom Water (AABW) and transfers back to the North Atlantic Ocean (Fig. 8.10). The global thermohaline current cycle is composed of sinking waters in high latitudes, the bottom currents transferred to low latitude-transmission up (flip) in low latitudes, and the advection of ocean surface currents that flow to high latitudes. This global circulation phenomenon is illustrated by the so-called "conveyor belt" mode in Fig. 8.11.

The density of seawater is a function of temperature and salinity, which is more sensitive to changes in salinity than in temperature under low temperatures, such as seawater in the deep formation area. The shrinkage and expansion of cryosphere can lead to the release or storage of a large volume of cold freshwater into the sea or land. This process affects both the ocean temperature and salinity, which

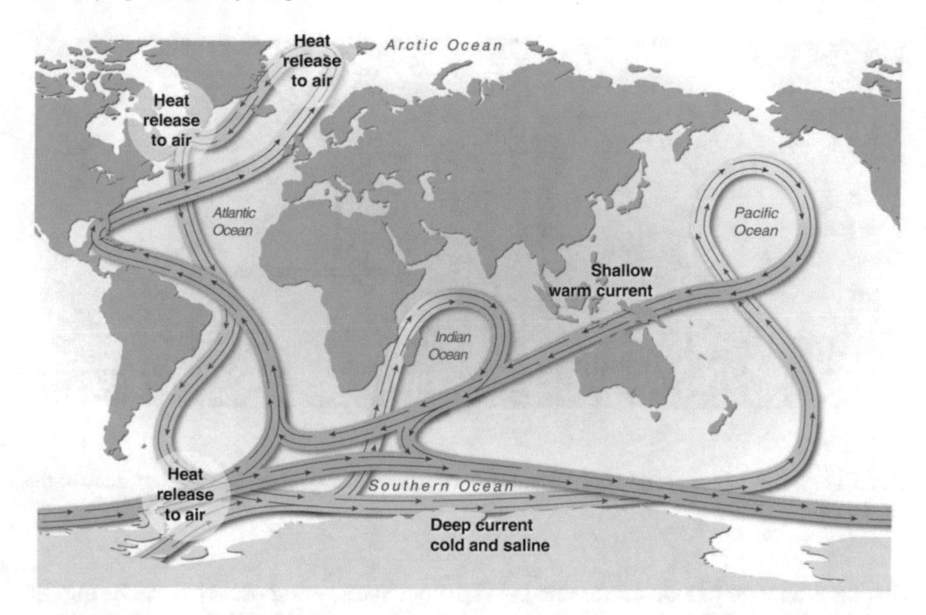

Fig. 8.11 Scheme of Broecker great ocean conveyor belt. Stephen modified the relationships between ocean and basins leads by the high-density sink water near the Antarctic and Antarctic circumpolar current (Broecker 1982; Stephen 2013). Red is the upper stream, blue is the deep water

further influence the thermohaline processes. When the amplitude of temperature or salinity is large enough, it can change the direction of Meridional Overturning Circulation (MOC), leading to an abrupt climate change (Fig. 8.11). MOC is the most common hypothesis to explain the Quaternary glaciation. A large volume of freshwater is stored in ice sheets and mountain glaciers in middle and high latitudes of the continent throughout the Quaternary. The fluctuation of terrestrial glaciers is equivalent to dozens of meters of sea level change due to the release of freshwater into the ocean from land. Therefore, many studies have investigated the role of freshwater perturbations on the MOC stability. Based on ice core records, early studies revealed that sudden warming events (interstadials) caused climate change of millennial-scale amplitude, which lasted for hundreds to thousands of years. These events, known as the Dansgaard-Oeschger (D-O) cycle or D-O fluctuations, which are also found in the sedimentary record of the North Atlantic, reflect the role or response of the oceans.

8.3.3 Cryosphere and Sea Level

1. Main factors affecting sea level change

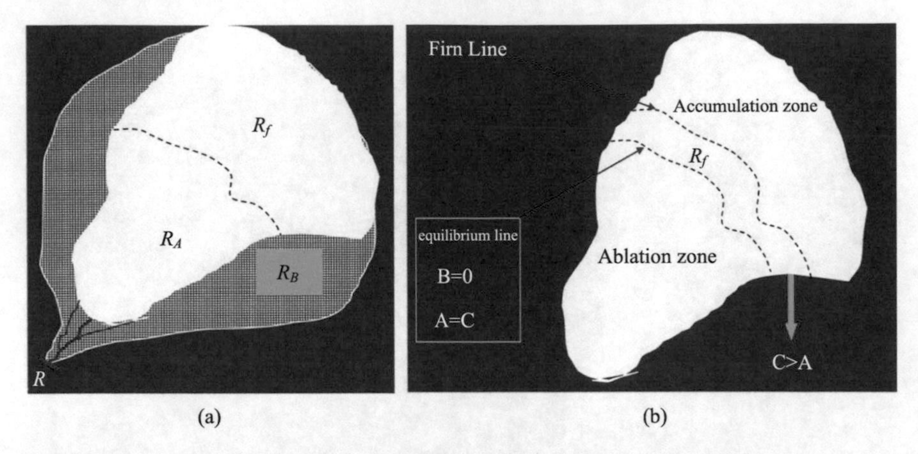

Fig. 8.12 The components of glacier runoff (**a**) and accumulation zone and ablation zone of the glacier (**b**)

Sea level change can be represented at different spatio-temporal scales. During the geological period (100 Ma ago), the largest global mean sea level change in the range of 100–200 m had been occurred. It was mainly caused by geological formation processes. With the formation of continent ice sheets (for example, Antarctic ice sheet formed 35 Ma years ago), the global mean sea level decreased by 60 m. The changes of earth's orbit and eccentricity led to glacial/interglacial cycles since 3 Ma ago, and quasi-periodic fluctuations (on a 10 ka scale) of ice caps in the Northern Hemisphere had a significant impact on the global mean sea level change as much as 100 m. In the period of centuries to thousands of years, sea level fluctuations was mainly affected by natural forces (solar radiation and volcanic eruptions) and changes in the climate system (atmosphere–ocean oscillations, such as El Niño-Southern Oscillation (ENSO), the North Atlantic Oscillation (NAO), and the Pacific Decadal Oscillation (PDO)). Meanwhile, sea level rise was significantly impacted by global warming as a result of anthropogenic emissions since industrialization. Over the past century, the change in global mean sea level was observed due to anthropogenic influence on climate warming, and an accelerated sea level rise has been observed recently. The oceanic thermal expansion and intensified melting of cryosphere are the main contributors of sea level rise nowadays, which are the focus of this section.

2. Estimated cryosphere's contribution to sea level change in the past

Global Mean Sea Level (GMSL) will rise by about 7 m if the Greenland and the west Antarctic ice sheet would fully have melted, respectively. Thus, even small decreases of ice volume in the Greenland and Antarctic ice sheet will have a substantial impact on GMSL change. The recent accelerated ice melting of Greenland and Antarctic ice sheet has compensated the decreasing contribution of slower ocean thermal expansion to GMSL rise (Cazenave et al. 2009). Gravity Recovery and Climate Experiment

(GRACE) data reveal that ice loss of Greenland and Antarctica had a significant increase trend with a rate of 392.8 ± 70.0 Gt a^{-1} during 2003–2010 (Cazenave and Llouel 2010), which equals to a GMSL rise of 1.09 ± 0.19 mm a^{-1}. Although estimates vary, ice loss was suggested to be responsible for 25% of observed GMSL rise during 2003–2010.

The authoritative assessment of the impact of cryospheric changes on GMSL rise was initiated after the release of the IPCC 5th Assessment Report (AR5) (Table 8.1). IPCC AR5 points out that the contribution rate of all mountainous glaciers (including glaciers around the Greenland and Antarctic ice sheets) to GMSL rise was 0.71 (0.64–0.79) mm a^{-1} during 2003–2009. Because the contribution of glaciers around ice sheets are difficult to separate from that of ice sheets, the global glacier

Table 8.1 The global mean sea level budged of observation and simulation in the different periods (mm yr^{-1}) (Church et al. 2013)

Source	1901–1990	1971–2010	1993–2010
Observed contribution to global mean sea level (GMSL) rise			
Thermal expansion	–	0.8 [0.5–1.1]	1.1 [0.8–1.4]
Glaciers except in Greenland and Antarctica	0.54 [0.47–0.61]	0.62 [0.25–0.99]	0.76 [0.39–1.13]
Glaciers in Greenland[a]	0.15 [0.10–0.19]	0.06 [0.03–0.09]	0.10 [0.07–0.13][b]
Greenland ice sheet	–	–	0.33 [0.25–0.41]
Antarctic ice sheet	–	–	0.27 [0.16–0.38]
Land water storage	– 0.11 [–0.1 to –0.06]	0.12 [0.03–0.22]	0.38 [0.26–0.49]
Total of contributions	**–**	**–**	**2.8 [2.3–3.4]**
Observed GMSL rise	**1.5 [1.3–1.7]**	**2.0 [1.7–2.3]**	**3.2 [2.8–3.6]**
Modelled contribution to GMSL rise			
Thermal expansion	0.37 [0.06–0.67]	0.96 [0.51–1.41]	1.49 [0.97–2.02]
Glaciers except in Greenland and Antarctica	0.63 [0.37–0.89]	0.62 [0.41–0.84]	0.78 [0.43–1.13]
Glaciers in Greenland	0.07 [–0.02 to 0.16]	0.10 [0.05–0.15]	0.14 [0.06–0.23]
Total including land water storage	**1.0 [0.5–1.4]**	**1.8 [1.3–2.3]**	**2.8 [2.1–3.5]**
Residual c	**0.5 [0.1–1.0]**	**0.2 [–0.4 to 0.8]**	**0.4 [–0.4 to 1.2]**

Notes

[a]Data for all glaciers extend to 2009, not 2010

[b]This contribution is not included in the total because glaciers in Greenland are included in the observational assessment of the Greenland ice sheet

[c]Observed GMSL rise-modelled thermal expansion-modelled glaciers-observed land water storage Uncertainties are 5–95%. The Atmosphere–Ocean General Circulation Model (AOGCM) historical integrations end in 2005; projections for RCP4.5 are used for 2006 –2010. The modelled thermal expansion and glacier contributions are computed from the CMIP5 results, using the model proposed by Marzeion et al. (2012a) for glaciers. The land water contribution is due to anthropogenic intervention only, not including climate-related fluctuations

contributions to GMSL rise excluding glaciers around the ice sheets during four different assessment periods were 0.54 (0.47–0.61) mm a^{-1} (1901–1990), 0.62 (0.25–0.99) mm a^{-1} (1971–2009), 0.76 (0.39–1.13) mm a^{-1} (1993–2009), and 0.83 (0.46–1.20) mm a^{-1} (2005–2009), respectively. The contribution pathways of the Greenland and Antarctic ice sheets to GMSL rise are slightly different. The mass balance of the Greenland ice sheet is determined by its surface mass balance and ice flow out to the sea, while the Antarctic mass balance is mainly influenced by the accumulation, calving, and basal melt of the ice shelf. The observations by satellite and aerial survey of ice contribution to GMSL change were started nearly 20 years ago. There are three main methods to estimate the ice sheet mass balance, including the field observed mass balance, repeat altimetry, and gravity measurements. Observations showed that the contributions of Greenland Ice Sheet to GMSL increased from 0.09 (–0.02 to 0.20) mm a^{-1} during 1992–2001 to 0.59 (0.43–0.76) mm a^{-1} during 2002–2011. The contribution rate of Antarctic Ice Sheet to GMSL increased from 0.08 (–0.10 to 0.27) mm a^{-1} during 1992–2001 to 0.40 (0.20–0.61) mm a^{-1} during 2002–2011. During 1993–2010, the total annual contribution of the ice sheets was 0.60 (0.42–0.78) mm a^{-1} (Table 8.1). The estimated ice sheet contribution was significantly larger than that estimated in IPCC AR4 0.21 $\pm$ 0.07 and 0.21 $\pm$ 0.35 mm a^{-1} during 1993–2003 for Greenland and Antarctic, respectively.

Overall, excluding the effects of terrestrial water storage change, the influence on GMSL rise by cryosphere conditions and thermal expansion via ocean warming has been nearly the same since industrialization (in the early 1900s), whereas the projected contribution of the cryosphere, mainly impacted by cryosphere change, will exceed than that of thermal expansion.

8.3.4 Cryosphere and Terrestrial Hydrology

1. Glacier hydrology

Glacier hydrology is the cross-discipline of glaciology and hydrology, which includes the generation and confluence processes, characteristics of changes, and hydrological roles of glacier runoff. It mainly concerns hydrological phenomena, processes, and basic laws from glacier ablation to runoff.

(1) Glacier melting and runoff

The basic concepts of glacier accumulation and ablation were introduced in the previous chapter. This section introduces some knowledge of glacier ablation again to link with glacier hydrology. The related concepts about ablation include total ablation and net ablation. Total ablation refers to all lost water on glacier surface in a year, including pure ice loss, snow melting and rainfall in summer, ablation (sublimation) in winter, and avalanche loss of snow and ice. Net ablation refers to the amount of pure ice lost from a glacier within a year.

Radiation is the key factor in determining glacier ablation, and the relationship between ice surface ablation and radiation balance can be expressed as:

$$Q = aB^n \quad (8.1)$$

where Q is the average of daily runoff (l s^{-1}) or ablation depth (mm) from runoff field over the ice surface; B is radiation balance (J cm^{-2} h^{-1}); a is coefficient; n is the power; and a and n vary with the glacier surface characters and climate at different regions.

Energy balance is the most accurate method for calculating glacier ablation. However, it is generally difficult to directly measure each energy component on the glacier surface. Because air temperature is a good and easy-to-access indicator that reflects energy balance, it is the most commonly used index to estimate total glacier ablation by establishing a relationship between air temperature and glacier ablation.

The glacier runoff can be determined by glacial ablation. However, glacier runoff processes are also related to hydrological systems, such as the stream system, supraglacial lake, and the coverage of moraine. Glacier runoff include surface runoff, englacial runoff, and subglacial runoff. The ablation of snow and ice is the source of glacier surface runoff, while englacial runoff mainly comes from the infiltrated glacier surface ablation through englacial channels. Subglacial runoff is formed through subglacial channels at the interface between a glacier and bedrock and comes from ice surface runoff and englacial runoff.

Glacier discharge represents the flowing out of glacier runoff through the upstream system, which then converges to the river channel. It also includes rainfall runoff from the bare rock in the surrounding mountainous areas (Fig. 8.12a). The glacier runoff R can be represented as:

$$R = R_{\mathrm{f}} + R_{\mathrm{A}} + R_{\mathrm{B}} \quad (8.2)$$

where R_{f} is runoff generated from snow and firn at the accumulation zone; R_A is runoff generated from the ablation zone; and R_B is runoff generated from the bare rock. The R_{f} mainly occurs between the equilibrium line and firn line in summer when the air temperature is high (Fig. 8.12b). The R_{f} is nearly negligible for continental type glacier due to high altitude, low temperature, and low energy, where snowmelt in the glacier accumulation zone is very limited, thus the contribution of R_{f} to glacier runoff is quite weak. The R_B from bare slopes includes current rainfall (liquid precipitation) runoff and meltwater from ground ice because the bare rock is always covered by permafrost or seasonal frozen ground. The R_A of the ablation zone can be calculated as follows:

$$R_{\mathrm{A}} = R_{\mathrm{w}} + R_{\mathrm{s}} + R_{\mathrm{I}} + R_{\mathrm{m}} \quad (8.3)$$

where R_{w} is the seasonal snowmelt runoff in the glacier ablation zone in winter and spring; R_{s} is the summer rainfall (including solid and liquid precipitation) runoff in the glacier ablation zone (mm); R_I is the runoff generated from melting of pure ice, including bare ice of glacier surface, englacial runoff, and subglacial runoff (mm); and R_m is the meltwater of buried ice (mm).

Since it is very difficult to accurately observe glacier runoff, the observed runoff at the terminal of glaciers generally contains more or less rainfall runoff from bare rock, which prompted the concern of whether this observed runoff fully represents the glacier runoff. The following paragraphs describe various definitions of glacier runoff R_g.

Observed runoff at the glacier terminal: Glacier runoff represents by the observed runoff of the areas above glacier terminal, including runoff generated from the glacier ablation zone, accumulation zone, and bare slopes above glacier terminal. The observed runoff is nearly equal to the glacier runoff when the ratio of the bare rock area to the basin area above the glacier terminal is small. It is obviously not suitable for the definition of glacier runoff due to it include the rainfall-runoff generated from the bare rock area, especially if the ratio of bare rock area to basin area is relatively large. The definition of R_g is a maximum. It can be represented as:

$$R_g = R_f + R_A + R_B \tag{8.4}$$

Runoff generated from the whole glacier: This definition of glacier runoff is equal to the observed runoff subtracted rainfall runoff generated from the bare rock. It includes meltwater generated from seasonal snow in winter and spring R_w, solid and liquid precipitation in summer R_s, melting of bare ice in the glacier accumulation and ablation zones, englacial runoff, subglacial runoff R_I, and meltwater from buried ice R_m. Runoff generated from the bare rock is considered as mountain snowmelt runoff. This is the most common definition of glacier runoff. It can be represented as:

$$R_g = R_f + R_A = R_f + R_w + R_s + R_I + R_m \tag{8.5}$$

$$R_g = R - R_b \tag{8.6}$$

Runoff generated from the glacier, excluding summer precipitation: The definition of glacier runoff excludes runoff generated from the bare rock and runoff from summer precipitation that hasn't transformed into ice in the current year. It can be represented as:

$$R_g = R_f + R_w + R_I + R_m \tag{8.7}$$

Runoff generated only from firn and ice: The definition of glacier runoff only refers to the meltwater generated from firn and ice. The runoff generated from precipitation over glaciers, whether it is in the form of rainfall or snow, is considered as the mountain snowmelt runoff. The definition of glacier runoff can be represented as:

$$R_g = R_f + R_I + R_m \tag{8.8}$$

Runoff generated only from ice: The definition of glacier runoff only includes runoff generated from ice, it can be represented as:

$$R_g = R_I + R_m \tag{8.9}$$

Obviously, the estimated volume of glacier runoff varies with the definitions applied. The first two definitions are rather wider and overestimated the role of glacier meltwater. The third and fourth definitions consider the formation of glacier ice, which classifies precipitation on mountain glaciers as snow. They don't exclude the role of precipitation in glacier melting, which seems quite reasonable to evaluate the role of glacier runoff in river runoff. However, there are difficulties in the estimation of ice melt due to limited data availability. The fifth definition ignores the role of snow transformed into glacier ice. To simplify calculations, the second definition of glacier runoff is generally used.

Glacier runoff is directly associated with ablation (Fig. 8.13). According to the principles of water balance, the relationship between glacier runoff (R_g) and ablation can be represented as:

$$R_g = (A_f - (E_f + \Delta A_f) + (A_a - E_a) \tag{8.10}$$

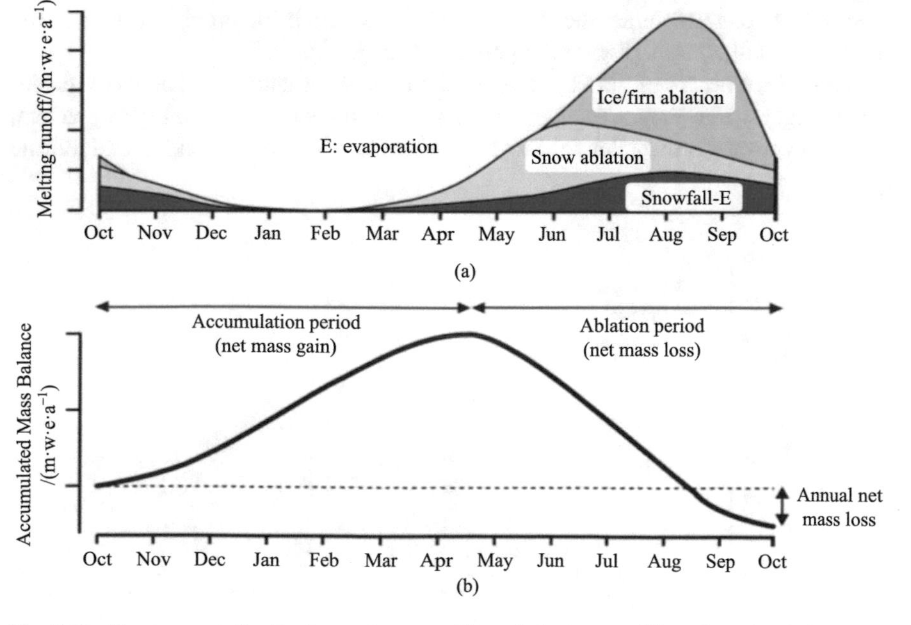

Fig. 8.13 The seasonal distribution and composition of glacier runoff based on the definition of runoff generated from the whole glacier (Radic and Hock 2014)

where A_f is ablation in the accumulation zone (mm); E_f is evaporation in the accumulation zone (mm); $\triangle A_f$ is the refreezing of meltwater in the accumulation zone (mm); A_a is ablation in the ablation zone (mm); and E_a is evaporation in the ablation zone (mm).

(2) Characteristics of glacier runoff

Diurnal cycle: The diurnal cycle of glacier runoff is characterized by one peak and one valley, whether the type of glacier is continental (polar) or marine (temperate). The lag time of the peak of glacier runoff behind the air temperature depends on the type of glacier, features of the glacial drainage system, basin size, the distance between hydrological observation section and glacier terminal, and other factors.

Seasonal distribution: The seasonal distribution of glacier runoff is associated with the length of melting period and glacier types. The hydrological year of continental type glaciers is defined from October to September in the subsequent year, while the hydrological year of marine type glaciers is consistent with the calendar year in China. Generally, glacier discharge mainly happens between June and August, which accounts for about 85–95% of the annual glacier discharge, and the discharge disappears in December, January, and February (Fig. 8.14).

Runoff processes affected by air temperature: The glacier runoff is significantly affected by air temperature due to that ablation is mainly dictated by heat. The solid precipitation suppresses ablation since fresh snow has a higher albedo, leading to a decrease in air temperature and eventually the glacier runoff. In individual cases, precipitation can accelerate the ablation of the glaciers if the precipitation is in the form of rainfall, with higher air temperature (Fig. 8.15).

Annual variances of glacier runoff: The annual variance coefficients of glacier runoff suggest that they are relatively larger for smaller continental type glaciers, while they are much smaller for larger continental, the subcontinental, and marine type glaciers.

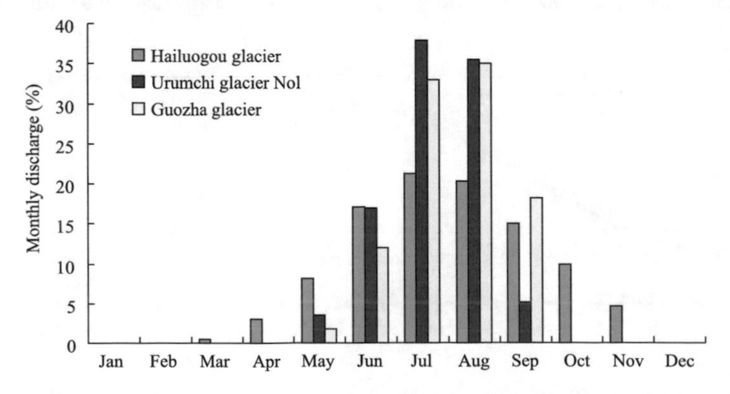

Fig. 8.14 Seasonal distribution of glacier runoff of Hailuogou, Urumchi glacier No. 1, and Guozha glacier in China (Cao 1998)

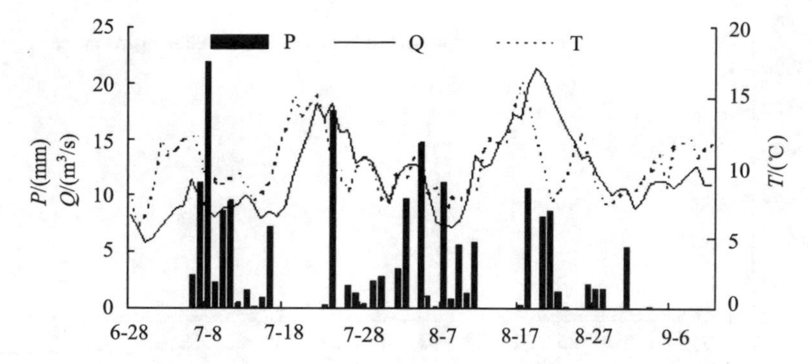

Fig. 8.15 The relationship between air temperature, precipitation and glacier runoff of Gongga glacier in China

(3) Main indexes of glacier runoff

Glacier runoff modulus (M_g). M_g is the glacier runoff at per unit time per unit area. It is an index of the volume of glacial runoff, and generally uses the unit L/(s km^2):

$$M_g = \frac{W_g}{F_g \cdot t_g} = \frac{W - W_B}{F_g \cdot t_g} = \frac{R \cdot F - R_B \cdot F_B}{F_g \cdot t_g} \times 100\% \tag{8.11}$$

where W_g is glacier meltwater volume (m^3); W is the runoff volume of basin (m^3); W_B is the runoff volume for the bare rock (m^3); F_g is the area of glacier (km^2); F is the area of basin (km^2); R is the glacier runoff depth (mm); R_B is the runoff depth of the bare rock (mm); and t_A is the melting period of the glacier. In general, the melting periods of continental, subcontinental, and marine glaciers are May–September, April–October, and March–November, respectively.

Glacier runoff depth (*R*). *R* is another form of glacier runoff, which is defined as the annual water production (W) of the glacier area (S), usually expressed in mm.

Runoff coefficient (α). α can be expressed as:

$$\alpha_g = R/P \tag{8.12}$$

where R is the runoff depth (mm); and P is the average rainfall over the watershed (mm). When the glacier mass balance is positive, the glacier accumulation is greater than ablation: $R < P$, $\alpha_g < 1.0$. Conversely, $\alpha_g > 1.0$ when the glacier mass balance is negative, indicating that $R > P$. It suggests that the glacier runoff R not only includes runoff generation from the precipitation over the glacier but also the mass loss of the glacier itself during dry years.

(4) Glacial floods

Glacial lake outburst flood (GLOF). GLOF was defined by International Association of Hydrological Sciences (IAHS) in 1974. It is that kind of flood which

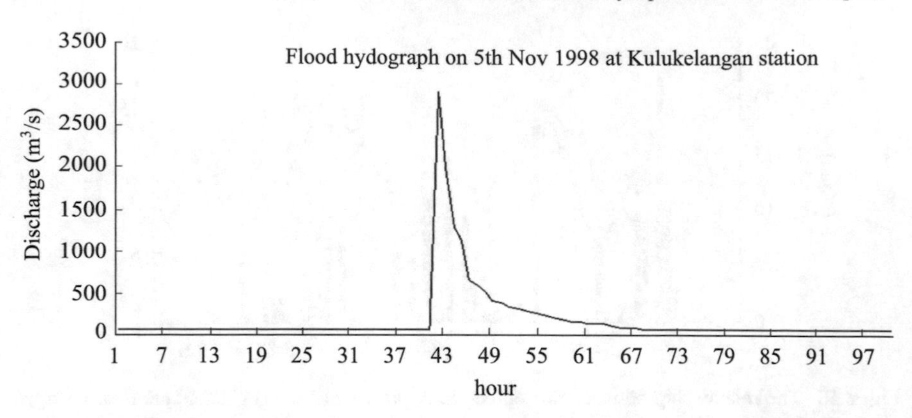

Fig. 8.16 Glacier lake outburst flood on 5th Nov 1998 at Kulukelangan station of Yarkant River in Karakoram Mountains

occurs very suddenly and is generally unpredictable with a short flood peak time and larger peak value. In short, it is a flash flood usually caused by glacial lake outburst. The glacial lakes generally include glacier-dammed lakes, moraine-dammed lakes, supraglacial lakes, subglacial lakes, and so on. GLOF also referred to as Jökulhlaup (in Icelandic), can be divided into two types. One is a glacier-dammed lake outburst, or glacier runoff outburst due to the sub-glacial or englacial plug, and is also called glacial lake outburst in short. The second is the glacier terminal moraine-dammed lake outburst, also called a moraine lake outburst in short. In most cases, GLOF induces glacial debris flow.

Characteristics of GLOF. GLOFs usually have a high peak and low flood volume. There are sharp rises and falls in the GLOF peaks, and it displays a single thin peak in the hydrograph (Fig. 8.16). The occurrence time of the flood has a large uncertainty, and the occurrence frequency is high. The flood volume has little relationship with antecedent precipitation and glacier runoff. It directly depends on the lake volume and scale of the dam break.

Mixing flood. Meltwater from ice or snow can cause immense floods under the influence of continuously high temperatures and/or rapid warming. In addition, floods mixed by snowmelt, ice meltwater, and rainfall runoff always developed under earlier high air temperature along with larger precipitation, or earlier larger precipitation along with high air temperature.

Glacial debris flow. Glacial debris flow is one kind of special flow surrounding modern glacier regions, which contains lots of soil, sand, stones, and other loose materials. Solid materials in the flow mainly come from fresh and old glaciers formed from modern glaciers and paleo-glaciation, and the water mainly comes from the strong melting of glaciers and snow, icefall, avalanche, etc.

2. Glacier water resources

(1) Method of calculating glacier ablation

As mentioned earlier, the glacier ablation calculation methods can be divided into two categories: one is energy balance or related to energy balance, which mainly based on energy balance equations. Another category is the statistical method. The main statistical method of the calculation of the glacier mass balance and ablation is the use of relationship between air temperature and glacier ablation.

Observations of glacier runoff are essential for calculation of glacier ablation both for the energy balance and statistical methods. To obtain glacier runoff, hydrologic sections are needed to set up at the glacier terminal and at bare rocks around the glacier for parallel comparison.

Relationship between glacier runoff and air temperature. The relationship between air temperature and glacier runoff can be established by analyzing the observed glacier runoff and air temperature during different time periods (daily, ten dailies, monthly, etc.). The air temperatures are generally observed at meteorological stations or hydrological stations. The glacier runoff can be extended by using long-term observed air temperatures. It provides a basis for analyzing the glacier runoff. However, the method has some limitations. The established relationship only fits for the observed glaciers, so it is difficult to directly apply it to other glaciers.

Glacier runoff modulus method. The glacier runoff modulus has a significant spatial pattern. The pattern can be obtained from limited information of the observed glacier runoff at some observed sites. Then, the glacier runoff modulus in the ungauged basins can be interpolated from the pattern.

Linear reservoir method based on energy balance. This method applies energy balance to calculate glacier meltwater from observed sites. Then, the glacier meltwater is used as the input data to obtain the runoff process by a linear reservoir model. The principle of the linear reservoir is that glacial runoff is proportional to the size of glacier meltwater storage:

$$V(t) = KQ(t) \tag{8.13}$$

$$Q(t) = \int_0^t \frac{R(t)}{k} e^{(\tau - t)/t} d\tau + Q_3(t) + Q_4(t) \tag{8.14}$$

where $V(t)$ is the runoff, K is a constant, $Q(t)$ is the inflow to the reservoir, τ is the time, t is the time between inflow to outflow, $Q_3(t)$ and $Q_4(t)$ are base flows, respectively.

Hydrological model. A hydrologic model is a basic method of calculating glacier runoff. However, due to data constraints, hydrological models which consider both glacier runoff and rainfall runoff process at the basin scale are still in the developing stages. Currently, the most popular method is embedding the glacier modulus into the distributed hydrological model, where the glacier modulus always applies the degree-day model, which is based on air temperature. In general, the simulation accuracy of the hydrological model is always lower when considering the process in more detail or with more complex model structure due to the limited observations of glacier runoff.

(2) Glacier meltwater resources in China and their distribution

The amount of glacier meltwater at basinal, regional, or Chinese scales can be estimated by the above-described methods. The glacier meltwater resources in China are mainly calculated by the runoff modulus method in the 1990s, and by the hydrological models based on the degree-day modulus in the 2010s.

The spatial pattern of glacier runoff modulus. The meltwater runoff modulus generally decreases with the increase of aridity. The maximum glacier runoff modulus appears in marine type glaciers in southeastern TP, which are affected by the southwest monsoon. The maximum glacier runoff modulus is 196.7 $L/km^2\ s^{-1}$ of the Guxiang glacier in the eastern part of Nyenchen Tanglha range. The modulus decreases from southeastern Tibet towards the west and northwest regions, including the northern interior part of the TP, Pamirs Plateau, and the western part of Qilian Mountain, which decreases by 7.7–43.1 $L/km^2\ s^{-1}$.

The spatial pattern of glacier runoff depth. The spatial pattern of glacier runoff depth is like that of the glacier runoff modulus, which decreases from marine type glaciers in southeastern TP to continental glaciers in western and northwestern TP. For instance, the glacier runoff depth is above 3,000 mm at Guxiang glacier in southeastern TP, which is about 2,037 mm at Gongba glacier in Gongga Mountain. To the westward direction in the northern slope of the Himalayas and to the southwest direction in western parts of Qilian Mountains and Pamir Plateau, the meltwater runoff depth is approximately 400–550 mm. The lowest runoff depth is about 200 mm in the southern slope of western of Kunlun Mountains.

The average annual glacier runoff of China. The estimated annual glacier runoff in China was 56.33 billion m^3 in the 1980s by integrating the glacier runoff modulus method, the relationship between glacier runoff and air temperature, and contrasting experiments in the field. The number was updated to 60.465 billion m^3 in 2006 (Table 8.2). The annual glacier runoff in the 1980s was about 2.2% of annual river runoff (2.7115×10^{12} m^3) in China, more than the Yellow River inflow into the sea, which is also equivalent to 10.5% of the total river runoff of four provinces, including Gansu province, Qinghai province, Xinjiang Uygur Autonomous Region (XUAR), and Tibet Autonomous Region (TAR) in western China (5.76×10^{11} m^3). In addition, there is some glaciers runoff in Sichuan and Yunnan provinces. If we compare glacier runoff in the different mountain ranges, the maximum volume of glacier runoff is found in the Nyenchen Tanglha range, which accounts for 35% of the total glacier runoff in China, followed by 15.9% and 12.7% of the Tian Shan and Himalayas, respectively. The minimum ratio is at the Altai range, which only accounts for 1%. When comparing the volumes of glacier runoff in different provinces, TAR ranks first, which accounts for 58% of the total glacier meltwater in China, followed by 33% in XUAR. About 60% of the glacier runoff in China contributes into the outflow rivers, while 40% flows into inland rivers. However, in terms of the glacier area, the

Table 8.2 Glaciers and meltwater runoff in China in 2006

Mountain range	Glacier area/km^2	Glacier meltwater/10^8 m^3	Percentage
Qilian Mountains	1930.51	11.32	1.9
Altai Mountains[a]	296.75	3.86	0.6
Tianshan	9224.80	96.30	15.9
Pamir	2696.11	15.35	2.5
Karakoram Mountains	6262.21	38.47	6.4
Kunlun Mountains	12,267.19	61.87	10.2
Himalaya Mountains	8417.65	76.60	12.7
Qiangtang Plateau	1802.12	9.29	1.5
Kailas Range	1759.52	9.41	1.6
Nyenchen Tanglha range	10,700.43	213.27	35.3
Hengduan Mountains	1579.49	49.94	8.3
Tanglha range	2213.40	17.59	2.9
Altun Mountains	275.00	1.39	0.2
Total	59,425.18	604.65	100.0

[a]Including 16.84 km^2 glacier area at Moose Ridge

total glacier area accounts for 60% in inland basins, while 40% in the outflow river basins.

(3) Supply role of glacier runoff in river runoff

The proportion of glacier runoff in river runoff varies in different basins. It depends on the glacier coverage above a certain hydrological section. Inland river basins in China have a higher percentage of glacial runoff supply proportions. For example, in western China, the largest supply proportion of glacier runoff to rivers is in the Tarim inland basin in XUAR, which accounts for 25.4%, followed by 8.6% in TAR. Gansu province provides the least in western China with just 3.6%. In terms of spatial distribution, the pattern of the supply proportion increases with the glacier area and arid climate from the perimeter to the center of Tibetan Plateau. To inland river basins, the supply proportion of glacier runoff is about 14% in the Gansu Hexi corridor, Junggar basin, and increases by 38.5% in the Tarim inland basin. In the Hexi corridor region, the glacier runoff supply proportion is only 4% in the Shiyang River basin at east of the corridor and is 8% and 32% in the Heihe River basin and Shule River basin in the middle and west parts of the corridor, respectively. The spatial pattern of the supply proportion is the same in the outflow river basins, which increases with the glacier area and aridity. The supply proportion is less than 10% in the upper reach of Lancang River basin and Ganges River basin in southeast Tibet but increases to nearly 40% in the upper reach of the Indus River, including the Shiquan River basin and Sutlej River basin.

(4) The regulation of glacier runoff to river runoff in China

Glaciers have a regulation role in river runoff on an annual scale. In humid years with lower air temperatures, there is not enough energy for melting, which causes more glacier accumulation. Conversely, in drought years, there are more sunny days and stronger glacier melting, thus releases larger amount of glacier meltwater. Therefore, there is no shortage of water in drought years in the basins with higher glacier runoff supply proportions located in the mountainous areas in western China. In contrast, there is less water in humid years, which decreases the difference between the high flow and low flow years. For example, the glacier covers an area of 37.95 km^2 in the upper reach of the Urumqi River basin (above heroes station), which is only 4.1% of the total basin area. According to the observed runoff in glaciated and non-glaciated areas, the average supply proportion of glacier runoff was 11.3% from 1982 to 1997. However, the glacial runoff supply proportion increased to about 28.7% in drought years, such as in 1986, while it was only 5.1% in humid years, such as in 1987. These results prove that glacier runoff has a strong role in the regulation of river runoff changes as a "solid reservoir."

In addition, the annual variation coefficient of runoff is small in rivers with a higher glacier meltwater supply proportion. Statistics of the ratio of Coefficient of Variations (CV) of annual river runoff to CV of annual precipitation of main basins in northwest China (Fig. 8.17) reveal that they are less than 0.5 in basins with rich glacier runoff (glacier runoff supply proportion greater than 30%), while they are greater than 1.0 in basins without glacier runoff supply. The variation of river runoff significantly decreases when the glacier coverage is more than 5%, which further proves that the glacier has a strong annual regulation on river runoff. Threats of droughts and floods are relatively small in the basins with rich glacier runoff, which is very important for stable and sustainable development of agriculture in arid regions of western China.

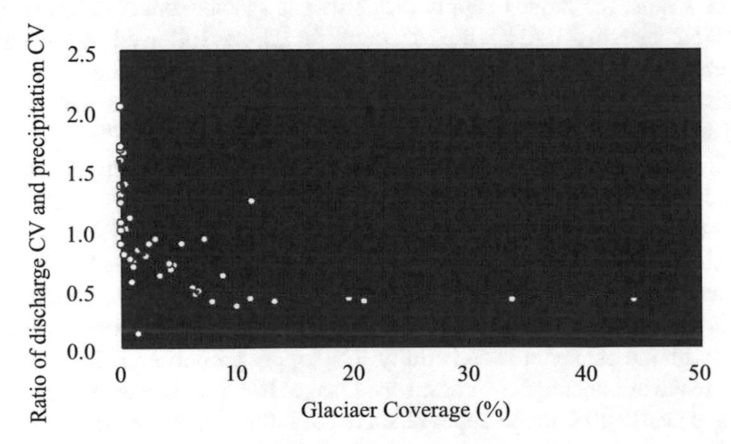

Fig. 8.17 The relationship between glacier coverage and the ratio of the coefficient of variation of runoff to the coefficient of variation of precipitation (Cvr/Cvp)

3. Snow hydrology

(1) Snow hydrology and snowmelt resources

Snow hydrology is a discipline of snow melting and snowmelt runoff generating and confluencing processes which have impacts on water resources and the environment, which is important to understand the global water cycle and water resources. Global snow contributes 5.95×10^{12} m^3 freshwater to terrestrial land, and snow storage (water equivalent) in the Northern Hemisphere can amount to 2×10^{12} m^3 in winter. For many large rivers in Asia, Europe, and North America, including source reaches of Yangtze River and Yellow River in China, the spring water supply mainly comes from snowmelt. Especially in the arid and semi-arid regions of the world like northwest China, the water supply for agriculture is highly dependent on winter snow in mountainous areas. Spring snowmelt in northeast China, XUAR, TAR, and other provinces leads to spring floods, which timely meet the urgent need for spring irrigation and provide abundant water resources for agriculture. Winter wheat which grows in the Yellow River basin, drinking water and grazing of livestock in pastoral areas in XUAR, Qinghai, and TAR are highly dependent on snowmelt water.

Snow water resources can be classified into supply resources, storage resources, and runoff. Accumulated snow comes from snowfall. Due to there are no separate snowfall observations after 1979, the snowfall record in China only before 1979. Therefore, the amount of snow supply in China is estimated from the observations at 2,300 meteorological stations during 1951–1979. It suggests that the annual average snowfall in China is 36.00 mm, which is approximately 5.7% of the annual precipitation. The estimated annual snowfall is 3.45×10^{11} m^3, among which 78.2% of the total is on three regions, including the TP, XUAR, and Northeast-Inner Mongolia, accounting 1.39×10^{11} m^3, 5.61×10^{10} m^3, and 7.49×10^{10} m^3, respectively. Snowfall influenced by mountainous terrain is far greater than the influence by plains. However, the meteorological stations are scarce in mountainous areas, especially in alpine areas, which lead to underestimation of snowfall. In addition, due to wind disturbance on precipitation gauges, the observed snowfall is much lower than the real snowfall. Therefore, the estimated annual snowfall and snowmelt discharge may be much lower than the real values.

(2) Snow cover formation and melting process of snow

Snowfall is one form of precipitation under certain weather conditions, which can be affected by many factors, such as air temperature, precipitation formation height, etc. Generally, precipitation is in rainfall form at the average air temperature above 2 °C, whereas snow is formed at the air temperature less than –2 °C. When the air temperature varies between –2 and 2 °C, the rainfall probably formed in mixture of rainfall and snowfall. Whether the snowfall accumulated as snow cover depends on both the atmospheric condition and surface characteristics. In atmospheric condition, the air transparency affects the amount of solar radiation that reaches the snow surface, clouds emit long-wave radiation, and turbulence affects the sensible heat and latent heat fluxes from air to the snow surface. The surface characteristics (including land surface temperature and terrain shadow) have impacts on wind force and solar radiation, while the canopy of trees reduces the incident of shortwave radiations onto the snow surface and increases the long-wave radiations from the snow surface to air.

Snowmelt is a process from crystal snow to liquid water. During this phase change process, water absorbed a large amount of energy as latent heat. The melting energy comes from different sources, which include radiation melting and advection melting.

Radiation melting process: The melting energy mainly comes from solar radiation, and melting processes have obvious diurnal and seasonal variations. The reflectance of snow determines the solar shortwave radiation received by the snow surface. The high albedo of snow and low solar elevation angle in winter limits the absorption of solar radiation of snow, which eventually leads to the snow accumulates. In the spring season, the snow albedo decreases with snow age, and solar elevation angle gradually increases, thus solar radiation received by the snow surface significantly increases. Meanwhile, as the daytime hours are prolonged in spring and the solar incident radiation increases, snow starts melting. The diurnal cycle of the melting process leads to the diurnal cycle of snowmelt.

Advection melting process: The advection of snow melting is determined by movements of warm and humid air masses. When warm and humid air mass moves over snow, the air mass transfer energy to the snow surface in the following ways: the emission of long-wave radiations from warm vapours and precipitable cloud base, the energy of rainfall itself, and the turbulent heat fluxes produced by strong winds blown across the rough ground surface. The strong turbulence can destroy the stable (neutral or inversion) atmospheric structure lead by warm air over the 0 °C snow surface, which will be more pronounced in the presence of trees. Energy carried by rain is generally 40–60 J/g, which is a very limited amount compared with the latent heat of melting (335 J/g). Although the direct energy of per gram of rain can melt only 0.2 g of snow, the decrease in the snow albedo led by rainfall is more important.

The snowmelt runoff is generated after meltwater infiltrates and saturates the soil. When the snow cover starts melting, the melting water over the surface infiltrates into the interior of the snow layers. The infiltration process of the snowmelt or rainfall in the snow is similar to the process of water infiltration into the soil, which can be described by similar dynamic equations. The main difference between water movement in the snow and in the soil is the mutual phase transformation between liquid water and ice crystal in the freezing and thawing processes. Liquid water freezes into ice crystal, while ice crystal can melt into liquid water, which constitutes a much more complex system than soil infiltration process.

(3) Snowmelt runoff calculations

Snowmelt runoff calculations include the snow melt speed at a point and watershed scales, and snowmelt runoff generation and routing, etc. The snowmelt speed at the point or basinal scales depends on the energy balance and spatial distribution of snow cover. The calculation of the generation and routing of snowmelt runoff is the same as that for rainfall runoff. The energy balance method and degree-day model are the primary methods to calculate the snowmelt rate at point scale. The principle of energy balance is to determine each component of the energy budget of snow cover and the heat used for melting, in which an accurate estimation of snow surface albedo is very crucial. Air temperature is the most available data, which is also the main factor that influences snowmelt. Therefore, the degree-day model, in most cases, which uses air temperature as the heat index, is widely used. Moreover, various equations have been established and widely applied for the estimation of snowmelt speed and sunshine duration, on the basis of air temperature, wind speed, relative humidity, radiation, precipitation, cloud cover, and other meteorological variables. Such as proposed by the United States Army Corps of Engineers (USACE) are as follows:

Open area:

$$M = 0.6(T_{\mathrm{m}} - 24)\ \mathrm{mm/d} \tag{8.15}$$

$$M = 0.4(T_{\max} - 27)\ \mathrm{mm/d} \tag{8.16}$$

Forest land:

$$M = 0.5(T_{\mathrm{m}} - 32)\ \mathrm{mm/d} \tag{8.17}$$

$$M = 0.4(T_{\max} - 42)\ \mathrm{mm/d} \tag{8.18}$$

where air temperature is measured in Fahrenheit. These equations can be applied when daily average air temperature T_{m} is within the range of 34–66 °F and daily maximum air temperature $T_{\max}$ is within the range 44–76 °F.

However, it is quite difficult to calculate snowmelt runoff in one basin. On one hand, due to the uncertainty in estimation on snowmelt infiltration and retention processes in the snow, the snowmelt runoff model has many difficulties in efficient simulation of the early snowmelt runoff processes. On the other hand, accurate measurements and estimates of snow cover on the basinal scale are still urgently needed. especially in the TP, where the snow cover is mainly thin and patchy. The interpretation of mix pixels of snow cover from remote sensing data remains a key issue in current research on snow cover over the TP.

The Snowmelt Runoff Model (SRM) is a commonly used method to calculate snowmelt runoff. SRM was developed on a small watershed in Europe in 1975, it has been successfully applied in 60 basins in more than 20 countries with a basin area ranging 0.76–122,000 km^2 and altitude ranging 305–7690 m a. s. l. The readers who are interested in SRM can get more details in literature.

(4) Characteristics of snowmelt runoff in China

Periods of snowmelt runoff: The snowmelt runoff in China mainly occurs in spring. However, due to the uneven spatial distribution of snow cover, there is a large difference between the periods of snowmelt runoffs. The daily discharge series in a whole year for representative rivers in western China, from the Irtysh River in the north to the Brahmaputra river in the south, are given in Fig. 8.18. Overall, the snowmelt runoff occurs from May to June in the north and from March to April in the south of western China. Most of the snowmelt runoff occurs in June at Kaqun station, which originates in the northern slope of Karakoram Mountain and flows into the Yarkant River. This snowmelt runoff is so late because it is mainly formed from glacier and snow in the alpine areas of the basin, which causes the period of snowmelt runoff later than that of the Irtysh River located north of the Yarkant River. It can be seen in Fig. 8.19 that the snowmelt runoff proportion is relatively large in the Irtysh River Basin, which has abundant snow cover. However, it is relatively small in other river basins. Snowmelt runoff in the spring, which is the planting season in northwest China, has important roles in agricultural production.

Changes in snowmelt runoff: Changes in snowmelt runoff are mainly controlled by air temperature, and global warming has impacted the changes of snowmelt runoff worldwide. For example, a study on discharge of the Irtysh River in China suggested that the spring snowmelt runoff has significantly increased, and the seasonal distribution of snowmelt process has been predated (Fig. 8.19).

4. Permafrost hydrology

1) Characteristics of permafrost hydrological processes

The soil temperature gradient drives the movement state of soil moisture when the active layer over permafrost freezes and thaws, causing frost heaving. The migration of soil moisture is mainly concentrated near the freezing front, which is the impermeable layer for rainfall or snowmelt water. The characteristics of hydrological processes in permafrost regions are:

Higher direct runoff coefficient: Due to the impermeable layer, most of the snowmelt and rainfall directly becomes runoff.

Higher and shorter flood peak: There is little winter runoff (if the watershed is entirely distributed 100% by permafrost then the winter runoff is 0), which increases the steep slope and peaks in the summer (Fig. 8.20).

Significant hydrological effects of the active layer: The direct runoff coefficient varies with freezing and thawing of the active layer over permafrost, leading to the different regulation of active layer to river runoff. The groundwater table is influenced by snowmelt runoff, rainfall runoff, and the thawing depth of the active layer. Therefore, the groundwater table is primarily controlled by changes of the active layer.

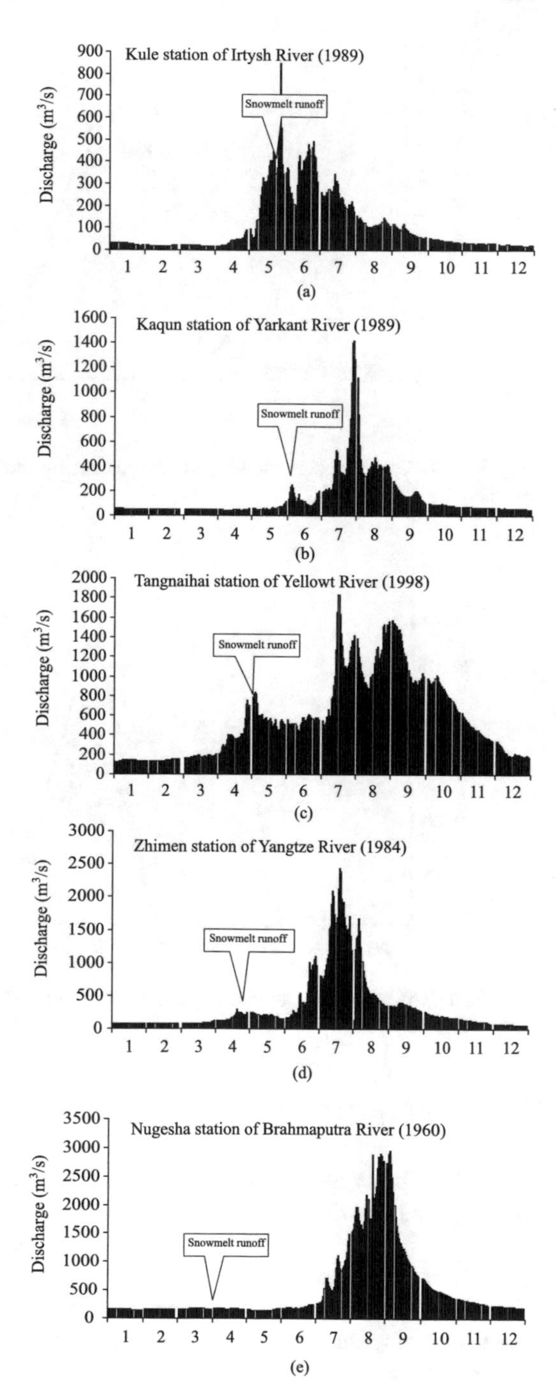

Fig. 8.18 Daily discharge series of representative rivers from north to south in Western China

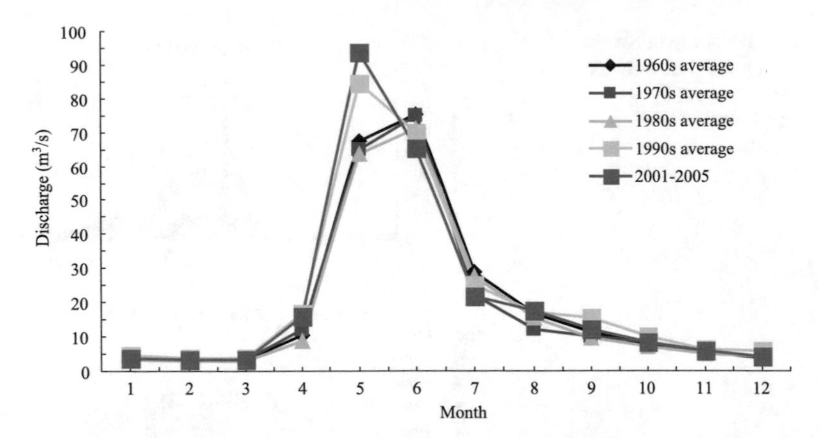

Fig. 8.19 Decadal average monthly discharges during the year from 1959 to 2005 in Kelan River in Irtysh Basin

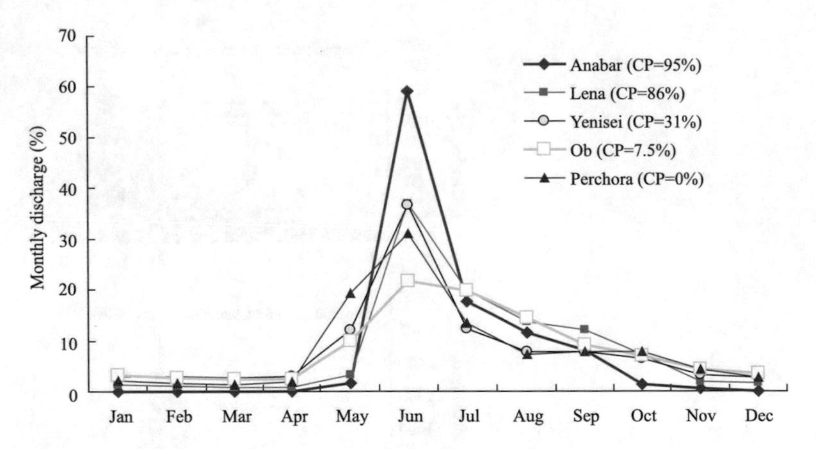

Fig. 8.20 The monthly river discharge in the different permafrost coverage basins of in Siberian Arctic regions

2) Hydrological processes of permafrost

Permafrost hydrological processes depend on the impermeability of frozen ground, freezing and thawing processes of active layer over permafrost, release and storage of ground ice, and water storage in the basin, which includes snow accumulation, meltwater infiltration in the basin (active layer), and development of ground ice. Therefore, the regulation of permafrost on runoff has three effects, including the regulation of ground ice, freezing and thawing of the active layer, and impermeability of permafrost.

(1) Ground ice

Due to the impermeability and moisture transfer are forced by the temperature gradient of permafrost, there is a large amount of ground ice near the upper limit of permafrost. It is estimated that the volume of ground ice exceeds the ice storage of mountain glaciers, which accounts for only 0.12% of global freshwater, while permafrost ground ice accounts for 0.86%. However, accurate estimations the volume from ground ice to river runoff still cannot be obtained due to lack of observation data.

(2) Active layer

Due to the impermeability of permafrost, groundwater sinks at the bottom of the active layer when permafrost melts, which directly impacts its hydrological processes and surface soil moisture. Meanwhile, the high latent heat is required for freezing and thawing in the active layer, which impacts soil temperature, hydrological cycle processes and ecosystems. Observation results indicate that soil moisture of the active layer has obvious storage or release process accompany with annual freezing and thawing of the active layer.

The freezing and thawing of active layer of permafrost depend on the process of soil temperature and water heat transfer. As observed soil temperature and moisture at Wudaoliang station in the TP, the freeze–thaw process can be divided into 4 stages: Summer Thawing (ST), Autumn Freeze (AF), Winter Cooling (WC), and Spring Warming (SW) (Fig. 8.21). The water heat coupling process in the active layer is very complex during ST and AF, and the volume of soil moisture migration is high. There is little moisture transfer in the active layer during WC and SW, and the heat is mainly transmitted by conduction. The water-heat coupling processes in the

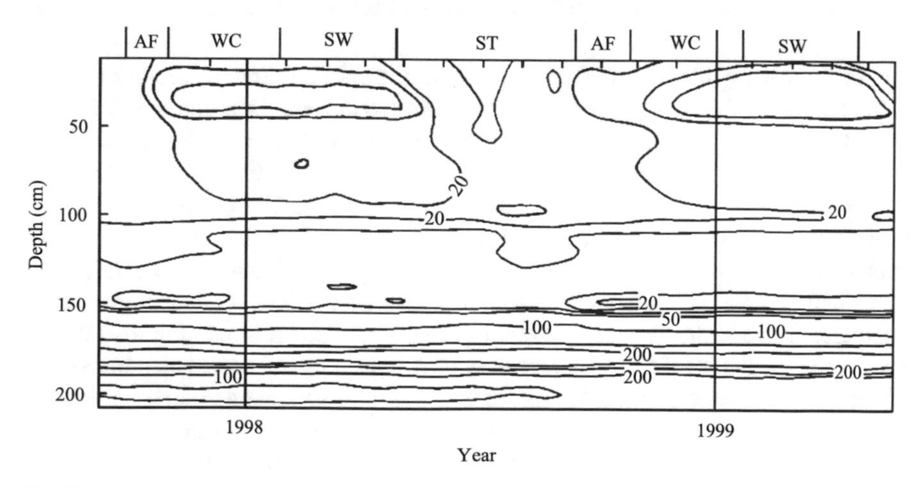

Fig. 8.21 Soil moisture contour in the active layer over permafrost at Wudaoliang station in the Tibetan Plateau

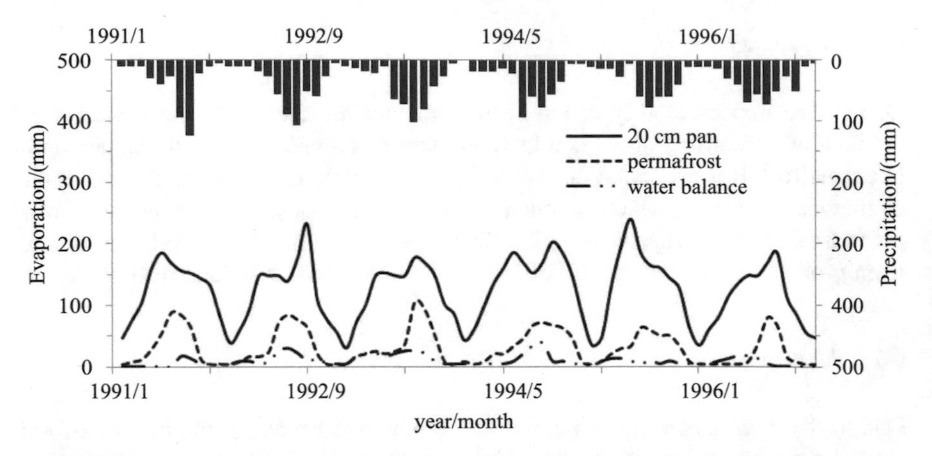

Fig. 8.22 Simulated evaporation with only water balance and with permafrost thawing and freezing process by VIC model in Maduo of Qinghai Province

active layer vary with the form of water transport at different stages of the freezing and thawing processes. The soil moisture at the upper limit of permafrost tends to increase after freezing and thawing, which is the main reason for the development of thick ground ice layer in the upper limit of permafrost.

The freezing and thawing processes release or absorb large amounts of latent heat, and affect the water and energy balance of permafrost regions. Simulations suggest that most of the energy is used for melting frozen ground in the summer, particularly in permafrost, leading to reduced energy for evaporation and transpiration. Therefore, frozen ground obviously suppresses evaporation, especially in summer, leading to an increase of runoff (Fig. 8.22).

(3) Influence of spatial distribution of permafrost on river runoff

In permafrost regions, most of the snowmelt and rainfall directly becomes runoff due to impermeability of permafrost, which leads to the larger coefficients of surface runoff. There is less groundwater recharge to winter river runoff, and even no runoff during some winters. Observation data of main rivers in the Arctic region and upper reaches of the Yellow River and Yangtze River in China suggest a significant correlation between the proportion of minimum and maximum monthly runoff and permafrost coverage (Fig. 8.23). The proportional contribution to the total runoff has small changes in a lower (<40%) coverage of permafrost, while it significantly increases in the higher (>60%). These results indicate that future permafrost degradation caused by global warming probably have important influences on the seasonal distribution of runoff in the basins with high permafrost coverage. In other words, permafrost degradation will lead to a large change in seasonal runoff distribution in high permafrost coverage basins but will have less impact on runoff in lower permafrost coverage basins.

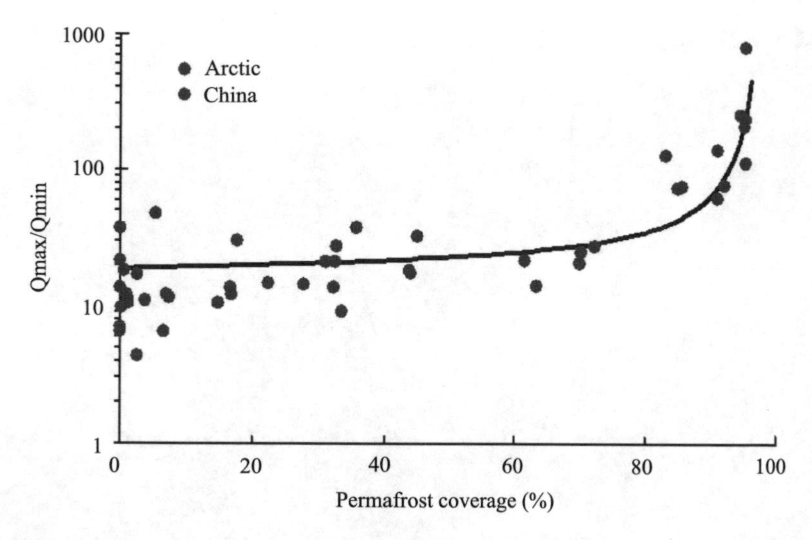

Fig. 8.23 The relationship between permafrost coverage and proportion of minimum and maximum monthly runoff in Arctic rivers, Yellow River, and Yangtze River

5. Hydrological modelling in the cryospheric basin

As mentioned above, the basic objective of cryospheric hydrology research is to scientifically understand the role of each cryospheric component in the whole basin. The hydrological model, which consider each individual cryospheric component, is an important mean for understanding the cryospheric hydrology. The characteristics of hydrothermal properties of ice, snow, and frozen ground and their roles in river runoff generation are key issues in hydrological studies in cold regions. The most recent hydrological models have not fully used for frozen ground and glaciers. Thus, to develop the distributed hydrological models which involve process of glacier changes, snow accumulation and melting, and freezing–thawing of frozen ground, are crucial for improving cryosphere hydrological models.

In order to modify the hydrological model to consider the function of cryosphere, the glacier components were introduced in the VIC (Variable Infiltration Capacity) model as an example, which is a grid-based spatially distributed hydrological model. The original VIC model considers the processes of snow and frozen ground but does not include glacier. Therefore, a scheme which consider glaciers as a special land cover type in a grid (Fig. 8.24) was developed and applied if there are glaciers in the grid, and calculated glacier ablation and runoff were calculated, and it then coupled with the original VIC model.

Simulations by the modified VIC model at the Aksu River basin in China (Fig. 8.25) revealed significant improvements (R^2) compared to the scheme that does not consider glacier runoff (old R^2). Therefore, in the cryosphere basin, the hydrologic simulation must be taken into account the influences of different cryosphere components.

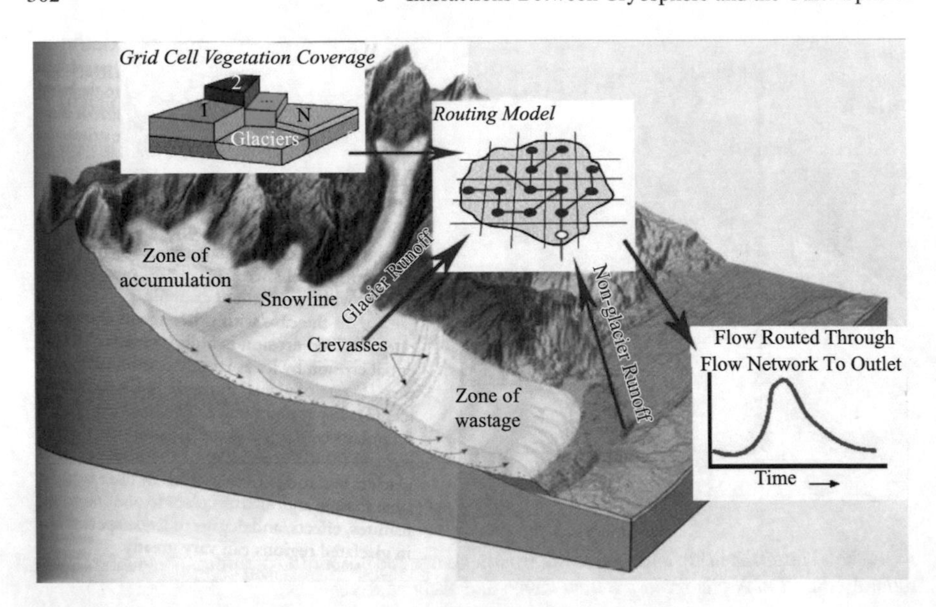

Fig. 8.24 Scheme of the glacier modulus coupled with the VIC model

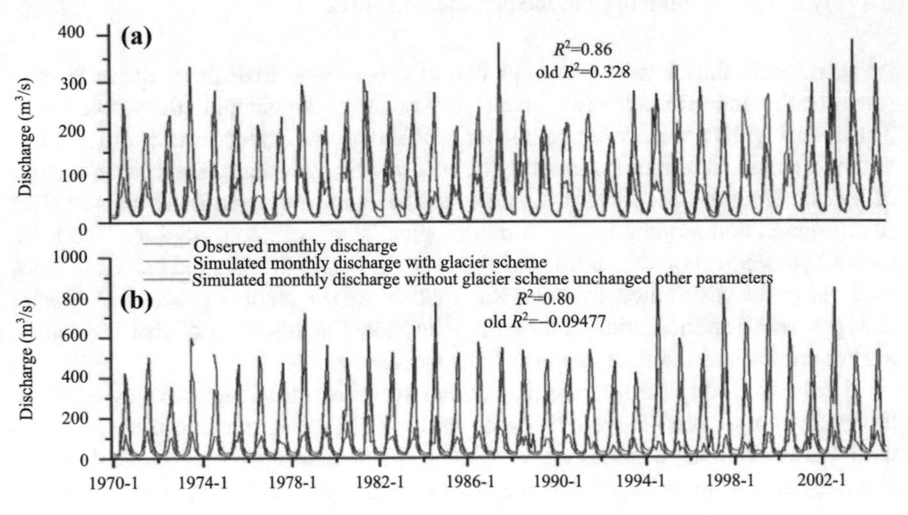

Fig. 8.25 Simulated and observation monthly discharge at **a** Toxkan river, **b** Kunma Like river in Aksu River Basin

8.4 Cryosphere and Lithosphere

The global pattern and regional difference in lithological structures can be determined by Earth's internal dynamics. The cryosphere is the combinations of complex physical geographical elements under the control of Earth's exterior, which has a global distribution pattern. Another way of looking at the interaction between cryosphere and Lithosphere is as a coupling between climatic and tectonic processes. The lithosphere's impact on cryosphere is its influence on glacier and permafrost (periglacial) systems. Therefore, this section focuses on the lithosphere roles in glacial and periglacial erosion, transportation, and deposition.

8.4.1 Glacial Erosion, Transportation, and Deposition

1. Glacial erosion

1) Glacial erosion processes

There are two processes in glacial erosion: ploughing and grinding. Glaciers exert enormous pressure on their underlying bedrock, sometimes, the pressure is great enough to break up the bedrock. The repeated freezing and thawing processes in the bedrock's joints can also fragment the bedrock. As the glacier flows downward, it carries away the bedrock debris. This whole process is called ploughing. The debris produced by the ploughing process are relatively coarse and contain gravels, pebbles, boulders, etc. Direct erosion of glaciers tends to occur principally at the interfaces between the ice and rock. As the glacier flows, rocks broke up by weathering, avalanches, meltwater, gravity, and wind. This debris are materials of moraine complex when they enter into the glaciers. The debris at the bases of glaciers act as a tool that shaves, grind, and polishes the bedrock when they pass over the bedrock, and this process is called grinding. The debris, produced by grinding, are usually in finely-powdered sand and clay forms, and sometimes as course as sand and gravel.

2) Glacial erosional landforms

Large-scale glacial erosional landforms include horns (peaks), arêtes, cirques, and troughs (U-shaped valley) (Fig. 8.26). As global sea level rise, it causes submergence of glacier canyon in the sea (which are called fjords), such as those along the coasts of the Scandinavian peninsula.

Glacial grinding occurs at the interface between the glacier base and the under bedrock, which forms glacier pavement or polish surfaces with striae, chute and crescent-shaped crevasses. Glacial striations are usually several millimeters to several centimeters wide, several millimeters deep, and several meters long and extend horizontally in the direction of glacier flows (Fig. 8.27c). There are different kinds of plastic deformed gouges and cracks on the polish surface. Rôche moutonnées are also a common geomorphological feature of glacial erosional bedrocks (Fig. 8.27d). Potholes are produced when glacier meltwater grind the bedrock under the glacier. All of these landforms are small-scale glacial erosional landforms. Regions covered by ice sheet or ice cap are characterized by erosional plains and/or erosional plateaus,

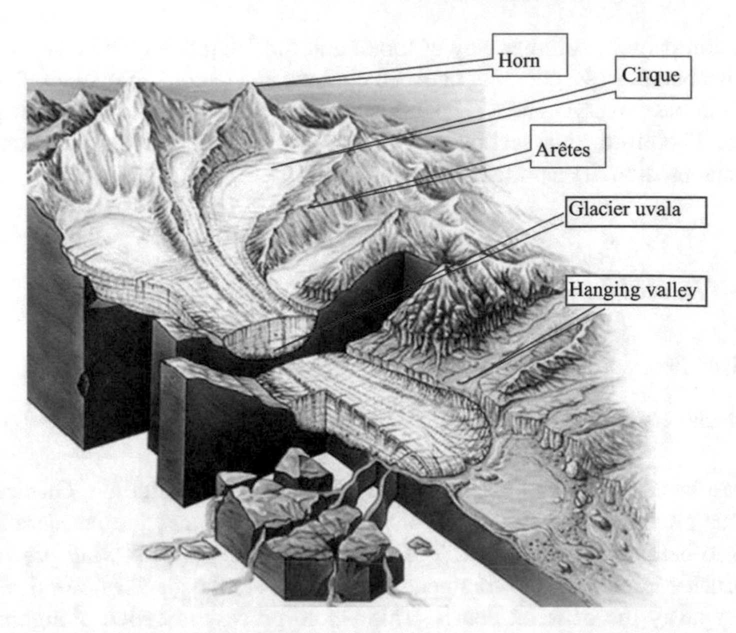

Fig. 8.26 The landscapes produced by glacial erosion process

e.g., the Great Eastern Plains of Canada in North America is produced by the Laurentide Ice Sheet, and the Haizishan Plateau in western Sichuan, China is produced by the Daocheng Ice Cap.

2. Glacial transportation

Glaciers have huge transportation abilities to carry massive glacial erratic boulders over long distances, up to several hundreds of kilometers. A mass of glacially eroded materials can be transported by moving glaciers, and the tills form moraines when deposited. The debris deposited along two sides of a glacier form lateral moraines. When two branches of a glacier converge, their lateral moraines form medial moraines. Englacial till refers to debris in the glacier and the basal till refers to debris at the base of glaciers. Debris that fall on a glacier surface from the sidewalls of a cirque or valley form superglacial moraines. The inner moraines, which ablate and emerge at a glacier's terminus, can also be termed as another type of superglacial moraine. When the above-mentioned moraines reach a glacier's terminus, they form terminal moraines.

3. Glacial deposits

A glacier's transportation abilities may decrease as a result of ablation or overloading of glacier's tills, leading to the dump of debris transported by the glacier. This process is called glacial deposit.

Fig. 8.27 Landscapes formed by glacial erosion. **a** Peak, arête, cirque. **b** U-shaped valley. **c** Glacial polish surface. **d** Rôche moutonnée

1) Glacial depositional landforms.

The glacial deposit landforms can be subdivided into two broad categories: mountain glacier landform and ice sheet landform. Another commonly formed glacial deposit landform is the end moraines, which deposits at a glacier's terminus. If the glacier retreats intermittently, it can produce a series of end moraines. The debris scattered between different terminal moraines are called hummocky moraines. Lateral moraines are always formed along both sides of a mountain glacier (Fig. 8.28a). If the ice sheet flow encounters topographical obstacles beneath them, the ground tills are prevented from being transported further and produce drumlins. Drumlins are commonly composed by mud gravel with rôche moutonnée-type bedrocks as their cores. In other words, drumlins are the product of both glacial erosional and depositional processes. Landforms produced by an ice sheet include eskers, ground moraines, flutes, etc. (Fig. 8.28b).

The deposited debris carried by glaciers and transported by meltwater are called glaciofluvial deposits (Fig. 8.28c). Glaciofluvial deposits are characterized by a mixture of glaciation, such as boulders with polish surfaces and striations, and sorted, rounded, and formed layers under fluvial.

Fig. 8.28 Types of glacial deposit landforms. **a** Moraine complex of a mountain glacier. **b** A diagram of an ice sheet deposits. **c** Glaciofluvial fan. **d** Ice-raft deposit. **e** Supraglacial and subglacial melt-out tills. **f** Varve of a glacial lake

2) The types of glacial deposits

The glacial deposits include lodgment till, melt-out till, deformation till, and glaciofluvial deposits etc.

Lodgment till: Lodgment till forms at a glacier base with high pressure, where the movement of till is blocked by multiple ways and, therefore, comes to a lodgment. The long axis of gravel in till is usually parallel to the movement of the glacier, and the till's flat surfaces are parallel to the ice bed (Fig. 8.28d). The particles of lodgment till generally have a sub-round, striations, or ironstone shape, fine-grained matrixes to coarse-grained clasts size, and strong fabric and appear highly compacted with mainly local lithology.

Melt-out till: Melt-out till can be divided into subglacial and supraglacial melt-out till, which is composed of debris dumped by meltwater (Fig. 8.28e). The processes associated with subglacial melt-out till (ground moraines) are related to the stagnate ice. These tills probably deformed with the flow after accumulation. The characteristics of subglacial melt-out till are similar to that of the lodgment till, but its structure is relatively weak and poorly compacted. Supraglacial melt-out till (also known as ablation till) is the product of ablation from the top surface down to the base of the glacier. Particles of melt-out till have sub-angular shape with remarkable transportation character, coarse-grained with a certain degree of sorting and with relatively low content of sand and clay grains, weak consolidation, and complicated lithologies, often include the debris from both sides of the valley.

Deformation till: Subglacial deformation till (also known as deformation moraines and deformation ground till) is formed by the fragmentation and assimilation of glacier compression, shearing of bedrock, and/or moraines at the base of a glacier. Deformation till has similar characteristics of shape, size, and rock property with that of subglacial melt-out till, sometimes strong but usually weak and tightly compacted. The deformation till mostly occurs beneath the ice caps.

Glaciofluvial deposits can be divided into three different types: pre-glacial type, ice-contact type, and marine type.

(1) The pre-glaciofluvial deposit occurs beyond a glacier and is deposited by meltwater, which includes glacial lakes, glaciofluvial fans, and outwash plains. Kames and glaciofluvial terraces can be found on both sides of the valley.

Glacial lakes: sometimes, the terminal moraines could block the meltwater and form glacial lakes. In the warmer spring and summer seasons, meltwater transports abundant amount of sediments into these glacial lakes, gravel and coarse sands deposit at the lakefront. Fine-grained sand, silt, and clay are carried into the center of the lake as suspended sediment load. The fine-grained sands and coarse-grained silts are relatively easily depositable, whereas the clay can be suspended for a long period of time. In the autumn and winter seasons, in the frozen lakes, no sediments load transported to the glacial lakes, therefore, clay is continuously deposited on the top of earlier-deposited sandy layers, which causes the formation of varve. The varve has an alternate layer with coarse and fine grains. The coarse grains layers are of light colour, whereas, the fine grains layers are of dark colour (Fig. 8.28). The granularity of varve vary from coarse to fine over a year, the rate and age of accumulation can be determined from the varve strata.

When overloaded meltwater enters a glacial lake, it causes the accumulation of sediments formation of deltas. In a profile, bottomsets, foreset and topsets could be observed. Commonly, the glaciofluvial fans look like fan-shaped and formed beyond the terminal moraines. When several glaciofluvial fans converge, it formed an outwash plain with a gentle slope.

Fig. 8.29 **a** Ice wedge. **b** Ice wedge mold. **c** Pingo. **d** rock glacier

(2) Ice-contact deposits: commonly, cross-bedding or horizontal bedding in non-stratified deposits could be found in ice-contact deposits, including eskers and kames. Eskers are long and curved embankments formed by glaciofluvial deposits. The length of an esker vary from several kilometers to several hundreds of kilometers, sometimes, it has tributaries. The direction of an esker could be used to indicate the glacier flow direction. Kame has hummocky topography, single kame or several kames in a group could be observed in the glacial areas. They originate from the sediments of supraglacial lakes. When the glaciers melt and the sediments deposited on the bedrocks, a kame or a group of them will be formed. The sediments have a concentric lamellar structure. The kame has a sorted and horizontal bedding of gravel and sand, therefore, the bedding of a kame is different from that of the original supraglacial lakes.

(3) Glacial marine deposits: A kind of sediments is formed by the interactions between glacier and ocean in the marine environment. They could provide important information about the evolutionary history of an ice sheet and paleoclimate variation. There are many kinds of sediments in the glacial marine deposits. The development of a glacial marine deposit is constrained by the interaction between ocean and ice sheet, where dominant factors are the discharge of meltwater, the stability of a tide glacier, the melting speed of icebergs, and marine biological processes. Marine glacial deposits are also influenced by the debris composition, the extent of sea ice, and intensity of ocean currents, gravity, mass movements, turbidity currents, and other depositional phenomena. The ice-raft debris and drop-stones are two very important evidences of glacial marine deposits.

8.4.2 Permafrost and Surface Layers of Lithosphere

The freeze–thaw is the main process in the periglacial environment, where freeze–thaw of soil moisture is of critical importance, and frost weathering, stream, wind, gravity, ice, and snow etc. can also affected. It is one of the most fundamental Earth's exterior dynamic processes on the surface of Lithosphere. On one hand, the processes of weathering, erosion, transportation, and deposition in periglacial environments contribute to the recycling of materials from Earth's surface. On the other hand, permafrost constitutes a unique hydrothermal "sub-sphere" on the surface of Lithosphere, which influences the composition and spatial patterns of Earth's exterior dynamic processes. In terms of area, the influential extent of permafrost is beyond that of the glacial "sub-sphere." In other words, the periglacial areas are not just part of the cryosphere, the periglacial processes are one of forces that shapes the geomorphology of cold regions.

The main periglacial processes and geomorphological landforms include the ice wedges, pingos, and rock glaciers associated with permafrost, as well as the patterned ground, peat mounds, and seasonal ice mound with the active layer over permafrost.

1. Periglacial processes and landforms related to permafrost

Ice-wedge appears as a polygon in surface geometry, and as wedge-shaped in profile geometry and in downward spike (Fig. 8.29a). Ice wedges are formed during the winter when the frigid temperature causes the rock surface to form a narrow gap, then these gaps are filled with water in summer. The veins of ice are formed in the early stages of winter, establishing a basis for ice wedges. When the soil temperature falls below 0 °C, new frigid-induced gaps are produced. As this process repeats, ice wedges are formed. The increase in the temperature degradation in the permafrost cause the thawing of the ice wedges and filling of the original ice-occupied wedges with sand grains, forming "ice wedge cast", which is also called "sand wedge cast" or "sand wedge mold" (Fig. 8.29b). Ice wedge cast are reliable indicators of the existence of paleo-permafrost, therefore, they are being widely used in both paleoenvironmental and paleoclimatic reconstructions.

Pingos are ice mounds that developed in permafrost regions and appear as hilly, bulge-shaped formations. They have ice cores formed from either invasive or coagulated ice and can be classified into annual and perennial types. Annual pingos are relatively small, several tens of centimeters to over 1 m high. Perennial pingos form downward can even reach permafrost and can be with relatively large size, like those found on the mountain passes of the Kunlun Mountains in China, which are 20 m high, 75 m long, and 35 m wide (Fig. 8.29c). Pingos can be classified into hydrostatic closed-system pingo and hydraulic open-system pingo by their hydrological characteristics. hydrostatic closed-system pingos are formed by the coagulation of ground ice within permafrost. In continuous permafrost zones, pingos are well developed in alluvial and lacustrine alluvial plains which have rich groundwater. Hydraulic open-system pingos formed by invasive ice, as pingos observed in the Kunlun mountain passes, where the groundwater seeps into the permafrost along cracks.

In permafrost regions, the mixtures of ice and rock creep slowly along valleys or slopes are called rock glaciers, which blocky moved as a whole. Most rock glaciers are composed of frozen gravels, it flows similar as glaciers. The surface speed of rock glacier may be as slow as 1–2 m a^{-1}, which is less than that of a glacier. Unique ditches and ridges resulting from its movement and the internal thawing of ice are found on a rock glacier's surface. Debris of rock glaciers are classified into two types, till and weathered talus, thus rock glaciers can be divided into till-type (Fig. 8.29d) or talus-type rock glaciers. Rock glaciers can also be categorized as tongue-shaped rock glacier and lobate-shaped rock glaciers.

2. Periglacial processes and landforms related to surface layer of lithosphere

Patterned grounds are typical landforms in periglacial environments, which always found under cold climates. The clearly sorted patterned grounds are generally located

(a) (b)

Fig. 8.30 Patterned ground. **a** Pattern ground at ice-free cirque in the source of the Urumqi River in the Tianshan Mountains. **b** Measurement of the two years' frost heave at the center of a stone cirque

in cold climate zones where the air temperature often falls below freezing point. Frost weathering and ground ice are two basic factors which control the formation of patterned ground. Patterned ground appears in multiple morphologies and it can be displayed as active and relict morphology. The relict patterned ground can be used to interpret paleoclimates, and their morphology and degree of sorting are the most important distinguishing features. Their most important morphology is stone cirques, which can appear individually or in groups. Sorted stone cirques are often present as a border of coarse gravel, surrounding a center composed of finer materials (Fig. 8.30a). Stone cirques can also form in polar and alpine areas; therefore, they are not limited to permafrost regions. Unsorted stone cirques can even appear in unfrozen environments, such as in arid areas of Australia. It has been pointed out that patterned ground formations can have many different causes, the primary processes include sorting due to frost heaving, and cracking accompanying dehydration and fluvial (Fig. 8.30b). The process, morphology, and size of patterned grounds are also influenced by the number of freeze–thaw cycles, soil texture, soil moisture, and topography.

Block stripes appear as alternatives of coarse and fine gravels and extend downhill. In coarse stripes, which are usually 20–35 cm wide, the coarser material collects in grooves with 7.5 cm depth, and finer material forms little ridges. The maximum course stripes may be as wide as 1.5 m and extend more than 100 m downslope, whereas their depth depends on their size.

3. Periglacial processes and landforms related to slope processes

Gelifluction (or solifluction) is a blocky movement of periglacial environments. It occurs on the surfaces of slopes covered by weathered debris (Fig. 8.31a). The conditions which are advantageous to gelifluction include: ① weathered debris which are disturbed in a number of ways by the freeze–thaw process; ② meltwater from snow, ice, or ground ice causes the debris to become wet and, thus, facilitates their movement; ③ the underlying frozen surface prevents any water in the thawing surface from migrating downward; and ④ there is limited vegetation cover, which allows a blocky

Fig. 8.31 **a** Gelifluction lobe. **b** Stratiform colluvium

downslope movement to occur even on gentle slopes. Gelifluction (or solifluction) process include gravity, frost creep, cryoturbation, needle ice, etc.

The frequent cycles of freeze–thaw supply frost-weathered debris. The stream slope then modifies the supplied debris which eventually accumulate at foot of the slope to form stratified sedimentary material, known as stratified slope deposits (Fig. 8.31b). This phenomenon is distributed both in modern or paleo-periglacial environments. Owing to the unique topographical and climatic conditions under which such deposits forms, it can be used in paleoenvironmental reconstructions. Chinese scientists early noticed this phenomenon and called it "stratified chippings."

Due to the frost weathering, gravity, stream, block slopes, block streams, and talus scree are always formed on slopes in a periglacial environment.

4. Frost weathering and cryoplanation

Frost weathering is dominated by physical weathering, and chemical weathering is comparatively weak. Frost weathering provide large amounts of raw materials for other periglacial phenomena, such as block fields, block streams, and talus scree. The debris which accumulate on the horizontal or gently slope are termed as block fields, felsenmeers, or blockmeeres. The formation processes of block fields are principally related to freezing–thawing condition. Block fields are mainly composed of local lithology, sometimes contain rocks transported by glaciers over quite long distances. Cryoplanation is formed by the combination of frost weathering, nivation, and gelifluction under the influence of stream and wind. Cryoplanation present as flat ground or stepped terrain in ridges or high platforms. The residual outcrops of bedrock are called periglacial rock columns (tor).

5. Periglacial aeolian

Periglacial aeolian forms widely-distributed periglacial aeolian erosion/deposit landforms. Aeolian produces blowout, niches, grooves, ventifacts, stony, gravelly deserts, and other landforms. Aeolian deposits give rise to sandy lands, sand dunes, and periglacial loess deposits. Every type of periglacial aeolian erosion/deposit landforms is widely distributed in the foothills of the Altai, Tianshan, Qilian, and Kunlun mountain ranges, as well as on the TP in China.

6. Thermokarst landform

The landforms of sinking, collapse, and erosion formed by permafrost thermokarst are collectively called thermokarst landforms. Thawing of the active layer over permafrost, as well as the erosion of sea coasts and shorelines resulting from the mechanical and thermal effects of sea wave, can cause shorelines to retreat. If there are rivers in permafrost regions, the mechanical and thermal effects produced by fluvial can lead to fluvial incision and riverbank retreat. With global warming, the temperatures of permafrost slowly increased, leading to intense thermokarst. Swamps have appeared and expanded, and thermally undercutting of swamps become

Fig. 8.32 Thermokarst. **a** Swamp at Aiken Daban, Tianshan Mountains. **b** Thermokarst collapse on a road through a Bayanhar Mountain pass, Qinghai Province

more powerful (Fig. 8.32a). The impact of this on the engineering construction and ecological environment of cold regions is becoming increasingly evident (Fig. 8.32b).

The thermokarst landforms of China are distributed principally in the permafrost regions of the TP. It varied greatly in different sub-regions of TP due to topography. Such as the areas surrounding the Tibetan Highway, the proportion of number and surface area of swamps in the Fenghuoshan Mountain region to the whole highway are relatively low, which mainly because the low-temperature permafrost in this high altitude area are not conducive to the formation of swamps. The Chumar River High Plains region is most densely populated with swamps on TP, followed by Beiluhe Basin as the second-most populated. Global climate change, permafrost degradation, and ecological environment deterioration, both have significant impacts on the formation and development of thermokarst landform. In turn, the development of thermokarst landform can lead to changes in permafrost environments, thereby influencing changes in the ecological environment and global climate.

Questions

1. What are main interactions between the cryosphere and the other spheres?
2. What are the impacts of cryosphere on climate? Explain applying an example.
3. Please explain hydrological roles of the cryosphere in details.
4. Please explain how frozen grounds may have impacts on the organisms and physical environment in details.

Extended Readings

1. *Permafrost hydrology*

Authors: Ming-ko W.

Publisher: Springer, 2012

Hydrology is by nature both scientific and applied, as is permafrost investigation. The development of permafrost hydrology benefits from the progress in other disciplines; among them are the atmospheric and climatic sciences, hydrogeology and pedology, geotechnical and environmental engineering, biology and forest sciences, periglacial, fluvial and glacial geomorphology. The interdisciplinary flavour of permafrost hydrology adds to its scientific strength and practical merit while its relevance to these other disciplines is reciprocal. The permafrost domain still encompasses many scientifically uncharted territories with innumerable hydrologic features yet to be discerned, many processes to be understood and pertinent new concepts to evolve. The excitement of discovery will continue to entice future investigators.

This book provides a survey of the status of progress. Through this book, I wish to share my experiences with professional and non-professional but interested readers, be they practitioners, researchers or students. The materials are presented in sufficient detail for the instruction of young permafrost hydrologists at a senior level, and broad enough to satisfy the needs of cross-disciplinary researchers and practitioners who can make use of the information without having to delve into the complexity of permafrost or hydrologic sciences. Emphasis is placed on discussion of permafrost and hydrologic processes with the premise that an understanding of the physical processes is fundamental to experimentation, theoretical and modeling work in permafrost hydrology.

2. *Geosystems: an introduction to physical geography*
Authors: Robert W. Christopherson
Publisher: Prentice Hall, 2011

Among the most highly regarded in physical geography, Robert Christopherson's best-selling texts are known for their meticulous attention to detail, currency, accuracy, and rich integration of climate change science. Geosystems: An Introduction to Physical Geography, Ninth Edition is uniquely organized to present Earth systems topics as they naturally occur: atmosphere, hydrosphere, lithosphere and biosphere. This interconnected and organic systems-based approach is highlighted in the strong pedagogical tools, structured learning path, and up-to-date information found in the text. This new edition presents bold new features that cultivate an active learning environment both in and outside the classroom. The Ninth Edition is available with MasteringGeography™, the most effective and widely used online tutorial, homework, and assessment system for the sciences. This program will provide an interactive and engaging learning experience for your students. Here's how: Personalize learning with Mastering Geography: Mastering Geography provides students with engaging and interactive experiences that coach them through introductory physical geography with specific wrong-answer feedback, hints, and a wide variety of educationally effective content. Teach with current and relevant content. An emphasis on currency includes a new chapter on global climate change and provides students and instructors with the most significant and current information and applications for learning physical geography. Leverage strong pedagogical tools and a structured active learning path: The text reinforces central hallmark physical geography themes of Earth systems, human-Earth relations, and global climate change by providing a consistent framework for mastering chapter concepts.

Chapter 9
Cryosphere Change Impact, Adaptation and Sustainable Development

Lead Authors: Yuanming Lai, Shiyin Liu, Xiaoming Wang, Yongjian Ding

Contributing Authors: Jianping Yang, Fujun Niu, Zhijun Li, Lijuan Ma, Yiping Fang, Ninglian Wang, Shijin Wang, Jianming Zhang, Guoyu Li, Jiahong Wen, Cunde Xiao

Changes in the cryosphere impact on social and economic development at a global, regional or local scale. On the global scale, the expansion or shrinkage of the cryosphere affect water cycles in hydrosphere, including sea level changes and extreme weather events. On the regional scale, they have profound impacts on water resources, coastal environment and closely associated economy as well. On the local scale, changes in individual components of the cryosphere lead to water supply variability, floods, ice and snow disasters, causing interruptions to communities and damage to infrastructure. This chapter introduces the cryospheric hazards in relevance to cryosphere changes, as well as their consequent impacts on societies. It also intends to provide principles and approaches for adapting to the cryosphere changes, with an intention to meet the sustainable development goals under changing environment.

9.1 General Concepts for Cryospheric Effects and Sustainable Development

9.1.1 Impact and Service

The cryosphere has functions capable of provisioning water supply, regulating climate, supporting ecosystem, and serving the cultural needs, from which human being eventually benefits. At the same time, it may bring glacier lake outburst, avalanche, and snowstorm etc. that lead to adverse impacts on societies. Therefore, the cryosphere has both positive and negative effect on our societies, as shown in Fig. 9.1. In interaction with the other spheres, it generates adverse hazards but also beneficial stocks, causing risks and providing services to societies at the same time. Changes in the cryosphere may either diminish or enhance the effects, resulting in changes of risks and services. In particular, the reduction in services may add more risks when demand would not be met.

D. Qin et al. (eds.), *Introduction to Cryospheric Science*, Springer Geography,
https://doi.org/10.1007/978-981-16-6425-0_9

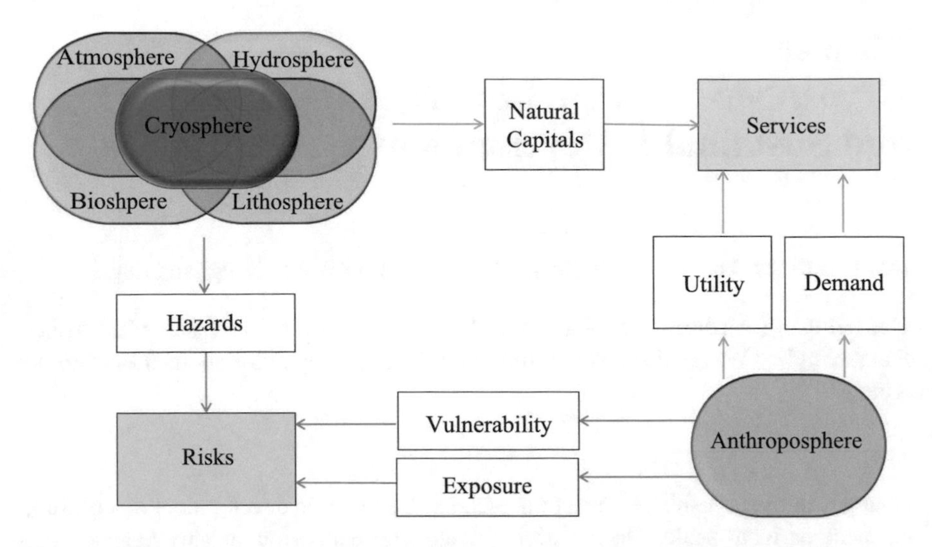

Fig. 9.1 Effects of the cryosphere interacted with the other spheres on the anthroposphere

More specifically, the risk represents the potential impact of cryospheric hazards on a concerned object. The object can be assets, people, infrastructure, communities, environment, economy or broad society at all scales. Risk is generally expressed by multiplication of the consequence of a cryosphere change event and the likelihood of event occurrence, or simply described by

$$\text{Risk} = \text{Likelihood} \times \text{Consequence} \tag{9.1}$$

In more details related to the cryosphere, the event is usually considered as a hazard, such as glacier lake outburst flooding (GLOF), avalanches, snowstorms, freezing and thawing, frost, permafrost erosions, coastal storm surges and so on. The consequence is considered as a result of the impacts of hazards, often in monetary or non-monetary terms of damages, loss or any form of adverse effects incurred by the concerned object, generally on social, economic and environmental aspect. In practice, concerning the object such as natural and built assets or population, the consequence is determined by both likelihood of exposure of the object to hazards as well as its vulnerability to the hazard. In this regard, the risk can also be described by

$$\text{Risk} = \text{Likelihood of Hazard} \times \text{Likelihood of Exposure} \times \text{Vulnerability} \tag{9.2}$$

In general, risk is the combined effect of hazards, exposures, and vulnerability of the concerned object, as shown in Fig. 9.1.

Vulnerability is deemed here to be the susceptibility of a concerned object and measured as a likely loss or any adverse effect in correspondence to a given degree of a hazard. For either built or natural assets, vulnerability can be described by the loss of functionality, serviceability or/and integrity given a exposure to hazards, for example, infrastructure damage loss in association with GLOF. It is often represented in a monetary term, although other measures may be used. Vulnerability assessment is the key step to understand the performance of concerned objects subject to cryosphere changes.

The degree of vulnerability is closely related to the capacity. The capacity is considered to be an inherent property and the ability to withstand or accommodate expected (future) adverse hazard impacts without loss of its functionality, serviceability or integrity. For example, the ability of road pavement to resist freeze–thaw cycles without damage.

In terms of the service as described in Fig. 9.1, in parallel to the risk, the service represents the potential benefit that human-being can obtain from the cryosphere. The benefit can be results of provisioning, regulating, supporting and culture provided by the cryosphere, as shown in Fig. 9.2.

A degree of service can be generally expressed by multiplication of the likelihood of available cryospheric natural capitals (or cryospheric capitals), the likelihood of societies' demand for the natural capitals, and the service utility that can be described in a monetary term. In this regard, the degree of the service can be represented by

$$\text{Service} = \text{Likelihood of cryospheric capital} \times \text{Likelihood of Demand} \times \text{Utility} \tag{9.3}$$

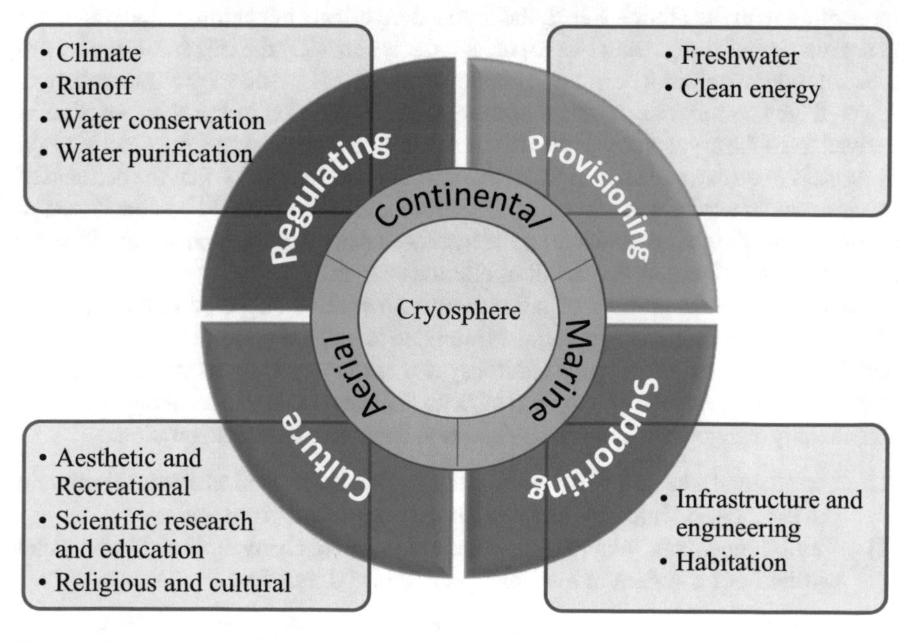

Fig. 9.2 Cryosphere service

The cryospheric capital is a term for the stock of renewable and non-renewable natural resources in the cryosphere, particularly relevant to snow, frozen ground and glaciers, which yields the service benefiting to societies.

The likelihood of demand, in another term, can be deemed as the likelihood of cryospheric capitals to meet societies' demand.

The utility measures the usefulness of cryospheric capitals, which societies derived from the service.

Estimation of the risk in Eqs. (9.1), (9.2) and (9.3) can be carried out by either qualitative or quantitative approaches. Quantitatively, the risk is often expressed as an expected loss considering uncertainties, with a reference to impact consequences of the hazards at categorically defined severity, such as a range from very frequent to very rare events. The impact consequence is normally related to economic loss, but it could also be described as broader socioeconomic and environmental loss at a scale from local to global level. On the contrary, the service can be qualitatively measured with a reference to the availability level of cryospheric capitals and the significance of utilities when societies rely on the capitals. Quantitatively, they can be estimated based on probability theory.

9.1.2 Adaptation to Changes

Interacting with the other spheres, the cryosphere affects the anthroposphere, either generating benefit or causing loss to societies. It is generally the task for researches to seek solutions to minimize risks and maximize services. Considering climate change, it would create more uncertainties of the effects. In addition to the changes of risks derived from the cryospheric hazards as a result of climate change, the deterioration in services may also add more risks as there would be a likelihood that the demand of services would not be met any more. Therefore, adaptation should be considered to accommodate the change and reduce the increased risk resulting from hazard impacts and service deterioration as a result of climate change.

Adaptation is the process of adjustment to actual or expected climate and its effects. In human systems, adaptation is intended to moderate, avoid harm or exploits beneficial opportunities. In some natural systems, adaptation by human intervention may facilitate adjustment to expected climate change and its effects. Adaptation can be generally categorized into incremental and transformational adaptation:

(1) Incremental adaptation: Adaptation actions where the central aim is to maintain the essence and integrity of a system or process at a given scale.
(2) Transformational adaptation: Adaptation that changes the fundamental attributes of a system in response to climate and its effects.

Adaptation can also be classified as autonomous and planned adaptation. The autonomous adaptation can happen in natural systems represented by ecological changes in response to environment change. The planned adaptation may include anticipatory, progressive and reactive approaches in terms of the timing of adaptation. Anticipatory or preventive adaptation acts before any expected change occurs; Progressive adaptation acts gradually in response to any expected occurrence of environmental changes; Reactive or corrective adaptation acts depending on the observation of any environmental change.

For adaptation in the cryosphere with its intention of moderating and avoiding harm or exploiting opportunities in mind, pathways to develop the adaptation to changes can be approached by reducing vulnerability and exposure of societies to hazards, as shown in Fig. 9.3. Though hazards are deemed to be natural phenomena that is hardly regulated, human-being may avoid deteriorating the environment such as carbon emission to limit the worsening of hazards in terms their frequency, intensity and duration.

To exploit more opportunities in align with the service provided by the cryosphere under climate change, potentially increased natural capital available to utilize would revitalize the service on one hand. On another hand, pathways to improve utility through effectiveness and efficiency while increasing demand may considerably enhance the service, as shown in Fig. 9.4.

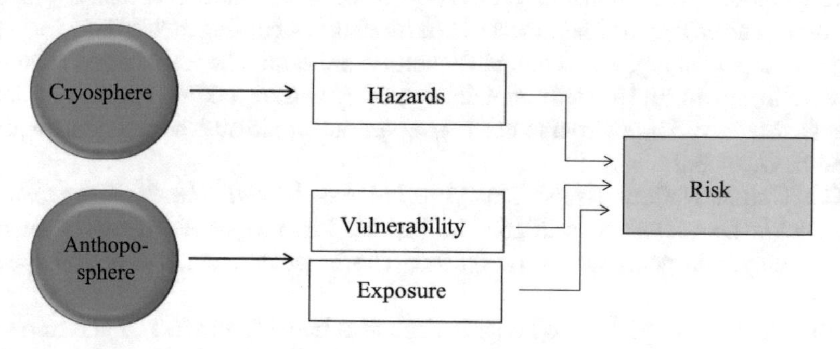

Fig. 9.3 A schematic description of risk in the cryosphere

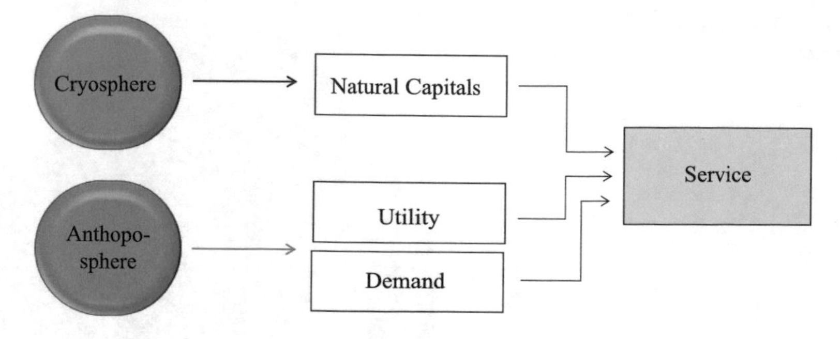

Fig. 9.4 A schematic description of service in the cryosphere

9.1.3 Sustainable Development

The question is how the service that benefit to the societies could be maintained for long-term prosperity, especially considering the threats from changes of cryospheric hazards. This leads to sustainable development.

The concept of sustainable development was initially raised in the Brundtland report (1987), which is defined as the development that meets the needs of the present without compromising the ability of future generations to meet their own needs. The sustainability is an attribute of a supply-consumption process that can be maintained over time in a system.

On the ecological term, the sustainability is defined as the way that human interacts with the biosphere to maintain its life support function on the aspect of biological diversity, ecosystem conservation and regional interconnectedness.

On the economic term, it is defined as the way that human manages their economy to preserve its productiveness and prosperity through efficiency, investment, diversification, and balance of internal demand and external supply now as well as in the future.

In general, the sustainability includes three key aspects, which are social, economic, and environmental aspects, in particular, concerning their balance now and future, as shown in Fig. 9.5. Economic development cannot be viable without considering environment, and hard to maintain equability without taking into account social aspects. Meanwhile, environmental conservation would not be bearable without social involvement.

The United Nations issued *Transforming our World: The 2030 Agenda for Sustainable Development* in 2015. The agenda has further identified 17 sustainable development goals shown in Fig. 9.6. The 17 goals include total 169 specific targets.

For examples, Goal 2 (Zero hunger) aims to achieve food security and improved nutrition, and to promote sustainable agriculture; Goal 6 (Water and sanitation)

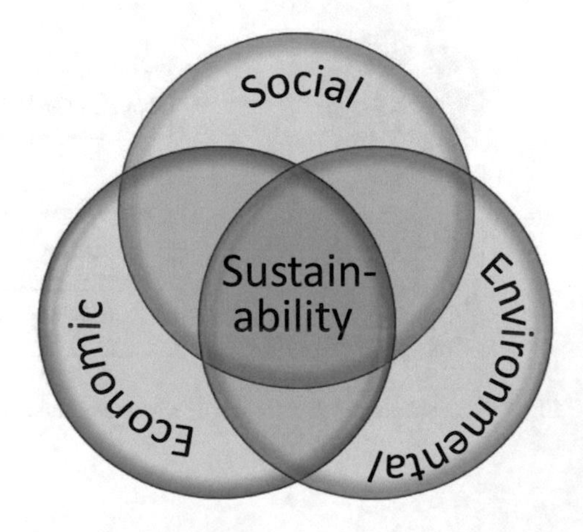

Fig. 9.5 Context of sustainability

Fig. 9.6 Sustainable Development Goals by 2030 (all logos sourced from www.un.org/sustainabledevelopment/development-agenda)

ensures availability and sustainable management of water and sanitation for all; Goal 9 (Infrastructure and industrialization) is to build resilient infrastructure, promote inclusive and sustainable industrialisation, and foster innovation; Goal 13 (Climate change) is to take urgent action to combat climate change and its impacts; Goal 14 (Oceans) is to conserve and sustainably use the oceans, seas, and marine resources for sustainable development; and Goal 15 (Biodiversity, forests and desertification) devotes to protect, restore, and promote sustainable use of terrestrial ecosystems, sustainably manage forests, combat desertification, halt and reverse land degradation, and halt biodiversity loss.

The cryosphere interacts with the other spheres including atmosphere, hydrosphere, lithosphere, biosphere and anthroposphere. It contributes to the well-being of societies as well as environmental conservation, directly or indirectly. Therefore, it closely contributes to, but not limited to, the sustainable development goals indicated in Fig. 9.7. Changes in the cryosphere may fundamentally affect the sustainable development. The changes can be shrinkage or retreat of the cryosphere such as glaciers and permafrost, potential sea-level rise caused by melting of ice sheets in the Arctic and the Antarctic, and other events closely associated with cryosphere changes such as redistribution of seasonal runoff, increased soil degradation by freezing and thawing cycles, the accelerated deterioration of soil mechanical properties required to sustain infrastructure and construction, and so on. Adaptation to the changes is prerequisite to maintain sustainable development.

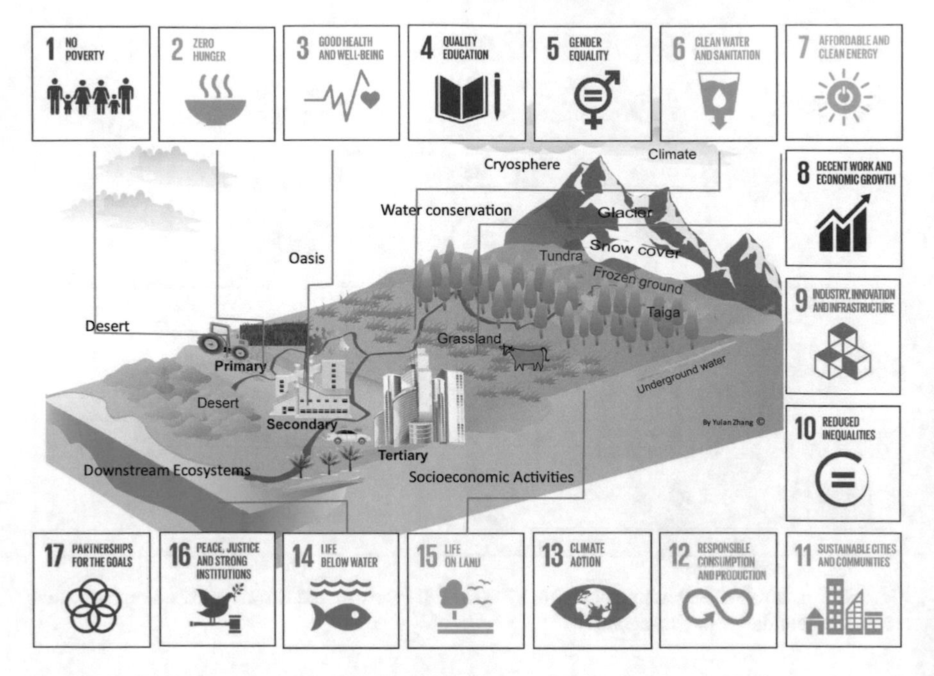

Fig. 9.7 Potential contributions of the cryosphere to sustainable development

9.2 Cryospheric Hazards and Disasters

There are various types of cryospheric hazards in association with cryospheric components. The hazards relevant to glaciers may include glacial lake outbursts, glacier flood and glacial debris flows. Those pertinent to snow are extreme snow cover, snowmelt floods, freezing rain and snow, avalanches, and snowdrift. Some linking to permafrost include frost heave, thaw or thermokarst settlement, creep deformation, and so on. Others also include river ice flooding, sea ice that block waterway, ports or wharfs, and any causing damages to societies such as engineering structures and aquaculture and so on.

Major cryospheric hazards in China are mainly distributed in the Qinghai–Tibet Plateau, Xinjiang, and northeast China regions (Fig. 9.8). More details about cryospheric hazards are to be described below.

9.2.1 Glacier Hazards

Disasters can be caused by glacier hazards such as glacial lake outbursts, glacial floods, and glacial debris flows. The glacier lake outburst refers to massive drainage from a glacier lake or water drainage caused by sudden collapse of a moraine lake.

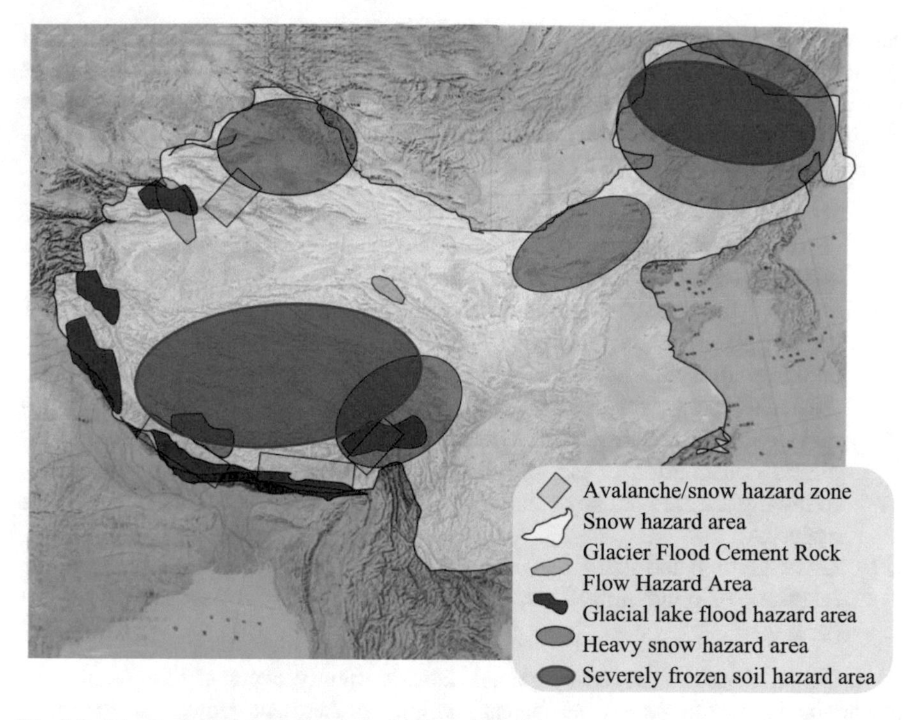

Fig. 9.8 Distribution of major cryospheric hazards in China

Moraine lakes generally refer to the lakes formed as a result of glacial meltwater accumulation due to the blockage caused by moraines, which came from the existing extra glaciers in the Little Ice Age (LIA). The size of moraine lakes varies with changes of glaciers, which are distributed extensively in China. The outburst of moraine lakes is a common cause leading to flood disasters. It can generally be classified into the following three types.

Glacier collapse or rockfall: Water surges or a sharp rise of water level is caused by collapse of steep glacier tongues of mother glaciers or significant rockfalls around a lake basin, subsequently impacting on terminal moraine and eventually causing collapses that lead to a sudden water drainage or floods.

Glacier lake overflow or piping dam break**:** The terminal moraine dams formed in the LIA has a high water permeability, often with water seepage at their bottoms. On one hand, an increase in glacier melting and a rise of lake surface result in overflow. On another hand, the piping is intensified or dead ice thaws and collapses beneath the moraine dam under the effect of hydrostatic pressure. All increase the risk of dam break.

Glacier lake: A lake occurs either by an advancing glacier that blocks a main waterway. For example, the Kyagar Glacial Lake was created by the Kyagar Glacier in the upstream region of the Yarkant River in the Karakoram Mountains that blocked the Shaksgam Valley. It also happens when a main glacier blocks a glacier-carved valley, which is evacuated by a glacier branch after its fast recession from the main glacier, for example, Merzbacher Lake in the upstream region of the Kunmalike River watershed of the Aksu River, Tianshan Mountains. In both circumstances, water is blocked and stored by glacier ice acting as a dam. It depends on the following factors for the occurrence of a sudden outburst of glacier lakes.

(1) Hydrostatic pressure: When the depth of a lake reaches 90% of the height of the ice dam, the ice dam is then forced by the hydrostatic pressure of the lake water, causing dam break and mass drainage.
(2) Drainage channels: When the thawing of ice dams intensifies and the water level of glacier lakes rises, hydraulic connections by drainage channels inside ice dams are established. As a result of hydraulic pressure and thermodynamics, glacier lakes are drained through the channels, and the channels are continuously enlarged causing even more rapid drainage. Meanwhile, influenced by plastic deformation of the glacial ice, the channel section may also shrink following reduced drainage until it is completely closed, eventually ending the drainage.
(3) Other factors: Ice dams may collapse or thaw due to earthquake, volcanic eruption, or geothermal heat, resulting in glacier lake outburst or sudden drainage.

In China, disasters caused by glacial hazards mainly occur at the precincts of mountains in the Qinghai–Tibet Plateau, which has frequent neotectonic activities and high topographical relief. The hazards related to moraine lake mainly occur in the middle Himalayas, the great canyon region of the Yalung-Zangbo River. The glacial lake outbursts occur most frequently in the Karakoram Mountains and western regions of the Tianshan Mountain, where mountains are high, valleys are deep, and glacier lakes lie above the headwater with steep slopes at their two sides of, mainly at altitudes of 4500–5200 m. A large glacier lake may cover an area up to 3–4 km^2, and a small one covers an area of 0.01 km^2.

9.2.2 *Snow Hazards*

The snow hazard is a natural phenomenon caused by large-scale snow cover after continuous heavy snowfalls. It often occurs in regions with stable snow cover and mountainous areas with unstable snow-cover. According to the characteristics in occurrence, the snow hazard can be divided into avalanche, snowdrift, snow hazard in pastoral areas, etc. Among them, snowdrift may lead to an avalanche and form blizzard conditions. Avalanche, snowdrift, and blizzard often cause snow disasters in pastoral areas. In general, different snow hazards interact with each other, often occurring frequently and coincidently, resulting in a chain action of disasters. In particular, heavy snowfalls, deep snow cover, long duration, and snowmelt flooding in spring may cause considerable threats on agriculture, livestock, regional transportation,

snow disaster in northern Gansu in March 2011

snow disaster in Inner Mongolia in January 2010

snow disaster in Europe in November 2010

snow disaster in mid-east in December 2008

Fig. 9.9 Snow disaster events

communication, road infrastructure, electric transmission lines, potentially causing impacts on sustainable economic and social development in the region (Fig. 9.9).

Snow disasters in pastoral areas mainly refer to the loss of livestock due to food shortage caused by heavy snowfalls and deep snow covers with a long duration. The degree of snow disasters in pastoral areas not only depends on snow precipitation, air temperature, snow cover depth, snow duration (days), slope gradient and orientation, grassland type, pasture height, and other natural factors, but are also closely related to herd structure, feed reserves, emergency response funds for snow disasters, and the level of regional economic development. Snow disasters mostly occur in Altay, the Three-River Headwater region, the Naqu and Xilinguole Leagues in Western China, and large-scale pastoral areas in Mongolia.

Snowdrift disasters are caused by strong winds that transport snow. Snowdrift is one of cryospheric hazards that may threaten agriculture, livestock, transportation, industries and mining. It is also called a "snow-driving wind disaster". According to the height and strength of blowing snow grains and their effects on visibility, it can be divided into low snowdrift, high snowdrift, and blizzard. Snowdrift, as the source of alpine glaciers, polar ice sheets, and avalanches, can induces such as ice and snow floods, avalanches, debris flows, and landslides, leading to disasters and causing serious loss of economy, human life, and properties. Snowdrift is a relatively complex and unique form of fluid. Snowfalls and snow covers are the sources of snowdrift, and wind drives it. It can be classified into multi-year and seasonal one according to its occurrence duration. In China, extreme snowdrift hazards largely occur in northwest China, the Qinghai–Tibet Plateau and its mountainous precincts, Inner Mongolia, and the northeast mountainous regions and plains. They may lead to disasters with significant interruptions to air traffics and impacts on industrial, agricultural, and pastoral production.

Freezing rain and snow disasters refer to significant impacts on people and socioeconomic systems by low-temperature rain and snow in winter and spring. It is one of meteorological or cryospheric hazards that are developed in a freezing process. The climatic conditions leading to severe freezing rain and snow hazards include: ① simultaneous occurrence of snowfall, freezing rain, and rainfall, with freezing rain being the main cause of disasters; ② occurrence of low temperature, rainfall and snowfall with a climate for strong freezing rain; ③ occurrence of low-temperature, rainfall and snowfall with a long duration of freezing climate. Low-temperature freezing rain and snow disaster is a consequence of the interaction of multiple meteorological and non-meteorological effects in the same period and same region. Such disasters often occur in the economically developed regions in central and eastern China, causing considerable losses and damages in agriculture, forestry, transportation, power transmission, communication, and aviation.

Blizzard disasters are caused by extreme weather conditions with wind speed greater than 15 m/s, duration longer than 3 h, and visibility less than 400 m due to continuous snowfall or snowdrift. It is one of the most common snow disasters to societies. The occurrence of blizzard is often accompanied by strong wind and heavy snowfall, steep decline in temperature, poor visibility, and deep snow cover. Snow dump localized on urban roads may interrupt traffics including expressway close, flight delay or cancellation. Large-scale blizzards, snow cover and cold weather in pastoral and agricultural regions often cause a massive loss of livestock due to freezing conditions and hunger, also result in freezing damage to agricultural crops together with other severe adverse consequences.

In China, the regions with a frequent occurrence of snow disasters agree with the regions with continuous snow cover. In other words, snow disasters mainly occur in northeast China, northwest China, and the Qinghai–Tibet Plateau regions. They are mainly concentrated in the Xilinguole League in the Inner Mongolia Autonomous Region, Altay and Ili regions in the Xinjiang Uygur Autonomous Region, Three-River Headwater region in the Qinghai–Tibet Plateau, and northern regions of Tibet (Fig. 9.10).

9.2.3 Avalanche

When the snow cover on slopes loses its stability, i.e. the ground frictional force cannot withstand the downward force of snow on the slope, it slides causing massive ice and snow collapse. This natural phenomenon is called avalanche.

In 1973, the International Avalanche Classification Team of the International Snow and Ice Committee proposed the international avalanche classification, i.e., shape–cause classification (Table 9.3).

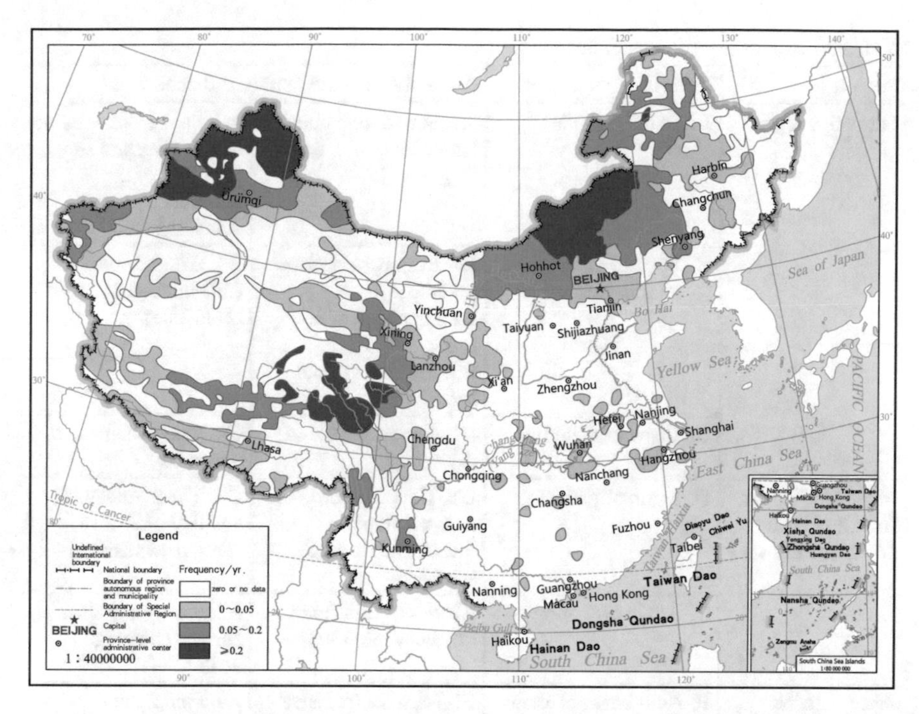

Fig. 9.10 Distribution of snow disasters in China. The Source Region of the Yangtze River (Yellow River)

Avalanche is closely associated with the gradient of slopes. It tends to happen when the slopes are in the range of 30°–40°. According to the observation data in Tianshan Mountains, with the slopes ranging from 25° to 35°, an increase of gradient is correlated with the increase in the risk of avalanche. However, the probability of avalanche sharply decreases when the gradient is higher than 45°. By analyzing 140 avalanche events in the southeast Tibet, it was found that avalanches occur mostly on the slopes with a gradient of 30°–45°. It is difficult for avalanche to occur if the gradient is higher than 50°.

Studies have also reported that a 50 cm thick fresh snow layer (density = 0.08 g/cm^3) in a continental climatic zone can slide on a slope of 25°. In oceanic avalanche regions, there is snow cover thickness exceeding 70 cm on a slope of 40° and good development of a deep frost layer at the bottom of the snow cover. The whole deep frost layer may collapse with a slight increase of snow thickness. A snow layer of 100 cm thickness, composed of refrozen moderate snow grains, coarse snow grains, and deep frost, will not slide on the slope of 37°, where the cohesive force of snow is from 2000 to 4000 Pa. In the period of climate warming up, melting water starts penetrating into different snow layers, which become wet snow, resulting in a sharp reduction of cohesive force and the frictional force as well. In this case, a full-layer wet avalanche may occur even on a slope of 25° (Table 9.1).

Table 9.1 Avalanche shape classification

Region	Criteria	Alternative feature, name, and code	
Formation region	A. Starting mode	A1. Start at one point (snow loosening type)	A2. Start from one line (snow plate and snow slide) A3. Soft A4. Hard
	B. Position of the sliding surface	B1. In snows (surface avalanche) B2. Breakage of new snow B3. Breakage of old snow	B4. Ground surface (full-thickness avalanche)
	C. Water content in snow	C1. No (avalanche of dry snow)	C2. Yes (avalanche of wet snow)
Movement region (free flow and decelerated flow)	D. Form of pathway	D1. The pathway is on an open slope (avalanche on slope surface)	D2. The pathway is in a valley or groove (groove avalanche)
	E. Mode of motion	E1. Snow dust cloud (staublawine)	E2. Flow on ground surface (flowing avalanche)
Accumulation region	F. Roughness of snow surface	F1. Massive (massive accumulation) F2. Snow blocks with edges F3. Rounded snow blocks	F4. Fine grains (stacking of fine grains)
	E. Water content in snow blocks	G1. No (avalanche of dry snow)	G2. Yes (avalanche of wet snow)
	F. Pollution of snowdrift	H1. Slight pollution (clean avalanche)	H2. Pollution (polluted avalanche) H3. Stones and mud H4. Branches and trees H5. Construction debris

When the snow temperature is higher than –5 °C, snow crystals grow quickly, and a deep frost layer forms when the temperature gradient exceeds the critical value (–0.2 °C/cm). The occurrence of a deep frost layer below snow cover is a sign of a large-scale full-layer avalanche. Temperature rises quickly in spring and surface snow melts. The melting water penetrates downward quickly through loose snow layers, and the temperature of the whole snow layer approaches to 0 °C. In this case, the snow strength drops sharply and the cohesive force declines. In particular, the deep frost layer at the bottom of a snow cover is reduced by melting water. When melting water occurs on the sliding surface, a full-layer wet avalanche could easily happen.

A thick cornice may form on the leeward slope of a watershed. Influenced by snowdrift or snowfall, the cornice may collapse when its dead load exceeds the breaking strength of the snow in the cornice, thus causing snow sliding on the lower slope and further causing avalanche. For a snow plate with weak cohesive force between it and the underlying surface, further influenced by snowfalls and dramatic temperature changes or other external factors (walking of humans and livestock or rolling stones), there would be cracks developing quickly on the snow plate surface, resulting in a snow plate avalanche.

9.2.4 Spring Floods

Spring flood refers to the phenomenon of river water level rise in spring. With climate warming in spring, melting snow and ice in the upstream of a river basin, together with spring rainfalls, can cause significant rise of river water level. Rivers, such as the Irtysh, Ob, Yenisei, and Lena, and others on plains in Canada, flows through regions with snow covers. In spring, the warming climate leads to the snow in low-altitude regions quickly melting and feeding into rivers, finally causing snowmelt floods or nival flood.

The spring is the largest flood season in some river regions in western China and may sometime cause flooding disasters. However, the spring flood is also a valuable source of water for farmland irrigation in the most regions of northern China. The volume of snow cover in winter and the scale as well as time of spring floods after snow melting closely affect agricultural and animal husbandry production in north China. Overall, there is less snow cover in China than many other countries, thereby, less spring floods.

9.2.5 Ice-Jam Hazards

Most rivers in northern China freeze in winter. Ice-jams mainly occur during the unstable ice periods, e.g. the freeze-up or breakup of rivers. There are detailed records of river ice hazards in the northeast and northwest China, and more detailed records of ice-jam hazards of the Yellow River.

In northern China, which is affected by Mongolian high pressure, the low temperature leads to frozen rivers and lakes. The rivers in northeast and northwestern China, as well as the section of the Yellow River in the Inner Mongolian, usually freeze early and breakup late and thus have a long freeze period. The maximum ice thickness is typically 0.8–1.63 m, which results in ice accumulation and ice-jams. Rivers in the northern China and the Central Plains have relatively stable ice conditions in winter, and the maximum ice thickness is 0.25–0.61 m, with small runoffs and weak ice-jams, but there is a high risk of ice-jam hazards in built channels. In the middle and lower reaches of the Huaihe River, the Yihe River, the Shuhe River, and the northern

Jiangsu section of the Beijing–Hangzhou Grand Canal, there are frozen rivers or drift ice when cold weather occurs in winter, affecting shipping. The maximum ice thickness is 0.25–0.46 m,

There are several types of ice-jam hazards:

(1) Ice-dam flood. An ice dam is developed when a large number of ice accumulating in rivers, resulting in a decrease in river cross-section, an increase in flow resistance, a rise in water level, and then overflow, causing flooding.
(2) Ice-flower jam. Drifting ice flowers stick to the surface of cold solid objects, layers by layers, becoming more and more thick, which reduces the river cross-section and eventually blocks it. As an example, a power station cannot operate when its fence barriers for debris are blocked by ice-flower jams. At the same time, the upstream of the power station becomes overflowed due to a high level of water causing ice-jam floods.

The ice-jam hazards may considerably impact on:

(1) Shipping and construction safety: Drifting ices can generate a great impact force, when colliding with ships and other structures. They make shipping impossible in winter and cause damages to any structures they meet.
(2) Embankment and hydraulic structures: The expansion of ice sheets creates a large static pressure that can damage river embankment and hydraulic structures as well, such as intake water towers, bridge columns and abutments. The Yongding River Diversion Canal has suffered from the impact of ice-jam hazards. The canal was built in 1957, and it is 26 km long. The design maximum flow rate is 35 m^3/s. Since the Guanting Reservoir Power Station is for peak regulating, there is sometime tailwater and sometime not. It increases the amount of ice produced in the adjustment pool of the upstream, resulting in more ice damage.

9.2.6 Sea-Ice Hazards

Sea ice usually indicates the saltwater ice as a result of frozen seawater, but it also includes the freshwater ice from rivers and icebergs from glaciers. Sea-ice hazards may cause disasters, in which the sea ices cause significant adverse impacts on societies and economy. The severity of sea-ice disaster is not only related to the intensity of hazards, but also depends on the impacted infrastructure or economic activity. For example, the severity of sea-ice disasters in the Bohai Sea is proportional to the frequency of industrial activities and inversely proportional to engineering measures or countermeasures. A low sea-ice hazard does not mean that sea-ice disaster is reduced when its potential consequence is significant. In the context of climate change, the prevention capacity for sea-ice disasters should be strengthened.

The risk of sea-ice hazards in the Bohai Sea includes the destruction of marine engineering structures and facilities, blocking of water intakes, damage to ships, blockage of harbour channels, and damage of aquaculture facilities. Sea-ice hazards may also cause oil spill accidents when offshore oil production, storage, and transportation facilities are damaged, causing serious pollution of marine environment and possibly endangering human life in addition to economic loss.

At present, there are more coastal and offshore economic activities in the frozen area of the Bohai Sea and the northern part of the Yellow Sea. Those include offshore projects, mainly offshore oil exploration and development projects, and also coastal nuclear power plants, port terminals, cross-sea bridges, coastal energy facilities, such as wind power, thermal power, hydropower and so on. Others are coastal base facilities, such as petrochemical storage and refining, steel etc.

9.3 Vulnerability, Risk and Adaptation Assessment

9.3.1 Vulnerability Assessment

As discussed early, vulnerability is the susceptibility of a concerned object and can be measured as a degree of loss or any adverse effect in correspondence to a given magnitude of a hazard. The vulnerability can emerge at different scales, for example, vulnerability to changes of sea level and ocean circulation at a global scale, to changes of monsoons at a regional scale, and to water supply shortage, floods, debris flows and so on at a local scale.

The concerned object can be physical or social or their integrated systems at different scales. For physical systems, they may include buildings, roads and bridges, power transmission and other infrastructure. For social system, they may include communities, population and the population with special needs, institutions and organizations and so on. They may include more complex systems such as cities, coastal and high-mountain ecosystems, environment, and human-earth systems in general. More specifically, the vulnerability assessment is here to understand the fragility of our concerned objects to the changes of cryosphere, and to some extents, their ability to resist the adverse impact of cryosphere changes.

The changes impact on agriculture and ecosystems, and eventually affect economy and society. It demonstrates that agriculture, industries, ecosystems and cities (or towns) as well as their down streams are vulnerable to the changes of cryosphere. To fully understand the impact, there are needs to assess the vulnerability along the impact chain.

The vulnerability can be considered to be proportional to exposure and sensitivity, and inversely proportional to adaptability to a hazard. Among them, exposure refers to the possibility that the system is subject to adverse impacts of the hazard. Sensitivity is the degree of response of the system to the effect of the hazard, and adaptive capacity is the ability to adjust the system to respond to the external pressure resulting from the hazard. In this regard, given a hazard, the vulnerability can be described by:

$$V = (E \times S)/A \text{ or } V = (E - A) \times S \quad (9.1)$$

where V is vulnerability, E is exposure, S is sensitivity, and A is adaptive capacity. When the degree of exposure is higher, the sensitivity is stronger, or the adaptability is smaller, then the vulnerability is greater. On the contrary, when the degree of exposure of the system is lower, the sensitivity is weaker, or the adaptability is greater, and then the degree of vulnerability of the system is smaller. This kind of approach is normally known as composite index.

The composite index combines individual indictors into one to simplify a complex multi-dimensional issue, and is useful in identifying the trend of the issue. Therefore, construction of a composite index has to be carefully designed, and may include the following steps:

(1) Developing a theoretical framework: a structure of composite index should be clearly defined, linking various sub-groups and their underlying variables that should be independent of each other.
(2) Selection of the underlying variables: guided by the developed framework, the selected variables should be able to describe the concerned issue on the basis of relevance, analytical, soundness, timeliness, accessibility. Proxy measures can be applied when the data of desired variables are not available.
(3) Imputation of missing data: The missing data can be categorised into missing at random and not at random. To deal with missing data, approaches are to be taken either by case deleting, single imputation or multiple imputation. The first is to ignore the missing record, but may cause incomplete samples; the second is to impute data by such as mean and median substitution, or regression and the third is to do by such as Monte-Carlo simulation, which may likely cause issues of a false perception of complete data achieved, leading to misinterpretation. Therefore, all should be fully checked statistically.
(4) Multivariate analysis: There are likely many variables that may affect the concerned problem, which is intended to be described by a composite index. While it is not realistic to include all variables that may affect the problem, it is important to identify the variables that are critical. The approaches can be taken by methods such as principle component analysis (PCA) and factor analysis (FA). The analysis is essential to develop a structure of a composite index.
(5) Normalization: Considering that all variables may be physically or categorically different, normalisation of all variables involved is essential before their integration into a composite index. The technique may include ① ranking; ② standardization by converting the variables into the scale with zero mean and standard deviation of one; ③ min–max normalisation by converting the variable into an interval of [0, 1] via subtracting the minimum and then dividing it by the range of the variables; ④ Normalised distance to a reference, for examples, external benchmark, the average of a concerned group; ⑤ categorical scale that assigns a score to the variable; ⑥ anomaly that are often applied to describe the yearly change with respect to the average over a specified period.

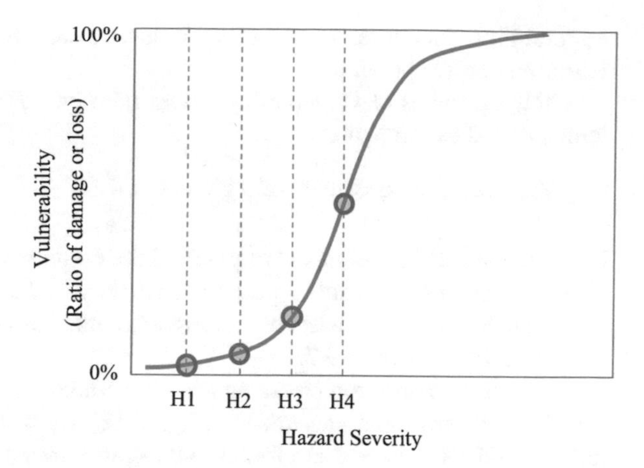

Fig. 9.11 Vulnerability

(6) Weighting and aggregation: To consider the significance or contribution of each variable to the composite index, weightings are applied in regard to each variable in aggregation. Though many simply rely on equal weighting, the technique of weighting includes statistical models such as FA, or participatory approach such as analytic hierarchy process (AHP) and conjoint analysis (CA).

However, different from models based on scientific principles, the development of composite index is more or less subjective. Due to the reason that it is not a physical variable and has difficulty to properly take into account uncertainties, it could even lead to over-simplistic conclusions.

Considering that the vulnerability is essentially the susceptibility measured as a degree of loss or any adverse effect in correspondence to a given magnitude of a hazard, it can be quantitatively formulated in association with a concerned object (or system) and hazards that affect the object. For physical assets, vulnerability can be described by the loss of functionality, serviceability or/and integrity of the assets (see Fig. 9.11), for example, building structure damage in association with GLOF.

The degree of vulnerability is closely related to the adaptive capacity of a system of interest. Adaptive capacity is considered as an inherent system property that enables adjustments of its capacity or capability threshold to accommodate expected (future) adverse hazard impacts without loss of its functionality and integrity, which may lead to disasters. It can generally be described by social, human, financial, environmental and physical capitals though there are many other representations. Enhancement of the adaptive capacity could also be beneficial to immediate disaster mitigation.

To be stricter, vulnerability is one of characteristics that a system holds to give its response to external stimuli such as hazards, Vulnerability assessment is the key step to understanding how an asset physically functions given its exposure to hazards. It can be done by either modeling or damage survey data.

It should also be pointed that there are more types of vulnerability definitions in literature as well as its assessment, often based on different interpretation of vulnerability. Therefore, it is important to have a clear understanding of any type of definition before applying it.

A few examples of vulnerability assessment based on the composite index are demonstrated as following:

1) Vulnerability assessment of regions to GLOF

The vulnerability to the impact of glacial lake outburst flood (GLOF) can be determined by degrees of hazard, exposure, sensitivity, and adaptive capacity. It is evaluated qualitatively on the basis of a composite index aggregated from the indicators as demonstrated in Table 9.2.

According to the composite index, the vulnerability to GLOF disasters in Himalayas is evaluated as shown in Fig. 9.12. The zones with very high vulnerability to GLOF disasters are found to be concentrated mainly in Nyalam, Tingri, Dinggyê, Lhozhag, Kangmar and Zhongba, and the mid-eastern Himalayas, whereas zones with very low vulnerability are located in the eastern Himalayas. The zones

Table 9.2 Indicators for the composite index of vulnerability to glacial lake outburst flood (GLOF) disasters

Objective layer (*A*)	Principle layer (*B*)	Index layer (*C*)	Unit
GLOFD vulnerability index	Hazard (B_1)	Number of glacial lakes (C_1)	–
		Area of glacial lakes (C_2)	km^2
		Area change of glacial lakes (C_3)	%
	Exposure (B_2)	Population density (C_4)	Persons/km^2
		Livestock density (C_5)	10^4
		Cultivated area (C_6)	km^2
		Density of road network (C_7)	km/km^2
		Density of agricultural economy (C_8)	Yuan/km^2
	Sensitivity (B_3)	Proportion of rural population (C_9)	%
		Percentage of small livestock (C_{10})	%
		Percentage of national and provincial roads (C_{11})	%
		Farmers' income (C_{12})	10^4 Yuan
	Adaptive capacity (B_4)	Regional GDP (C_{13})	10^8 Yuan
		Percentage of financial revenue in GDP (C_{14})	%
		Density of investment in fixed assets (C_{15})	10^4 Yuan/km^2

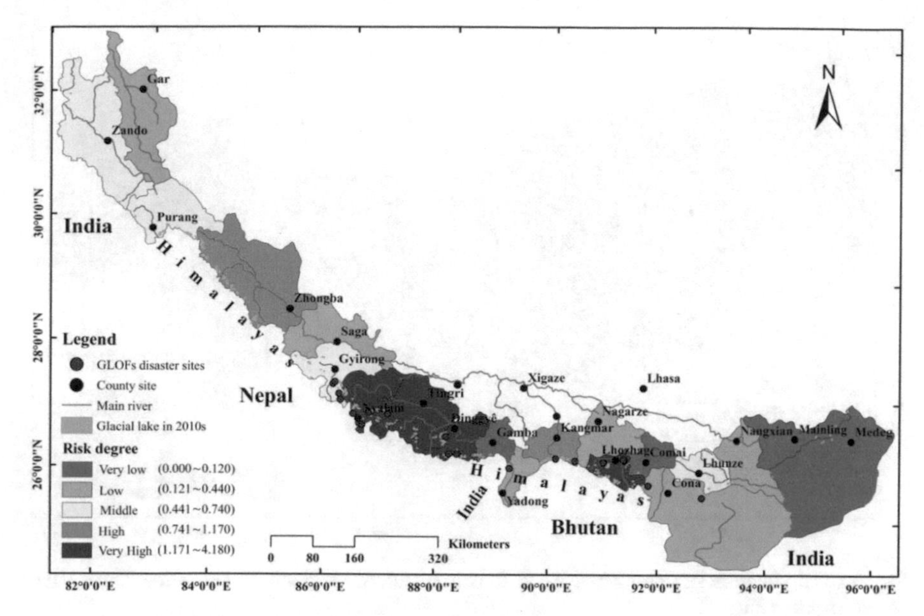

Fig. 9.12 Vulnerability to GLOF in the Himalayas

with higher vulnerability are characterized by a higher degree of hazard, exposure and sensitivity as well as lower adaptive capacity.

2) Vulnerability to snow hazards

Snowstorm hazard is represented by heavy snowfall over a long duration, and can be divided into light, medium, and heavy categories:

(1) light snowstorm: the snowfall in winter and spring is equivalent to 120% of the average annual snowfall;
(2) medium snowstorm: the snowfall in winter and spring is equivalent to 140% of the average annual snowfall; and
(3) heavy snowstorm: the snowfall in winter and spring is equivalent to 160% of the annual average snowfall.

The snowstorm hazard can be indicated by quantities, such as snow depth, density, temperature, etc., which are easy to obtain and use. The risk to snowstorms is also related to the exposure of local economic development and the adaptability of infrastructure.

Taking an example for the area in the region of Three Rivers Headwater, 10 factors are selected in relation to hazard, exposure, sensitivity, and adaptive capacity to assess the vulnerability to snowstorm. As shown in Fig. 9.13, the zones with extremely high vulnerability to snowstorm are mainly located in the Yushu, Chengduo, Zaduo, and Nangqian counties of the southern Bayan Har Shan, the Bayan Har Shan, A'nyemaqen Shan, Gander, and nearby counties, while the areas with very low vulnerability are

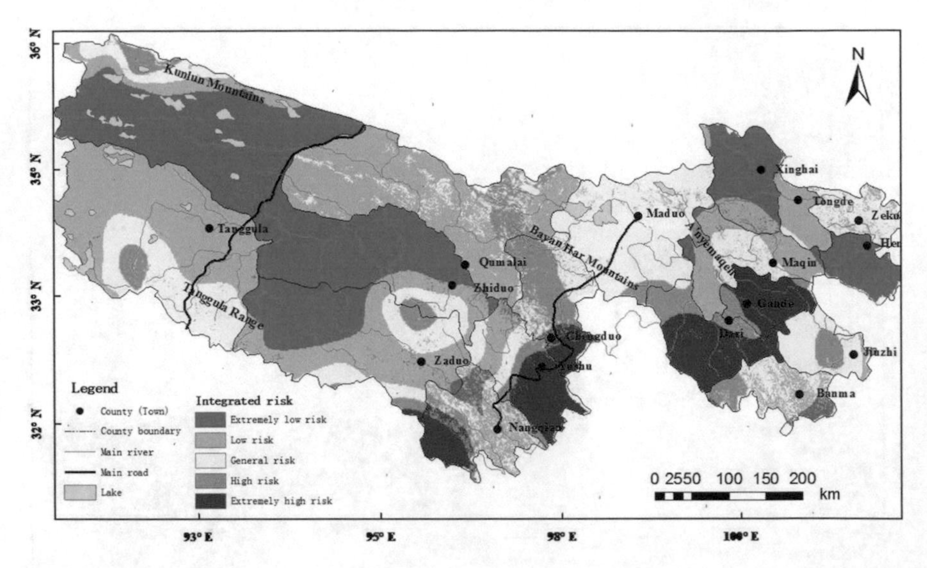

Fig. 9.13 Snow disaster comprehensive risk assessment and regionalisation in the Three River source region, China

in the West Maqin land where there are no inhabitants, as well as most of the Jiuzhi County Basin.

3) Assessment of freeze–thaw impact

Disaster in the permafrost region is mainly caused by thermokarst hazards in the process of permafrost degradation. In polar regions, thermokarst subsidence, thermokarst lakes and ponds, thermokarst mud flows, and thaw slump are amid the main permafrost hazards.

Although there is low population in permafrost areas, there are more and more engineering constructions in recent years for resource development and roads. They are increasingly impacted by the hazards as a result of permafrost degradation. In this regard, by considering the effects of climate change on permafrost regions, Nelson et al. evaluate the sensitivity of thermokarst subsidence and thermokarst development, and also the effects of engineering construction on permafrost, which in turn affect the safety of construction. In this case, it was reported that there are 20 deaths caused by thermokast subsidence building difference in Norilsk, and about 300 buildings in the Siberian permafrost region were destroyed.

Meanwhile, hazard zoning and assessment in the Arctic permafrost region was done by the JAKUZK considering different climate scenarios. The evaluation is mainly based on the settlement index (I_s) with an aid of ECHAM1-A and UKTR climate models. The settlement index is described by

$$I_{\mathrm{s}} = \Delta Z_{\mathrm{al}} \times V_{\mathrm{ice}} \tag{9.2}$$

where Z_{al} is the increase in active layer thickness and V_{ice} is the ratio of ice volume in the ground to that in the near surface soil. The results showed that the thermokarst hazards in the permafrost regions of the Northern Hemisphere may potentially affect engineering structures, such as those in Alaska, Siberia, Canada (Norman Pipelines), as shown in Fig. 9.14.

Therefore, the assessment of vulnerability to freeze–thaw hazards is important for engineering planning and management. Factors are selected more or less pertinent to the causes to vulnerability, and a combination of expert opinions and field investigation is utilized to acquire relevant data. The analytic hierarchy process (AHP) is

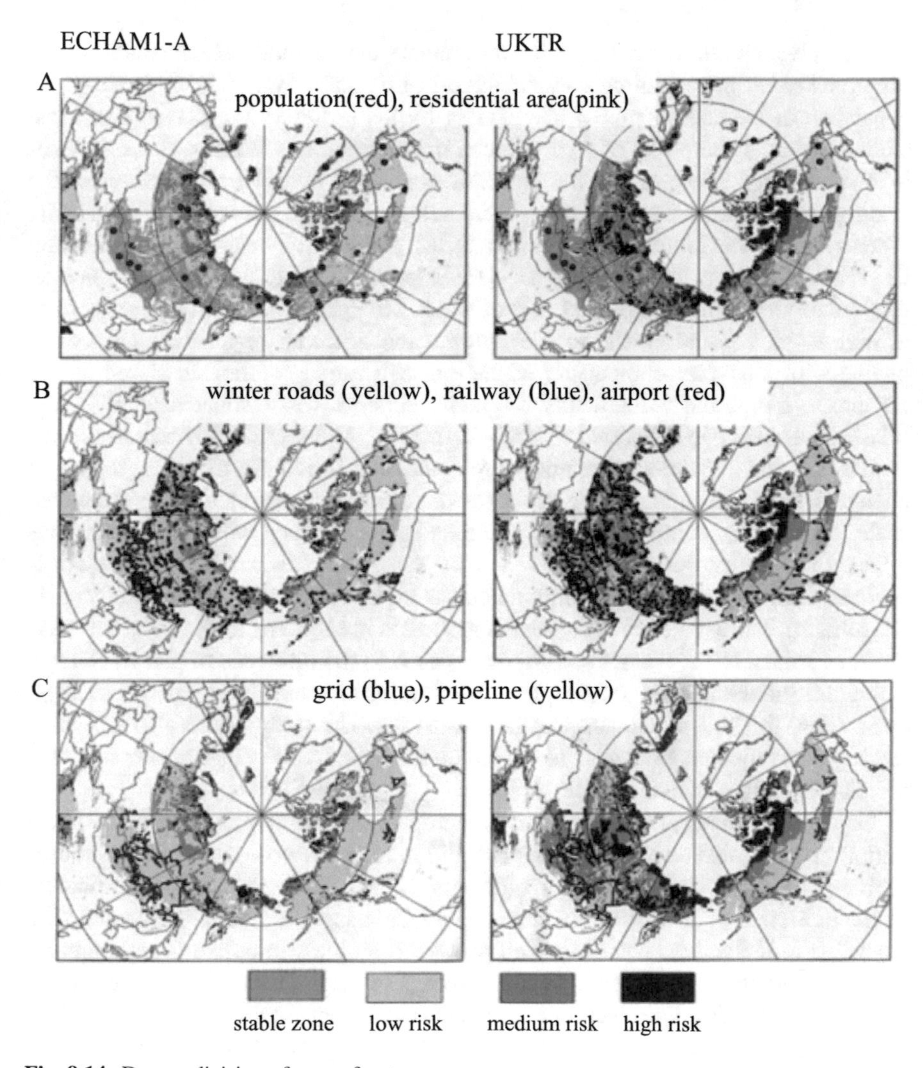

Fig. 9.14 Danger division of permafrost

applied to obtain the weights of each factor. The factors include frozen soil distribution, ice content, soil temperature, soil types, surface conditions, slope, groundwater, and other regional geological and geomorphic factors. The soil distribution, ice content and soil temperature are mainly acquired by field drilling, geophysical exploration, and spatial modelling; quantification of the regional geology, geomorphology, and vegetation cover affecting permafrost development is obtained mainly by on-site investigation, geological exploration, extraction and remote sensing as well as analysis of historical geological data.

4) Vulnerability of ecosystem to cryospheric hazards

Glacier advance and retreat, snow cover change, and permafrost degradation play important roles in regulating and stabilizing oasis stability and lake shrinkage or expansion in the ecosystems of inland river basins in the arid region of northwest China, which is composed of cryosphere in the mountainous areas, oases, and lakes in lower reaches. The glacier is the source of freshwater to support life of oases in the arid region of China. Sixteen indices were selected for the vulnerability composite index system. These indices are runoff modulus, glacial meltwater recharge ratio and its variation, dryness, average oasis area, GDP, population density, urbanization rate, total grain yield, GDP output per unit of water, NPP, labor productivity, proportion of high water consumption industries, proportion of tertiary industries, number of people with nine-year compulsory education, and Engel coefficient. Based on the composite index, the vulnerability to glacial changes was evaluated for the Hexi Inland River Basin in the period of 1995–2009 at the scales of watershed and county.

It was found that the vulnerability of the oasis social-ecological system to glacier change in the Hexi Inland River Basin has an increasing tendency at the watershed scale (Fig. 9.15a). Among the three watersheds, the Shiyang River Basin is mostly affected by the glacier change, followed by the Shule River and the Heihe River basins. Meanwhile, the oasis economic belt is highly vulnerable at the county scale (Fig. 9.15b). The expansion of the oasis and growth in economy and population cause the Hexi Inland River Basin to be highly exposed to the influence of glacier change. The grain yield and GDP output per unit of water are sensitive to glacier change, and so does the high water consumption industries. The study results also show that the impact of socioeconomic development on this vulnerability has far surpassed the effects of changes in natural factors.

Cryosphere changes have an important impact on ecosystems. The wide distribution of the alpine swamp wetland and alpine meadow ecosystems in the source regions of rivers, where annual precipitation is less than 400 mm, is attributable to the existence of permafrost over the Tibet Plateau. The unique water and heat exchange in the active layer of the soil produced by the permafrost is key to maintaining the stability of the alpine ecosystems. The permafrost and its upper alpine marsh wetland and alpine meadow ecosystems have a remarkable water conservation function, which is an important factor for stabilizing the water cycle and river runoff in the source regions of the rivers. In recent decades, ecological degradation and significant changes in the hydrological environment, such as rivers, lakes, swamps,

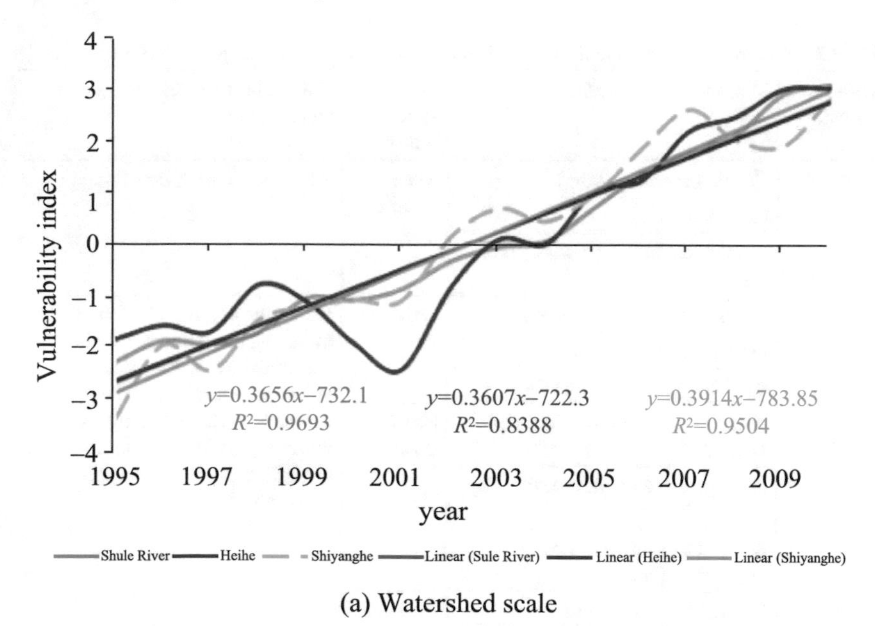

(a) Watershed scale

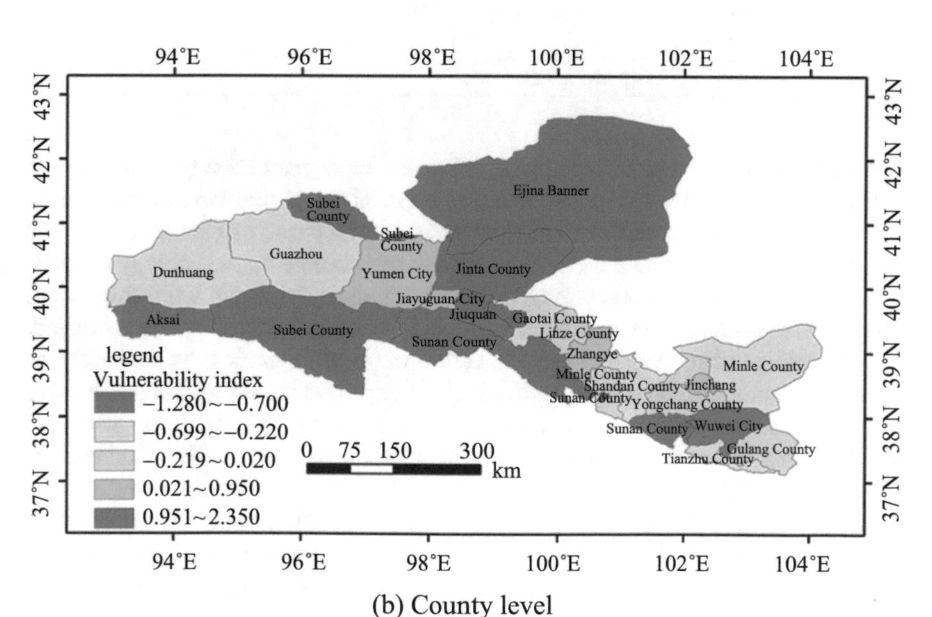

(b) County level

Fig. 9.15 Temporal and spatial variation of vulnerability of the oasis system affected by glacier change in the inland river basins of the Hexi region, northwest China. **a** at the watershed scale; **b** on the country scale

Table 9.3 Vulnerability assessment indicators and methods for the alpine grassland ecosystem

Assessment dimension	Measurement indicator	Spatial state matrix	Vulnerability expression model and method
Grassland quality	Usable grassland (km^2)	Mass matrix	Structural dynamics model
	NPP (kg/hm^2)		
	Heavily degraded grassland area (km^2)		
	Grassland fence area (hm^2)	Damping matrix	Under the action of different state matrices, the vulnerability is expressed by the displacement of grassland ecosystem. The greater the displacement, the higher the vulnerability, the smaller the displacement, the lower the vulnerability
	Livestock shed (10^4 m^2)		
	Artificial grassland area (hm^2)		
Grassland potential	Average precipitation from Apr. to Oct. (mm)	Stiffness matrix	
	Average air temperature from Apr. to Oct. (°C)		
	Active layer thickness of permafrost (cm)		
Grassland pressure	Population density/km^2	Pressure matrix	
	Economic density (ten thousand RMB/km^2)		
	Livestock density (sheep unit/km^2)		

and wetlands, in the source regions of the rivers have been closely related to the freeze–thaw cycle changes in soil and the degradation of permafrost. An evaluation index system of grassland ecological vulnerability based on permafrost change was developed to evaluate the effects of permafrost change on the ecology over the Tibet Plateau (Table 9.3). According to the Ricardian equation, a relationship model between grassland ecosystem vulnerability and frozen soil change was constructed, and the change characteristics and the future trend of ecological vulnerability of alpine grassland were revealed quantitatively. The projection showed that the active layer of permafrost will increase, the ecological carrying capacity of the grassland will decrease, and that the range of reduction will increase with the quickness of increase of the active layer in the source region of the Yellow River. The preliminary results of the above assessment show that the degradation of permafrost may offset the effects of ecological restoration. Therefore, both the above ground grassland and the underground frozen soil are protected ecological engineering in future.

9.3.2 *Risk Assessment*

Risk refers to the potential loss or consequence caused by hazards including extreme events, described in the early sections. As mentioned, the risk is determined by two basic elements, i.e. the impacts and the likelihood of their occurrence. In more details, natural disaster risks in the cryosphere depend on cryopsheric hazards, exposures and vulnerability, and its assessment generally include cryospheric hazard assessment, exposure assessment and vulnerability assessment. It can be assessed qualitatively or quantitatively by approaches, either based on a composite index or probabilistic theory.

The *cryospheric hazard assessment* includes: ① cryospheric hazard identification, a process to identify hazards that could cause damages and losses at different scales, such as individual, local (regional) and national scales; ② hazard information acquisition, a process to acquire historical hazard information, and if future outlook is considered, projection information from modelling given different scenarios, such as climate and landscape changes; ③ hazard modelling, in qualitative approach, to rank hazards into a class, such as severe, strong, medium, weak, very weak, and describe the likelihood of the hazard either in a categorical approach or ARI as shown in Table 9.4 in quantitative approach, to model the hazard in terms of severity at different average recurrence interval (ARI) or return periods, which can be depend on location and time (see Fig. 9.16); ④ hazard mapping, if required, mapping the hazards of different average recurrence or return periods across a scale as required, such as local, regional and national scales, at different time horizons (if future environmental changes are considered). Hazard maps can be developed for each occurrence frequency of the hazard, as shown in Fig. 9.17 The high frequency hazard event shows less magnitude than the low frequency event. It implies that rare hazard event has a large magnitude of hazards.

The *exposure assessment* includes: ① identification of points of interest (POIs): identify points of interests, such as physical assets, communities and natural resources, which distribution generally has the nature of spatiality and temporality, for example, urban sprawl and population growth; ② hazard exposure analysis, to evaluate the exposure of POIs to the hazards in association with the severity of hazards, for example, the houses may not be exposed to the low-depth of flooding, but exposed to severe flooding.

Table 9.4 Descriptive examples of cryospheric hazard magnitude and frequency

ARI (years)	Descriptions	Hazard magnitude (e.g. severe, strong, medium, weak, very weak)
1	Average occurrence of once every year	
10	Average occurrence of once every ten years	
100	Average occurrence of once every a hundred of years	
500	Average occurrence of once every five hundreds of years	
1000	Average occurrence of once every a thousand of years	

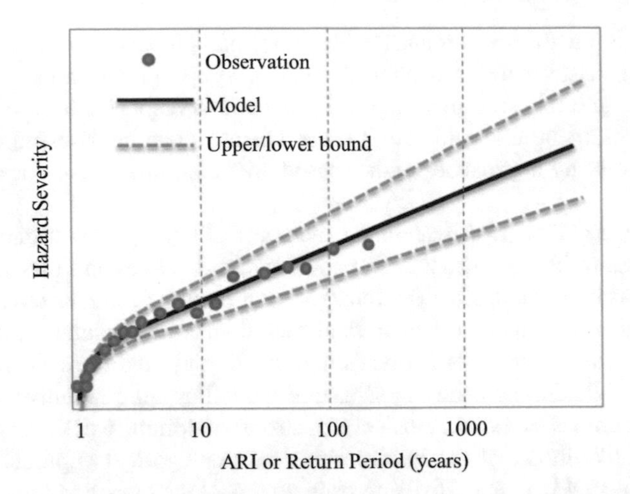

Fig. 9.16 Illustrative description of quantitative representation of a cryospheric hazard through statistical modelling

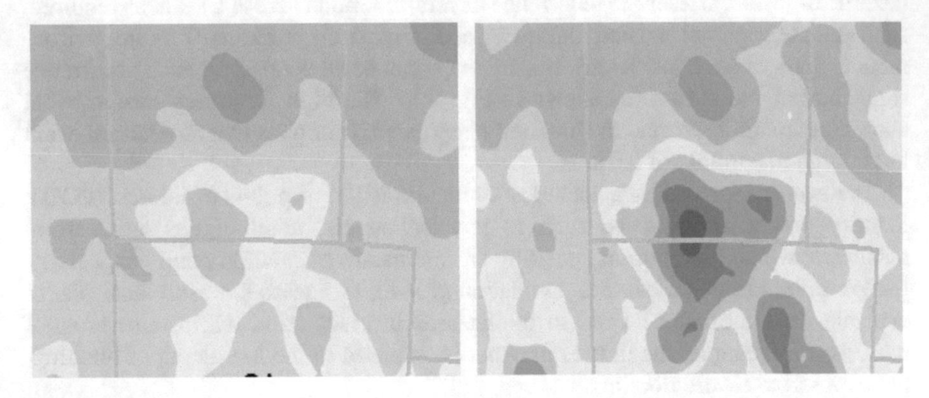

Fig. 9.17 Illustrative maps of cryospheric hazard magnitude at a high occurrence frequency (left) and low occurrence frequency (right). Red and blue colour indicates high and low intensity, respectively

The *vulnerability assessment* includes: ① qualitative approach, for example, ranking the vulnerability to different level of hazard (if exposed) in terms of consequences, such as catastrophe, significant, moderate, small, and very small; ② simplified vulnerability curve approach (quantitative approach), by taking the steps of developing damage/loss data inventory in association with the severity of hazards, developing collective performance of POIs subject to the impacts of identified hazards at different severity, or fragility curves that give a distribution of four physical damage states (slight, moderate, extensive and complete) in relation to hazard severity, as shown in Fig. 9.18; developing vulnerability curves based on fragility curves (see Fig. 9.18); ③ detailed system analysis (quantitative approach). The vulnerability

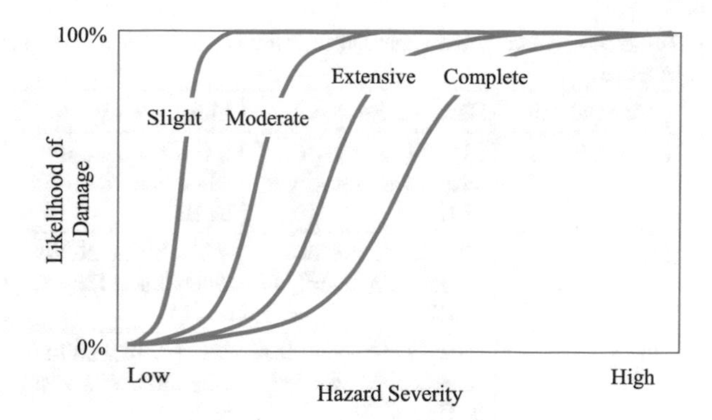

Fig. 9.18 Illustrative fragility curves that describes the relation between cryospheric hazard severity and the likelihood of damage that in one of the damage categories including slight, moderate, extensive and complete

assessment is one of key steps in risk analysis, but its concept is often expanded as described in the early section of vulnerability assessment.

Risk evaluation, in a qualitative approach, to rank the risk based on hazard severity and likelihood together with its consequence; and in a quantitative approach, to estimate the risk based on the hazard, exposure and vulnerability quantified in other early assessment.

For the qualitative approach, the risk to a specific severity of hazard can be measured on the basis of Table 9.5. For multiple hazards or hazard at various severities, the overall risk is considered as the highest among all risks.

For example,

(1) Hazard 1, its likelihood is very often and the consequence is small, hence the risk is 'high'.
(2) Hazard 2, its likelihood rare and its consequence is moderate, hence the corresponding risk is moderate.

Therefore, the overall risk is the highest among them, which is high.

For a hazard event with a discrete severity variable, the quantitative approach using vulnerability curves to measure the risk can be done following the steps described in Table 9.6 which has assumed that the objects of interest are exposed to the hazard. The total likelihood of the hazard at all severities should be equal to one.

Table 9.5 Qualitative measurement of risks based on hazard likelihood and corresponding consequences

Hazard Likelihood	Consequences (damage or loss)				
	Minor	Small	Moderate	Significant (extensive)	Catastrophic (complete)
Very often	Moderate	High	High	Extreme high	Extreme high
Often	Moderate	Moderate	High	High	Extreme high
Occasional	Low	Medium	Moderate	High	High
Rare	Low	Low	Moderate	Moderate	High
Very rare	Low	Low	Low	Moderate	Moderate

Table 9.6 Quantitative approach to estimate risks for a cryospheric hazard event with discrete severities

Hazard severity	Hazard Likelihood	Vulnerability	Risk
H1	L1 = likelihood of hazard with severity to be H1	V1 = Vulnerability when hazard severity to be H1	R1 = L1 * V1
H2	L2 = likelihood of hazard with severity to be H2	V2 = Vulnerability when hazard severity to be H2	R2 = L2 * V2
H3	L3 = likelihood of the hazard with severity to be H3	V3 = Vulnerability when hazard severity to be H3	R3 = L3 * V3
…	…	…	…
Total risk	Sum(L1, L2, …) = 1		R = Sum(R1, R2, R3 …)

For the hazard in a continuous severity variable (e.g. Snow, GLOF), Table 9.7 can be applied to calculate the risk as one in addition to many other theoretical approaches. In this case, the severity of hazard is firstly converted into discrete intervals, separated by values such as H1, H2 and H3 etc. in Table 9.6 As results, the hazard likelihood is described as the likelihood of the hazard event with its severity within an interval, and the vulnerability can be expressed as average damage or loss due to the hazard within the interval.

Table 9.7 Quantitative approach to estimate risks for a cryospheric hazard event with continuous severities (when vulnerability is 100% or 1, it is complete damage or total loss)

Hazard severity	Hazard Likelihood	Vulnerability	Risk
<H1	L1 = likelihood of hazard with severity less than H1	Average for V < V1, or V1/2	R1 = L1 * V1/2
≥H1 and <H2	L2–L1 = likelihood of hazard with severity large/equal H1 and less than H2	Average for V between V1 and V2, or (V2 + V1)/2	R2 = (L2 – L1) * (V1 + V2)/2
≥H2 and <H3	L3–L2 = likelihood of hazard with severity large/equal H2 and less than H3	Average for V between V2 and V3, or (V3 + V2)/2	R3 = (L3 – L2) * (V2 + V3)/2
…	…	…	…
≥Hn (last discrete point)	1-Hn	(1 + Vn)/2	Rn = (1 – Hn) * (1 + Vn)/2
Total risk			R = Sum(R1, R2, R3…, Rn)

9.3.3 *Risk Management and Adaptation*

In general, the development of adaptation aims to reduce potential impacts of cryospheric hazards or adverse effect as a result of service deterioration, another kind of hazards. From the prospect of risks, adaptation is fundamentally developed on three principles, i.e.

(1) Reduce the vulnerability of all relevant institutional levels to cryospheric hazards at relevant spatial and temporal scales, and
(2) Reduce the likelihood of the occurrence of and exposure to cryospheric hazards, but it should be aware that reducing the occurrence likelihood of natural hazards is in most cases difficult to achievable.
(3) Reduce any residual adverse consequences as a result of the impact of hazards

More specifically, it could be implemented through three steps as shown in Fig. 9.19.

Risk minimization is to reduce/avoid the manageable adverse consequence as a result of cryospheric hazard impacts; risk sharing is to share the inevitable (residual) impacts as a result of cryospheric hazard attacks, to reduce the corresponding adverse consequences incurring on each of individuals; and impact management is to manage the inevitable adverse consequence as a result of cryospheric hazard impacts for recovery.

Risk sharing can be implemented by distributing the consequence of impacts among multiple parties, such as private and public, individuals and institutions, communities and government, through instruments such as insurance, regulation and government incentives, to redistribute climate risks. It is particularly important to balance the needs of socially disadvantaged groups who disproportionally incur high risks to the climate impacts.

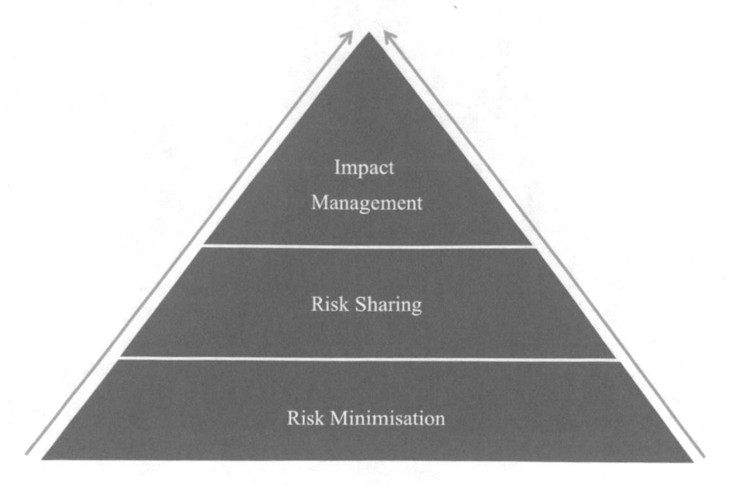

Fig. 9.19 Three principles for adaptation to reduce risks

The impact management is implemented to minimize the hazard-induced adverse consequence, particularly to reduce those subsequent downstream impacts caused by the direct hazard impacts. It normally covers the short-term emergency responses and humanitarian disaster recovery, but it is also associated with long-term reconstruction and recovery of, not only physical assets and services, but also local/regional economy, social systems, environment and community livelihood etc.

As mentioned, adaptation can generally be developed by minimizing risk, which depends on its three key factors i.e. hazard, exposure and vulnerability. More specifically, the risk can be managed by reducing the effects of those factors.

Climate change is related to all spheres and their interactions, which are normally assessed by earth system models, but will not be discussed in this chapter. As shown in Fig. 9.20, the hazard is related to cryosphere change that depends on climate change. Meanwhile, it is also affected by the other spheres interacting with the cryosphere. Therefore, changes in other spheres may alter the exposure and vulnerability, eventually either exacerbate or reduce the risk. Interventions to reduce the risk or development of adaptation to changes can be approached at different levels, such as by improvement in design, planning and policies, which intend to address risk issues at local, regional and national scales, respectively. It should be noted that all measures developed in adaptation may also more or less changes the dynamics in the other spheres, which may in turn affect the risk indirectly.

While it is important to develop adaptation for the sake of reducing risk of climate change induced hazards, the adaptation should also be looked after to address the change in service as a result of climate change, leading to risks such as shortages in service. In fact, climate change may potentially vary natural capitals required for services, while affecting demand and utility indirectly. Adaptation measures can be

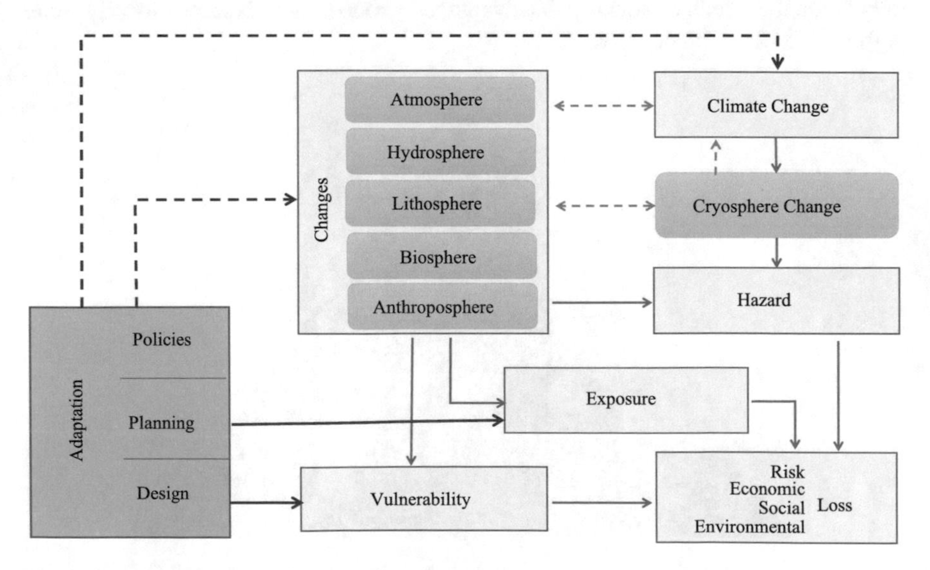

Fig. 9.20 Development of adaptation by reducing risk

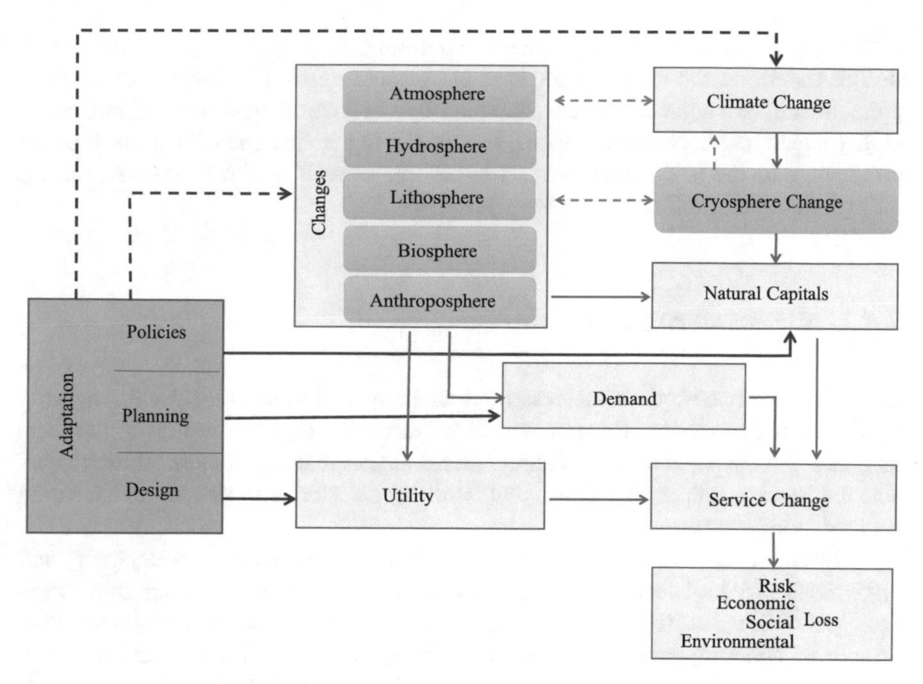

Fig. 9.21 Development of adaptation by reducing risk caused by the change in service

identified by the approaches aiming to improve the utility, efficient use of the natural capitals, and management of the demand, similarly through design, planning and policies.

It should be noticed, the approach given in Fig. 9.21 only provides potential options for adaptation, the decision has to made on which option is more practical and effective. One of methods is applied by comparing cost and benefit of those potential options. The benefit can be considered as the risk reduction resulting from adaptation, while the cost is related to the implementation of the adaptation.

9.4 Cryosphere Services and Their Value

The cryosphere stores a huge amount of water, energy, and gas resources and carries endemic biological species and indigenous cultural structures, so it is not only an irreplaceable resource but is also a special material and cultural basis for the sustainable development of populations, resources, environments, and social and economic systems in the high-altitude and polar regions. Therefore, the cryosphere has unique service functions. The cryosphere service function (CSF) is an important part of the general ecosystem and its healthy development; it should be considered an important component of ecosystem services, and its values should be measurable and calculable.

As a special sphere, the cryosphere is natural entity studied exclusively for its service functions and values. There are huge population and economic capacities in polar, alpine, and adjacent regions, and the life and development of people there are highly dependent on water resources, suitable living environments, various tourism products, and the indigenous cultural structure, as well as the habitats of unique biological populations related to the cryosphere.

9.4.1 Cryosphere Services

Cryosphere service means the benefits that people obtains from the cryosphere. Similar to the ecosystem function and its services, the cryosphere function reflects its characteristic capable of providing natural stocks that are in demand by people, and the service reflects the flows generated by the stocks to people and creating socioeconomic benefits.

Human demands can be classified as anything for livelihood, development, and enjoyment, which all depend on environmental resources. People's recognition, attention, and willingness to pay for cryosphere service will increase gradually with the growth of economy and improvement of living standards. Having said that, not all cryosphere functions can produce services. Cryosphere services includes provisioning, regulating, culture, supporting services as indicated in early section. Habitat service was also applied instead of the use of support service in more general term (Xiao et al. 2015). All of these services can provide benefits to the livelihoods and wellbeing of people.

As shown in Fig. 9.2 of the Sect. 9.1, the provisioning services of the cryosphere include freshwater and clean energy. The cryosphere is considered as a solid freshwater reservoir, continuously supplying societies with melting freshwater and showing to tremendous value to people. Clean energy related to the cryosphere consists mainly of hydropower using glacial melting and existence of natural gas hydrates. The runoff over steep terrains serviced a good condition to produce hydropower. The main component of natural gas hydrates is also known as solid methane, commonly known as combustible ice. The natural gas hydrate is normally distributed in permafrost areas, submarine sediments of the continental shelves, and deep lake sediments, a potential source of clean energy.

The cryosphere's regulating services include climate regulation, runoff regulation, and water conservation and ecological regulation. As one of land covers, the cryosphere plays a critical role in global and regional climate systems through its high albedo and cold water that drives ocean circulations, leading to comfortable habitation and a stable planetary ecosystem. It can be said that the cryosphere is crucial in the regulating global the climate system. The cryosphere is the important source for many rivers in middle-latitude and low-latitude mountains, and it also regulating runoff to reasonably maintain its annual volume. Compared to other areas where precipitation is the main source of freshwater, the cryosphere runoff is mainly affected by temperature. When temperature rises, more melting of ice, snow, and permafrost occurs, leading to increased runoff. The ice, snow, and frozen ground all contribute to the regulation of runoff. The water conservation function of the cryosphere is also remarkable. Among all components, the frozen ground closely affects terrestrial ecology. Because of water impermeability of the frozen ground and moisture transfer with temperature gradient, there is an amount of groundwater stored above the upper layer of permafrost, supporting the need of vegetation. The frozen ground is important for maintaining the stability of ecosystem in cold regions. Without the water conservation and hydrothermal effects of the frozen ground, there had been only a desert on the Qinghai–Xizang Plateau, rather than the large area of alpine meadow and alpine wetland ecosystems. In the pan-Arctic regions, due to significant hydrothermal effects of permafrost, it exhibits typical polygonal tundra and taiga ecosystems.

The cryosphere's cultural services include aesthetic and recreational, scientific research and environmental education, and religious and cultural services. The aesthetic value, as a fundamental value of cryosphere tourism resources. It mainly refers to artistic characteristics (shape, colour, etc.), status, and significance (such as diversity, specialty, pleasure, and entirety) of the cryosphere landscape, which are main reasons to attract cryosphere tourism. The uniqueness of the cryosphere landscape is that it cannot be copied or transferred (Litzinger 2004; Steinberg 2008; Wang and Qin 2015). In regard to the service for scientific research and environmental education, it are primarily reflected in the increase in knowledge that may help the development of national economy and improvement of human welfare.

Over a long period of historical and cultural development, many snow-covered and glacierised mountain peaks have been given spiritual significance and cultural implication. They are considered to be the manifestation of spirits, which forms a unique cultural structure and a specific religion for mountain residents. In the cryosphere region, it may include some unique cultures. For example, Eskimos and Sami live all time of a year in the Arctic and central Arctic cryosphere. People who lived in the cryosphere a generation by a generation is in possession of characteristics, inextricably and closely related to the cryosphere. There is no doubt that a unique culture of those minorities is related to the cryosphere and its unique landscapes.

The cryosphere supports habitations where ecosystem relies on to survive. The cryosphere provides a large area of habitation services for settled and migratory populations in the worldwide cold regions, in addition to human beings. For the associated terrestrial and marine species, it offers diverse living spaces and biodiversity, giving a variety of habitats, shelters, feeding and breeding grounds. It is a also place to home some rare and endangered wildlife. In the polar and sub-polar regions, in addition to supporting indigenous people's livelihood, the cryosphere also provides crucial habitats for microbes, algae, worms, crustaceans, sea birds, penguins, seals, walruses, polar bears, and whales (UNEP 2007).

9.4.2 Cryosphere Service Value

The diversity of cryosphere functions and services determine the various representations and approaches in evaluating their values. In association with the path, extent, and duration, by which human benefit from the services, the cryosphere service value can be divided into use value and non-use value. The use value includes a direct use value and an indirect use value, and the non-use value contains option value, existence value, heritage value, and so on. The direct use value refers to the value of market products in the cryosphere services to meet the human demand of production or consumption, such as freshwater and clean energy, or non-market products, such as aesthetics and recreation, scientific research and environmental education, religious and cultural values, etc.

Compared with the direct use value, the indirect use value means the benefits obtained indirectly from the cryosphere, such as climate regulation, runoff regulation, water conservation, ecological regulation services, etc. It does not enter into the processes of production or consumption in the cryosphere service. The option value refers to the potential value of cryosphere services. The heritage value is a manifestation of some use value and non-use value of cryosphere services reserved for descendants, namely the values we are willing to leave to descendants who will use the cryosphere services in the future. The existence value, also known as the intrinsic value, is the willingness to pay to ensure that cryosphere services can exist persistently. The existence value is the inherent value of the cryosphere, and it does not concern about the use of cryosphere services now or in the future. All are summarized in Table 9.8.

9.5 Adaptation of Major Engineering Constructions in the Cryosphere

There is a close relation between the cryosphere and engineering constructions, for example, in China, such as Qinghai–Tibet Railway, the Qing–Kang Highway, the North-Eastern Railway, the Western Route of the South-to-North Water Supply Project, the West–East National Gas Pipeline Project, the China–Russia Crude Oil Pipeline, the Lanzhou–Lhasa Optical Cable Project, the Golmud–Lhasa Oil Pipeline Project, the Sanjiangyuan Ecological Conservation Project, the Qilian and Tianshan Mountains Ecological Protection Project and so on. Their design and construction have to be supported by the understanding of Cryospheric Science. This support will be even more important for the development of Western China and the progress in the Belt and Road Initiative.

Table 9.8 Cryosphere service value, type, and evaluation methods

Service function classification		Evaluation method of service function value			
		DUV	IUV	NUV	Difficulty of evaluation
Supply service	Freshwater	MPM			Relatively easy
	Clean energy	RCM			Relatively difficult
Regulating service	Climate regulation		RCM, WTP, HPM		Difficult
	Runoff regulation		SEM		Difficult
	Water conservation and ecological regulation		SEM, RCM, MPM		Relatively difficult
Social and cultural services	Aesthetic and recreational service			HPM, WTP, TCM	Relatively easy
	Scientific research and environmental education			CAM	Relatively easy
	Religious and cultural service			CAM, WTP	Relatively difficult
Habitat services	Providing habitat	OCMCVM			Relatively difficult

Abbreviations Direct use value, DUV; Indirect use value, IUV; Non-use value, NUV; Replacement cost method, RCM; Wish to pay, WTP; Hedonic pricing method, HPM; Shadow engineering method, SEM; Market price method, MPM; Expenditure method,EM, or cost analysis method, CAM; Travel cost method, TCM; Contingent valuation method, CVM; Opportunity cost method, OCM

9.5.1 The Construction of Railways and Highways in the Cryosphere

1. Road engineering and construction

From an engineering point of view, the isotherm of mean air temperature at 0 °C during the coldest month of the year is used to define the boundary of cold regions, a part of the cryosphere. Cryospheric engineering is a general term for engineering construction in the cryosphere. It mainly includes ice engineering, prevention and control engineering for drifting snow, and permafrost engineering. The core issue of cryospheric engineering is usually related to the freeze–thaw effect in soil layers and the ice–snow variation during both construction and operation periods. The design, construction, and maintenance operations should consider factors such as climate, geology, geography, ecology, hydrology, and so on. To analyze and solve the problems, many disciplines would be applied, including geography, remote sensing, physics, chemistry, and mathematics. Therefore, cryospheric engineering is a very wide field. Special considerations in design are required due to unique environmental conditions in cold regions.

Permafrost engineering includes the construction of railway, highway (embankments, tunnels, bridges, and culverts), oil/gas pipeline, power transmission line, optical cable, hydraulic facilities, and buildings. The frost heave, thaw settlement, and salt heave are the three major problems threatening the stability of structures and diminishing their service life. Specifically, the freezing–thawing effect can result in the extra bending of a bridge pile due to uneven frost jacking of the bridge pile foundation, which may cause cracking, dislocation of a lining canal, and even collapse. It can also result in inclination of a retaining wall structure under horizontal frost heave force, and cause the settlement, waving, and longitudinal cracking of an embankment. The freezing–thawing action of soil is a complex process of hydro-thermal–mechanical interaction. Methods to control the structural deformation caused by freezing–thawing action form a basic principle in the design aspect of permafrost engineering. Some famous road engineering projects built in cold regions are described in the following.

The Canadian Pacific Railway, with a total length of 4667 km, is one of the Canada's Class I railroads. The railway, which was constructed from 1881 to 1885, extends from Vancouver in western Canada to Montreal in eastern Canada. It was the first transcontinental railway in Canada, and it has contributed to the development of both eastern and western Canada.

Construction of the first Trans-Siberian railway in Russia was started in 1891 and completed in 1905. The total length of this railway is 9446 km, making it the world's longest railway. The second Trans-Siberian railway, the Baikal-Amur Mainline (BAM), has a length of 4275 km. The project, which was conducted to accelerate the economic development of Siberia, has a reported cost of 15 billion dollars. The designed transport capacity of this railway is 35×10^6 tons per year. The BAM was completed in 1985. The construction of this railway not only shortened the transport distance from western Russia to Pacific coast ports but also alleviated the transportation intensity of the first Trans-Siberian railway.

The Qinghai–Tibet Highway in China has a length of about 1150 km, of which 550 km is in continuous permafrost regions. Work on the Qinghai–Tibet Highway started in 1950. Due to the lack of both scientific research and engineering practice at this time, no specific techniques were used to protect the underlying permafrost during the first construction. Thaw bulbs formed under the embankments and thaw settlement continued to increase from its construction. In 1956 and 1972, the highway was reconstructed. During the second reconstruction, black asphalt pavement with strong heat absorbing ability was added. However, problems still existed, and several maintenance efforts were conducted.

Existing engineering projects in permafrost regions have all experienced various problems due to the complexity of permafrost and the lack of knowledge about environmental conversion and permafrost engineering problems. The engineering damage rates of the permafrost railway in Russia and the Qinghai–Tibet Highway reached nearly 30% and 33%, respectively. Additionally, the damage rates of the Yakeshi–Mangui Railway and the Nenjiang–Gulian Railway in the Greater and Lesser Khingan Range regions in northeastern China exceeded 30%. Thus, the maintenance costs of these engineering projects are very high. For example, the reconstruction of the railway from Gilham to Churchill in Canada, including settlement alleviation, maintenance of bridges and culverts, recovering of longitudinal slopes, and replacement of sleepers, cost more than 30 million US dollars from 1978 to 1983.

The Golmud–Lhasa section of the Qinghai–Tibet Railway is one of the most recent major engineering projects in a permafrost region. To adapt to climate change, adverse effects caused by climate warming were carefully considered in the design of the Qinghai–Tibet Railway. A series of innovative design methods and technical measures were employed to eliminate or reduce the adverse impacts induced by climate warming.

2. Design principles of roadways in cold regions

The melting of ground ice is considered to be the immediate cause of thaw settlement. However, both natural and artificial factors can result in the thawing of permafrost. The natural factors mainly include climate warming, permafrost degeneration, and rainwater infiltration. The main human factors are unreasonable engineering design, construction, and management.

The design principles of permafrost engineering mainly refer to keeping permafrost frozen, allowing permafrost to thaw gradually, and controlling the thaw rate and the preconstruction thaw method. The design principle of maintaining permafrost in a frozen state is generally used in low-temperature and ice-rich permafrost regions. If the heat-absorption capacity of the engineered structure is weak, engineering techniques can be effectively employed in combination to control the heat transfer condition and thus keep the permafrost foundation frozen. Thus, the strength of the permafrost can supply the bearing capacity of the foundation. The engineering effect is to adjust the artificial permafrost table in a certain depth. Meanwhile, the underlying permafrost cannot be allowed to thaw or warm to ensure the stability of the construction. In contrast, the design principle of controlling the thaw rate is often applied to warm and ice-poor permafrost regions. For engineering structures with strong heat-absorption ability, the design principle allows the permafrost stratum to thaw gradually within a certain limit of the operation condition and safety limits. The engineering effect of this principle is to allow the permafrost stratum to warm, the artificial permafrost table to move down, and the permafrost to thaw slowly. On the contrary, the preconstruction thaw method can be used in warm and ice-rich permafrost regions. In addition, the method is suitable for engineering structures with strong heat-absorption capacity. In this method, the permafrost layer with a high thaw-settlement coefficient is melted before construction. The design principles used in normal regions can then be used in the thawed layer. Overall, the main research content for the permafrost foundation issue is aimed at forecasting the thermo-mechanical state reasonably, determining the interaction between permafrost stratum and construction, the use of proper design methods and engineering techniques, and evaluating the operating conditions and service performance.

3. Passive insulated embankment

The basic construction principle for permafrost engineering in China is to protect the permafrost foundation. However, due to economic limitations, engineering requirements, and a lack of understanding of permafrost, the traditional protective measures are mainly either to increase the height of the embankment or to apply heat insulation materials. Previous research has shown that although increasing the height of the embankment and applying heat insulation materials can raise the permafrost table by increasing the thermal resistance under some conditions, the ground temperature of the permafrost layers will be raised simultaneously. This method can delay permafrost degradation under the scenario of climate warming. However, it cannot prevent the permafrost from degrading. Hence, it is a passive method protecting permafrost only by increasing thermal resistance. The method cannot effectively ensure the long-term stability of a permafrost embankment, especially in warm permafrost regions. Numerous engineering studies have indicated that the traditional methods have great limitations.

Installing insulation materials (Fig. 9.22a) and appropriately increasing the height of an embankment can reduce the heat absorption of the soil and decrease the maximum seasonal thaw depth in warm seasons. Thus, these measures can maintain the thermal stability of a permafrost embankment. Expanded polystyrene (EPS) thermal insulation material was employed early in permafrost engineering practice. However, some new materials have appeared recently. These include polyurethane (PU) board and extruded polystyrene (XPS) board. These thermal insulation materials have been used widely in permafrost regions along the Qinghai–Tibet Highway, the Qing–Kang Highway, and the Qinghai–Tibet Railway.

The layout of the insulation berm can adjust the heat absorption of the slope and prevent water erosion, which is also done to tackle embankment problems. The insulation berm can reduce the disturbance to the embankment toe and the natural ground surface during construction. It can also decrease the adverse effect of water infiltration into the underlying permafrost, and it can strengthen slope stability (Fig. 9.22b). However, significant settlement can occur within the insulation berm when the permafrost foundation is degraded intensively. Observations of the Qinghai–Tibet Highway have illustrated that the insulation berms can increase the thaw depth and annual ground temperature, despite shrinking the thaw bulb in the permafrost layer. Consequently, many limitations remain in the use of insulation berms in permafrost engineering.

4. Proactive cooling embankment

According to analysis of the traditional methods for permafrost embankment problems based on increasing thermal resistance, the traditional methods cannot fully

Fig. 9.22 Highways and railways in permafrost regions. **a** XPS insulation embankment of the Qinghai–Tibet Railway; **b** Insulation berm of the Qinghai–Tibet Railway; **c** Shading board of the Qinghai–Tibet Railway; **d** Duct-ventilated embankment of the Qinghai–Tibet Railway; **e** Crushed-rock based embankment of the Qinghai–Tibet Railway; **f** Thermosyphons embankment of the Qinghai–Tibet Highway; **g** Cut slope with grass cover of the Qing–Kang Highway; **h** Dry bridge of the Qinghai–Tibet Railway

prevent embankment problems, although they can alleviate the extent of the problems by delaying permafrost degradation. Road construction in permafrost regions started more than 100 years ago. Based on experiences and lessons from engineering practices around the world, Cheng Guodong and other scientists proposed a proactive cooling approach in the design and construction of the Qinghai–Tibet Railway. At present, the approach has been used widely in permafrost engineering.

The "roadbed cooling" methods include, in the main, adjusting solar radiation by the use of shading sheds; adjusting heat convection by the use of ventilation ducts, thermosyphons, and crushed-rock bases and revetments; adjusting heat conduction with thermal-semiconductor material, as well as combining these methods to enhance the cooling effect. Long-term field observations on the geotemperature and deformation of the Qinghai–Tibet Railway embankment have shown that the above measures can effectively cool the underlying permafrost and consequently maintain embankment stability.

(1) Shading shed

The use of a shading shed can reduce the solar radiation received by the ground and lower the ground temperature. Thus, it can cool the embankment. According to field data from the Fenghuo Mountain experimental site on the Qinghai–Tibet Plateau, the ground surface temperature under the shed is at least 5 °C lower than that outside of the shed. The maximum temperature difference can reach 15–20 °C. Currently, there is only one embankment section using this method on the Qinghai–Tibet Plateau, at the transition section from the cutting to filling zone in "no man's land" in Tongola (Fig. 9.22c).

(2) Duct-ventilated embankment

Field experiments on the duct-ventilated embankment were performed at the Qingshuihe and Beiluhe testing sites during the initial construction period of the Qinghai–Tibet Railway (Fig. 9.22d). The main mechanism of the duct-ventilated embankment is to use the low-temperature and strong-wind climate of the Qinghai–Tibet Plateau fully. The intense heat convection processes occurring within the duct can cool the embankment. Long-term field observations have demonstrated that the soil temperatures under the duct-ventilated embankment are lower and that the permafrost table is significantly raised. Therefore, the thermal stability of the embankment is improved. Additionally, related numerical studies have shown that duct-ventilated embankments can ensure the thermal stability of underlying permafrost in the regions with mean annual air temperature lower than − 3.5 °C under the scenario of a 2.0 °C warming in 50 years.

(3) Crushed-rock embankment

The crushed-rock embankment mainly includes the crushed-rock-based embankment, crushed-rock revetment embankment, and U-shaped crushed-rock embankment (Fig. 9.22e). The crushed-rock embankment can increase the porosity of the embankment and allow air convection to occur due to the unstable air pressure/temperature gradients in cold seasons, thus accelerating heat dissipation. In warm seasons, it can provide a stable layer of insulating air and solar shading. Thus, under certain conditions, the crushed-rock embankment acts as a thermal semiconductor, which can cause interannual heat dissipation and cool the underlying permafrost. Temperature distributions within the crushed-rock embankment are usually symmetric. Additionally, the geotemperature will also be reduced in the late period following construction. Research has also found that placing rock revetments with different thicknesses on the two embankment slopes can reduce the difference of the thawing depth. In addition, the U-shaped crushed-rock embankment can both cool the permafrost stratum and maintain the symmetric thermal distribution. Therefore, it is one of the most effective embankment structures in permafrost engineering practice.

(4) Thermocosyphons embankment

The gravity heat-pipe embankment is one of the cooling embankments in which heat pipes are typically embedded within the embankment or at the shoulders (Fig. 9.22f). At present, the most widely used heat pipe is the closed two-phase (vapour and liquid) thermosyphon (TPCT). Energy within permafrost can be transferred via latent heat. Heat pipes are highly efficient heat-transfer devices that have been applied widely in permafrost embankments, especially in the strengthening and maintenance of permafrost engineering constructions, such as the Qinghai–Tibet Railway and Highway. In addition, a special pile foundation embedded with heat pipes is also in widespread use due to its cooling performance in permafrost engineering, e.g. the thermal pile foundation of the Trans-Alaska Oil Pipeline.

(5) Other measures

The natural peat and humus layer can also protect underlying permafrost because its thermal conductivities under frozen states are considerably larger than those under unfrozen states, resulting from the fact that the thermal conductivity of ice is about four times larger than that of water. The ratio of thermal conductivity under the unfrozen state to that under the frozen state can reach 0.33, which can be utilised advantageously in the permafrost embankment. Planting grass or transplanting turf onto slopes is a recognised way to adjust the thermal state and prevent wind and water erosion (Fig. 9.22g). In addition, this measure can help to protect and beautify the environment.

In engineering construction, we can use a single control measure or use comprehensive techniques. For instance, dry bridges can lower the ground temperature because they can both shield the ground from the sun and act as an air duct (Fig. 9.22h). They can also support heavy loads. In the Qinghai–Tibet Railway, dry bridges with a total length of 11.7 km were built to cross the warm and ice-rich permafrost. However, this approach is usually used in warm and ice-rich permafrost regions because of its high cost.

To control the thaw settlement of permafrost embankments, reasonable embankment structures should be used. In addition, we should prevent surface water and recover vegetation.

9.5.2 Oil and Gas Pipelines in Cold Regions

With world economic development and increasing energy demand for oil and gas, the speed of oil and gas exploration and the development and utilisation of the cryosphere have been accelerated. A number of oil and gas transmission pipelines have been constructed. Thaw settlement caused by climate warming and engineering disturbance is a major issue for the construction of pipelines in the cryosphere. Frost heave in cold winters is also a significant problem for such engineering. Frost heave and thaw settlement can affect the stability of pipelines and also threaten their accessory structures such as pump stations. However, pump stations are generally less affected by frost heave and thaw settlement because their site selection is relatively flexible, which means that they can be constructed in areas with better engineering geological conditions. To prevent thaw settlement and frost heave damage of pipelines and their accessory structures, researchers have proposed a series of measures. At present, some major oil and gas pipeline projects in the cryosphere are the Trans-Alaska Oil Pipeline in Alaska, the Norman Wells Oil Pipeline in Canada, the Nadym-Pur-Taz network of natural gas pipelines in Siberia, and the China–Russia Crude Oil Pipeline in northeastern China. All of these pipelines have contributed to local and national economic and social development.

The Trans-Alaska Oil Pipeline, with a total length of 1280 km, was built in 1977. The diameter and wall thickness of the pipelines are 122 cm and 13 mm, respectively. The oil temperature ranges from 38 to 63 °C. The pipeline is a typical long-distance, large-calibre and high-temperature pipeline. About three-quarters of the pipeline route traverses permafrost regions, and at least half of the pipeline is located in ice-rich permafrost regions. To avoid pipe cracking resulting from differential thaw settlement, thermal piles (heat pipes and pile foundations) were used to support 676 km of the pipeline above the ground in the ice-rich and unstable permafrost regions (Fig. 9.23). The above-ground pipeline is used to prevent the underlying permafrost foundation from thawing due to the high oil temperature. The thermal piles are used to cool and protect the underlying permafrost and thus to support the pipeline. The thermal piles can prevent permafrost degradation and thaw settlement under the pipeline even if problems caused by thaw settlement exist in some ice-poor permafrost regions. The damage caused by the construction is relatively low.

Fig. 9.23 The Tran-Alaska Oil Pipeline across ice-rich permafrost regions

9.5.3 Ports in Sea-Ice Zones

There are a number of shipping ports in the Bohai Sea and the northern Yellow Sea. Large-scale ports that operate in winter include the Dandong Port, Dalian Port, Yingkou New Port, Jinzhou Port, Bohai Shipyard, Qinhuangdao Port, Tianjin New Port, Longkou Port, and Yantai Port. Some small-scale ports and satellite ports have also been built around these large-scale ports.

During the construction and management processes of these coastal ports, ice is a common concern because seawater freezes in the winter. Although global warming has mitigated the ice problem in recent years, the design and management of ports in the ice zone cannot be evaluated by large-scale concepts in the geosciences. The construction of ports should consider the recurrence interval, which is usually taken as 50 years. The recurrence interval should be considered if ice blockage events have occurred in recent history. The safe operation of ports depends on the real-time ice situation. Generally, the designed ice-resistance ability of the Bohai Sea ports is relatively high. Ice damage does not occur because the ice situation cannot exceed their ice-resistance ability. However, problems may still occur at ports operating in ice zones, including human errors in ship manoeuvring.

The Bohai Sea features a regular semidiurnal tide with a tidal range nearly 3 km in most areas. Much of the floating sea ice accumulates as thick ice near the coast and exists in the form of thin drifting ice far from the coast. Sea ice near the coast flows slowly, while sea ice far from the coast flows quickly. The moving sea ice can exert force on port constructions. The magnitude of the force is determined by the speed of movement and the size of the floating sea ice. When the kinetic energy of the moving ice is less than the energy required for the ice to reach its failure limit, the moving ice exerts an impact force on constructions. Conversely, crushing failure of the moving ice occurs when the kinetic energy of the moving ice is greater than the energy required for the ice to reach its failure limit. Thus, constant ice extrusion pressure is applied to constructions. Making the structure more resistant to external forces than ice squeezing force can solve the problem.

Construction at ports around the Bohai Sea usually involves a breakwater, which can obstruct the floating sea ice. The mobility and kinetic energy of the ice inside the port are therefore small, and the ice cannot produce a large force on the constructions. A breakwater is a gravity-type structure with good stability, and it can withstand the force caused by floating ice. Wharf structures for docking include gravity-upright wall structures and long-pile wharf structures. The stability of the gravity-upright wall structure is high enough to resist floating ice, so no severe ice damage can occur. The long-pile wharf structure should strengthen the ice resistance ability. A breakwater should be built as a sloping structure because the force under bending failure for ice is lower than that under crushing failure.

The force of ice on a long-pile wharf can be determined according to a physical model between ice and structures when the calculation method for ice force on a simple structure cannot be adapted to validate the structural stability. With this aim, Tianjin University maintains a refrigerated breakable ice laboratory and the Dalian University of Technology has a non-refrigerated breakable ice laboratory.

9.6 Cryosphere Services in Tourism

9.6.1 Cryosphere Tourism

Cryosphere tourism is an activity combined with other forms of tourism such as sightseeing, experience, adventure, scientific research, and education. It mainly depends on various forms of natural landscape of the cryosphere, complex and varied meteorological and climatic resources, and profound cultural antecedents. Among these, mountain glaciers, remnants of glaciation, ice sheets, ice shelves, sea ice, freezing and thawing, frost heaving, snow cover, glaze, rime, and other related aesthetic and cultural landscapes and characteristics are important tourist attractions of the cryosphere.

Modern cryosphere tourism originated from the mountaineering, expeditioning, and early pilgrimage activities of the early 1800s; it developed to a form of mass tourism in the 1900s, and started to become popular in leisure experience tour activities in the 1980s. With improvement of the economy and living standards and the increase of leisure time, cryosphere tourism has become a new tourism project and has been developed vigorously in various countries. It plays an important role in increasing regional economic benefits and in enhancing regional tourism and its popularity, thus promoting the sustainable development of regional economies and societies.

9.6.2 Characteristics of Cryosphere Tourism Resources

Cryosphere tourism resources have obvious value functions, such as aesthetic appreciation, popular science education, and tourism experiences. Compared with other tourism resources, cryosphere tourism resources also have obviously unique characteristics. The main elements of the cryosphere are located mostly in the high-latitude regions of the north and south poles but are also scattered in high-altitude regions in the middle and low latitudes, all of which are typically far from human settlements. These regions are far from source markets, their locational advantages are not obvious, and their accessibility is relatively difficult. However, for these reasons, it also peaks the curiosity of people and produces a strong attraction. The elements of the cryosphere have many landscape phenomena, but these are extremely sensitive to climate change. Cryosphere tourism resources are relatively vulnerable, and their tourism capacity is relatively small. In the future process of this tourism development, the environmental capacity of cryosphere tourism should be a major concern. Of course, the small environmental capacity has also promoted cryosphere tourism as a high-end and in-depth type of tourism. The special climatic conditions and geographical characteristics of the cryosphere determine the unique ethnic characteristics of the residents in some areas. Within and outside of the Arctic Circle, there are a large number of unique indigenous nations, including the Eskimos. The Arctic region is vast, so the Eskimos are highly scattered. Their cultural differences are significant at the regional scale. The unique national characteristics of the cryosphere provide a solid cultural foundation for cryosphere tourism development.

9.6.3 Overview of International Cryosphere Tourism Development

The immense economic benefits of glaciers and snow cover as a major tourist resource for the cryosphere have prompted the political and academic communities of many countries to pay a great deal of attention to glacier and ski tourism development. As early as 100 years ago, the mountain glaciers and snow cover in foreign countries began to be used as tourist resources. Currently, many scenic locations relying on glaciers and snow-covered landscapes have become popular tourist destinations favoured by tourists from all over the world. At present, more than 100 glacial tourist attractions have been developed around the world and more than 6,000 ski resorts have been built. Among these, some tourism destinations have been included in the World Biosphere Reserve and the UNESCO World Heritage List due to their unique and spectacular landscape value and environmental significance in response to climate sensitivity. Glacier and ski tourism are two mature types of cryosphere tourism. Tourism destinations are concentrated mainly in the Rocky Mountains of North America and Alaska, the Alpen Hills in Europe, and in East Asia. China's glacier and ski destinations are mainly concentrated in the Hengduan Mountains

of north and northeast China. World famous glacier and ski tourism destinations include the Saint Moritz ski resort in Switzerland, the Chamonix ski area in France, the Ilulissat Ice Bay in Denmark, the Los Glacier National Park in Argentina, and the Fuji Zen Mountain Ski Resort in Japan.

Questions

1. How are the effects of the cryosphere changes assessed?
2. What are the major social and economic impacts of cryosphere changes? Provide an example.

Extended Readings

Classic Works

1. *Climate Change 2014: Impacts, Adaptation, and Vulnerability*
   Author: IPCC
   Publisher: Cambridge University Press, Cambridge, United Kingdom, 2014
   Content introduction: *Climate Change 2014: Impacts, Adaptation and Vulnerability* is the second volume of the IPCC AR5, which focuses on the effects that have already occurred and the potential risks of the future. The content involves both natural and human systems, as well as regional aspects. The report consists of two parts, A and B. Part A consists of the background, natural resources and systems, human settlements, industries and infrastructure, human health, welfare and safety, adaptation and influence, risk, and vulnerability and opportunities. Part B is composed mainly of regional classification topics. The report suggests that the increase in the temperature increase range of climate change will aggravate the risk of extensive, serious, and irreversible impacts to natural and human systems. To mitigate the adverse effects of climate change and reduce the vulnerability of natural and human social systems, the initiative of human society to adapt to climate change should be carried out jointly at the global and regional scales. Strengthening the management of disaster risk and enhancing the ability to restore the human social system are an effective way to adapt to climate change and reduce the impacts of extreme climate events. A society of sustainable development requires a combination of adaptation and mitigation, and a transition from economic, social, technical, and political decision-making and action to a climate recovery capacity path.

2. *Extreme Weather and Climate Events and Disaster Risk Management and Adaptation National Assessment Report in China*
   Author: Qin D. H., Zhang J. Y., Shan Q. C., Song L. C.
   Publisher: Science Press, Beijing, 2015
   Content introduction: risk management of extreme climatic and weather disasters has become an important sector of the international community to adapt to climate change. Under global warming, China will face frequent and serious extreme weather and climate hazards in the future, and the exposure and vulnerability of disasters will become increasingly prominent. Food, water resources, ecology, energy, urban and rural environment, economy, as well as security in China will encounter the thread

from these extreme climatic hazards. The assessment report has been contributed jointly by over 100 researchers from governmental departments, including the China Meteorological Administration, research institutions and universities by integrating their most recent progress in the field of extreme weather and climate hazards and the risk management. The report can be a value-added guidance to sectors adapting to the adverse impact of climate change.

Chapter 10
Field Observations and Measurements for Cryospheric Science

Lead Authors: Shichang Kang, Xin Li, Bo Sun

Contributing Authors: Junying Sun, Baiqing Xu, Qingbai Wu, Tao Che, Jianping Yang

The rapid development of cryospheric science has benefited from innovations in methodologies and techniques for field observations and laboratory experiments. Various types of data from field observations and laboratory analyses help us to understand the cryospheric processes and their interactions with the other spheres. This chapter describes general techniques and methodologies for meteorological and hydrological observations in the cryosphere, including drilling and pitting techniques and geophysical methods such as electromagnetism and ground-penetrating radar. Specific monitoring techniques for cryospheric components (e.g., glaciers, snow cover, permafrost, sea ice, river/lake ice) and surveys of socioeconomic status in cryospheric regions are also described. In addition, mechanical, thermology, and optical methods for laboratory analyses of the physical structure, chemical compositions, and chronology of the cryosphere are introduced. In view of the extensive use of remote sensing techniques in cryospheric science over the last several decades, this chapter briefly introduces the principles and applications of optical and microwave remote sensing, altimetry, radio-echo, and gravity satellite methods.

10.1 The Role of Observations and Experimental Technology in Advancing Cryospheric Science

Early cryospheric studies relied on simple manual observations and survey analyses of individual components of the cryosphere, such as glaciers (ice sheets), snow cover, permafrost, river ice, lake ice, sea ice and fossil preglacial geomorphology. For example, observations of terminal variations, area, mass balance, temperature and movement of glaciers in the Alps have been manually performed since the late 19th century. Observations of the physical parameters include snow cover days, snow depth and snow density; freezing (breakup) period, area, thickness and density of river (lake) ice; and sea ice type, sea ice concentration, sea ice thickness, inter-sea ice river and polynya. However, manual observations are time-consuming and labor-intensive, and they can only be carried out in limited areas because of the remoteness

D. Qin et al. (eds.), *Introduction to Cryospheric Science*, Springer Geography,
https://doi.org/10.1007/978-981-16-6425-0_10

of some regions. These characteristics of manual observations have greatly limited the comprehensive understanding of the processes and mechanisms of the elements of the cryosphere.

Since the 1970s, the application of remote sensing technology has improved the understanding and research of the cryosphere. Observations using aerial and satellite remote sensing, including visible light, near-infrared, thermal infrared, microwave, laser, radio and gravity technologies, can efficiently obtain the geometric mass and energy parameters at large scales and high resolutions. In addition, the applications of other advanced technologies in field observations, such as drilling and pit techniques, ground penetrating radar, high-density electrical methods, transient electromagnetic methods, and frequency domain electromagnetic methods, have improved field observations at the microcosmic to macroscopic scales. The combination of remote sensing and field observations further enhances the measurement accuracy. Automatic field observations (such as automatic weather stations, eddy covariance systems, and automatic photography) have also greatly increased the efficiency and accuracy of observations of the mass and energy processes in the cryosphere. In recent years, cryosphere theme groups have launched cryospheric theme reports initiated by cooperating international organizations, such as the Integrated Global Observing Strategy (IGOS), which focuses on global environmental changes. The cryospheric components and their variations have been observed using integrated "ground-air-space" observation systems. These have facilitated the establishment of a complete, collaborative and comprehensive cryospheric observation system on a global scale and have increased the observation data and information available for fundamental studies of cryospheric science and routine services. In addition, the World Meteorological Organization (WMO) has launched the Global Cryosphere Watch (GCW) program, which will set the standard for cryospheric monitoring and achieving data sharing. These initiatives have advanced the development of cryospheric science.

Over the past several decades, laboratory analyses technique for various environmental components of the cryosphere, including sampling, preprocessing, laboratory analysis theory and technology, have improved. Advanced theories and methods, such as mechanics, thermology, optics, physical structure and electromagnetism, are applied to analyze the physical and chemical parameters of samples. Improvements in detection accuracy, the rapid upgrades of instruments and equipment, and innovations in analytical methods have led to new opportunities in cryospheric science. In addition, model simulations have been widely applied to the simulation, attribution and prediction of the changes in cryospheric components, including glacier mass balance models, glacier (ice sheet) dynamics models, frozen soil models, snow cover models, sea ice models, and river (lake) ice models. Each type of model can be coupled to a climate system model to predict the dynamic changes of the cryosphere under different climate scenarios. The models applied in cryospheric science are gradually becoming more complex. Greater numbers of physical, chemical and biological processes are being integrated into models. Moreover, models are being developed with longer integration times, higher spatial resolutions and more comprehensive descriptions of each subsystem. In summary, the application of technology and the development of cryospheric models have improved the integration and rapid development of cryospheric science.

10.2 Field Observations, Survey Methods and Techniques

10.2.1 General Methods and Techniques

1. Meteorological observations

The World Meteorological Organization has developed detailed meteorological standards, such as snow cover and permafrost. However, meteorological observations of the cryosphere are different from those from ordinary weather stations. The arrangement of observations should be flexible, the instruments should be portable, and the meteorological parameters should be observed with unique specifications. The objects of the meteorological observations of the cryosphere are as follows: ① routine meteorological factors affecting glaciers, snow cover, permafrost and meltwater runoff from snow and ice; ② micro-climatic characteristics of glaciers, snow cover and permafrost regions; ③ the energy-mass exchange characteristics on the surfaces of glaciers and permafrost. Cryospheric meteorological observations are mainly carried out by automatic weather stations (AWSs) (Fig. 10.1). The observed parameters include air temperature, relative humidity, wind speed, wind direction, air pressure, four-component radiation, snow depth, total precipitation (T200B or the national standard total rain and snow), and evaporation (from evaporation dishes). In addition, the energy transmission across the snow (ice)-air and ground-air interfaces can be observed using eddy motion systems. The sensors for meteorological observation must be characterized by low temperature resistance, a wide measurement range, high precision and easy maintenance. The working principles and application scopes of meteorological measurement sensors are shown in Table 10.1.

2. Hydrological observations

Hydrological processes and their influence on the cryosphere are unique, and the main objective of the observations is to determine and measure the runoff from glaciers, permafrost and snowmelt. The basic elements of hydrological monitoring in the cryosphere include the water level, flow velocity, water temperature, water chemistry and isotopes; thus, attention should be paid to extreme factors such as low temperatures and gravel in channels. Therefore, the sensors have higher requirements (Table 10.2).

Observations of river runoff in alpine regions are the same as conventional hydrological observations. Because of the small areas covered by glaciers and permafrost, river weirs can be used as hydrological observation sections. A water level gauge or

Fig. 10.1 Automatic weather stations for the cryosphere. **a** Surface automatic weather station (1. Lightning rod; 2. Wind speed and direction sensor; 3. Snow depth sensors; 4. Data collector; 5. Battery; 6. Temperature and humidity sensors; 7. Solar panels; 8. Four component radiation). **b** Ice eddy dynamic system (1. Lightning rod; 2. Temperature and humidity sensor; 3. Data logger; 4. Data processor; 5. Three-dimensional ultrasonic wind speed wind direction sensor; 6. Batteries). **c** T200B rain gauge (1. Solar panel; 2. Data logger; 3. Rain and snow meter container; 4. Wind protection fence). **d** Permafrost meteorological observation field (1. Lightning rod; 2. Solar panels; 3. Wind speed and direction sensor; 4. Data logger; 5. Snow depth sensors; 6. Temperature and humidity collector; 7. Windbreak fence; 8. Snow and rain gauge container; 9. Data processor; 10. CO_2/H_2O analyzer; 11. Three-dimensional ultrasonic wind speed and direction sensor). **a** and **b** were provided by Jizu Chen, **c** was provided by Guoshuai Zhang, and **d** was provided by Erji Du

water level meter is set up at the side of the stream section to measure the depth of runoff, and water level flow relationship curves are established at low, medium and high water levels. In larger river basins, it is possible to choose straight natural river sections or the upper and lower river sections near a highway bridge as the stream section and to set up a water ruler and self-recording water level gauge near the river bank. Rivers in glacial areas are generally intermittent rivers that freeze during the winter months and thaw in the summer. The runoff depth is measured at the beginning of the thawing period. During the summer, the water discharge is large, and it is difficult to conduct flow measurements. If the section is damaged by flash floods, the flood mark method can be used to estimate the peak flood flow.

Isotopic hydrology is a principal topic of the water cycle and is based on the abundances of stable isotopes and radioisotopes in natural water. The most commonly

Table 10.1 Automatic weather observation systems for the cryosphere

Classification	Observed parameters	Description	Scope of applications
Automatic weather station	Temperature and humidity	The sensor is a digital probe with 0–1 V linear output signals for temperature and humidity. The D/A converter used to generate the analog output signals has 16-bit resolution. The default configuration is for temperature –40 to +60 °C, and 0–100% relative humidity	Temperature: Temperature measurement range: –50 to +100 °C (default –40 to + 60 °C); Accuracy at 23 °C: ±0.1 °C with standard configuration settings; Long term stability: <0.1 °C/year Humidity: Measurement range: 0–100% non-condensing; Accuracy at 23 °C: ±0.8% RH with standard configuration settings; Typical long term Stability: <1% RH per year
	Wind speed and direction	Wind speed is measured with a helicoid-shaped, four-blade propeller. Rotation of the propeller produces an AC sine wave signal with frequency proportional to wind speed Wind direction: Lead lengths for the Wind Monitors are specified when the sensors are ordered. Vane position is transmitted by a 10 kΩ potentiometer. With a precision excitation voltage applied, the output voltage is proportional to wind direction	Wind speed: Measurement range: 0–100 m/s; Accuracy: ± 0.3 m/s (±0.6 mph) or 1% of reading Wind direction: Measurement range: 0°–360° mechanical, 355° electrical (5° open); Accuracy: ±3°

(continued)

Table 10.1 (continued)

Classification	Observed parameters	Description	Scope of applications
Automatic weather station	Four component radiation	The radiometer consists of a pyranometer pair, one facing upward, the other facing downward, and a pyrgeometer pair in a similar configuration. The pyranometer pair measures short-wave solar radiation, and the pyrgeometer pair measures long-wave far infrared radiation. The upper long-wave detector has a meniscus dome to ensure that water droplets roll off easily while improving the field of view to nearly 180°, compared with a 150° for a flat window. All four sensors are integrated directly into the instrument body, instead of separate modules mounted onto the housing. Each sensor is calibrated individually for optimal accuracy	Field of view Upper: 180°; Lower: 150°; Net-irradiance: –250 to +250 W/m^2; Non-stability: <1% (sensitivity change per year) Uncertainty in daily total: <10% (95% confidence level) indoor calibration; Typical signal output for atmospheric application: ±5 mV
	Snow depth	The sensor measures the distance from the sensor to a target, which is based on a 50 kHz (Ultrasonic) electrostatic transducer. The Sensor determines the distance to a target by sending out ultrasonic pulses and listening for the returning echoes that are reflected from the target. The time from transmissions to return of an echo is the basis for obtaining the distance measurement	Measurement range: 1.6–32.8 ft (0.5 to 10 m); Accuracy: ±0.4 in (±1 cm) or 0.4% of distance to target, whichever is greater, Accuracy specification excludes errors in the temperature compensation; Resolution: 0.01 in (0.25 mm)

(continued)

Table 10.1 (continued)

Classification	Observed parameters	Description	Scope of applications
	Air pressure	This barometer is encased in a plastic shell (ABS/PC blend) fitted with an intake valve for pressure equalization. The sensor outputs a linear 0–2.5 VDC signal that corresponds to 500–1100 mb. It can be operated in a shutdown or normal mode. In the shutdown mode the datalogger switches 12 VDC power to the barometer during the measurement. The datalogger then powers down the barometer between measurements to conserve power	Range: 500–110 hPa; Calibration accuracy: ± 0.07; Total accuracy: ±0.25; Long-term stability (hPa/a): ±0.1; Working temperatures: –40 to 60 °C
Frozen ground observation system	Frozen ground temperature	Thermistor sensors are used to measure the soil resistance through a data collector or high-precision multimeter, and the temperature is obtained using a calibration equation in the laboratory	Measurement range: +30 to –30 °C; Accuracy: ±0.05 °C
Frozen ground observation system	Soil heat flux	The Soil Heat Flux Plate uses a thermopile to measure temperature gradients across its plate. Operating in a completely passive way, it generates a small output voltage that is proportional to this differential temperature	Measurement range: ±2000 W m^{-2}; Sensitivity (nominal): 50 μV W^{-1} m^{-2}; Expected typical accuracy (12 h totals): within –15 to +5% in most common soils; Nominal resistance: 2 W; Sensor thermal resistance: <6.25 × 10^{-3} Km^2 W^{-1}
	Soil moisture	The soil moisture sensor provides simultaneous measurement of soil moisture, salinity, and temperature using a unique patented design	Temperature measurement range: –10 to 65 °C; ±0.1 °C; Soil electrical conductivity: 0.01–1.5 S/m; ±2.0% or 0.005 S/m; Temperature environment: –10 to 65 °C, 0 to 65 °C

(continued)

Table 10.1 (continued)

Classification	Observed parameters	Description	Scope of applications
	Precipitation (bucket type rain and snow meter)	This telemetry instrument consists of a sensor and a signal recorder. Rain passes by the top bearing of the gate into the water funnel and then into the skip. When the water reaches a certain height, it overturns the skip, and the switch circuit sends a pulse signal to the recorder, which records the rainfall	Rain mouth diameter: Φ ~ 200 mm Instrument resolution: 0.2 mm (1 type); Working environment temperatures: –20 to 70 °C
	Precipitation (T200B total pluviometer)	The T-200B Series Precipitation Gauges are weighing bucket precipitation gauges. The collection container in the T-200B Series gauge is suspended from three points, each supporting 1/3 of the weight. With this type of set-up there are options available to measure precipitation with up to triple redundancy. With equal load distribution, 1, 2 or 3 VW transducers, form now on referred to as sensors, can be used to measure total precipitation and rate of precipitation	Volume: 600 mm; Collection area: 200 cm^2; Sensitivity: 0.05 mm; Temperature range: – 40 to 60 °C
	Evaporation	The evaporation capacity is calculated by measuring the change of water level in the evaporation dish	Accuracy: 0.25%; Operating temperature range: – 40 to 60 °C
	Albedo	Albedo is the ratio of reflected short-wave radiation to incoming short-wave radiation	Accuracy: ±1%
	Infrared radiation temperature	The Infrared temperature sensor provides a non-contact means of measuring the surface temperature of an object. It includes a thermopile for measuring a millivolt output dependent on the target to sensor body temperature difference	Measurement range: – 40 to 70 °C; Accuracy: <±0.1 °C

(continued)

Table 10.1 (continued)

Classification	Observed parameters	Description	Scope of applications
Eddy covariance system	Three-dimensional wind speed, CO_2 and H_2O flux	The gas analyzer provides measurements of absolute densities of carbon dioxide and water vapor, while the sonic anemometer measures orthogonal wind components. It has been optimized for remote eddy-covariance- flux applications, addressing issues of aerodynamics, power consumption, spatial displacement, and temporal synchronicity	CO_2 density equation: mg m^{-3} = 0.38632 · (mVout) – 102.59, Full scale range: –103 to 1829 mg · m^{-3}; H_2O density equation: g · m^{-3} = 0.00865 · (mVout) – 2.26, Full scale range: –2 to 41 g · m^{-3}

Table 10.2 Hydrologic measurements in glacier areas

Water level and water temperature	The collector has temperature and pressure sensors; the temperature sensor measures the ambient temperature, and the pressure sensor measures the water level	Water level: Accuracy: ±0.3 cm Resolution: 0.14 cm Blasting pressure: 310 kPa; Water temperature: Range: –20 to 50 °C Accuracy: ±0.37 °C Error: 0.1 °C
Velocity	This sensor uses sound waves from a source (ultrasonic) to a receiver (suspended solids) in the water. It uses the received body acoustic wave frequency and the difference between the sound source frequencies to calculate the flow velocity	Range: ±10 m/s Accuracy: 1% of reading or ± 0.5 cm/s Operating temperature: –4 to 30 °C

used stable isotopes are δD and δ^{18}O, and the radioactive isotopes are tritium (T) and ^{14}C. Since the 1970s, isotope technology has been applied to studies of runoff formation, flow paths and drainage time. The emergence of isotope technology also provided an ideal physical basis for the division of flow process lines. The process of precipitation runoff is analyzed based on changes of δ^{18}O, δD and water chemistry, and the two water flow process lines are divided based on the equations of mass balance and concentration balance. With the joint application of isotope and geochemical tracers, the δ^{18}O values of three water sources (snow melt, precipitation, and groundwater) from glaciers and hydrochemistry ion tracers are used to model the contributions of the water flow process lines (i.e., three water flow process line segments). The number of water sources determined by this method need not be determined artificially. The number of water sources, the amounts of water and the flow processes of the different water sources can be determined automatically based on the differences in the flow processes of the water sources in the outlet section of a river basin. Because this method is based on the laws of physics, its rationality and effectiveness are widely accepted.

3. Drilling and pit techniques

The cryosphere (such as glaciers and permafrost) contains a large amount of paleoclimatic and environmental information. Drilling and pit techniques are the fundamental requirements to obtain this information. Drilling technology is applied to reach a certain depth in a material to obtain samples for measuring various physical and chemical parameters in the laboratory. Borehole or vertical profiles can also be obtained by drilling to make observations. Pit techniques are applied to observe and sample from shallow profiles. In general, drilling and pit techniques are mainly applied in permafrost and glacier surveys. Drilling and pit technologies can differ in terms of the specific research objects and purposes.

1) Drilling and pit in permafrost

Pitting in permafrost requires digging a pit with a certain width and depth by manual or mechanical means based on the requirements of the field observations. Physical and chemical parameters of the permafrost profile can be surveyed and measured in the pit. In addition, various observation instruments (such as soil temperature and moisture sensors) can be arranged at different depths within a pit to perform long-term observations. Drilling technology can be applied to obtain a permafrost borehole to a certain depth (approximately several tens of meters) using manual or mechanical power, to install sensors and obtain samples for physical and chemical parameters analysis. After drilling and sampling, a string of thermistors that measure soil temperature are usually installed in the borehole for long-term observations.

2) Drilling and pit in glaciers

Drilling and pitting in a glacier mainly include digging a shallow pit or mechanical or thermal drilling. Shallow pit digging is generally carried out on firn/snow on the surface of a glacier. Manual or miniature electric tools are used to dig to a certain depth and produce an observation profile. The physical parameters of the snow layers (e.g., snow density and thickness, temperature, firn fabrics, dust layers) can be observed (Fig. 10.2a). In addition, snow samples can be collected at certain intervals along the vertical profile to analyze their physical and chemical parameters in the laboratory.

In glaciers, boreholes and ice cores can be obtained by drilling to a certain depth using manual, mechanical or thermal power. The borehole can be used to collect different types of parameters. Manual and mechanical drills use manual and mechanical power, respectively, to rotate the drill head and cylinder. Snow/firn or ice layers can then be sliced by the cutter on the drill head, and the snow/firn or ice samples retained in the cylinder can be extracted to measure the physical and chemical parameters (Fig. 10.2b). Mechanical drills can integrate electric systems into the drill head, which is connected to electric power by a cable. The drill head can be powered to cut, move up and down and collect ice core samples. Thermal drills use electric power to heat the annular head of the drill cylinder and melt the snow/firn and ice layers for

Fig. 10.2 Snow pit sampling and ice coring

vertical drilling. The other structural designs of thermal drills are similar to those of mechanical drills.

In addition, thermal drilling technology can be used to drill a borehole on a glacier without collecting samples. This method uses a drill to spray hot water (hot water drills) or high-pressure steam (steam drills) to melt the ice layers downward to create the borehole. In mountain glaciers, steam drills can be used to drill boreholes to depths of 10–20 m. Large hot water drills, which are used in polar regions, can drill to depths of hundreds of meters.

Generally, manual drill sets are used to collect shallow ice cores, and mechanical drill sets are used to collect deep ice cores with cold ice temperatures. Thermal drills are commonly used to collect deep ice cores in warm ice layers (approximately 0 °C).

4. Electromagnetic methods

1) Electrical resistivity tomography (ERT)

The principle of ERT is to measure the current intensity with two current electrodes at the ground surface and measure the potential difference with two other electrodes. The apparent resistivity is then calculated from the ratio of the potential difference to the current intensity multiplied by a k-factor related to thc electrode array and topography. The apparent resistivity measured by ERT is not the true electrical resistivity structure unless the subsurface is an isotropic uniform half-space. Inversion calculations are indispensable for deriving the subsurface resistivity distribution. When soil is frozen, it contains a large amount of ice, and ice is nonconductive; therefore, the electrical resistivity of thick ground ice will be tens or hundreds of times that of thawed soil. The electrical resistivity structure of permafrost provides a good prerequisite for ERT investigations because of the presence of ground ice or ice lenses near the permafrost table. Therefore, ERT has been widely used to detect the depth of the permafrost table, the ground ice distribution and the permafrost thickness.

2) Time-domain electromagnetic (TEM)

The basic principle of the TEM method is the law of electromagnetic induction. When the electrical conductivity of a geological body is high under the action of a step pulse, the intensity of electromagnetic eddy currents is high, which causes a greater intensity of the secondary electromagnetic field. The primary electromagnetic field is generated by shutting off the current in ungrounded or grounded loops. In TEM data collection, the changes of the secondary eddy current field over time are analyzed to investigate the electrical resistivity of the medium. TEM has been widely used in studies of glacial and periglacial environments, such as for detecting ground ice in rock glaciers in the Rocky Mountains and determining the distribution of permafrost, the depth and base of the permafrost table, and the permafrost thickness in thermal spring area on the Tibetan Plateau.

3) Frequency-domain electromagnetic (FEM)

The FEM method, which is similar to the TEM method, generates a primary magnetic field using electrified loops while the current is changed sinusoidally at a certain frequency rather than being shut off rapidly. The received signal includes primary and secondary fields, and the frequency of the field is identical to that of the emissions current. The primary field is excited by the emissions source and exists when there is a nonconductive medium in the ground. The secondary field results from the eddy current induced by the primary field. The phase of the secondary field lags behind that of the primary field, and the frequencies of the two fields are identical. The electrical resistivity structure of the ground can be studied by analyzing the characteristics of the secondary field. FEM uses the skin effect of electromagnetic induction to change the working frequency from high to low and explore the geological target from shallow to deep. This method has been successfully used in studies of glacial and periglacial environments in polar regions, such as detecting ground ice in rock glacier, shallow ground ice in the European Alps, and the lower limit of permafrost in the high mountains of Norway.

5. Penetrating radar technology

Penetrating radar technology is based on permittivity contrasts in the ground. The spatial distribution, pattern and physical properties of geological bodies in the cryosphere can be detected and identified by analyzing the dynamics, travel time signal processing and imaging of reflected electromagnetic waves. These methods can be divided into two categories based on the principles of the radar sounding techniques.

1) round penetrating radar (GPR)

GPR is a single pulse radar system. It transmits and receives high frequency electromagnetic waves (MHz) and uses the propagation time and amplitude of the electromagnetic wave in the medium to obtain the permittivity of the stratum or target. The permittivity can then be explained from a geological perspective. The receiving antenna receives the direct wave, reflected wave and refracted wave, which are emitted by the transmitting antenna, pass through the shallow ground surface and are reflected or refracted at the interface between different permittivities in the subsurface. The conditions and potential investigation depths of the three kinds of GPR waves are different. The reflected wave method is the most commonly used method in permafrost surveys because it more intuitively represents the reflective surface of the underground medium, and the data processing is simpler.

2) Ice radar

Ice radar is also known as radio echo-sounding (RES) or ice-penetrating radar and is a frequency-modulated pulse compression radar system. It has better penetration capability than GPR, which can be used to measure the thickness of polar ice sheets, their internal structure, the subglacial topography, the ice-bedrock interface and the environment features in the detection areas. Ice radar data are mainly presented as radar images (radargrams). There are two main types of radar images: single-channel waveforms (A-scope) and multichannel images (Z-scope). The platforms used to collect ice radar observations can be loaded on vehicles and aircraft. Vehicle ice radar coverage is small, but the detection accuracy and positioning accuracy are high; therefore, it is suitable for small-scale ice sheet surveys with complex ice conditions. Airborne ice radar has wide coverage and high detection efficiency. However, it has the disadvantages of poor attitude stability and weak positioning ability. It is commonly used for large-scale investigations of ice sheets.

10.2.2 Observations of Cryospheric Components

There are several differences in field observations and survey techniques for cryospheric components, such as glaciers (ice sheets), snow cover, permafrost, river ice, lake ice, and sea ice. The specific methods and techniques are shown in Table 10.3.

1. Glacier observations

Glacier observations are the basis for studying the changes in glaciers, understanding the responses of glaciers to climate change and predicting future glacier changes. Glacier observations mainly include the glacier mass balance, glacier terminal, area and volume.

1) Glacier mass balance

Glacier mass balance is the algebraic sum of glacier accumulation and ablation and is the most direct indication of the glacier's response to climate change. Each component of the mass balance is generally expressed in terms of the water mass or water depth per unit area (g/cm^2 or mm, respectively). Mass balance and its accumulation and ablation are the integrals over a certain time period. The basic time unit of the glacier mass balance is the hydrological year. The hydrological year in the northern hemisphere for continental glaciers is from October 1 of a year to September 30 of the following year. The observation methods of glacier mass balance mainly include direct observation methods (stakes and snow pits), repeated ground stereo photogrammetry, the water balance method and flight surveying methods.

(1) Measurement of stakes and snowpit: The method uses stakes placed directly on a glacier to regularly measure snow accumulation. Observations of each measured point are used to calculate the mass balance and its components of the entire glacier or a part of the glacier over the entire year or a certain time period. In this method, stakes are inserted vertically into the ice in the glacier

Table 10.3 Field observation and survey technique for the cryosphere

Components of the cryosphere	Observation items	Observation methods and techniques
Glacier	Mass balance	Stake and snow pit method, repeat ground stereo photogrammetry, water balance method, remote sensing technology
	Glacier area	Ground photography, aerial photogrammetry, remote sensing technology
	Glacier thickness	Hot drilling, seismic method, gravity methods, electrical measurements and theoretical estimate, remote sensing technology
	Glacier surface velocity	Theodolite intersection methods, GPS measurement, repeat ground stereo photogrammetry, remote sensing technology
	Glacier temperature	Borehole temperature measurements, non-contact radiation thermometer measurements
	Ice formation	Snow pit level observations, ice core observations
Ice sheet	Stratigraphy	Snow densification, visible reference observations, logging of ice cores
Snow cover	Snow depth	Stakes, time-out snow probes
	Snow density	Volume measurements and weighing method
	Snow water equivalent	Snow pillow, cosmic-ray meter
	Snow particle size	Optical microscopes, grid paper, CT scanners
	Snow hardness	Punching hardness tester
	Liquid water content	Snow characteristic analyzer
	Snow temperature	Thermal infrared thermometer, needle thermometer
	Impurity elements	Filtration, weighing, chemical composition analysis
Components of the cryosphere	Observation items	Observation methods and techniques
Permafrost	Seasonal freezing and thaw depth	Ground penetrating radar, mechanical detection methods, ground temperature methods
	Frozen soil temperature	Thermistor method with temperature probes, thermocouple method with temperature probes, distributed fiber optic thermometers
	Frozen soil moisture	Oven-drying method, dielectric constant method [time domain reflection (TDR) or frequency domain reflection (FDR) moisture sensors], resistance method, tension meters, neutron scattering method, gamma ray method
	Permafrost table	Ground penetrating radar, high density electrical method, pitting exploration, drilling exploration, ground temperature method

(continued)

Table 10.3 (continued)

Components of the cryosphere	Observation items		Observation methods and techniques
	Permafrost thickness		Ground temperature method, high density electrical method, drilling exploration, electrical conductivity imaging method
	Ground ice		Pit exploration, drilling exploration, high-density electrical method
River ice, lake ice	Freezing period	Ice thickness	Borehole measurements, ice radar and fixed-point automatic monitoring instruments
		Ice volume	Visual measurements
		Frazil	Ice core sampling
		Ice flower thickness	Ice flower ruler
		Ice jam	Underground electrical method
	Thawing period	Ice concentration	Ice core sampling and weighing
		Flowing ice area	Visual measurements and image methods
		Flowing ice velocity	Current meters
	Ice-flowing period	Flowing ice concentration	Visual estimation, statistical methods and photographic methods
		Ice volume	Visual measurements
	Lake ice	Thickness	Borehole measurements, ice radar and fixed-point automatic monitoring instruments
Sea ice	Sea ice extent		Satellite remote sensing
	Sea ice thickness		Ship (aircraft) loaded EM, borehole measurements, ice radar, fixed-point monitoring instruments
	Intensity		Satellite remote sensing
	Density		Sampling and weighing
	Salinity		Sampling analysis, electrical conductivity measurements

ablation areas, and the height differences of the glacier surface in reference to the stakes are observed. When the glacier surface is covered with snow (firn) and superimposed ice, their thickness (h) and average density (ρ) are recorded separately. The glacier accumulation areas mainly consist of snow and firn layers, and snow pits are mainly used, whereas stakes are used as a supplemental method. The total accumulation of the layer is calculated from the density and thickness measurements from a snow pit.

(2) Water balance method: When a glacier covers a large area and the terrain is complex, direct observations are difficult to make. In these cases, the hydrology

method can be used to estimate the accumulative mass balance of the entire glacial basin. The basic principle is the basin water balance formula:

$$\begin{aligned} B &= P - R - E - I \\ B_g &= B/k \end{aligned} \tag{10.1}$$

where *B, P, R, E* and *I* are the water balance, average precipitation, average evaporation, average runoff and water permeability of the entire basin, respectively; B_g is the mass balance of all of the glaciers and snow cover in the basin; $k = S_g/S$ is the glacier coverage ratio; S_g is the glacier area in the basin; and S is the total basin area.

Using the observed average precipitation, runoff depth, evaporation, water permeability and glacier area in the basin, the glacier mass balance can be calculated from the water balance. When the glacier melts, the water permeability is small, so it is often ignored. In marine glaciers, sublimation and condensation often compensate for each other and are negligible. In continental glaciers, evaporation can be estimated by experiments. In subcontinental glaciers, the correction coefficient for the runoff coefficient is approximately 0.95; in polar continental glaciers, it is 0.90.

(3) Survey method: This method can be used to estimate the glacier mass balance in a region. Aerial photographs during late summer or satellite image data are used to determine the equilibrium line altitude (ELA) and calculate the change in the accumulation area ratio (AAR). When the ELA is high and the AAR is small, the glaciers in the area may be in a negative mass balance, and vice versa. This method is suitable for marine glaciers because the equilibrium line is consistent with the firn line and it is easy to identify on aerial photographs.

(4) Remote sensing technology: Satellite altimetry and satellite gradiometry can be used to monitor changes in glacier mass. The former is sensitive to elevation changes. It mainly determines the ground elevation through optical stereo photogrammetry, Interferometric Synthetic Aperture Radar (InSAR) and laser altimeter data and then monitors the elevation changes of the glacier surface over different periods to estimate the changes in glacier ice volume and mass. Satellite gradiometry is sensitive to changes in mass. It can be used to analyze mass changes in the study area by monitoring the changes in the gravitational field. The advantage of satellite gradiometry is that it can monitor mass changes over areas of several hundred kilometers or more, such as the Antarctic and Greenland ice sheets. However, it cannot determine the specific location where a mass loss occurs, nor can it distinguish between the mass changes on the ground and underground.

(5) Mass balance observations of polar ice sheets: The component method is a common method to calculate the mass balance of polar ice sheets. Almost all of the mass of the polar ice sheet comes from solid precipitation. The average mass accumulation and its interannual variations can be calculated by stake and snow pit methods and sometimes from shallow ice core records, such as

from the peaks of radioactive isotopes (such as tritium). The expenditures of the ice sheet mass can be divided into the disintegration of icebergs, surface ablation, blowing snow, ice beds and the ablation at the bottom of the ice shelf. The condensation and evaporation on the ice sheet are considered to be mutually offsetting and are negligible. Measurements of iceberg calving are mainly conducted using aerial photographs or satellite images, and the thicknesses and survival durations of icebergs are also estimated. It is difficult to estimate the ablation at the bottom of an ice shelf. The ablation of an ice sheet only occurs on its margins, and a considerable amount of meltwater reinfiltrates into the firn layer and becomes internal recharge.

The ice discharge method indirectly calculates the amount of iceberg calving from the margin of an ice sheet. By multiplying the ice thickness by the average flow velocity of the ice at a particular point, the ice discharge per unit width of ice edge can be obtained. Furthermore, by comparing it with the accumulation of the ice sheet, the mass balance of the ice sheet can be obtained.

This method does not estimate the gain and loss of mass separately; however, it can directly measure the change in volume of the ice sheet. The change in the ice sheet elevation is measured by satellite altimetry, and the surface equilibrium adjustment or the vertical motion related to the structure is then estimated. Finally, the change in volume of the ice sheet is estimated. In addition, the mass balance of an ice sheet can be estimated directly by the gravity method. The development of space technology and the application of new technology have provided broad prospects for this method.

2) Glacier area

The most effective way to observe the glacier area is terrestrial stereo photogrammetry. Geodetic measurements and terrestrial stereo photogrammetry are used to measure the size (length, area, volume) and morphological changes of the entire glacier. However, ground measurements and mapping of glaciers are more complex and costly. Currently, this method is only applied to research of glaciers. Its precision is higher than aerial surveys and satellite remote sensing images.

Satellite remote sensing has been widely used to measure glacier areas. Satellite remote sensing monitoring methods can be divided into two categories: information extraction based on visual interpretation and computer-assisted classification. The accuracy of visual interpretation is high, but it is time-consuming and labor-intensive, so it is mainly applied to tasks that require higher precision. Computer-assisted classification is also relatively mature. Commonly used methods for extracting glacier boundaries from remote sensing image data include the ratio threshold, unsupervised classification, supervised classification, principal component analysis, the snow cover index, and fuzzy mathematics based on geographic information systems (GIS) and digital elevation models (DEMs). None of these methods can be used to extract the boundaries of debris-covered glaciers, so a semiautomatic classification based on the combination of remote sensing and DEMs has been proposed. Although this method has achieved good results in some research fields, it is difficult to apply to other areas. Currently, there is no general and sophisticated method for extracting the boundaries of debris-covered glaciers. In actual operations, several methods can be tried and compared to select an appropriate technology.

3) Ice flows

The transit method is the traditional method used to measure glacier velocity. In recent years, the global positioning system (GPS) has been applied to observe glacier velocity in a static or dynamic mode. The static mode with repeated measurements can improve the measurement accuracy. Dynamic models calculate the change in glacier velocity from the interval between known reference points and the measured distance. Repeated terrestrial stereo photogrammetry can also characterize glacial movements.

4) Glacier thickness

Several methods have been used to measure glacial thickness. The most direct and effective methods of obtaining the thickness distribution of a single glacier are through thermal drilling and seismic, gravity and electric measurement methods. The most accurate method is thermal drilling. However, drilling is difficult, and the glacier thickness can be obtained only at a limited number of discrete points. The seismic method is used to observe the thickness and structure of glaciers based on the distribution of elastic vibrations in ice. It is simpler and cheaper than drilling. However, the devices used for seismic methods are heavy and complex, and explosive material must be used. This method is difficult to use if the glacier is thin. The gravity method considers the glacier thickness as a function of the negative gravity anomalies obtained on the glacier. Its precision is not as high as that of the seismic method, but it is simpler. While calculating the glacier thickness of a section using the gravity method, the ice thickness at a certain point on the section must be known in advance. Currently, the most widely used method is electrical measurements, or radio echo detection (radar detection), which is better than the seismic method. For ice below the melting point, the attenuation of electromagnetic waves during propagation is very small, which implies strong penetration ability. Therefore, radar (including ground-penetrating radar and airborne sounding radar) is widely used to measure the thickness of glaciers and ice sheets. The glacier thickness data obtained by these methods can be used to determine the thickness of the entire glacier through spatial interpolation. These methods have obtained good data, especially in measurements of the thickness of continental glaciers.

5) Glacier temperature

Noncontact radiation thermometers cannot directly measure the surface temperature of a glacier accurately. Generally, these methods refer to conventional surface temperature observations at meteorological stations. First, relatively flat areas without a coarse surface are chosen to measure the glacier surface temperature, and the temperature sensor of precision thermistor is set facing east. Half of the wire is embedded into the snow and ice or fine surface till, and the other half is exposed at the surface. The probe must be pressed against the underlying surface with no gaps. Temperature measurements of snow pit profiles are taken using a precision thermistor. Drilling ice temperature measurements use several ice temperature probes integrated into a cable, which are placed into a borehole in a PE plastic pipe that is sealed at the lower end of the ice hole.

2. Snow cover observation methods

Due to the discontinuity of the snowfall process, the role of wind and snow metamorphism, there are different physical characteristics in different regions and different snow layers in the same area.

1) Snow depth

Using a ruler is a direct and convenient method for measuring snow depth. To continuously measure snow depths at numerous locations, stakes can be set up in the measurement area, and the snow depth can be observed with visual observations or telescopes. This type of measurement is more common in mountainous areas. Ultrasonic snow depth monitors are a kind of ultrasonic telemetry to measure snow depth. They emit an ultrasonic pulse to the target, receive the echo, measure the propagation time, and then calculate the distance between the sensor and the target based on the ultrasonic velocity in the air. This method is suitable for unattended fields and can perform automatic continuous monitoring of snow depth.

2) Snow density

The snow density is the snow mass per unit volume (g/cm^3). It is usually measured with a volume gauge and snow fork. A volume gauge is an instrument for measuring the snow pressure (unit: g/cm^2) and snow depth (unit: cm), and the snow density can be obtained by dividing the snow pressure by the snow depth. A snow fork is an instrument for measuring the snow density and other characteristics of snow. Its main components include a data reader and a probe, which is a steel, fork-shaped microwave resonator. It measures the dielectric constant of snow, with which the snow density and liquid water content can be calculated using a semiempirical formula.

3) Snow water equivalent (SWE)

The snow water equivalent is the depth of water that would result if the snow mass of a given region or a confined snow plot melted completely, whether over a given region or a confined snow plot. The snow pillow is a traditional method to measure the snow water equivalent. The snow pillow is installed on the ground, flush with the ground, or is buried under a thin layer of soil or sand. The liquid static pressure in the snow pillow is used to measure the weight of the snow on the snow pillow. This liquid static pressure is measured by a buoy or pressure sensor so the snow water equivalent can be measured continuously.

A cosmic ray meter is an advanced instrument that can measure snow water equivalent in place of a snow pillow. It obtains the snow water equivalent by measuring the amount of cosmic rays (such as gamma rays) emitted by the ground that are absorbed by the snow layer. Therefore, the snow samples are not damaged. Its principle is that the snow layer can attenuate the cosmic rays emitted by radioactive elements in the topsoil. The amount of gamma rays naturally emitted from the ground depends on the water content of the medium between the radioactive source (ground) and the detector. Therefore, the greater the snow water equivalent of the snow layer is, the more the rays are attenuated.

SWE can also be calculated based on the snow density and snow depth, and the equation is SWE = snow density × snow depth. Therefore, the instruments described above for measuring the snow density, such as traditional volume gauges, weighted snow gauges and snow forks, can be used to calculate the snow water equivalent.

4) Snow particle size

The snow particle size is usually expressed by two values, the physical size and the optical size. The former can be measured with grid paper, an optical microscope, and a CT scanner. Grid paper is a simple and easy way to measure the snow particle size. The snow sample is placed on a millimeter grid, and the average and maximum mean snow particle sizes are estimated by comparing the space between the snow particle and the grid lines on the grid. The snow particle size can also be measured with an optical microscope or an electron microscope. The snow sample is placed on a paper, and a photo is taken. The size of the ice particles is measured using software that can measure the length of a target object. The accuracies of the first two methods are limited by subjective factors and are also influenced by the temperature. A CT scanner is currently the best method to measure the snow particle size. In a low-temperature laboratory, this technology not only can observe the snow particle size but also can obtain the three-dimensional shapes of snow particles and snow blocks. The volume and surface area of the snow grains can then be calculated. The optical particle size is a function of the distance between the scattering probability (e.g., bubble or inner particle boundary), including optical effective particle size and optical equivalent particle size. The optical effective particle size can be calculated from the albedo ratio of the NIR band to the visible band. The optical equivalent particle size can be represented by the ratio of the volume of the snow particle to its surface area.

5) Snow hardness

The snow hardness can be measured using a durometer. Active metal weights are located at the top of the durometer. The hardness of a snow layer can be calculated based on the weight and the descent height as well as the depth of the hardness tester into the snow layer.

Table 10.4 Classification of snow cover by liquid water content

Terminology	Description	Approximate range of liquid water content (%)
Dry	The snow layer temperature is usually less than 0 °C, and loose snow grains make the viscosity very low. Even when squeezed hard, it is difficult to make a snowball from dry snow	0
Slightly wet	The snow layer temperature is 0 °C. Even at 10 times magnification, water cannot be seen, and the snow compacts easily when squeezed gently	<3
Wet	The snow layer temperature is 0 °C. At 10 times magnification, half-moon water marks can be observed in the snow grains, but water will not be produced when the snow is squeezed gently	3–8
Very wet	The snow layer temperature is 0 °C. Water is produced when the snow is squeezed by hand, but the snow contains significant amounts of air	8–15
Extreme wet	The snow layer temperature is 0 °C. The snow is soaked in water, and the air content in the snow pores is only 20–40%	>15

6) Liquid water content

The liquid water content is the volume or mass of liquid water in a unit of snowpack. It can be measured with a snow fork. Table 10.4 shows the classification criteria for the snow liquid water content by volume fraction.

7) Snow temperature

The surface temperature of the snow cover is usually measured with a thermal infrared thermometer. The internal temperature of a snow layer can be measured by a needle thermometer and a temperature probe. These are based on thermistors, and the temperature is measured by the changes in the properties of metal conductors or semiconductors with changing temperature. Needle thermometers are compact and portable. Temperature sensor probes can be placed in the field for long-term observations of snow temperature.

8) Impurity elements

Common impurities in snow cover include dust, sand, soot, biomass, and organic matter. The type and quantity of impurities are generally measured by laboratory chemical analyses of snow samples collected in the field.

3. Frozen soil observation methods

Frozen soil observations mainly include the characteristic parameters of frozen soil, the thermal state of the active layer, and the thermal state of frozen soil. The characteristic parameters include the seasonal freezing and thaw depth, depth of zero annual ground temperature variation, annual mean ground temperature, permafrost base, and permafrost thickness.

1) Seasonal freezing and thaw depth

The seasonal freezing depth is observed in seasonally frozen soil regions, and the seasonal thaw depth is observed in permafrost regions. The seasonal thaw depth is also known as the permafrost table. Three kinds of observation methods are used: mechanical probes, soil temperature measurements and visual measurements.

The mechanical probe method is mainly performed using freezing tubes and thaw tubes. Freezing tubes are used to observe the seasonal freezing depth, and they consist of an outer tube with a diameter of 40 mm and an inner tube with a diameter of 30 mm. The outer tube is a rigid rubber hose with a 0 scale line, and the inner tube is a soft rubber tube with a centimeter scale (a chain or wire rope with a fixed icicle is located within the tube). The bottom of the tube is sealed, and the top of the tube is connected to the short metal tube, wood rod and iron cover. Local clean water (low salt content) is filled to the 0 scale of the inner tube. During the observation, the inner tube is extracted to read the corresponding scales on both ends of the ice. The seasonal thaw depth can be observed with thaw tubes. The structure of a thaw tube is similar to that of a freezing tube. Soil temperature observations are commonly obtained with thermistor sensors in the active layer, and the seasonal freezing and thaw depth are determined using calculations from a soil thermistor. Because the seasonal freezing and thaw depths will change within a few years, the general observation depth must exceed the maximum seasonal freezing depth or the maximum thaw depth.

Visual observations are mainly used to observe the maximum seasonal freezing and thaw depths using pits or drilling methods. This method is generally used between mid-March and early April in seasonally frozen soil regions, and it is used to determine the maximum thaw depth in thick ground ice near the permafrost table from late September to early October in permafrost regions.

2) Depth of zero annual ground temperature variation, annual mean ground temperature, permafrost base, and permafrost thickness

The depth of zero annual ground temperature variation is commonly obtained by calculations of the resistance values of a string of thermistors at a range of depths. This depth varies by region from 10 to 20 m. Therefore, the monitoring depth of the soil temperature must be greater than 10–20 m. Generally, the depth of zero annual variation can be determined from continuous observations of soil temperature over a year, and the annual mean ground temperature (ground temperature at the depth of zero annual variation) can then be determined. In addition, the depth of zero annual variation can be estimated from calculations based on temperature measurements taken during drilling. To obtain the permafrost base or the permafrost thickness from the soil temperature, the observed depth of the permafrost should be greater than that of the permafrost base. However, the permafrost base or permafrost thickness can be estimated using the temperature gradient of the permafrost from temperature observations.

3) Thermal state of frozen soil

The thermal state of frozen soil includes the spatial and temporal variations at various depths, which can be obtained by measuring thermistor values at different depths. A string of thermistors at different intervals and different depths is developed based on the observation goals. In general, two methods of thermistor observations are used. First, high-precision multimeters with a resolution of $\pm 1\ \mu V$ can be used for manual observations, which are then converted to soil temperatures based on a calibration equation. The second method is by automatically recording the soil temperatures at a regular time interval with multichannel data loggers.

4) Hydrothermal process of the active layer

The soil temperature can be measured by thermistors, platinum resistance or thermocouples. Water can be monitored using time domain reflection (TDR), frequency-domain reflection (FDR) or other types of moisture sensors. Due to spatial and temporal changes in the shallow soil temperature and moisture, the depths of the soil water and temperature measurements in the active layer should be set at 5, 15, 30, 50, 80, 120, 180, 240, and 300 cm to the permafrost table. Readings are automatically recorded at a regular time interval of 30 min using multichannel data loggers.

4. River ice and lake ice observation methods

Observations of river ice and lake ice are divided into visual observations, manual measurements and automatic observations. Observations of lake ice mainly include the initial icing period, the complete occlusive phase, the ice melt period, the complete thaw period, and the ice thickness. The main content of river ice observations is the ice conditions. Different observation methods are used in the three transit periods. During the ice flow period, the characteristics of the river ice flows can mainly be observed, such as the ice density, ice flower, ice size, ice volume, and changes in shore ice change. During the freeze-up period, the main observations are of the location of the frozen river, length, width, and number of segments, the frozen state (e.g., flat seal, seal), ice thickness, ice volume, ice water area within the cross section, water ice, ice jam conditions, and the river channel storage capacity. During the thawing period, the main observations are of the ice quality, changes in ice color, ice removed from the shore, sliding, thawing mouth location, time, length, parts, ice area and speed, ice plugs, the accumulation of ice, ice dam locations, degree of obstruction, development and changes, and the river floodplain, such the level of water against dikes. Most of the elements measured during these stages require visual or manual measurements. With advances in science and technology, several prototype observation instruments have been developed. The SWIPS system measures the water temperature, the rate of the growth and ablation of sea ice and the temperature of the riverbed. The advantage of this system is that it can be used to measure the formation of ice in the river and the water during the thawing period, the development of floating ice, the prototype and the observations under severe conditions under sea ice.

1) River ice and lake ice thickness

The thickness of river ice and lake ice can be measured by magnetostrictive ice thickness measurement sensors and resistivity ice thickness measurement sensors. A magnetostrictive ice thickness measurement system consists of two pieces of hardware: an instrument box and a measuring rod. During the measurement, the lower magnetic loop moves downwards under the weight of a heavy hammer and is placed on the ice/snow surface. The motion of the lower magnetic ring is controlled by a pneumatic control. The lower magnetic ring has an airbag that is connected to the cylinder through a guide pipe. The airbag expands as the cylinder compresses the air as it driven down by the motor. The force of the magnetic loop mechanism and the force of gravity pull it upward to contact the surface of the ice. The magnetostrictive sensor fixed on the circular and annular core distance are then used to obtain the positions of the ice/snow surface and bottom surface, and the measured values are stored in a data recorder.

2) Concentration of flowing ice

Image methods are used to automatically monitor the density of flowing ice by setting up a high resolution camera on the shore to collect images. The remote data are then transmitted to the monitoring center. Finally, the density of ice can be analyzed through image processing. The density of the flowing ice increases with decreasing temperature.

3) Ice flow rate

Image methods can also be used for real-time monitoring of the ice flow velocity, but it is difficult to monitor the flow rate at night because a method for monitoring the velocity using infrared remote sensing images has not been developed. Several new velocity monitoring instruments have been developed, such as the ADCP (acoustic Doppler current profiler). In addition, several small monitoring sensors are available, such as the wireless monitoring ZigBee technology that carries a speedometer.

5. Sea ice observation methods

Sea ice observations include the extent, intensity, thickness, shape and type of sea ice. Sea ice field observation technology (Fig. 10.3) is mainly used on research vessels, ice stations and buoys.

1) Ship-based sea ice observations

Because it is a mobile platform, sea ice observations collected by an icebreaker during navigation are beneficial for collecting data over wide areas and can guarantee the observation precision. These data are a bridge linking satellite remote sensing data and ice observations. Sea ice observations based on research ships mainly include

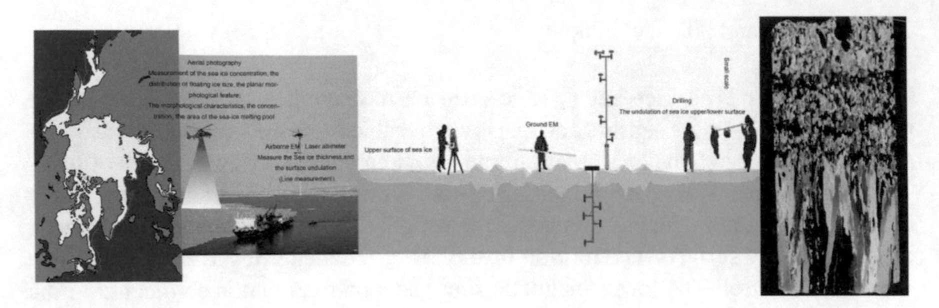

Fig. 10.3 Sea ice observation systems based on remote sensing, shipping, helicopters, stations, and ice coring

observations of morphological parameters, such as the sea ice concentration, sea ice thickness, the coverage of melt ponds and the distribution of ice ridges. The main observation techniques include manual observations based on observation criteria, electromagnetic induction technology for sea ice thickness and image recognition for sea ice morphology.

The WMO defines parameters for the classification and morphology of sea ice and uses the egg code to record the sea ice in three categories. Ship-based observations and records are helpful for obtaining the spatial distribution of the basic physical parameters of sea ice, optimizing deciphering algorithms for satellite remote sensing products, providing feedback to sea ice forecasting systems, and improving their prediction accuracy. The standard for the classification of sea ice age by the WMO oval recording method is shown in Table 10.5.

In addition to manual observations, a series of external equipment can be used on research vessels to observe the physical characteristics of sea ice. As shown in Fig. 10.4, the Chinese "Xuelong" research vessel carries an infrared temperature measuring instrument for sea ice and to measure sea surface temperatures and uses an outward tilting automatic camera to monitor the sea ice concentration and surface morphology. It also uses a downward-looking video recorder to monitor the icebreaker pressure over the sea ice section (the section is determined by comparing the thickness and suspension markers for the sea ice thickness and ice) and uses an EM-31 electromagnetic induction measuring instrument to measure the sea ice thickness.

Table 10.5 Classification of sea ice

Classification	Frazil	Shuga/Grease	Nilas	Pancakes	Young gray ice
Thickness	/	<0.05 m	0.05–0.10 m	<0.30 m	0.10–0.15 m
Classification	Young gray-white ice	Thin first-year ice	First-year ice	Thick first-year ice	Multiyear ice
Thickness	0.15–0.30 m	0.30–0.70 m	0.70–1.20 m	>1.20 m	>2.50 m

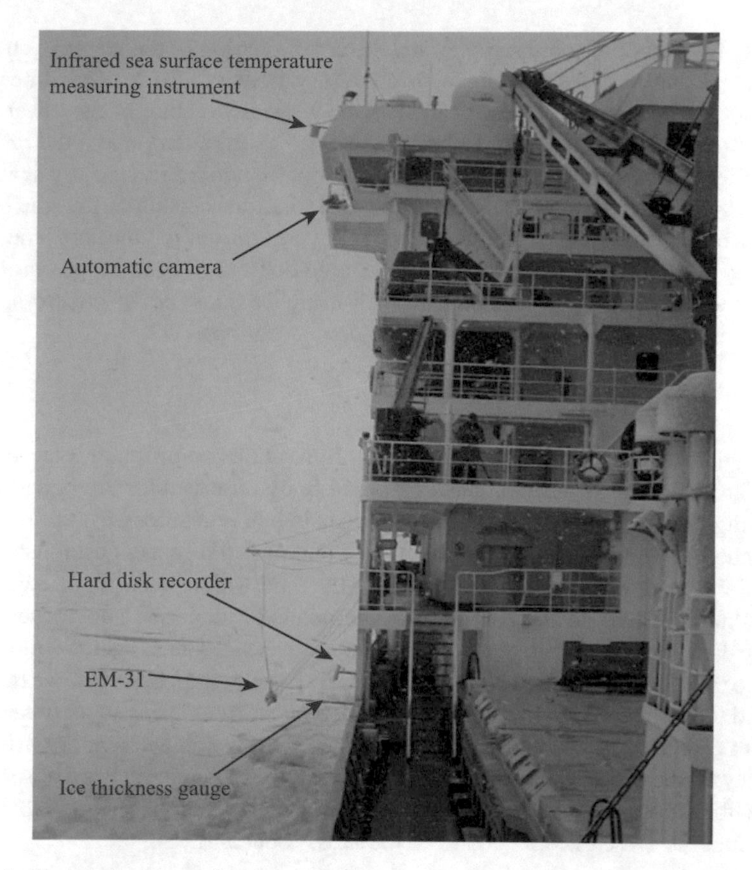

Fig. 10.4 Observation system for sea ice on the Snow Dragon vessel during the Chinese Antarctic Expedition

Electromagnetic induction has been widely used for sea ice thickness observations. Electromagnetic induction technology is nontouch, and observations are easy to implement with high data accuracy. It is suitable for many purposes, including ice observations, shipborne suspension observations and airborne observations. Underwater elevation sonar is also one of the main automated techniques for observing sea ice thickness. Observations from submarines and underwater robots are similar to ground-based and airborne observations; they are snapshot observations. Submarine observations are collected at the scale of the ocean basin, so they are conducive to analyzing the large-scale distribution of sea ice thickness. Underwater robot observations are generally limited to 100 m but are useful for obtaining the three-dimensional structure of the ice bottom.

2) Ice-based seawater observations

Observations of the base of sea ice are mainly used to observe multiple elements of sea ice through the establishment of ice stations. Short-term ice station observations focus on the collection of ice core samples and the determination of the physical structure of the ice. Long-term ice stations focus on observations of air-ice-ocean interaction processes. Long-term observations include the vertical structure of the lower atmosphere, the structure of the atmospheric boundary layer, radiation and the turbulent flux of the air-ice interface, mass balance of snow cover and sea ice, the physical structure of the snow cover and sea ice, shortwave radiation transmission at the bottom of the ice, the flow field in the upper ocean layer, and sea ice motion. The corresponding observation techniques include moored boats/GPS probes, meteorological gradient towers, EM-31 instruments for measuring ice thickness by electromagnetic induction, spectral flux instruments, and underwater robots.

3) Ice-based buoy observations

Ice-based buoy observations are similar to observations from floating ice stations, which are Lagrangian observations. Ice-based buoy observations are beneficial for collecting data regarding to the key air-ice-ocean interaction processes. The buoys are unattended, which greatly reduces the human and material costs of establishing and maintaining floating ice stations. Therefore, they are widely used for sea ice observations in the Antarctic and Arctic. The parameters of ice-based buoy observations include the atmospheric boundary layer, snow cover-sea ice mass balance, movement of sea ice and ice deformation, turbulence at the bottom of the ice, shortwave radiation flux and the upper ocean stratification structure and current. Movement of sea ice in the 20th century was generally observed by ARGOS, which had a poor positioning accuracy on the order of 100 m. Since 2000, buoys have generally been observed by GPS with an accuracy of 10–20 m. Sea ice movement is the most observable of the parameters, so these historical data are the most abundant.

6. Methods of social economic surveys in the cryosphere

Methods of social economic surveys in the cryosphere refer to investigators using specific methods and means in core areas, functional areas and impact areas of the cryosphere to identify relevant cryospheric changes and their effects on humans, livelihoods, resources, infrastructure, society, the economy, traditional knowledge, and the adaptation of existing information. These methods include reviews, sorting, analysis, and interpretation.

1) Principles of social economic surveys in the cryosphere

(1) Objectivity: The influences of changes in the cryosphere on the natural environment and economic society are complex and diverse, and there are significant regional differences. Different surveys are performed based on the focus of the research and its intended purpose. Survey areas are chosen based on the situation and the objective of the investigation. To be objective, it is necessary to explore the causal relationships between the changes in the cryosphere and the social economy.

(2) Authenticity: Whether research is scientific or not depends mainly on the authenticity of the collected data. The cryosphere areas in China are mainly located in the western high mountains and the plateau region, which are not easy to reach and are mostly inhabited by minority nationalities. These factors, including language barriers, significantly affect the investigation process and the authenticity of obtaining information. Therefore, while carrying out social investigations, it is better to anticipate various situations that may occur in the study areas and to prepare as early as possible. For example, a questionnaire can be translated into the regional language of the survey area or through the local government by using translators to obtain data.

(3) Accuracy: Subjectivity is considered to be a major flaw in social research studies. To overcome this problem, we must be realistic in our social investigations and accurately describe the investigation.

2) Common methods of social surveys in the cryosphere

(1) Methods to determine the subject of the investigation include general investigations, typical investigations, sample surveys, and case investigations. Sample surveys are mainly used in social investigations of the cryosphere.

(2) Investigation and data collection: questionnaire and interview methods are used to perform social surveys in the cryosphere. Questionnaires are designed to fully consider the purpose of the survey, the content of the survey, the nature of the samples, the data processing and analysis methods, the financial resources, manpower and time, and the use of the questionnaire. The questionnaire mainly includes a cover letter, guidance language, questions and answers, and coding. Questionnaire surveys can be divided into a preparation stage, investigation stage, analysis stage and summary stage.

10.3 Experimental Techniques in the Laboratory

10.3.1 Mechanics

1. Uniaxial test

Uniaxial tests are mainly used to perform unconfined compression strength tests on materials such as frozen soil and ice. There are two kinds of samples: disturbed samples and undisturbed samples. The uniaxial test equipment used for frozen soil and ice is usually modified from equipment used for materials testing, and it includes a temperature control chamber, axial compression equipment, and an axial stress and deformation system. Different loading control methods are used according to the test requirements. Strength tests usually use a constant strain rate, whereas creep tests use a constant load. For uniaxial strength tests, the uniaxial compressive strength (unconfined compressive strength) is obtained. For uniaxial creep experiments, three elements of creep (failure time, failure strain and minimum creep rate of frozen soil), long-term strength curves and long-term strength limits can be obtained.

2. Triaxial test

Triaxial tests are mainly used to perform triaxial compression tests on materials such as frozen soil and ice. The samples can be disturbed samples and undisturbed samples. There are two unique requirements of the triaxial equipment used for frozen soil. One is that the temperature in the pressure chamber must be controlled, and the other is that the axial compression and confining pressure provided by the equipment should be greater than those for soil; for example, the maximum axial forces are 5 t, 10 t and 15 t, and the maximum confining pressures are 5, 10 and 20 MPa. The triaxial test equipment mainly includes a pressure chamber, axial pressure system, confining pressure system, counter pressure system, pore water pressure measurement system, and an axial deformation and volume change measurement system.

10.3.2 Thermology

Thermology is a branch of physics concerned with heat and temperature. The main parameters are the thermal conductivity, thermal diffusivity, and specific heat capacity (or specific heat), and they are closely related {$\alpha = \lambda / \rho C$, where α is the thermal diffusivity (m/h); λ is the thermal conductivity [W/(m K)]; ρ is the density (kg/m^3); and C is the specific heat capacity [J/(kg K)]}. As long as two parameters can be measured, the third parameter can be calculated.

1. Thermal conductivity

Thermal conductivity is the property of a material to conduct heat. In the laboratory, steady-state and nonsteady-state methods are mainly used to measure the thermal conductivity of a material. In steady-state methods, the sample is heated, and the temperature difference inside the sample causes heat to conduct from high temperature to low temperature. Therefore, the temperature at each point changes with the heating rate and the heat transfer rate. At a controlled temperature, the heat transfer process reaches an equilibrium state and forms a stable temperature distribution. The thermal conductivity can be calculated according to the relationship between the heat, sample thickness, sample area, time and temperature. Steady-state methods include the longitudinal heat flow method, radial heat flow method, direct electric heating method, thermoelectric method and thermal comparison method. Intelligent dual-plate thermal conductivity detectors measure the thermal conductivity of thermal insulation material using steady-state methods (longitudinal heat flow method- absolute method). In nonsteady-state methods, the sample is heated for a short time to instantaneously change the sample temperature. The thermal conductivity of the sample is obtained by solving the differential thermal conductivity equation based on the temperature change and heating time. The transient heat flow method (hot wire method), which is a nonsteady-state method, can be used to measure the thermal conductivities of thermal insulation materials, rock and soil. The transient heat flux method (laser flash analysis) can be used measure the thermal conductivities of metals, rock, soil, liquids and powders.

2. Thermal diffusivity

The thermal diffusivity can be obtained from the thermal conductivity and specific heat capacity, and it can also be directly measured. The cylinder transient heat flow method is generally used in the laboratory, whereas the temperature wave method and thin plate method are used in the field. The normal state technique is used for rock and soil in the laboratory.

3. Specific heat capacity

The specific heat capacity is usually measured in the laboratory using a traditional specific heat capacity tester, plate thermal conductivity tester and thermal analysis. The thermal analysis method is used most often. The Q2000 differential scanning calorimeter, which uses the thermal analysis method, can measure the specific heat capacities of metals, rock-soil, liquids and powders.

10.3.3 Optical Methods

1. Single particle soot photometer (SP2)

The single particle soot photometer (SP2) utilizes the high optical power available intracavity from an Nd-YAG laser as an analytical technique. Light absorbing particles, mainly black carbon (BC) or elemental carbon (EC), absorb sufficient energy and are heated to the point of incandescence. The energy of this incandescence is measured and used to quantitatively determine the mass of black carbon. Particles that do not absorb light will still scatter light according to the Mie theory. The SP2 operates in a single particle mode to count and measure the light scattering or incandescence of each particle.

An air jet containing a sample aerosol intersects an intense Nd-YAG intracavity continuous laser beam pumped by a diode laser. Aerosol particles elastically scatter light in the laser beam and, if they contain light-absorbing material at the wavelength of the incident light, such as BC, they are heated as they absorb the laser radiation and eventually incandesce and vaporize due to evaporation at high temperatures. The mass of BC with diameters of approximately 150–1000 nm is derived by measuring the intensity of the laser-induced incandescence signal from individual particles. The size of nonlight absorbing aerosols with diameters of 250–1000 nm can be measured by the scattering intensity. BC particles coated by nonlight absorbing material scatter laser light more efficiently than uncoated BC due to the larger scattering cross section.

This technique is designed to measure the mass and mixing state of individual BC of the ambient aerosol. It can also be used to analyze the black carbon content of water samples after the water sample is ultrasonically atomized and dried.

2. Laser particle size measurement technology

Since the back-scattering of particles in a uniform liquid solution is linearly related to the particle size and concentration, the particle concentration and size distribution in the solution can be determined by measuring the intensities of the scattering particles at different scattering angles. Light from a He–Ne laser is expanded by a beam expander and then irradiated onto the tested particles. The light scattered by the particles is focused by a lens, received by a photo-detector, and converted into an electrical signal, which can be related to the particle diameter and size distribution. This technique is mainly used to analyze the particle concentration and size distribution in snow and ice samples and water samples.

3. X-ray diffraction technique

The X-ray diffraction technique is used to determine the atomic and molecular structure of crystals, in which the crystalline atoms cause a beam of incident X-rays to diffract into many specific directions. By measuring the angles and intensities of these diffracted beams, a crystallographer can produce a three-dimensional picture of the density of electrons within the crystal. Crystals are regular arrays of atoms, and X-rays can be considered waves of electromagnetic radiation. Atoms scatter X-ray waves, primarily through the atoms' electrons. The crystal lattice (the mean positions of the atoms, regular arrays of atoms) as well as the unit cell, lattice defects, shape, and other information can be determined from the electron density.

X-ray diffraction analysis is simple, fast, economic, stable, and does not cause damage to the sample. It has wide applications to the mineral crystallization process, mineral surface analysis, mineral quantitative analysis, and mineral crystal structure determination.

4. Optical physical imaging technology

Microscopes are instruments used to see objects that are too small to be seen by the naked eye. Microscopes include optical microscopes, electron microscopes (transmission electron microscopes and scanning electron microscopes), stereoscopes, fluorescence microscopes, and various types of scanning probe microscopes. Electron microscopy and electron micro-probes use a beam of accelerated electrons as a source of illumination to obtain the surface micro-scopic characteristics of a sample and can quantitatively determine the distribution of micro-elements when coupled with a wavelength dispersive spectrometer or energy dispersive spectroscope. A scanning electron microscope (SEM) produces images of a sample by scanning the surface with a focused beam of electrons, which interact with the atoms in the sample and produce various signals that contain information about the sample's surface microstructure and composition. An SEM is equipped with two detectors (a secondary electron detector and a back-scattered electron detector). The elastic scattering of electrons generally occurs within a few hundreds of nanometers of the surface of the sample. Because its yield increases with the sample's atomic number, it can be used to analyze not only the topography but also the composition of the sample. The technology is mainly used in mineralogy, rock topography, structure and structural analysis.

10.3.4 Microphysical Structure

Based on the classification of the cryosphere components, the techniques for analyzing the physical structure can be divided into two categories: ① technologies for snow and ice (including snow, glacier ice, river ice, lake ice and sea ice); and ② technologies for frozen soil.

The micro-physical structure of snow and ice mainly include the crystal morphology, structure and grain size. Morphology detection also includes the firn porosity, ice formation depth, density of snow and ice, concentration of impurities, and the number, size, shape and distribution of the bubbles that form in snow during glaciation. The equipment for determining the microphysical structure of snow and ice mainly include weighing meters, magnifying glasses, digital image equipment, high magnification microscopes, electron microscopes and laser particle size analyzers. Ice structure analysis utilizes the anisotropy of ice under visible light to obtain the statistics of the ice crystal size and crystallographic orientation and to analyze the direction of the crystal axis, which are used to determine the growth directions and subsequent changes of ice crystals. Ice core physical parameter scanning technology utilizes monochromatic light of a characteristic wavelength to scan an ice core to obtain a complete image and reproduce the physical characteristics of the ice core to clearly reflect layers of dust, white ice, ice lenses and firn. The ice crystal's grain size reflects the temperature during ice formation, and the ice crystal's C-axis can reflect various processes of ice formation.

The microphysical structure of frozen soils mainly includes the lithological characteristics, moisture content, density, color and thickness of frozen soil as well as the mixed structure of rock and soil, the distribution of unfrozen water in frozen soil, the structure of the frozen fringe, the type of soil and its cryostructure, the morphology, granular structure, and porosity of soil particles and the structure of segregated ice. The main instruments used are digital image equipment, magnification microscopes, electron microscopes, laser particle size analyzers, pulsed nuclear magnetic resonance, and automatic specific surface area and porosity analyzers.

10.3.5 Chemical Composition

1. Ion chromatography (IC)

Ion chromatography is a liquid chromatographic method for the analysis of ions and polar molecules that separates ions and polar molecules based on their affinity to the ion exchanger. Based on the separation mechanism, ion chromatography can be divided into high-performance ion exchange chromatography (HPIC), ion exclusion chromatography (HPIEC) and ion-pair chromatography (MPIC). HPIC is the most commonly used type of ion chromatography; it usually uses an ion exchange resin with low capacity to separate ions. The most important component of an ion chromatograph is the separation column. Ion chromatography can simultaneously detect a variety of ions in a sample rapidly with good selectivity and high sensitivity. The average analysis times for anions (F^-, Cl^-, Br^-, NO_2^-, NO_3^-, SO_4^{2-}) and cations (Li^+, Na^+, NH_4^+, K^+, Mg^{2+}, Ca^{2+}) are less than 15 min. The detection limit for anions is usually less than 10 μg/L. Ion chromatography is widely used in the analysis of soluble species in snow and ice samples.

2. Inductively coupled plasma mass spectrometry (ICP-MS)

Inductively coupled plasma mass spectrometry is a highly sensitive analytical technique that combines the high temperature ionization characteristics of inductively coupled plasma with the high sensitivity and fast scanning of a mass spectrometer. Quadrupole mass spectrometry is usually used as a mass analyzer, as well as high resolution dual-focus magnetic field mass spectrometry and time-of-flight mass spectrometry. The main features of this technology are high sensitivity, short analysis time (one can quantitatively analyze several tens of elements in a few minutes), simple spectra with low interference, a wide linear range of up to 7–9 orders of magnitude, and simple sample preparation. This technique can be used not only for elemental analysis but also to determine the isotopic composition, and it has been widely used in analyses of inorganic elements and their isotopes in snow and ice.

3. Thermal-optical analysis of organic carbon (OC)/elemental carbon (EC)

Thermal-optical carbon analysis is based on the preferential oxidation of organic carbon and elemental carbon compounds at different temperatures. It relies on the fact that organic compounds can be volatilized from a sample in helium gas at low temperatures, whereas elemental carbon is not oxidized and removed. In helium gas, the sample is heated to liberate the organic compounds at increasing temperatures to convert these compounds to CO_2 by passing them through an oxidizer (heated manganese dioxide, MnO_2). By adding 2% oxygen to the helium gas and increasing the temperature, the elemental carbon will be oxidized to CO_2. CO_2 is produced to methane by passing the flow through a methanator (hydrogen-enriched nickel catalyst), and the CH_4 is quantified by a flame ionization detector (FID). In the course of analyzing organic carbon in the helium gas, some organic materials in the samples are pyrolyzed to elemental carbon. The correction for the pyrolysis of organic carbon compounds to elemental carbon is carried out by continuously monitoring the reflectance and transmittance using a helium–neon laser and photo detectors. This method can effectively measure between 0.2 and 750 μg C/cm^2 with lower quantifiable limits of 0.82 μg C/cm^2 for the total organic carbon and 0.20 μg C/cm^2 for the total elemental carbon. This technology can be applied to the analysis and detection of organic carbon and elemental carbon in water and sediments.

4. Isotope-ratio mass spectrometry (IRMS)

Isotope-ratio mass spectrometry (IRMS) is a specialization of mass spectrometry, in which mass spectrometric methods are used to measure the relative abundances of isotopes in a sample. Most of the instruments used for the precise determination of isotope ratios are of the magnetic sector type. The core parts of the IRMS are the ion sources, mass analyzers and ion detectors. The instrument operates by ionizing the sample under high vacuum conditions, accelerating it over a potential in the kilovolt range, and separating the resulting stream of ions according to their mass-to-charge ratios (m/z) in the mass analyzer. After separation, the beams with lighter ions bend at smaller radii than those with heavier ions. The current of each ion beam is then measured using a 'Faraday cup' or multiplier detector. This technique has been widely used in the earth and environmental sciences, such as the nuclear sciences, geological dating, isotope dilution mass spectrometry, and isotope tracer analysis.

5. Cold vapor atomic fluorescence spectroscopy (CVAFS)

Cold vapor atomic fluorescence spectroscopy is a type of atomic fluorescence spectrometry that is used to quantitatively measure trace amounts of volatile heavy metals by measuring the fluorescence intensity of elemental atomic vapor. This method has advantages of high sensitivity, low detection limits, good stability, a wide linear range and a simple spectral line. It is the most commonly used method for the determination of mercury (Hg) in various environmental samples. The different forms of mercury in the sample are converted into atomic mercury, which is loaded into the system with high purity argon gas. The ground-state mercury atoms are excited by ultraviolet light with a wavelength of 253.7 nm, and the excited mercury atoms emit fluorescence of the same wavelength. The proportional relationship between the intensity and mercury content is used to determine the mercury content of samples with a detection limit of less than 1 ppt. Cold vapor atomic fluorescence spectrometry can detect the mercury concentrations in various forms of snow and ice.

6. Cavity ring-down spectroscopy (CRDS)

Isotope analysis technology based on cavity ring-down spectroscopy (CRDS) has the advantages of fast measurement, high sensitivity and a wide measurement range. The main components of the CRDS are the laser source, a pair of highly reflective mirror-formed optical resonators and photodetectors. In CRDS, the beam from a single-frequency laser diode enters a cavity defined by two or more high-reflectivity mirrors. When the laser is on, the cavity quickly fills with circulating laser light. A fast photodetector senses a small amount of light leaking through one of the mirrors to produce a signal that is directly proportional to the intensity in the cavity. When the photodetector signal reaches a threshold level (in a few tens of microseconds), the laser is abruptly turned off. The light already within the cavity continues to bounce between the mirrors, but because the mirrors have slightly less than 100% reflectivity, the light intensity inside the cavity steadily leaks out and decays to zero exponentially. This decay, or "ring-down", is measured in real time by the photodetector. The CRDS technique has been used in the high precision measurement and quantification of stable isotope ratios for liquid water, water vapor, and greenhouse gases with small gas-phase molecules. For example, water isotopes have a unique and differential absorption spectrum that is used to quantify the stable isotope ratios (e.g., $^{18}O/^{16}O$ and $^{2}H/^{1}H$).

7. Continuous flow analysis (CFA)

Continuous flow analysis is a well-established method that combines an on-line melting ice core technique with on-line analytical equipment or a collection of other on-line sample preparation devices. By optimizing the melting speed of the ice core sample, CFA can provide high temporal resolution records and very efficient sample decontamination because only the inner part of the ice sample is analyzed. This method greatly improves the efficiency of batch ice core sample analysis, which is more suitable for analyzing ice core samples at drilling sites and avoiding possible contamination during transport and post-processing. Flow injection analysis is a kind of analytical technology for automatic on-line processing and measurement of solutions that was developed in the mid-1990s. It can be classified into flow injection spectrophotometry, flow injection atomic spectrometry, flow injection electrochemical analysis, flow injection enzyme analysis, and flow injection fluorescence and chemiluminescence, which can be used to identify a variety of ions and elements in snow and ice.

8. Gas chromatography (GC)

Gas chromatography is a common type of chromatography used for separating and analyzing compounds that can be vaporized without decomposing. In gas chromatography, the mobile phase (or "moving phase") is a carrier gas, which is usually an inert gas such as helium or an unreactive gas such as nitrogen. The stationary phase is a microscopic layer of liquid or polymer on an inert solid support. The process of separating the compounds in a mixture is carried out between a liquid stationary phase and a gas mobile phase based on the difference in the partition coefficients of the mobile phase and stationary phase.

When a mixture of different chemical constituents of a sample enters the column, they pass through the column in a gas stream (mobile phase) at different rates depending on their chemical and physical properties and their interactions with the stationary phase. The components are separated in the column, which causes each to exit the column at a different time (retention time). As the chemicals exit the end of the column, they are detected and identified electronically.

The most commonly used detectors are the thermal conductivity detector (TCD), flame ionization detector (FID), electron capture detector (ECD), and mass spectrometer (MS). Mass spectrometer detectors are a common type of mass detector. Gas chromatography-mass spectrometry (GC–MS) analysis combines high chromatographic separation with the high sensitivity of mass spectrometry and strong structural identification. A GC can be used for the analysis and detection of organic matter in snow and ice samples and other environmental samples.

10.3.6 Methods of Dating

Age dating methods are the basis for analyzing the long-term evolution of the cryosphere. The chronologies of river ice, lake ice and sea ice are generally shorter than those of ice cores. Generally, cryospheric dating techniques mainly refer to the dating of ice cores and sediments (or deposits).

1. Ice core dating

Five technical methods are currently used in ice core dating, including the seasonal parameter method, reference horizontal position method, radioisotope method, theoretical model method and comparative analytical method through climate event. The seasonal parameter method has the highest precision and is widely used to date the upper layers of ice cores. The main parameters are the hydrogen and oxygen stable isotope ratios, the soluble ionic concentration and the insoluble particle content. The method is based on the significant seasonal variations of the parameters, which can be determined by counting layers. The reference horizontal position method is aimed at particular years, for which nuclear tests (tritium content and beta activation degree) and volcanic events provide references that augment other continuous ice core dating methods (such as the seasonal parameters method). This method is not sufficient for the determination of the ages of ice cores over long time scales, and it is generally used for absolute age dating to less than 100 years in specific intervals of ice cores. Radioactive isotopes can be generated by cosmic radiation and nuclear tests. The age of an ice core can be determined by measuring the radio-isotope strengths of different layers of the ice core and from the decay cycles of each radioactive nuclide. For example, ^{210}Pb has been successfully used to study the changes in the accumulation of ice cores over the past 100–200 years. ^{10}Be has also provided good results in the Antarctic ice cap. Theoretical models are widely used in studies of deep ice cores. The principle is that snow will gradually accumulate after settling on the surface of the glacier and then move toward the lower part of the glacier or away from the watershed over time. If ice is assumed to be incompressible, the snow in the glacier experiences only plastic deformation after it becomes ice; that is, the vertical pressure causes the ice to expand horizontally. The pressure in the upper part of the ice body can be calculated based on the variation of the measured density with depth. Because of thinning, ice in the glacier moves down every year. In a relatively stable ice cap or at the center of an ice cap, the annual vertical velocity of points in the ice mass must be equal to the ice's equivalent annual thickness. The current theoretical model is the Nye time scale. In the case of large deviations of the ice core model and no absolute dating of specific layers, the comparative analytical method through climate event can be used to determine the age. Based on the indications of extreme weather events in the ice cores, the cores can be compared with other media (e.g., other ice cores, lake and deep sea sediments, cores, stalagmites, trees) in which the same events have already been calibrated and dated and establish the overall time sequence of the ice cores. Dating using this method has a low resolution; therefore, it is only used to determine the approximate ages of changes in climate due to events.

2. Paleoglacial dating

Dating glacial landforms is a fundamental requirement for studying paleoglaciations, especially in Quaternary glaciation research and reconstructing past climates. In recent decades, numerous dating techniques that can potentially directly determine the ages of glacial landforms, including cosmogenic radionuclides (CRN), terrestrial in situ cosmogenic nuclides (TCN), optically stimulated luminescence (OSL) and electron spin resonance (ESR), have been developed and applied widely in paleoglaciology research. These dating techniques can be combined with lichenometry, traditional ^{14}C and accelerator mass spectrometry (AMS) ^{14}C, $^{40}K/^{40}Ar$, $^{40}Ar/^{39}Ar$, U-series, paleomagnetism and thermoluminescence (TL) techniques, which has allowed for significant advances in research on the patterns and timing of Quaternary glaciations throughout the world. Each technique has its own optimum dating range. Sometimes, all or part of their dating ranges overlap; therefore, if suitable dating materials can be found in a glacial complex, several dating techniques can be applied to improve the dating accuracy and reliability.

1) Lichenometry

Lichenometry is an effective dating technique to determine the ages of fresh glacial and periglacial deposits because some pioneering plants, such as moss and lichen, will colonize the bedrock or fresh glacial complex after glacier retreat. Several kinds of lichen, such as *Rhizocarpon geographicum (L.)DC.*, can be used to determine the timing of these events. Two parameters must be known: the growth rate and the maximum diameter of the lichen in a selected quadrat. The age of glacial retreat or of a fresh glacial complex can then be obtained. The dating range covers several to hundreds of years, and the upper limit can sometimes reach 5 ka. Therefore, this dating technique can be applied to determine the neoglaciation and Little Ice Age (LIA) in the Holocene. Its main advantage is to date the LIA, which can make up for the dating gap of 200–500 years ago.

2) ^{14}C dating

^{14}C dating is the earliest developed and most commonly applied radioactive dating technique. Most researchers consider it to be a mature and reliable dating technique, especially in determining the ages of organic materials. Its outstanding characteristics are its accuracy, use of ubiquitous organic and inorganic materials (e.g., rotted wood, charcoal, peat deposits, pollen, shells, furs, bones, secondary carbonate) and easy sample preparation. With progress in science and technology, the AMS ^{14}C method was developed and applied in the late 1970s. It has several advantages over traditional ^{14}C dating, such as requiring only small samples, high sensitivity of measurements, an increased upper dating limit (theoretically up to 100 ka), and shorter measurement time. Glaciers form at high elevations and high latitudes, and these environments restrict the growth of plants; therefore, the biological productivity in these areas is low, and the amount of organic material that is preserved in the glacial deposits is small and difficult to find. These factors restrict the use of this dating method in paleoglaciology research. Abundant ^{14}C results from glacial and glaciofluvial deposits have been reported in previous studies, but they are indirect ages rather than direct ages. Another disadvantage is the short half-life of ^{14}C ($T_{1/2} = 5730$ years), which limits the use of this dating technique to determine the ages of sediments and events during the late stage of the last glacial cycle.

3) Optical dating

Optical dating methods include thermo luminescence (TL) and optically stimulated luminescence (OSL). The former was developed earlier and is stimulated by heat, whereas the latter is stimulated by different wavelengths of light (e.g., blue, green, and infrared light). The basic principle is that the dating signals are reset by heat, light, weathering, erosion, and exposure to light during transportation. After sediments are deposited, the dating signals will accumulate again due to irradiation by α, β, and γ rays from internal and external minerals and cosmic rays from space. The signal intensity will increase with time before reaching saturation. The accumulated signals in some minerals, such as quartz and feldspar, will be released by heat or light of a specific wavelength, and the signal intensity is proportional to the total accumulation dose (TD). In addition, the annual dose rate (D) can be measured in situ or calculated from the radioactive elements in the minerals and their surroundings. This dating technique has been applied successfully in Quaternary glaciation research. Theoretically, this dating technique could be used to determine the ages of glacial complexes since the penultimate glaciations. Light should be absolutely avoided during sample collection. However, due to the potential partial bleaching of glacial deposits, more work is required.

4) Electron spin resonance dating

Electron spin resonance is also known as electron paramagnetic resonance (EPR). The principle of ESR dating is similar to that of optical dating. The dating signals can be completely or partially bleached by several mechanisms, such as shear stress from tectonism (e.g., fault activity), collisions (e.g., debris flows), sunlight, heat (e.g., geothermal, volcanism, natural and anthropic fire) and mineral recrystallization. After deposition, the dating signals will accumulate again due to irradiation by α, β and γ rays produced by the decay of radioactive elements (U, Th, ^{40}K, etc.) in the minerals and their settings as well as from high-energy cosmic rays. Under these conditions, some free electrons and vacancies will form, and these free electrons can easily be trapped by impurities (e.g., germanium (Ge), titanium (Ti), aluminum (Al)) and lattice defects to form impurity centers and defect centers in some minerals. These centers are paramagnetic; therefore, they are called paramagnetic centers. These paramagnetic centers can be identified by an ESR spectrometer. The number of paramagnetic centers is proportional to the geological time; the longer the time since deposition is, the more paramagnetic centers there are prior to saturation of the dating signals.

The number of paramagnetic centers is proportional to the total irradiation dose that was received by the minerals, and the accumulation dose is closely related to the signal intensity. Therefore, the age can be determined if the accumulation dose can be determined and if the annual dose can be calculated by analyzing the radioactive elements and the contribution of cosmic rays. The ESR dating technique has several advantages, such as a wide dating range (from several hundred to a billion years), ubiquitous datable materials (e.g., quartz, gypsum, carbonate, evaporites), and a relatively easy preparation procedure, and the samples can be prepared in natural light. However, sunlight must be avoided during sample collection. Glacial deposits could be dated using ESR dating techniques, especially using Ge centers, which are sensitive to light and grinding in glacial quartz. Significant progress has been made in Chinese Quaternary glaciation research due to the application of ESR dating techniques.

5) Cosmogenic radionuclide dating

Cosmogenic radionuclide (CRN) dating is also known as terrestrial in situ cosmogenic nuclide dating (TCN). It is a radioactive isotope dating method, and significant achievements have been made with the development of high energy accelerator mass spectrometers over the past three decades. Compared with traditional ^{14}C and AMS ^{14}C, OSL, ESR and other dating techniques, CRN can determine the exposure age as well as the burial age of terrestrial sediments.

The earth is bombarded by protons, helium nuclei and heavy nuclei, which are the primary high-energy nuclei, and secondary neutrons, photons, and mesons can be produced when these primary nuclei interact with the atmosphere. The secondary nuclei bombard the earth's surface and produce some CRNs. Commonly, the CRN production rate is proportional to the intensity of the cosmic rays and the concentrations of the target elements. Therefore, the longer the exposure time is, the more CRNs are produced. The ages of samples can be calculated from the concentrations of the CRNs and their production rates. Several CRNs, such as ^{3}He, ^{10}Be, ^{14}C, ^{21}Ne,

^{26}Al and ^{36}Cl, have been used to reconstruct geochronological frameworks. The dating range is from 10^3 to 10^7 years.

CRN dating is one of the most successful methods in Quaternary glaciation research because it can be used to determine the exposure ages of erratics (or glacial boulders) and eroded bedrock as well as the burial age of tills. In addition, quartz is a ubiquitous mineral in nature; it can be found in granite, gneiss, and sandstone. Therefore, this dating technique can be applied to many chronological problems.

10.4 Remote Sensing Technology

The harsh environment of the cryosphere makes it tremendously challenging to conduct field observations, which makes airborne and satellite-based remote sensing widely accepted and promising technologies for cryospheric studies. The application of remote sensing to cryospheric science covers almost all remote sensing bands, including the visible and near-infrared (VIS/NIR), thermal infrared (TIR), and microwave bands. In addition, innovative techniques such as gravity satellite observations, spaceborne/airborne radio echo detection, and LiDAR measurements have been rapidly developed and applied in cryospheric science. Due to the different characteristics and requirements of different studies, the comprehensive use of various remote sensing technologies has become a new trend in cryospheric remote sensing to improve the monitoring accuracy of cryospheric elements (Fig. 10.5). Satellite remote sensing imagery has become an important tool in cryospheric studies because it provides repeated observations and overcomes the small area and repeatability limitations of aerial remote sensing images. Table 10.6 lists the main cryospheric components and related satellite remote sensing missions/platforms.

10.4.1 Optical Remote Sensing

1. Visible and Near-Infrared (VIS/NIR) remote sensing

VIS/NIR sensors remotely sense the characteristics of surface targets by analyzing the VIS reflectance. The reflectances of key cryospheric components (snow and ice) in the VIS/NIR bands are higher than those of the surroundings. This principle is the foundation of cryospheric remote sensing. Therefore, satellite sensors operating in the VIS/NIR bands are widely used to detect cryospheric characteristics. The main applications of this method include ① retrieving parameters such as the snow cover extent and snow cover fraction at the subpixel scale; the surface albedo of snow and snow grain size; and the river, lake, and sea ice concentrations; ② mapping glaciers, permafrost and periglacial geomorphology, and sea ice; and ③ monitoring frozen soil deformation (frost heaving, thaw settlement and creep), sea ice motion, and glacial lake outburst flooding.

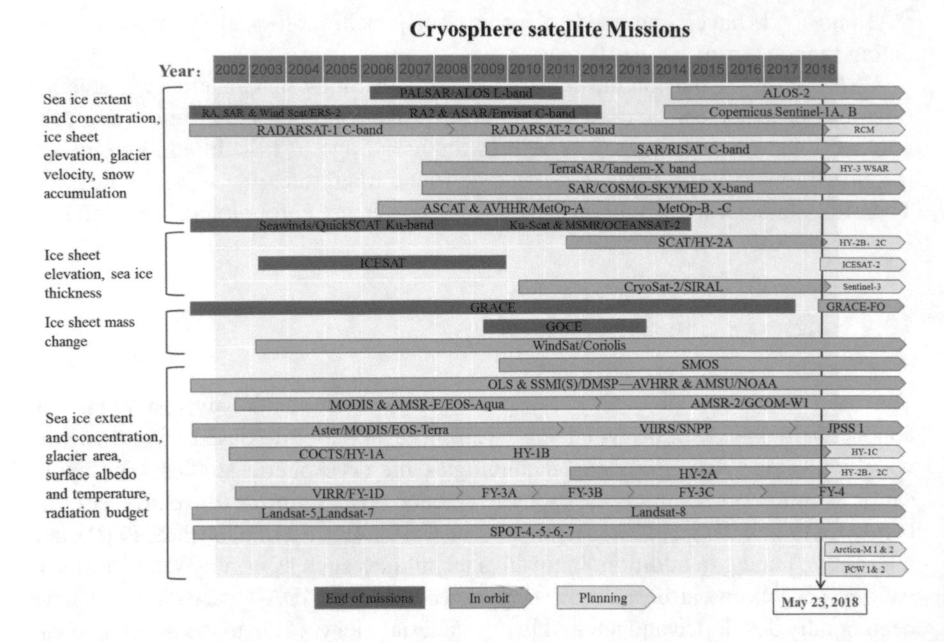

Fig. 10.5 Satellite-based cryospheric remote sensing projects and missions (http://globalcryospherewatch.org/satellites/overview.html)

1) Glacier mapping

Glaciers have high reflectivity in the visible band and a significant contrast with their surrounding areas. Based on the angular relationship between the main optical axis and the vertical direction, remote sensing acquisition modes can be classified into vertical photogrammetry and tilt photogrammetry. Vertical photogrammetry has high geometric accuracy and is widely used. Based on multi-temporal single images of the study area or the stereophotography method, remote sensing images are used for multiple applications, including mapping the topography of glacier surfaces, monitoring changes in ice and snow, and estimating glacier mass balance. Digital image processing techniques such as image enhancement and false-color composites can help identify differences in the spectral characteristics of ground objects, highlight subtle structures of high-brightness areas of ice and snow, and easily identify glacier surface sediment and ablation conditions, which are important steps in accurately extracting cryospheric elements.

The global high-resolution satellites that are commonly used to map glaciers include IKONOS, QuickBird, GeoEye, WorldView, and ZY-3 (Table 10.7). The mapping of glaciers by VIS/NIR remote sensing is based on the principle that glaciers have higher reflectance than the surrounding areas and are thus more easily detected in VIS/NIR images. VIS/NIR remote sensing images were first applied to study glacier inventories and area changes in the 1980s. Remote sensing image processing

Table 10.6 Remote sensing measurements of cryospheric components and parameters

Components	Parameters	Remote sensing technology
Glaciers	Albedo, area, inventory, surface temperature	VIS/NIR, TIR
	Volume	VIS/NIR photogrammetry, radar/laser altimetry
	Mass balance	VIS/NIR, thermal infrared, LiDAR, gravity satellite
	Elevation, thickness, surface morphology, motion	VIS/NIR photogrammetry, SAR, InSAR, radar/laser altimetry, radio echo detection
Snow cover	Extent, snow cover fraction, albedo, surface temperature	VIS/NIR, thermal infrared
	Grain size	VIS/NIR, high spectral
	Snow depth, snow water equivalent	Active and passive microwave, LiDAR
	Wetness	Active and passive microwave
	Density	Active microwave
Permafrost	Surface freeze–thaw state	Active and passive microwave
	Cartography, deformation, active layer thickness	VIS/NIR, thermal infrared, SAR
River and lake ice	Concentration, area	VIS/NIR, active microwave
	Thickness	Active and passive microwave, LiDAR
	Freeze/breakup dates	VIS/NIR, thermal infrared, active and passive microwave
	Temperature	Thermal infrared
	Surface roughness	LiDAR
	Ice jams and ice jam flooding	VIS/NIR and SAR
Sea ice	Extent mapping	VIS/NIR, radar altimetry
	Surface temperature, albedo	VIS/NIR, thermal infrared remote sensing
	Density, motion, thickness	VIS/NIR, thermal infrared, active and passive microwave, altimetry
	Ice jam and ice jam flooding	VIS/NIR and SAR

methods, such as band ratios, unsupervised classification, and supervised classification, have gradually replaced traditional visual interpretation using topographic maps and aerial photographs.

Table 10.7 Frequently used high-resolution satellite sensors

Satellites	Multispectral/panchromatic spatial resolution (m)	Stereoscopic observation capacity	Revisit interval/day	Launch date
Landsat MSS	78	N	18	July 1972
Landsat TM	30	N	16	July 1982
Landsat ETM	30/15	N	16	April 1999
Landsat OLI	30/15	N	16	February 2013
SPOT 1–4	20/10	Y	26	February 1986
SPOT 5	10/2.5	Y	5	May 2002
SPOT 6	6/1.5	Y	5	September 2012
CBERS-02B	19.5/2.4	N	26	October 1999
QuickBird	2.44/0.61	Y	1–6	October 2001
IKONOS	4/1	Y	3	September 1999
ASTER	30/15	Y	16	February 1999
IRS-P5	10/2.5	Y	5	May 2005
ALOS	10/2.5	Y	2	January 2006
WorldView-2	1.85/0.46	Y	3.7	September 2007
ZY-3	5.8/2.1	Y	5	January 2012
GF-1	8/2	N	4	April 2013
Sentinel-2A/B	10, 20, 60/no panchromatic	N	5	June 2015, March 2017

2) Glacial lake monitoring

Glacial lake outburst floods (GLOFs) are among the most serious natural disasters. High-resolution digital elevation models (DEMs), terrain data and the interpretation of remote sensing images have become the most effective means of investigating and monitoring glacial lakes. The enhancement of remote sensing imagery is helpful in interpreting the state and variations of glaciers and glacial lakes. In VIR satellite images, GLOFs can be identified using image information, such as brighter tones and riverbank erosion and accumulation. Additionally, historical GLOFs can be recognized from certain distinguishing features; for example, after collapse, moraine dams usually become end moraines separated by a small hill, and some moraines are accompanied by small ponds and rough textures.

3) Monitoring snow cover extent

Snow cover extent is usually mapped using VIS/NIR sensors such as NOAA AVHRR, GOES, Landsat TM/ETM+/OLI, MODIS, and SPOT-VEGETATION. Several approaches have been used to identify snow-free and snow-covered pixels, including the snow cover index method, the brightness temperature threshold method, supervised classification, manual interpretation and radiative transfer models. The first three methods are often used in practice, particularly the snow cover index method. The normalized difference snow index (NDSI), which utilizes two bands (one is the visible band with the high reflectance, and one in the near-infrared or short-wave infrared bands with high absorption), is a commonly used index and is defined as follows:

$$\mathrm{NDSI} = \frac{\mathrm{CH}(n) - \mathrm{CH}(m)}{\mathrm{CH}(n) + \mathrm{CH}(m)}$$

where n and m are the numbers of bands. Take example for the Landsat TM image, n and m are the band2 and band5.

The methods used to calculate the snow cover extent at the global and large scales are complicated but fully developed. Methods for identifying the snow cover extent at the local scale should be further developed to improve their accuracy. Recently, a new algorithm that uses the ratio of one visible band to two near-infrared bands was proposed to extract the snow cover extent from Landsat-5/TM data. The proposed algorithm has been demonstrated to improve the overall accuracy of snow mapping in vegetated regions.

4) Permafrost monitoring

The principle of mapping permafrost by remote sensing is based on establishing a morphogenic indicator system to identify cryogenic formations and interpret cryogenic development. Mapping permafrost by remote sensing involves three aspects: ① delimiting the permafrost extent by geomorphological and vegetation characteristics such as snow cover, bare bedrock, frost weathering clastic deposits, and alpine meadows in permafrost regions; ② interpreting thermokarst regions based on fracture structures in satellite remote sensing imagery; and ③ identifying periglacial landforms such as rock glaciers using aerial photography.

5) Sea ice monitoring

The albedo of sea ice is higher than that of open water in the VIS/NIR bands. Remote sensing is mainly used to retrieve the physical parameters of sea ice, such as the extent of the sea ice, sea ice types (one-year ice, multiyear ice and even finer types of sea ice), sea ice concentration, sea ice thickness, and the distribution of ice leads. The sea ice extent can be determined directly using cloud-free VIS/NIR images. Under the same sunlight and ice surface pollution conditions, different types of sea ice have different reflectances in the spectral range of 0.4–1.1 μm, especially 0.4–0.7 μm, due to the different structures and surface roughnesses. Reflectance images can be converted to grayscale, and the grayscale thresholds can be determined to discriminate sea ice types combined with auxiliary information such as baseline values.

2. Thermal infrared remote sensing (TIR)

TIR, which uses wavelengths of 8–14 μm, is mainly used to detect the thermal emissivities of objects and the surface temperature. It does not depend on sunlight and can work at night. TIR remote sensing is used in the estimation of land surface temperature, surface radiation, glacier mass balance, extents of snow, sea ice and permafrost, and the onset of melting of sea ice. TIR data can not only assist in identifying the boundaries of snow regions but also assist in evaluating the surface temperature of snow. Using MODIS data, local split-window algorithms are applied to simultaneously extract the surface emissivity and surface temperature. These algorithms have been accepted as the baseline algorithms for NASA's MODIS land surface temperature (LST) products.

10.4.2 Microwave Remote Sensing

Microwave remote sensing detects geometric and physical information by actively transmitting or passively receiving scattered or emitted radiation from objects in the microwave band. Microwave remote sensing is influenced less by sunlight and weather conditions, such as clouds and mist. This approach allows for observations in all weather and at all times, which is its advantage over optical/thermal remote sensing. In addition, microwave remote sensing has such a high penetration depth into snow and ice that it can monitor information within the snowpack and can obtain the thickness of glaciers, which cannot be obtained by optical/thermal remote sensing. Microwave remote sensing methods are divided into active and passive microwave remote sensing. The former actively transmits energy and receives the echoes backscattered by the objects, whereas the latter passively receives energy emitted from objects. Passive microwave remote sensing is usually used to retrieve the snow cover extent, snow depth, snow water equivalent (SWE), sea ice extent and sea/river/lake ice phenology. Synthetic Aperture Radar (SAR) can be applied to detect sea ice motion, ice phenology, ice jams, glacial lakes, GLOFs, glacier zones, snow lines, snout ends, ice sheet front changes, ice surface lakes, crevasses, surface ice till, ice speed, strain rate and ice mechanics. Additionally, glacial areas can be mapped using SAR interferometry and radar altimetry.

1. Passive microwave remote sensing

1) Snow depth and snow water equivalent

Snow depth and SWE can be derived from passive microwave data based on the volume scattering of snow particles. Microwave emissions from snow-covered soil are attenuated and scattered by snow particles. The attenuation density is proportional to the number of snow particles, which means that a deeper snowpack or larger SWE, which contains a larger number of snow particles, will scatter more microwave signals

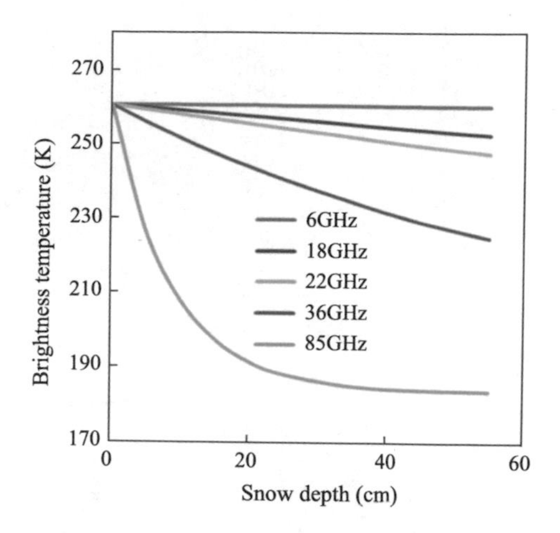

Fig. 10.6 Variations of brightness temperature with snow depth at different frequencies

from the soil. The scattering intensity also depends on the microwave frequency, and higher frequencies correspond to stronger scattering intensities. Therefore, the brightness temperature at a higher frequency is lower than that at a lower frequency, and their difference increases with increasing snow depth and SWE (Fig. 10.6). This relationship has been used to develop a snow depth retrieval method, the brightness temperature gradient method, which has been widely used to estimate snow depth and SWE at global and regional scales.

2) Surface freeze/thaw

Remote sensing perceives only the thin layer beneath the land surface; therefore, only surface freeze–thaw conditions can be detected. The dielectric constant of the soil is a key variable in the remote sensing detection of surface freeze–thaw. It is a function of the temperature, soil type, soil particle size, ice, free water, and bound water. During soil freeze/thaw processes, soil has different scattering and emissions characteristics. When the soil freezes, the soil emissivity increases, and its back-scattering coefficient decreases significantly. Therefore, the passive microwave brightness temperature of frozen soil is much higher than that of thawed soil. Volume scattering is another characteristic of frozen soil, which causes the brightness temperature gradient higher than that of thawed soil.

3) Sea ice concentration

The sea ice concentration can be efficiently derived from passive microwave remote sensing because the dielectric properties of ice and water are significantly different. When water freezes, the brightness temperature increases (Fig. 10.7). Therefore, the brightness temperature increases with the ice area in a passive microwave footprint. The frequency gradient, which is the ratio of the brightness temperature difference between 37 and 18 GHz for a vertical polarization to their sum, is commonly used to derive the sea ice concentration. Based on this method, the NSIDC has provided

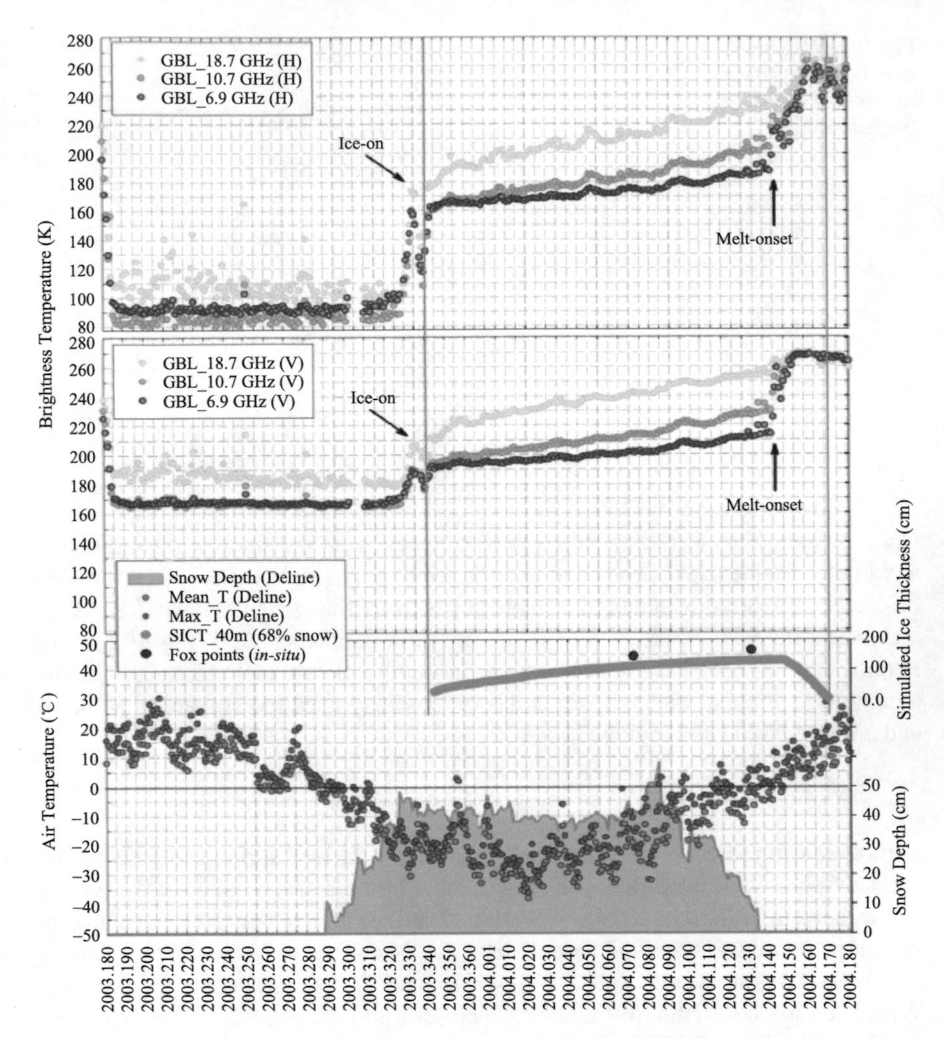

Fig. 10.7 Time series of H-pol (top panel) and V-pol (middle panel) brightness temperatures at 6.9, 10.7, and 18.7 GHz for the GBL (65°15′ N, 122°51.5 ′W) site during the 2003–2004 ice season. The time series of the maximum (Max_T, red) and mean (Mean_T, blue) daily air temperatures obtained at the Deline meteorological station are shown in the bottom panel along with snow depth (gray). Simulated ice thicknesses obtained with CLIMo are represented by the thick gray curve. The two red circles (Fox points) overlaid on the ice thickness curve correspond to in situ measurements made during field visits in 2004 at GBL (Kang et al. 2014)

daily and monthly time series of sea ice concentrations from SMMR, SSM/I and SSMI/S since 09 July 1987 for the northern and southern polar regions.

2. Active microwave remote sensing (AMRS)

Active microwave remote sensing actively transmits microwave signals and receives the scattered echoes from surface objects. In contrast to VIR/NIR remote sensing, active microwave remote sensing provides information about the cryosphere under all weather conditions and at all times. Therefore, this approach plays an important role in numerous applications, such as mapping mountain snow cover.

1) Synthetic Aperture Radar (SAR)

SAR has a similar detection capability to that of large-aperture antennas because its small aperture radar antenna is virtualized into a large-aperture antenna through the movement of the flight platform. It is used to detect objects from the back-scattering density, which is a function of the frequency, polarization, and incidence angle of incident electromagnetic waves. Thus, SAR can obtain rich information about ground objects by making full use of electromagnetic waves at different frequencies, polarizations and incidence angles.

SAR can be used to monitor ice phenology because the dielectric characteristics of ice and water are different, and the backscatter coefficient of ice is much higher than that of water.

2) nterferometric Synthetic Aperture Radar (InSAR)

InSAR represents an important field of research in SAR technology, and it is usually used to monitor the surface morphology and changes of glaciers (Rott 2009). By combining SAR imaging technology and interferometry technology, InSAR can be used to extract three-dimensional information (elevation or speed) from SAR data. By using two antennas to simultaneously observe one (single orbit double antenna mode) or two (repetitive orbital mode) near-parallel observations, two scenes of the same region are obtained. By acquiring the phase difference between two echoes corresponding to the same feature points, high-precision and high-resolution DEMs or surface deformation diagrams can be obtained.

10.4.3 Altimetry

Altimetry emits pulses and receives echoes from the ground surface, and the time interval between the emissions and the returns are measured. The distance between the aircraft and the pulse point is calculated based on the time interval of the echo signal, which thus measures the surface elevation. Altimetry methods can be divided into laser altimetry and radar altimetry. Both methods use the same basic principle, and both can be used to directly measure the absolute elevation of the surface. By combining altimetry data with the density distributions of ice and snow, the mass balance can be acquired using the quadrature technique. Therefore, airborne and spaceborne altimetry have become primary remote sensing methods for measuring the surface elevation of glaciers (ice sheets) and topography. The altimeters have a higher vertical accuracy (centimeter level) but lower horizontal resolution (kilometer level). Laser altimetry improves the horizontal resolution. Radar altimetry, especially airborne laser altimetry, has become the primary means of monitoring ice cracks.

1. Radar altimetry

Radar altimetry usually works in fairly large-wavelength bands. It can measure not only the absolute elevation of the ground but also topographical changes. Therefore, ice sheet ground lines can be detected if the ice surface topography is affected by lakes under ice sheets and water under ice. It also has ability to measure the three-dimensional distributions of snow and ice, such as the snow depth on sea ice, ice sheet thickness, and geomorphological features under ice.

2. Laser altimetry

Laser altimetry, which is also called LiDAR, uses energy in small wavelengths, such as visible green light, as the source signal. These signals in smaller wavelength can be reflected back by smaller objects. In addition, the area illuminated by a laser altimeter is much smaller than that illuminated by a radar altimeter for platforms at the same height. Therefore, more detailed information about ice surface changes can be obtained by LiDAR. GLAS loading on the Ice, Cloud, and land Elevation Satellite (ICESat-1) was the first spaceborne laser altimeter to provide laser echo point data in January 2003. It operates at a height of 600 km, and the diameter of the illuminated area is only 70 m. Its vertical accuracy is approximately 5 cm, which can provide accurate topographic information that satisfies the requirements of studies on the mass balance of ice sheets and dynamic modeling. From 2003 to 2009, this laser altimeter provided multiyear elevation data to determine ice sheet mass balances. ICESat was stopped operating on August 14, 2010, and the 2nd generation laser altimeter ICESat mission (ICESat-2) was launched in 2018.

10.4.4 Radio Echo Sounding

Radio echo sounding (RES) is also called ice radar. Based on the attenuation of different echoes within a layered or even ice sheet, the underlying interface parameters, such as ice thickness, subglacial relief, ice and ice-bottom conditions, and glacier flow velocity, can be detected effectively by RES. In the 1960s, ice radar was first introduced for Antarctic and Arctic Greenland ice sheet investigations. It is mainly used to produce topographic maps of ice thickness. The detecting modes have developed from monostatic, bistatic, and multistatic to multifrequency and multipolarization synchronous measurements.

10.4.5 *Gravity Satellites*

Gravimeters are used to reveal the mass changes of glaciers and ice sheets. Satellite gravimetry evaluates anomalies in gravity using observations at the satellite altitude. The mass balances of glaciers and ice sheets can be obtained directly based on Newton's law of universal gravitation with no assumptions (e.g., snow density). The appearance of satellite gravimetry technology began at the end of the 1950s and has progressed through three stages: ① optical technology, in which the global geoidal surface has one-meter accuracy; ② ground tracking and satellite observation techniques, which use the linear intersection method to measure the locations of satellite and measure elevations using satellite radar; and ③ orbit determination technology based on the precise orbit tracking by spaceborne GPS. In addition, satellite gravimetry is not significantly affected by the atmosphere. The precision of instrumental measurements can reach the centimeter level. The main low-orbit gravity satellites include CHAMP (DLR; July 15, 2000), GRACE (jointly developed by the United States and Germany; March 17, 2002) and GOCE (ESA; March 17, 2009). Based on a model of the Earth's gravitational field derived by gravity satellite missions, the mass changes of glaciers and ice sheets can be inferred.

Questions

1. Which key factors have stimulated the rapid development of cryospheric science in recent decades?
2. What are the challenges in the observations of ice sheet mass balance?
3. What kinds of remote sensing technology can be used to observe the cryosphere?

Extended Readings

1. Hauck C, Kneisel C. 2008. Applied Geophysics in Peri-glacial Environments. Cambridge University Press: Cambridge.
2. A Cryosphere Theme Report for the IGOS Partnership (IGOS-Cryo). For the Monitoring of our Environment from Space and from Earth. WMO/TD-No. 1405. 2007.
3. Key, J., M. Drinkwater and J. Ukita. 2007. Integrated Global Observing Strategy- Partnership (IGOS-P) Cryosphere Theme report 2007. Geneva, World Meteorological Association.

Cryosphere Theme Report.

Publisher: World Meteorological Organization (WMO) (2007), Integrated Global Observing Strategy.

Translator: Xiao C. D., Xie A. H., Ma L. J., et al.

Press: China Meteorological Press (Chinese version)

The cryosphere theme report summarizes the work of the Cryosphere Theme Team under the framework of the Integrated Global Observing Strategy (IGOS). Focusing on the cryospheric observation system (CryOS), the report defines the concept of the cryosphere and the main applications of the related observations and provides a concise presentation of the requirements of cryospheric observations, data and products. The report also provides a summary of the ability to observe and the requirements for fundamental weather variables in the main research domains of cryospheric science today. In addition, the report provides near-term, mid-term and long-term strategies. An integrated and coordinated observation system is constructed using ground-based, satellite-based and airborne observations, and the data management objectives are specifically illustrated. This report highlights the need for observations, data and data products and provides suggestions for the development and maintenance of the observation system. In particular, the report emphasizes the needs for data validation and coordination. With the advancement of observational techniques, the report suggests continuously improving the observation specifications and datasets. The appendix of the report has great reference value and provides the essential observation variables and precision requirements.

CryOS includes not only measurements of snow and ice properties but also satellite remote sensing instruments; ground observation network instruments; aircraft-based measurements; modeling, assimilation, and reanalysis systems; and a data management system. CryOS aims to foster the evaluation of the cryosphere in models, to reveal the role of the cryosphere in the climate system, to measure the predictability of the cryosphere in climate models and to stimulate improvements in the parameterization of cryospheric processes. The data and information management sectors must provide services in support of cryospheric science research, long-term scientific monitoring, and operational monitoring. These services must go beyond the traditional metadata services and web portals by encouraging the development of tools that combine all types of data, including model fields, from diverse and distributed data centers.

References

Aagaard K, Swift JH, Carmack EC (1985) Thermohaline circulation in the Arctic Mediterranean Seas. J Geophys Res 90(03):4833–4846
Allen PA, Etienne JL (2008) Sedimentary challenge to Snowball Earth. Nat Geosci 1(12):817–825
Anderson DM, Chamberlin GL, Guymon GL, Kane DL, Kay BD, Mackay JR, O'Neill K, Outcalt SI, Williams PJ (1984) Ice segregation and frost heaving. National Academy Press, Washington, p 72
Barber DG, Massom RA (2007) The role of sea ice in Arctic and Antarctic Polynyas. Elsevier Oceanogr Ser 74:1–54
Belmecheri S, Babst F, Wahl ER et al (2016) Multi-century evaluation of Sierra Nevada snowpack. Nat Clim Change 6(1):2–3
Belzile C, Gibson JE, Vincent WF (2002) Colored dissolved organic matter and dissolved organic carbon exclusion from lake ice: implications for irradiance transmission and carbon cycling. Limnol Oceanogr 47(5):1283–1293
Bennington K (1963) Some crystal growth features of sea ice. J Glaciol 669–688
Bond GC, Lotti R (1995) Iceberg discharges into the North-Atlantic on millennial time scales during the last glaciation. Science 267(5200):1005–1010
Bond G, Showers W, Cheseby M et al (1997) A pervasive millennial-scale cycle in North Atlantic Holocene and glacier climates. Science 278:1257–1266
Broecker WS, Peng TH (1982) Tracers in the sea. Eldigio Press, p 690
Brown RJE (1965) Factors influencing discontinuous permafrost in Canada. In: Abstracts, 7th international congress, International Association for Quaternary Research, Boulder, Colorado, p 47
Brown RJE (1966) The relationship between mean annual air and ground temperatures in the permafrost regions of Canada. Natl Acad Sci 241–246
Brown RJE (1966) Influence of vegetation on permafrost. Natl Acad Sci Publ 1287:20–25
Brown RE (1973) Influence of climatic and terrain factors on ground temperature at three locations in the permafrost region of Canada. Natl Acad Sci 27–34
Brown RJE, Péwé TL (1973) Distribution of permafrost in North America and its relationship to the environment: a review, 1963–1973. Natl Acad Sci 71–100
Bryan K (1922) Erosion and sedimentation in the Papago Country, Arizona: with a sketch of the geology. U.S. Geological Survey Bulletin
Bryant JP, Scheinberg E (1970) Vegetation and frost activity in an alpine fellfield on the summit of Plateau Mountain. Can J Bot 48:751–771
Cao ZT (1988) The hydrologic characteristics of the Gongba glacier in the Mount Gongga area. J Glaciol Geocryol 10(1):57–65
Cazenave A, Llovel W (2010) Contemporary sea level rise. Ann Rev Mar Sci 2:145–173

D. Qin et al. (eds.), *Introduction to Cryospheric Science*, Springer Geography,
https://doi.org/10.1007/978-981-16-6425-0

Cazenave A, Dominh K, Guinehut S et al (2009) Sea level budget over 2003–2008: a reevaluation from GRACE space gravimetry, satellite altimetry and Argo. Global Planet Change 65:83–88
Cheng GD (1983) The mechanism of repeated-segregation for the formation of thick-layered ground ice. Cold Reg Sci Technol 8:57–66
Cheng GD (2003) The impact of local factors on permafrost distribution and its inspiring for design Qinghai-Xizang Railway. Sci China (Ser D) 33(6):602–607
Cheng GD (2004) Influences of local factors on permafrost occurrence and their implications for Qinghai-Tibet railway design. Sci China (Ser D): Earth Sci 47(8):704–709
Cheng GD, Dramis F (1992) Distribution of mountain permafrost and climate. Permafr Periglac Process 3(2):83–91
Cheng GD, Wang SL (1982) On the zonation of high-altitude permafrost in China. J Glaciol Geocryol 4:1–16
Church JA, White NJ, Domingues CM et al. Sea-level and ocean heat-content change. International geophysics. Elsevier, New York, pp 697–725
Comiso JC, Parkinson CL, Green R et al (2008) Accelerated decline in the Arctic sea ice cover. Geophys Res Lett 35(01)
Cook FA (1955) Near surface soil temperature measurements at Resolute Bay, Northwest Territories. Arctic 8(4):237–249
Crowley T, Kim Y (1995) Comparison of proxy records of climate and solar forcing. Geophys Res Lett 23(4):359–362
Cuffey KM, Paterson WSB (2010) The physics of glaciers, 4th edn. Elsevier Science, Oxford
Cui ZJ, Zhao L, Vandenberghe J et al (2002) Discovery of ice wedge and sand-wedge networks in Inner Mongolia and Shanxi Province and their environmental significance. J Glaciol Geocyol 24(6):708–716
Echelmeyer K, Wang ZX (1987) Direct observation of basal sliding and deformation of basal drift at sub-freezing temperatures. J Glaciol 33(113):83–98
Ehlers J, Gibbard PL, Hughes PD (2018) Quaternary glaciations and chronology. In: Glacial environments, 2nd ed. Elsevier Publisher, Amsterdam, pp 74–101
EPICA community members (2004) Eight glacial cycles from an Antarctic ice core. Nature 429:623–629
Fedorova AP, Yankina AS (1964) The passage of Pacific Ocean water through the Bering Strait into the Chukchi Sea. Deep Sea Res 11:427–434
Fetterer F, Untersteiner N (1998) Observations of melt ponds on Arctic sea ice. J Geophys Res 103(11):24821–24835
Flavio L, Christoph CR, Dominik H et al (2012) The freshwater balance of polar regions in transient simulations from 1500 to 2100 AD using a comprehensive coupled climate model. Clim Dyn 39:347–363
Frakes LA (1979) Climates throughout Geologic time. Elsevier, Amsterdam, 310
Harry DG, Gozdzik JS (1988) Ice wedges: growth, thaw transformation, and palaeoenvironmental significance. J Quat Sci 3(1):39–55
Grenfeell TC, Maykut GA (1977) The optical properties of ice and snow in the Arctic Basin. J Glaciol 18(80):445–463
Groote PM, Stuiver M, White JC et al (1993) Comparison of oxygen isotope from the GISP2 and Grip Greenland ice core. Nature 366(6455):552–554
Harris SA (2010) Greenhouse gases and their importance to life. In: Global warming, Chapter 2. Sciyo Publishers, Rijeka, Croatia, pp 15–22
Harris SA (2013) Climatic change: casual correlations over the last 240 Ma. Sci Cold Arid Reg 5(3):259–274
Harris SA, Brouchkov A, Cheng GD (2017) Geocryology: characteristics and use of frozen ground and permafrost landforms (Series M). CRC Press, Boca Raton
Hays JD, Imbrie J, Shackleton NJ (1976) Variations in the earth's orbit: pacemaker of the ice ages. Science, 194: 1121-1132. (1976) Variations in the earth's orbit: pacemaker of the ice ages. Science 194:1121–1132

Hobbs PV (1974) Ice physics. Clarendon Press, Oxford
Hoffman PF, Kaufman AJ, Halverson GP et al (1998) A Neoproterozoic snowball Earth. Science 281:1342–1346
Hu GJ, Zhao L, Li R et al (2014) Characteristics of hydro-thermal transfer during freezing and thawing period in permafrost regions. Soils 46(2):355–360
Hu GJ, Zhao L, Zhu XF et al (2020) Review of algorithms and parameterizations to determine unfrozen water content in frozen soil. Geoderma 368:114277
IPCC (2013) Climate change 2013: the physical science basis. Contribution of Working Group I to the Fifth assessment report of the intergovernmental panel on climate change. Cambridge University Press, Cambridge
Ivanova EV (2009) The global thermohaline paleocirculation. Springer, Moscow
Johnsen SJ, Dahl-Jensen D, Gundestrup N et al (2001) Oxygen isotope and palaeotemperature records from six Greenland ice-core stations: Camp Century, Dye-3, GRIP, GISP2, Renland and NorthGRIP. J Quat Sci 16(4):299–307
Kämäräinen J (1993) Studies on ice mechanics
Kang S, Zhang Q, Kaspari S et al (2007) Spatial and seasonal variations of elemental composition in Mt. Everest (Qomolangma) snow/firn. Atmos Environ 41(34):7208–7218
Kang KK, Duguay CR, Lemmetyinen J et al (2014) Estimation of ice thickness on large northern lakes from AMSR-E brightness temperature. Remote Sens Environ 150:1–19
Karcher MJ, Oberhuber JM (2002) Pathways and modification of the upper and intermediate waters of the Arctic Ocean. J Geophys Res 107(C6):3049
Kawamura T, Shirasawa K, Ishikawa N et al (2001) Time-series observations of the structure and properties of brackish ice in the Gulf of Finland. Ann Glaciol 33:1–4
Kinnard C, Zdanowics CM, Fisher DA, Isaksson E, De Vernal A, Thompson LG (2011) Reconstructed changes in the Arctic sea ice over the past 1,450 years. Nature 479(7374):509–512
Kozlowski T (2004) Soil freezing point as obtained on melting. Cold Reg Sci Technol 38(2–3):93–101
Kwok R, Rothrock DA (2009) Decline in Arctic sea ice thickness from submarine and ICE Sat records. Geophys Res Lett 36(15):1958–2008
Lemke PJ, Ren RB, Alley I et al (2007) Observations: changes in snow, ice and frozen ground. In: Climate change 2007: the physical science basis. Contribution of Working Group I to the Fourth assessment report of the intergovernmental panel on climate change. Cambridge University Press, Cambridge
Li Z, Han T, Jin Z et al (2003) A summary of 40-year observed variation facts of climate and glacier No.1 at Headwater of Urumqi River, Tianshan, China. J Geophys Res 25(2):117–123
Li Z, Edwards R, Mosley-Thompson E et al (2006) Seasonal variability of ionic concentrations in surface snow and elution processes in snow-firn packs at the PGPI site on Urumqi glacier No. 1, eastern Tien Shan, China. Ann Glaciol 43(1):250–256
Li Z, Shen Y, Wang F et al (2007) Response of glacier melting to climate change-Take Ürümqi Glacier No.1 as an example. J Geophys Res 29(03):333–342
Li Z, Li C, Li Y et al (2007) Preliminary results from measurements of selected trace metals in the snow-firn pack on Urumqi Glacier No.1, eastern Tien Shan, China. J Glaciol 53(182):368–373
Li Z, Zhao S, Edwdrds R et al (2011) Characteristics of individual aerosol particles over Urumqi Glacier No.1 in eastern Tianshan, central Asia, China. Atmos Res 99(1):57–66
Li Z, Dong Z, Zhang M et al (2011) chemical characteristics and seasonal variation of snow on Urumqi Glacier No .1 of the Eastern Tianshan, China. Earth Sci J China Univ Geosci 36(4):670–678
Lindsay RW, Zhang J (2005) The thinning of Arctic sea ice, 1988–2003: have we passed a tipping point? J Clim 18(22):4879–4894
Mackay JR (1971) Origin of massive icy beds in permafrost, western Arctic coast. Can J Earth Sci 8:397–422
Mackay JR (1983) Downward water movement into frozen ground, western Arctic coast, Canada. Can J Earth Sci 20:120–134

Mackay JR (1984) The frost heave of stones in the active layer above permafrost with downward and upward freezing. Arct Alp Res 16:413–417
Marzeion B, Jarosch AH, Hofer M (2012) Past and future sea-level change from the surface mass balance of glaciers. Cryosphere 6:1295–1322
Maslanik JA, Fowler C, Stroeve J et al (2007) A younger, thinner Arctic ice cover: increased potential for rapid, extensive sea-ice loss. Geophys Res Lett 34(24)
Maslanik J, Stroeve J, Fowler C et al (2011) Distribution and trends in Arctic sea ice age through spring 2011. Geophys Res Lett 38(13)
Morgan VI, Budd WF (1978) The distribution, movement and melt rates of Antarctic icebergs. Iceberg Util 220–228
Oerlemans J (2005) Extracting a climate signal from 169 glacier records. Science 308:675–677
Overland JE, Francis JA, Hanna E et al (2012) The recent shift in early summer Arctic atmospheric circulation. Geophys Res Lett 39(19)
Parkinson CL, Cavalieri DJ (2008) Arctic sea ice variability and trends, 1979–2006. J Geophys Res 113(07)
Paterson WB (1994) The physics of glaciers, 3rd ed. Pergamon Press, Oxford
Petrich C, Eicken H (2010) Sea ice: growth, structure and properties of sea ice. Wiley
Radic V, Hock R (2014) Glaciers in the Earth's hydrological cycle: assessments of glacier mass and runoff changes on global and regional scales. Surv Geogr 35:813–837
Romanovsky V et al (2010) Thermal state of permafrost in Russia. Permafr Periglac Process 21:136–155
Rothrock DA, Yu Y, Maykut GA (1999) Thinning of the Arctic sea-ice cover. Geophys Res Lett 26(23):3469–3472
Rott H (2009) Advances in interferometri synthetic aperture radar (InSar) in earth system. Prog Phys Geogr 33(6):769–791
Schwerdtfecer P (1963) The thermal properties of sea ice. J Glaciol 4(36):789–807
Screen J (2017) The missing Northern European winter cooling response to Arctic sea ice loss. Nat Commun 8:14603
Screen J, Simmonds I (2010) The central role of diminishing sea ice in recent Arctic temperature amplification. Nature 464:1334–1337
Screen JA, Simmonds I (2010) Increasing fall-winter energy loss from the Arctic Ocean and its role in Arctic temperature amplification. Geophys Res Lett 37(16)
Serreze MC, Barry RG (2011) Processes and impacts of Arctic amplification: a research synthesis. Global Planet Change 77:85–96
Shi Y, Xie Z (1964) The basic characteristics of modern Chinese glaciers. Acta Geogr Sin 30(3):183–208
Shumskii PA (1964) Principles of structural glaciology. Dover Publications, New York
Skiles SM, Flanner M, Cook JM et al (2018) Radiative forcing by light-absorbing particles in snow. Nat Clim Change 8:964–971
Stephen RR (2013) Large-scale ocean circulation: deep circulation and meridional overturning. The encyclopedia of sustainability science and technology. Springer, Moscow
Stroeve J, Holland MM, Meier W, Scambos T, Serreze MC (2007) Arctic sea ice decline: faster than forecast. Geophys Res Lett 34(09)
Swift JH, Aagaard K (1981) Seasonal transitions and water mass formation in the Iceland and Greenland seas. Deep Sea Res 28:1107–1129
Taber S (1929) Frost heaving. J Geol 37:428–461
Taber S (1930a) The mechanics of frost heaving. J Geol 38:303–317
Taber S (1930b) Freezing and thawing of soils as factors in destruction of road pavements. Public Roads 11:113–132
Thomas DN (2017) Sea ice, second, edition. Wiley Blackwell Publishing, New York
Thompson LG, Davis ME, Mosley-Thompson E et al (2005) Tropical ice core records: evidence for asynchronous glaciation on Milankovitch timescales. J Quat Sci 20(7–8):723–733

Timco GW, Burden RP (1997) An analysis of the shapes of sea ice ridges. Cold Reg Sci Technol 25(1):65–77

Tsytovich NA (1975) The mechanics of frozen ground. McGraw-Hill Book Company, p 426

Ukita J, Honda M, Nakamura H, Tachibana Y, Cavlieri DJ, Parkinson CL, Koide H, Yamamoto K (2007) Northern Hemisphere sea ice variability: lag structure and its implications. Tellus A 59:261–272

Untersteiner N (1961) On the mass and heat budget of Arctic sea ice. Archives Meteorol Geophys Bioclimatol Ser A 12(2):151–182

van den Broeke, M (2005) Strong surface melting preceded collapse of Antarctic Peninsula ice shelf. Geophys Res Lett 32(12)

Vandenberghe J, French HM, Gorbunv A et al (2014) The Last Permafrost Maximum (LPM) map of the Northern Hemisphere: permafrost extent and mean annual air temperatures, 25–17 ka BP. Boreas 43(3):652–666

Wang YJ, Cheng H, Edwards DL et al (2005) The Holocene Asian monsoon: links to solar changes and North Atlantic climate. Science 308:854–857

Wang F, Li Z, Edwards R, Li H (2007) Long-term changes in the snow-firn pack stratigraphy on Urumqi glacier No. 1, eastern Tien Shan, China. Ann Glaciol 46(1):331–334

Wu ZW, Ma W (1994) Strength and creep of frozen soils. University of Lanzhou Press, Lanzhou, China

Xie ZC, Liu CH (2010) Induction to glaciology. Shanghai Popular Science Press, Shanghai, pp 135–188

Xie ZC, Zhou ZG, Li QY, Wang CG (2009) Progress and prospects of mass balance characteristic and responding to global change of glacier system in high Asia. Adv Earth Sci 24(10):1065–1072

Xu P, Zhu HF, Shao XM et al (2012) Tree ring-dated fluctuation history of Midui glacier since the Little Ice Age in the southeastern Tibetan Plateau. Sci China (Earth Sci) 55(4):521–529

Yang Z (1991) Glacier water resources in China. Gansu Science and Technology Press, Lanzhou

Yang SZ, Jin HJ (2011) $\delta^{18}O$ and δD records of inactive ice wedge in Yitulihe, Northeastern China and their paleoclimatic implications. Sci China (Earth Sci) 54(1):119–126

Yang XY, Fyfe JC, Flato GM (2010) The role of poleward energy transport in Arctic temperature evolution. Geophys Res Lett 37(14)

Yao TD, Duan KQ, Tian LD et al (2000) The accumulation rate record in the Dasuopu ice core and the variations of Indian summer monsoon precipitation over the past 400 years. Sci China (D) 30(06):619–626

Yen YC, Cheng KC, Fukusako S (1992) A review of intrinsic thermophysical properties of snow, ice, sea ice and frost. North Eng 23(4) & 24(1):53–74

Yershov ED (1979) Moisture transfer and cryogenic textures in fine grained soils. Moscow. Moscow University Publications, USSR: 214

You X, Li Z, Wang F (2005) Study on time scale of snow-ice transformation through snow layer tracing method—take the glacier No.1 at the Headwaters of Urumqi River as an example. J Geophys Res 27(6):853–860

Yu G, Xu J, Kang S et al (2013) Lead isotopic composition of insoluble particles from widespread mountain glaciers in western China: natural vs. Anthropocentric sources. Atmos Environ 75:224–232

Zachos JC, Pagani M, Sloan L et al (2001) Trends, rhythms and aberrations in global climate 65 Ma to present. Science 292:686–693

Zachos JC, Röhl U, Schellenberg SA, Sluijs A, Hodell DA, Kelly DC, Thomas E, Nicolo M, Raffi I, Lourens LJ, McCarren H, Kroon D (2005) Rapid acidification of the ocean during the Paleocene-Eocene thermal maximum. Science 308(5728):1611–1615

Zhang Y, Kang S (2017) Research progress of light-absorbing impurities in glaciers of the Tibetan Plateau and its surroundings. China Sci Bull 63(35):4151–4162

Zhang J, Lindsay R, Steele M, Schweiger A (2008) What drove the dramatic retreat of arctic sea ice during summer 2007? Geophys Res Lett 35(11)

Zhang Q, Huang J, Wang F et al (2012) Mercury distribution and deposition in glacier snow over western China. Environ Sci Technol 46(10):5404–5413

Zhao L (2019) Permafrost thermal state. Bull Am Meteor Soc 98:19–21

Zhao JP, Cao Y (2011) Summer water temperature structures in upper Canada Basin and their interannual variation. Adv Polar Sci 22(4):223–234

Zhao JP, Li T (2010) Solar radiation penetrating through sea ice under very low solar altitude. J Ocean Univ China (Ocean Coast Sea Res) 9(2):116–122

Zhao XT, Han YS, Li PR et al (1996) Regional coastal development, sea-level change and their geological records. China Sea Level Change. Shandong Science & Technology Press, Jinan, pp 52–70

Zhao L, Cheng GD, Li S, Zhao X, Wang S (2000) Thawing and freezing processes of active layer in Wudaoliang region of Tibetan Plateau. Chin Sci Bull 45(23):2181–2187

Zhao JP, Zhu DY, Shi JX (2003) Seasonal variations in sea ice and its main driving factors in the Chukchi Sea. Adv Marine Sci 21(2):123–131

Zhao JP, Li T, Zhang SG, Jiao YT (2009) The shortwave solar radiation energy absorbed by packed sea ice in the central arctic. Adv Earth Sci 24(1):35–41

Zhao L, Ding YJ, Liu GY, Wang SL, Jin HJ (2010) Estimates of the reserves of ground ice in permafrost regions on the Qinghai-Tibetan Plateau. J Glaciol Geocryol 32(1):1–9

Zhao L, Zou DF, Hu GJ, Du EJ, Pang QQ, Xiao Y, Li R, Sheng Y, Wu XD, Sun Z, Wang LX, Wang C, Ma L, Zhou HY, Liu SB (2020) Changing climate and the permafrost environment on the Qinghai–Tibet (Xizang) plateau. Permafrost and periglacial processes

Zwally HJ, Comiso JC, Gordon AL (2013) Antarctic offshore leads and polynyas and oceanographic effects. In: Jacobs SS (ed) Oceanology of the Antarctic continental shelf

Printed in the United States
by Baker & Taylor Publisher Services